Advances in Intelligent Systems and Computing

Volume 801

The series "Advances in Intelligent Systems and Computing" contains publications on theory, applications, and design methods of Intelligent Systems and Intelligent Computing. Virtually all disciplines such as engineering, natural sciences, computer and information science, ICT, economics, business, e-commerce, environment, healthcare, life science are covered. The list of topics spans all the areas of modern intelligent systems and computing such as: computational intelligence, soft computing including neural networks, fuzzy systems, evolutionary computing and the fusion of these paradigms, social intelligence, ambient intelligence, computational neuroscience, artificial life, virtual worlds and society, cognitive science and systems, Perception and Vision, DNA and immune based systems, self-organizing and adaptive systems, e-Learning and teaching, human-centered and human-centric computing, recommender systems, intelligent control, robotics and mechatronics including human-machine teaming, knowledge-based paradigms, learning paradigms, machine ethics, intelligent data analysis, knowledge management, intelligent agents, intelligent decision making and support, intelligent network security, trust management, interactive entertainment, Web intelligence and multimedia.

The publications within "Advances in Intelligent Systems and Computing" are primarily proceedings of important conferences, symposia and congresses. They cover significant recent developments in the field, both of a foundational and applicable character. An important characteristic feature of the series is the short publication time and world-wide distribution. This permits a rapid and broad dissemination of research results.

More information about this series at http://www.springer.com/series/11156

Sara Rodríguez · Javier Prieto
Pedro Faria · Sławomir Kłos
Alberto Fernández · Santiago Mazuelas
M. Dolores Jiménez-López
María N. Moreno · Elena M. Navarro
Editors

Distributed Computing and Artificial Intelligence, Special Sessions, 15th International Conference

Editors
Sara Rodríguez
BISITE Digital Innovation Hub
University of Salamanca
Salamanca, Spain

Javier Prieto
BISITE Digital Innovation Hub
University of Salamanca
Salamanca, Spain

Pedro Faria
GECAD - Instituto Superior de Engenharia
Porto, Portugal

Sławomir Kłos
Department of Computer Science and Production Management
University of Zielona Góra
Zielona Góra, Poland

Alberto Fernández
Computing Science and Artificial Intelligence
Rey Juan Carlos University
Móstoles, Madrid, Spain

Santiago Mazuelas
Basque Center for Applied Mathematics
Bilbao, Spain

M. Dolores Jiménez-López
Basque Center for Applied Mathematics
Universidad de Alcalá
Alcalá de Henares, Spain

María N. Moreno
Departamento de Informática y Automática
University of Salamanca
Salamanca, Spain

Elena M. Navarro
Departamento de Sistemas Informáticos
University of Castilla-La Mancha
Albacete, Spain

ISSN 2194-5357 ISSN 2194-5365 (electronic)
Advances in Intelligent Systems and Computing
ISBN 978-3-319-99607-3 ISBN 978-3-319-99608-0 (eBook)
https://doi.org/10.1007/978-3-319-99608-0

Library of Congress Control Number: 2018951889

This Springer imprint is published by the registered company Springer Nature Switzerland AG
The registered company address is: Gewerbestrasse 11, 6330 Cham, Switzerland

Preface

The 15th International Conference on Distributed Computing and Artificial Intelligence 2018 is an annual forum that will bring together ideas, projects, and lessons associated with distributed computing and artificial intelligence, and their application in different areas. Artificial intelligence is changing our society. Its application in distributed environments, such as the Internet, electronic commerce, environment monitoring, mobile communications, wireless devices, distributed computing, to mention only a few, is continuously increasing, becoming an element of high added value with social and economic potential, in industry, quality of life, and research. These technologies are changing constantly as a result of the large research and technical effort being undertaken in both universities and businesses. The exchange of ideas between scientists and technicians from both the academic and industry sector is essential to facilitate the development of systems that can meet the ever-increasing demands of today's society.

The present edition brings together past experience, current work, and promising future trends associated with distributed computing, artificial intelligence, and their application in order to provide efficient solutions to real problems. This conference is a stimulating and productive forum where the scientific community can work toward future cooperation in distributed computing and artificial intelligence areas. Nowadays, it is continuing to grow and prosper in its role as one of the premier conferences devoted to the quickly changing landscape of distributed computing, artificial intelligence, and the application of AI to distributed systems.

This year's technical program will present both high quality and diversity, with contributions in well-established and evolving areas of research. More than 120 papers were submitted to main and special sessions' tracks from over 20 different countries (Algeria, Angola, Austria, Brazil, Colombia, France, Germany, India, Italy, Japan, Netherlands, Oman, Poland, Portugal, South Korea, Spain, Thailand, Tunisia, UK and USA), representing a truly "wide area network" of research activity.

Moreover, DCAI'18 Special Sessions have been a very useful tool in order to complement the regular program with new or emerging topics of particular interest to the participating community. The DCAI'18 Special Sessions technical program has selected 64 papers and, as in past editions, it will be special issues in JCR-ranked journals such as Neurocomputing, and International Journal of Knowledge and Information Systems. Special Sessions that emphasize on multi-disciplinary and transversal aspects, such as Advances on Demand Response and Renewable Energy Sources in Smart Grids (ADRESS), AI–driven methods for Multimodal Networks and Processes Modeling (AIMPM), Social Modelling of Ambient Intelligence in Large Facilities (SMAILF), Communications, Electronics and Signal Processing (CESP), Complexity in Natural and Formal Languages (CNFL), Web and Social Media Mining (WASMM), have been especially encouraged and welcome.

This symposium is organized by the University of Castilla-La Mancha, the Osaka Institute of Technology, and the University of Salamanca. The present edition was held in Toledo, Spain, on June 20–22, 2018.

We thank the sponsors (IBM, Indra, IEEE Systems Man and Cybernetics Society Spain) and the funding supporting of the Junta de Castilla y León (Spain) with the project *"Moviurban: Máquina Social para la Gestión sostenible de Ciudades Inteligentes: Movilidad Urbana, Datos abiertos, Sensores Móviles"* (Id. SA070U16—Project co-financed with FEDER funds), and finally, the Local Organization members and the Program Committee members for their hard work, which was essential for the success of DCAI'18.

Sara Rodríguez
Javier Prieto
Pedro Faria
Sławomir Kłos
Alberto Fernández
Santiago Mazuelas
M. Dolores Jiménez-López
María N. Moreno
Elena Navarro

Organization

Honorary Chairman

Masataka Inoue	President of Osaka Institute of Technology, Japan

Program Committee Chairs

Sigeru Omatu	Osaka Institute of Technology, Japan
Sara Rodríguez	University of Salamanca, Spain
Fernando De la Prieta	University of Salamanca, Spain

Local Committee Chairs

Antonio Fernández	University of Castilla-La Mancha, Spain
Elena Navarro	University of Castilla-La Mancha, Spain
Pascual González	University of Castilla-La Mancha, Spain

Scientific Committee

Silvana Aciar	Instituto de Informática, Universidad Nacional de San Juan, Argentina
Naoufel Khayati	COSMOS Laboratory, ENSI, Tunisia
Miguel A. Vega-Rodríguez	University of Extremadura, Spain

Zita Vale	GECAD—ISEP/IPP, Portugal
Paulo Mourao	University of Minho, Portugal
Egons Lavendelis	Riga Technical University, Latvia
Mina Sheikhalishahi	Consiglio Nazionale delle Ricerche, Italy
Patricia Jiménez	Universidad de Huelva, Spain
Volodymyr Turchenko	Research Institute for Intelligent Computing Systems, Ternopil National Economic University, Ucrania
Olfa Belkahla Driss	University of Manouba, Tunisia
Amel Borgi	ISI/LIPAH, Université de Tunis El Manar, Tunisia
Heman Mohabeer	Charles Telfair Institute, Mauritius
Toru Fujinaka	Hiroshima University, Japan
Reza Abrishambaf	Miami University, EE.UU.
Luiz Romao	Univille, Brazil
Abdallah Ghourabi	Higher School of Telecommunications SupCom, Tunisia
Susana Muñoz Hernández	Universidad Politécnica de Madrid, Spain
Fabio Marques	University of Aveiro, Portugal
Ramdane Maamri	LIRE Laboratory UC Constantine2-Abdelhamid Mehri Algeria, Algeria
Julio Cesar Nievola	Pontifícia Universidade Católica do Paraná—PUCPR Programa de Pós Graduação em Informática Aplicada, Brazil
Reyes Pavón	University of Vigo, Spain
Raffaele Dell'Aversana	Research Center for Evaluation and Socio-Economic Development, Italy
Daniel López-Sánchez	BISITE, Spain
Emilio Serrano	Universidad Politécnica de Madrid, Spain
Amin Khan	INESC-ID, Instituto Superior Técnico, Universidade de Lisboa, Portugal
Daniel Hernández de La Iglesia	University of Salamanca, Spain
Alberto López Barriuso	USAL, Spain
Eleni Mangina	UCD, Ireland
Diego Hernán Peluffo-Ordoñez	Yachay Tech, Ecuador
Alberto Fernandez	CETINIA, Rey Juan Carlos University, Spain
Angel Martin Del Rey	Department of Applied Mathematics, University of Salamanca, Spain
María Navarro	BISITE, Spain
Muhammad Marwan Muhammad Fuad	Aarhus University, Denmark
Ichiro Satoh	National Institute of Informatics, Japan
Horacio Gonzalez-Velez	National College of Ireland, Ireland

Takuya Yoshihiro	Faculty of Systems Engineering, Wakayama University, Japan
Julio Ponce	Universidad Autónoma de Aguascalientes, Mexico
Antonio Pereira	Escola Superior de Tecnologia e Gestão do IPLeiria, Portugal
Jaime A. Rincon	Universitat Politècnica de València, Spain
Ali Selamat	Universiti Teknologi Malaysia, Malaysia
Ana Faria	ISEP, Portugal
José Luis Oliveira	University of Aveiro, Portugal
Shimpei Matsumoto	Hiroshima Institute of Technology, Japan
Masaru Teranishi	Hiroshima Institute of Technology, Japan
Sigeru Omatu	Osaka Institute of Technology, Japan
Paulo Vieira	Insituto Politécnico da Guarda, Portugal
Giner Alor Hernandez	Instituto Tecnologico de Orizaba, Japan
Michifumi Yoshioka	Osaka Prefecture University, Japan
Mohd Saberi Mohamad	Universiti Teknologi Malaysia, Malaysia
Jose M. Molina	Universidad Carlos III de Madrid, Spain
David Griol	Universidad Carlos III de Madrid, Spain
Vicente Julian	Universitat Politècnica de València, Spain
Zbigniew Banaszak	Warsaw University of Technology, Faculty of Management, Department of Business Informatics, Poland
Araceli Queiruga-Dios	Department of Applied Mathematics, University of Salamanca, Spain
Angélica González Arrieta	University of Salamanca, Spain
Stefania Costantini	Dipartimento di Ingegneria e Scienze dell'Informazione e Matematica, Univ. dell'Aquila, Italy
Juan Carlos Burguillo	University of Vigo, Spain
Álvaro Lozano Murciego	USAL, Spain
Elisa Huzita	State University of Maringa, Brazil
Rui Camacho	University of Porto, Portugal
Friederike Wall	Alpen-Adria-Universitaet Klagenfurt, Austria
Stefan-Gheorghe Pentiuc	University Stefan cel Mare Suceava, Romania
Zhu Wang	XINGTANG Telecommunications Technology Co., Ltd., China
Adel Boukhadra	National High School of Computer Science (Oued Smar, Algeria), Algeria
Carina Gonzalez	ULL, Spain
Francisco Garcia-Sanchez	University of Murcia, Spain
Irina Georgescu	Academy of Economic Studies, Romania
Ana Belén Gil González	University of Salamanca, Spain
Mariano Raboso Mateos	Facultad de Informática, Universidad Pontificia de Salamanca, Spain

Fábio Silva	University of Minho, Portugal
Li Weigang	University of Brasilia, Brazil
Nadia Nouali-Taboudjemat	CERIST, Algeria
Bozena Wozna-Szczesniak	Institute of Mathematics and Computer Science, Jan Dlugosz University in Czestochowa, Poland
Javier Bajo	Universidad Politécnica de Madrid, Spain
Gustavo Santos-Garcia	University of Salamanca, Spain
Jose-Luis Poza-Luján	Universitat Politècnica de València, Spain
Jacopo Mauro	University of Oslo, Norway
Worawan Diaz Carballo	Thammasat University, Thailand
Paulo Novais	University of Minho, Portugal
Peter Forbrig	University of Rostock, Germany
Nuno Silva	DEI & GECAD—ISEP/IPP, Portugal
Rafael Valencia-Garcia	Departamento de Informática y Sistemas, Universidad de Murcia, Spain
Yann Secq	Université Lille I, France
Tiago Oliveira	National Institute of Informatics, Japan
Rene Meier	Lucerne University of Applied Sciences, Switzerland
Aurélie Hurault	IRIT, ENSEEIHT, France
Fidel Aznar	University of Alicante, Spain
Paulo Cortez	University of Minho, Portugal
Leandro Tortosa	University of Alicante, Spain
Carmen Benavides	University of Leon, Spain
Rosalia Laza	Universidad de Vigo, Spain
Francisco A. Pujol	Specialized Processor Architectures Lab, DTIC, EPS, University of Alicante, Spain
Ângelo Costa	University of Minho, Portugal
Carlos Carrascosa	GTI-IA DSIC Universidad Politecnica de Valencia, Spain
Ana Almeida	ISEP/IPP, Portugal
Luis Antunes	GUESS/LabMAg/Universidade de Lisboa, Portugal
Ester Martinez-Martin	Universitat Jaume I, Spain
José Ramón Villar	University of Oviedo, Spain
Faraón Llorens-Largo	University of Alicante, Spain
Fernando Diaz	University of Valladolid, Spain
Jesus Martin-Vaquero	University of Salamanca, Spain
Maria João Viamonte	Instituto Superior de Engenharia do Porto, Portugal
Cesar Analide	University of Minho, Portugal
Pierre Borne	Ecole Centrale de Lille, France
Johan Lilius	Åbo Akademi University, Finland
Camelia Chira	Technical University of Cluj-Napoca, Romania

Organization of Special Session on Advances on Demand Response and Renewable Energy Sources in Smart Grids (ADRESS)

Smart grid concepts are rapidly being transferred to the market and huge investments have already been made in renewable-based electricity generation and in rolling out smart meters. However, the present state of the art does not ensure neither a good return of investment nor a sustainable and efficient power system. The work so far involves mainly larger stakeholders, namely power utilities and manufacturers, and their main focus has been on the production and grid resources. This vision is missing a closer attention to the demand side and especially to the interaction between the demand side and the new methods for smart grid management.

Efficient power systems require, at all moments, the optimal use of the available resources to cope with demand requirements. Demand response programs framed by adequate business models will play a key role in more efficient systems by increasing demand flexibility both on centralized and distributed models, particularly for the latter as renewable energy generation and storage are highly dependable of uncontrolled factors (such as wind and solar radiation) for which anticipated forecasts are subjected to significant errors.

The complexity and dynamic nature of these problems require the application of advanced solutions to enable the achievement of relevant advancements in the state of the art. Artificial intelligence and distributed computing systems are, consequently, being increasingly embraced as a valuable solution. ADRESS aims at providing an advanced discussion forum on recent and innovative work in the fields of demand response and renewable energy sources integration in the power system. Special relevance is indorsed to solutions involving the application of artificial intelligence approaches, including agent-based systems, data mining, machine learning methodologies, forecasting, and optimization, especially in the scope of smart grids and electricity markets.

Organizing Committee

Kumar Venayagamoorthy	Clemson University, USA
Zita Vale	Polytechnic of Porto, Portugal
Pedro Faria	Polytechnic of Porto, Portugal
Juan M. Corchado	University of Salamanca, Spain
Tiago Pinto	University of Salamanca, Spain

Program Committee

Bo Norregaard Jorgensen	University of Southern Denmark, Denmark
Carlos Ramos	Polytechnic of Porto, Portugal
Cătălin Buiu	Politehnica University Bucharest, Romania
Cédric Clastres	Institut National Polytechnique de Grenoble, France
Dante I. Tapia	Nebusens, Spain
Frédéric Wurtz	Institut National Polytechnique de Grenoble, France
Georg Lettner	Vienna University of Technology, Austria
Germano Lambert-Torres	Dinkart Systems, Brazil
Gustavo Figueroa	Instituto de Investigaciones Eléctricas, Mexico
Ines Hauer	Otto von Guericke University Magdeburg, Germany
Isabel Praça	Polytechnic of Porto, Portugal
István Erlich	University of Duisburg-Essen, Germany
Jan Segerstam	Empower IM Oy, Finland
José Rueda	Delft University of Technology, The Netherlands
Juan Corchado	University of Salamanca, Spain
Juan F. De Paz	University of Salamanca, Spain
Kumar Venayagamoorthy	Clemson University, USA
Lamya Belhaj	l'Institut Catholique d'Arts et Métiers, France
Nikolaus Starzacher	Discovergy, Germany
Nikos Hatziargyriou	National Technical University of Athens, Greece
Marko Delimar	University of Zagreb, Croatia
Nouredine Hadj-Said	Institut National Polytechnique de Grenoble, France
Pablo Ibarguengoytia	Instituto de Investigaciones Eléctricas, Mexico
Paolo Bertoldi	European Commission, Institute for Energy and Transport, Belgium
Pedro Faria	Polytechnic of Porto, Portugal
Peter Kadar	Budapest University of Technology and Economics, Hungary

Organization of Special Session on AI–Driven Methods for Multimodal Networks and Processes Modeling (AIMPM)

The special session entitled AI–driven methods for Multimodal Networks and Processes Modeling (AIMPM 2018) is a forum that will share ideas, projects, researches results, models, experiences, applications, etc., associated with artificial intelligence solutions for different multimodal networks-born problems (arising in transportation, telecommunication, manufacturing, and other kinds of logistic systems).

Recently, a number of researchers involved in research on analysis and synthesis of multimodal networks devote their efforts to modeling different, real-life systems. The generic approaches based on the AI methods, highly developed in recent years, allow to integrate and synchronize different modes from different areas concerning: the transportation processes synchronization with concurrent manufacturing and cash ones or traffic flow congestion management in wireless mesh and ad hoc networks as well as an integration of different transportations networks (buses, rails, subway) with logistic processes of different character and nature (e.g., describing the overcrowded streams of people attending the mass sport and/or music performance events in the context of available holiday or daily traffic services routine). Due to the above-mentioned reasons, the aim of the workshop is to provide a platform for discussion about the new solutions (regarding models, methods, knowledge representations, etc.) that might be applied in that domain.

Organizing Committee

Chairs

Sławomir Kłos	University of Zielona Góra, Poland
Izabela E. Nielsen	Aalborg University, Denmark
Paweł Sitek	Kielce University of Technology, Poland
Grzegorz Bocewicz	Koszalin University of Technology, Poland

Co-chairs

Peter Nielsen	Aalborg University, Denmark
Zbigniew Banaszak	Koszalin University of Technology, Poland
Paweł Pawlewski	Poznan University of Technology, Poland
Mukund Nilakantan Janardhanan	University of Leicester, Leicester, UK
Robert Wójcik	Wrocław University of Technology, Poland
Marcin Relich	University of Zielona Gora, Poland
Arkadiusz Gola	Lublin University of Technology, Poland
Justyna Patalas-Maliszewska	University of Zielona Góra, Poland

Organization of Special Session on Social Modelling of Ambient Intelligence in Large Facilities (SMAILF)

Ambient Intelligence (AmI) is intended to provide users with systems tightly integrated with their everyday environment and activities. The goal is minimizing the need of explicit actions by users, through the continuous and distributed orchestration of information and actuation devices. With the advances in the field, AmI is pursuing growingly ambitious goals in terms of the size and integration of its smart spaces, the number of served users, and the level of adaptation to them.

This special session was focused on the challenges and potential solutions that appear when AmI moves to Large Premises (LP). In this context, new requirements consider big groups of people moving in premises that fall beyond the classical closed and controlled environments of most AmI systems. The ways of interaction, the expected services, and the behavior of people acquire a new dimension and variability in those interconnected smart spaces. AmI systems need to adapt to the crowds using large numbers of multiple and heterogeneous AmI resources in distributed and frequently uncontrollable environments that cause unexpected dynamic changes in the system topology.

Organizing Committee

Alberto Fernandez	Rey Juan Carlos University
Jorge J. Gómez	Universidad Complutense de Madrid
Ramón Alcarria	Universidad Politécnica de Madrid
Álvaro Carrera	Universidad Politécnica de Madrid

Program Committee

Carlos A. Iglesias	Universidad Politécnica de Madrid, Spain
Geylani Kardas Ege	University International Computer Institute, Turkey
Gianluca Rizzo	HES SO Valais, Switzerland
Holger Billhardt	Rey Juan Carlos University, Spain
Iván García-Magariño	University of Zaragoza, Spain
Juan Pavón	Universidad Complutense de Madrid, Spain
Juergen Dunkel	FH Hannover, University for Applied Sciences and Arts, Germany
Marin Lujak	IMT Lille Douai, France
Paulo Novais	University of Minho, Portugal
Rubén Fuentes-Fernández	Universidad Complutense de Madrid, Spain
Sascha Ossowski	Rey Juan Carlos University, Spain
Tomás Robles	Universidad Politécnica de Madrid, Spain

Organization of Special Session on Communications, Electronics and Signal Processing (CESP)

Today's digital revolution, with millions of connected devices providing real-time information about cities, homes, buildings, vehicles, etc., would not have been possible without the great advances in communications, electronics, and signal processing of the last decades. This special session covers all aspects related with these three pillars: new communication approaches such as 5G, massive MIMO, network function virtualization (NFV), software-defined networks (SDN), or millimeter wave communications; novel results in the field of electronics such as new antennas design, emerging Li-Fi devices, micro-electromechanical systems, or nano-electronics devices; and prevalent signal processing methodologies such as adaptive filtering approaches, fusion techniques, navigation systems, or image and video processing.

Organizing Committee

Moe Win	MIT
Andrea Conti	University of Ferrara
Santiago Mazuelas	Basque Center for Applied Mathematics
Marco Chiani	University of Bologna
Javier Prieto	University of Salamanca
Soumya Prakash Rana	London South Bank University

Organization of Special Session on Complexity in Natural and Formal Languages (CNFL)

Complexity has become an important concept in several scientific disciplines. There has been a lot of research on complexity and complex systems in natural sciences, economics, and social sciences. Complexity has always been a central topic in area of formal languages and now also increasingly in natural language research. The main objective of this special session is to bring together researchers from different areas that have in common their interest on linguistic complexity, regarding formal and/or natural languages. We want to boost the interchange of knowledge and methods between specialists that have approached linguistic complexity from different viewpoints. In order to promote interdisciplinarity among researchers that are dealing with any type of linguistic (natural or formal) complexity, this special session was focused in contributions introducing methods, models, definitions, and measures to assess complexity.

Organizing Committee

M. Dolores Jiménez-López	Universitat Rovira i Virgili, Spain
Leonor Becerra-Bonache	University of Saint-Etienne, France
Adrian-Horia Dediu	Universitat Rovira i Virgili, Spain
Adrià Torrens-Urrutia	Universitat Rovira i Virgili, Spain

Organization of Special Session on Web and Social Media Mining (WASMM)

The Web has become an indispensable instrument in the daily life for business activities, learning, entertainment, communication, etc. Offer of products and services to Internet users is practically unlimited; nevertheless, this apparent advantage is also a great drawback due to the fact that the Web provides from multiple sources a great quantity of heterogeneous information difficult to handle and interpret. In this context, data mining methods arise as efficient tools for helping users in the recovery of suitable information, products, or services from the Web. For that reason, recommender systems have become very popular in recent years, mainly in the e-commerce sites, although they are increasing in importance in other areas such as e-learning, tourism, news pages.

Nowadays, social networks are big sources of data, from which valuable information can be extracted by means of data mining algorithms. Social media mining allows us to explore a wide range of aspects regarding users, communities, networks structures, information diffusion, and so on.

WASMM aims at providing a forum for the presentation and discussion of the advances achieved in the Web mining field.

Organizing Committee

María N. Moreno García	University of Salamanca, Spain
Ana María Almeida de Figueiredo	Polytechnic Institute of Engineering of Porto, Portugal

Program Committee

Harshavardhan Achrekar	University of Massachusetts Lowell, USA
Yolanda Blanco	University of Vigo, Spain
Rafael Corchuelo	University of Sevilla, Spain
Chris Cornelis	Ghent University, Belgium
María José del Jesús	University of Jaen, Spain
Anne Laurent	University of Montpellier 2, France
Vivian López Batista	University of Salamanca, Spain
Joel Pinho Lucas	Tail Target, Brazil
Constantino Martins	Institute of Engineering of Porto, Portugal

Organization of Doctoral Consortium Sessions

The aim of the Doctoral Consortium is to provide a frame where students can present their ongoing research work and meet other students and researchers, and obtain feedback on future research directions. The Doctoral Consortium is intended for students who have a specific research proposal and some preliminary results, but who are still far from completing their dissertation.

All proposals submitted to the Doctoral Consortium underwent a thorough reviewing process with the aim to provide detailed and constructive feedback.

The submissions should identify:

- Problem statement
- Related work
- Hypothesis
- Proposal
- Preliminary results and/or Evaluation plan
- Reflections

Doctoral Consortium Organizer

Antonio Fernández-Caballero University of Castilla-La Mancha, Spain

Contents

Special Session on Advances on Demand Response and Renewable Energy Sources in Smart Grids (ADRESS)

UCB1 Based Reinforcement Learning Model for Adaptive Energy Management in Buildings

Rui Andrade[1], Tiago Pinto[1,2](✉), Isabel Praça[1], and Zita Vale[1]

[1] GECAD – Research Group, Institute of Engineering, Polytechnic of Porto (ISEP/IPP), Porto, Portugal
{rfaar, tmcfp, icp, zav}@isep.ipp.pt
[2] BISITE Research Centre, University of Salamanca (USAL), Calle Espejo, 12, 37007 Salamanca, Spain
tpinto@usal.es

Abstract. This paper proposes a reinforcement learning model for intelligent energy management in buildings, using a UCB1 based approach. Energy management in buildings has become a critical task in recent years, due to the incentives to the increase of energy efficiency and renewable energy sources penetration. Managing the energy consumption, generation and storage in this domain, becomes, however, an arduous task, due to the large uncertainty of the different resources, adjacent to the dynamic characteristics of this environment. In this scope, reinforcement learning is a promising solution to provide adaptiveness to the energy management methods, by learning with the on-going changes in the environment. The model proposed in this paper aims at supporting decisions on the best actions to take in each moment, regarding buildings energy management. A UCB1 based algorithm is applied, and the results are compared to those of an EXP3 approach and a simple reinforcement learning algorithm. Results show that the proposed approach is able to achieve a higher quality of results, by reaching a higher rate of successful actions identification, when compared to the other considered reference approaches.

Keywords: Adaptive learning · Energy management in buildings EXP3 · Reinforcement learning · UCB1

1 Introduction

During the last decade a centralized approach is being used in energy (and more specifically, in electricity) markets. Energy consumers are only connected to energy producers and thus the energy distribution is all cantered around one production point [1]. Alternatives to this traditional energy market are emerging and future energy markets are evolving towards a more distributed model. The biggest difference is the decentralization of the energy production, which has originated a new type of role in

This work has received funding from the European Union's Horizon 2020 research and innovation programme under the Marie Sklodowska-Curie grant agreement No 641794 (project DREAM-GO) and from Project SIMOCE (ANI|P2020 17690).

S. Rodríguez et al. (Eds.): DCAI 2018, AISC 801, pp. 3–11, 2019.
https://doi.org/10.1007/978-3-319-99608-0_1

the market: besides energy producers and energy consumers, consumers who are also actively producing energy become part of the market, and are referred to as "prosumers" [2]. With the emergence of prosumers, new possibilities open in the market, enabling the emergence of distributed energy markets. These markets are categorized by connecting consumers and prosumers in a grid while still being connected to centralized producers.

The new characteristics, and consequently the new role, of consumers in the energy ecosystem, force these players to pursue more intelligent and adaptive energy management solutions, in order to be able to take as much advantage from the environment as possible. House or building Energy Management Systems (EMS) are designed to manage the energy consumption and generation within the buildings, respond to energy requests from the grid and minimize the energy bill, while at the same time taking into consideration the comfort levels within the users. The objective is that the EMS will use a minimal amount of energy and still keep the user satisfied [3].

Creating an EMS for smart-houses is the aim in [4]. The system takes into consideration five a total of five possible electrical loads: fixed loads, lights, dishwasher, washing machine and dryer. This EMS also considers the desired temperature for the house which is set by the user. In [5] a similar concept is explored. The proposed EMS has the ability to control the energy consumption within the building and aims to shift the electricity usage depending on the current electricity prices in order to lower the electric costs. The research presented in [6] proposes an EMS for a smart-house focused on managing renewable energy sources such as solar and wind energy generation, Hybrid electric vehicles with batteries, supercapacitors (SCs), and the house itself. The system makes use of maximum power point tracking (MPPT) algorithms to control and optimize the energy storage, solar generation and wind generation.

In order to improve its performance, EMS should collect data and make changes in its behavior when necessary [3]. Artificial intelligence, and machine learning in particular, are promising solutions to improve the processes of self-evaluation and adaptation [7]. Some relevant work has already been accomplished in this domain, e.g. by using reinforcement learning [8], but much has yet to be explored in order to enable an effective and dynamic adaptation of EMS to the constantly changing environment and uncertainty associated to energy resources, such as consumption habits, renewable generation and market prices.

This paper proposes a novel model based on a Markov decision process for decision-making in the context of a smart house. A reinforcement learning approach is presented, in which the goal is to learn the best action for the user to take, considering the expected state of energy resources at each time. An adaptation of the Upper Confidence Bound (UCB1) algorithm (a well-known algorithm for multi-armed bandits [9], is applied to solve de modelled problem. Results are compared to those achieved by the Exponential-weight algorithm for exploration and exploitation (Exp3), also a commonly used algorithm for adversarial bandits problems [10]; and by a simple reinforcement learning algorithm that simply updates the confidence value in each *action-state* pair according to the given reinforcement value at each time. A case study using real data is presented, and shows that the proposed UCB1 based algorithm is able to achieve better results than the other reference algorithms.

2 Proposed Approach

The proposed approach aims at enabling a house EMS to learn and adapt to the dynamic changes in the environment. The objective is to learn which is the best action a to perform at each time t, considering the current state s of the surrounding environment. The proposed model considers a generic set of actions, which can be instantiated depending on each specific application scenario, e.g. as presented in the case study. These may represent the action to consume the energy stored in the battery, to sell the generated energy to the network, etc. Performing an action in a current state results in a specific reward for time t, which represents the value that this action brings in the corresponding state. This process is called a Markov Decision Process (MDP) for decision-making.

Different reinforcement learning algorithms can operate on top of an MDP model. This process can be described in a simple number of steps. A state is given as input, an action is selected and performed, the reward given to the action is used to determine how good that action is in that state and the resulting state is given as the new input and the cycle continues [11, 12], this set of steps is shown in Fig. 1.

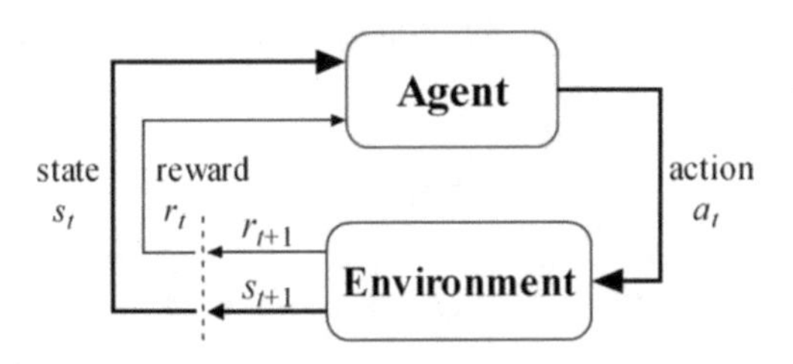

Fig. 1. Reinforcement learning algorithms learning process, [12]

Reinforcement learning algorithms are in constant learning and can adapt to changing environments, referred by [11] as non-stationary environments. This makes reinforcement learning an ideal option for problems that may need to deal with unpredictable changes, such as energy pricing.

There are many reinforcement learning algorithms, with different ways of processing the received rewards, which makes them have distinct results depending on the problem where they are being applied [12].

Multi-armed bandit algorithms [9] are reinforcement learning algorithms that try to find the best of playing in multiple slot machines, also known as “one-armed bandit” or simply “arm”. These may have a biased reward probability distribution picked a priori. Searching for the best arm is called the exploratory phase, and using that information to make the biggest possible profit is the exploitation phase. The aim on these algorithms is to try the different options until enough confidence is built on what option is the best. Upper Confidence Bound (UCB) algorithms are usually applied to solve this problem.

This work presents an adaptation of a UCB algorithm to solve the envisaged MDP problem. UCB1 combines the exploratory phase and an exploitation phase, in a way that the algorithm chooses one of those two approaches in each iteration depending on the received rewards. The algorithm also has a concept of a regret function that is used to try to find the loss correspondent with each arm. The arm with the lowest value in the regret function is considered the best option.

$$8.\left[\sum_{i:\mu_i<\mu^*}\left(\frac{\ln t}{\Delta i}\right)\right]+\left(1+\frac{\pi^2}{3}\right)\left(\sum_{i=1}^{K}\Delta_i\right). \tag{1}$$

Exponential-weight algorithm for exploration and exploitation (Exp3) is an algorithm for adversarial bandits problems, which is similar to the multi-armed bandit problem, but instead of fixed distributions, adversarial bandits follow the idea that an "adversary" is changing the rewards distributions in each time step algorithms [10]. The algorithm uses a parameter called egalitarianism, $\gamma \in [0, 1]$, this parameter is used to balance the exploration. The objective with this parameter is to determine the amount of time $(1 - \gamma)$, in which the algorithm is doing a weighted exploration/exploitation. The weighted exploration/exploitation is based on the current estimated reward, and the rewards received from the weighted exploration/exploitation are immediately used to update the correspondent arm's weight with (2) where i indicates the arm, and Pi represents the received reward for the arm, and (3) is used to calculate the current probability for each arm.

$$w_{i,t} = w_{i,t-1} \cdot e^{\gamma \cdot \frac{P_i}{p_{i,t} \cdot K}} \tag{2}$$

$$p_{i,t} = (1-\gamma)\frac{w_{i,t}}{\sum_{j=1}^{K} w_{i,t}} + \gamma \cdot \frac{1}{K}. \tag{3}$$

A simple reinforcement learning algorithm is also considered in this work, for benchmarking comparison purposes. This algorithm considers the updating of the confidence value in each action a in time t, through a direct increment of the confidence value C according to the reinforcement value R. The update of the values is expressed by (4).

$$C_{a,t+1} = C_{a,t} + R_{a,t} \tag{4}$$

3 Case Study

This case study aims at assessing the proposed approach and comparing the performance of the different reinforcement learning algorithms, by using a practical application case. The MDP model is instantiated as follows. 5 states are considered, combining different possibilities regarding the energy consumption, generation and retail market price, as shown in Table 1. The considered values for these three components are based on real data of a house studied in [13]. Each of the states is active during a specific period in each 24 h cycle. The probabilities of transitions between states at the end of each period are also specified and presented in Table 1. These define the probability of each state occurring, based on a random distribution.

Table 1. Considered states

State	Description	Transition states and probabilities
GgtC_PL	Generation greater than Consumption with low Price	GgtC_PH 50%
		GetC 50%
GgtC_PH	Generation greater than Consumption with high Price	GgtC_PH 66.7%
		GgtC_PL 33.3%
GltC_PL	Generation less than Consumption with low Price	GltC_PL 75%
		GltC_PH 25%
GltC_PH	Generation less than Consumption with high Price	GgtC_PL 10%
		GltC_PL 10%
		GltC_PH 80%
GetC	Generation equal to Consumption (Price independent)	GgtC_PL 50%
		GltC_PL 50%

Conversely, 5 possible actions are also considered. These are presented in Table 2, together with the considered rewards for each State-Action pair. The proposed MDP model is executed for 10000 iterations for each of the three considered algorithms.

Table 2. Possible actions and rewards for each *s-a* pair

	GgtC_PL	**GgtC_PH**	**GltC_PL**	**GltC_PH**	**GetC**
Buy	0.1	0	1	0.6	0.2
Sell	0.8	1	0	0.1	0.2
Store	1	0.8	0.1	0	0.2
Use_stored	0.2	0.1	0.8	1	0.2
Consume	0.4	0.4	0.5	0.8	1

Figures 2, 3 and 4 show the evolution of the rewards for each action over time, for the three algorithms. If the algorithm is able to learn ideally, all actions should converge to the reward value of 1, which is maximum reward value for each action.

From Fig. 2 it is visible that in the first iterations, actions appear to be chosen randomly, however as the iterations increase, the algorithm manages to learn when the action should be used and in that way the algorithm converges around iteration 8000, and all actions start being used ideally.

EXP3 shows a different trend of results. It appears that algorithm is converging to the 1.0 reward value, which would indicate that actions are often being used ideally, however there also low points throughout the entire graph, which indicates that the

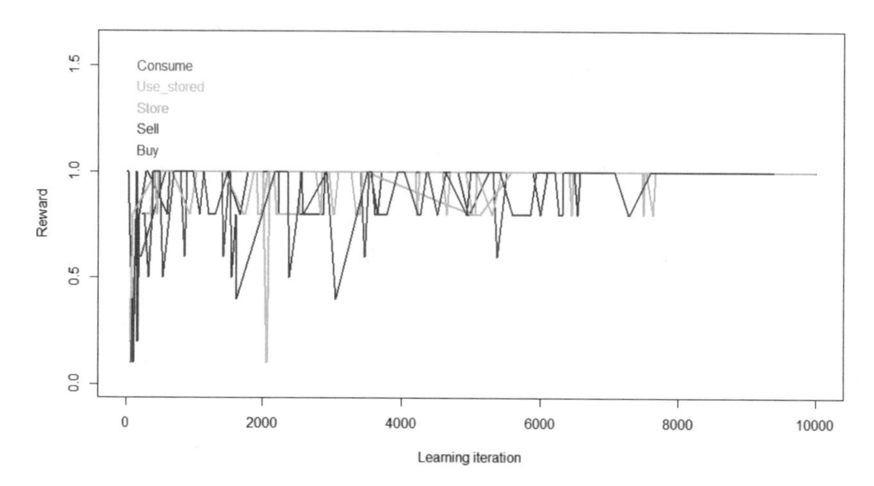

Fig. 2. Simple reinforcement learning algorithm

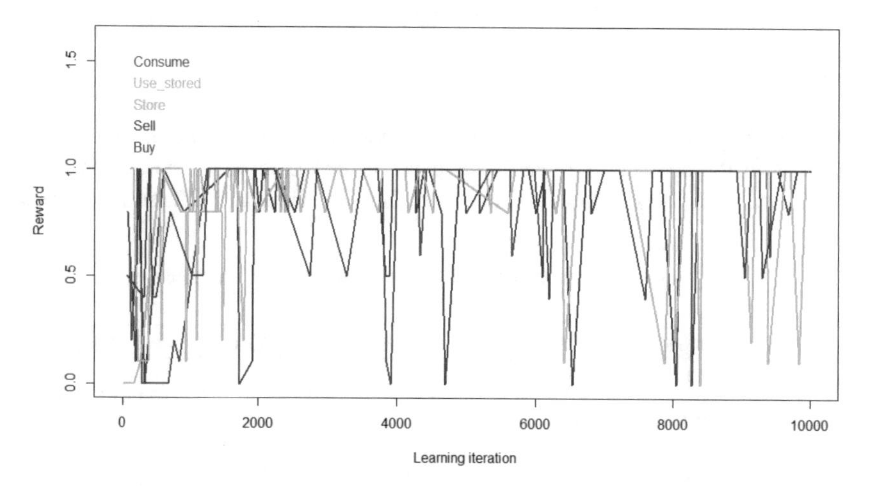

Fig. 3. EXP3

algorithm is constantly exploring and trying to adapt to possible changes to the environment. On the other hand, UCB1 shows a much more stable behaviour. Most of the time actions are chosen ideally (exploitation), with a convergence to the reward value = 1 around iteration 5000. However, the algorithm still continues to explore other actions in a very small frequency, trying to adapt to possible changes.

Table 3 shows the ideal action frequency, which should be achieved if the algorithms would only exploit the best actions and not explore possible alternative actions. Tables 4, 5 and 6 show the frequency of application of each action in each state, for all 3 algorithms. The highest frequency actions in each state are highlighted in green.

By comparing the three algorithms' action selection frequency in each state, it can be seen that the proposed UCB1 approach is able to achieve the best results, with the

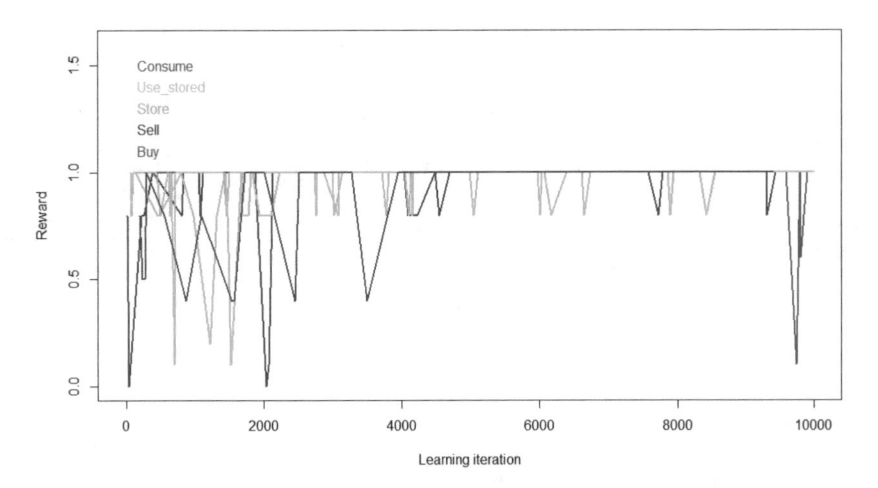

Fig. 4. UCB1

Table 3. Ideal action frequency

	Buy	Consume	Sell	Store	Use_stored
GetC	0	1	0	0	0
GgtC_PH	0	0	1	0	0
GgtC_PL	0	0	0	1	0
GltC_PH	0	0	0	0	1
GltC_PL	1	0	0	0	0

Table 4. Simple reinforcement learning algorithm action frequency

	Buy	Consume	Sell	Store	Use_stored
GetC	0.009	0.971	0.008	0.006	0.006
GgtC_PH	0.003	0.011	0.852	0.130	0.004
GgtC_PL	0.006	0.013	0.176	0.796	0.009
GltC_PH	0.043	0.119	0.004	0.002	0.831
GltC_PL	0.839	0.017	0.004	0.004	0.136

highest frequency of choice of the best action in 4 of the 5 considered states. Only for the state when the generation is equal to the consumption is the simple reinforcement learning algorithm able to reach a higher frequency for the Consume action, which means to simply consume the generated energy. The EXP3 algorithm reaches a low quality of results, giving a high priority to the exploration of alternative actions, and neglecting the exploitation of the best actions. The fact that the EXP3 algorithm does not consider the probability of transition between states, rather using a probability

Table 5. EXP3 action frequency

	Buy	Consume	Sell	Store	Use_stored
GetC	0.054	0.833	0.029	0.042	0.042
GgtC_PH	0.026	0.042	0.762	0.138	0.033
GgtC_PL	0.040	0.036	0.159	0.725	0.040
GltC_PH	0.028	0.037	0.017	0.023	0.895
GltC_PL	0.796	0.040	0.027	0.029	0.109

Table 6. UCB1 action frequency

	Buy	Consume	Sell	Store	Use_stored
GetC	0.022	0.914	0.022	0.022	0.022
GgtC_PH	0.007	0.017	0.878	0.090	0.008
GgtC_PL	0.011	0.021	0.109	0.845	0.013
GltC_PH	0.022	0.065	0.005	0.005	0.903
GltC_PL	0.896	0.017	0.005	0.006	0.075

distribution for each state independently from the possible transitions, makes this algorithm disregard important information about the problem, which may be one of the main causes for its lack of success in this problem. The UCB1 approach, on the other hand is able to learn that the best action is to buy when the price is low and the generation is lower than the consumption; to sell when there is more generation than consumption and the price is high, but to store instead when the price is low; and to use the stored energy when price is high and the generation is not enough to meet the consumption.

4 Conclusion

Energy management in buildings is a central priority worldwide, due to the incentives to the increase of energy efficiency and renewable energy sources penetration. The uncertainty associated to the different resources makes this a hard problem to solve while considering its dynamic characteristics, and constantly changing nature.

This paper addresses this problem by proposing a solution modelled as a Markov decision process. A reinforcement learning model is applied for the intelligent energy management in buildings. A UCB1 based algorithm is applied, and the results are compared to those of an EXP3 approach and a simple reinforcement learning algorithm.

The results from the presented case study show that the proposed UCB1 approach is able to achieve a higher quality of results, by reaching a higher rate of successful actions identification, when compared to the other considered reference approaches.

References

1. Borlase, S.: Smart Grids: Infrastructure, Technology, and Solutions (2012)
2. Rosen, C., Madlener, R.: Regulatory options for local reserve energy markets: implications for prosumers, utilities, and other stakeholders. Energy J. **37**, 39–50 (2016)
3. Fernandes, F., Morais, H., Vale, Z., Ramos, C.: Dynamic load management in a smart home to participate in demand response events. Energy Build. **82**, 592–606 (2014)
4. Acone, M., Romano, R., Piccolo, A., Siano, P., Loia, F., Ippolito, M.G., Zizzo, G.: Designing an energy management system for smart houses. In: 2015 IEEE 15th International Conference on Environment and Electrical Engineering (EEEIC), pp. 1677–1682 (2015)
5. Zhou, B., Li, W., Chan, K.W., Cao, Y., Kuang, Y., Liu, X., Wang, X.: Smart home energy management systems: Concept, configurations, and scheduling strategies. Renew. Sustain. Energy Rev. **61**, 30–40 (2016)
6. Afrakhte, H., Bayat, P., Bayat, P.: Energy management system for smart house with multi-sources using PI-CA controller. In: 2016 Iranian Conference on Renewable Energy Distributed Generation (ICREDG), pp. 24–31 (2016)
7. Pinto, T., Vale, Z., Sousa, T.M., Praça, I., Santos, G., Morais, H.: Adaptive learning in agents behaviour: a framework for electricity markets simulation. Integr. Comput. Aided. Eng. **21**, 399–415 (2014)
8. Li, D., Jayaweera, S.K.: Reinforcement learning aided smart-home decision-making in an interactive smart grid. In: 2014 IEEE Green Energy and Systems Conference (IGESC), pp. 1–6 (2014)
9. Burtini, G., Loeppky, J., Lawrence, R.: A Survey of Online Experiment Design with the Stochastic Multi-Armed Bandit. arXiv1510.00757 [cs, stat] (2015)
10. Bouneffouf, D., Féraud, R.: Multi-armed bandit problem with known trend. Neurocomputing **205**, 16–21 (2016)
11. Shimkin, N.: Reinforcement Learning – Basic Algorithms (2011)
12. Xu, X., Zuo, L., Huang, Z.: Reinforcement learning algorithms with function approximation: recent advances and applications. Inf. Sci. (Ny) **261**, 1–31 (2014)
13. Faia, R., Pinto, T., Abrishambaf, O., Fernandes, F., Vale, Z., Corchado, J.M.: Case based reasoning with expert system and swarm intelligence to determine energy reduction in buildings energy management. Energy Build. **155**, 269–281 (2017)

Electricity Price Forecast for Futures Contracts with Artificial Neural Network and Spearman Data Correlation

João Nascimento[1], Tiago Pinto[2(✉)], and Zita Vale[2]

[1] Energia Simples, Porto, Portugal
joao.nascimento@energiasimples.pt
[2] GECAD – Research Group, Institute of Engineering, Polytechnic of Porto (ISEP/IPP), Porto, Portugal
{tmcfp, zav}@isep.ipp.pt

Abstract. Futures contracts are a valuable market option for electricity negotiating players, as they enable reducing the risk associated to the day-ahead market volatility. The price defined in these contracts is, however, itself subject to a degree of uncertainty; thereby turning price forecasting models into attractive assets for the involved players. This paper proposes a model for futures contracts price forecasting, using artificial neural networks. The proposed model is based on the results of a data analysis using the spearman rank correlation coefficient. From this analysis, the most relevant variables to be considered in the training process are identified. Results show that the proposed model for monthly average electricity price forecast is able to achieve very low forecasting errors.

Keywords: Artificial neural networks · Electricity price · Forecasting Futures contracts · Spearman correlation

1 Introduction

The liberalization of the electric energy market has changed the paradigm of a sector whose organizational models were traditionally monopolistic, being mostly owned by the state [1]. This change has introduced the competitive factor in the sector, making it possible to reduce costs, improve quality and reliability of service, and even promote innovation in new products and services. However, this also brought instability to the prices practiced and a greater risk to the market agents of this sector [2].

The characteristics of the post-liberalization electricity market make forecasting price developments very complex, but important for the protection of those involved, be they producers, traders or consumers [3]. In the last decades several models have

This work has received funding from the European Union's Horizon 2020 research and innovation programme under the Marie Sklodowska-Curie grant agreement No 641794 (project DREAM-GO), from NetEfficity Project (P2020-18015) and from FEDER Funds through COMPETE program and from National Funds through FCT under the project UID/EEA/00760/2013.

S. Rodríguez et al. (Eds.): DCAI 2018, AISC 801, pp. 12–20, 2019.
https://doi.org/10.1007/978-3-319-99608-0_2

been proposed for predicting the price of electricity, however, there is still no consensus regarding the method that should be used for each of the three types of time horizons - short, medium and long term. Some of the most widely used forecasting methods are: statistical models, physical models and models based on computational intelligence [4].

Statistical models conceive forecasts of electricity prices based on a combination of historical price values and/or current or historical values of other variables, such as consumption, production, weather, among others. In this context, time-series techniques (e.g. ARIMA models [5]) and GARCH (Generalized autoregressive conditional heteroscedasticity [6]) models are highlighted.

Computational intelligence models are computational techniques designed to solve problems in which traditional methods do not provide an effective response. Models capable of dealing with complex and non-linear problems. One of the main models of this class is Artificial Neural Networks (ANN) [7], since it can be used for both short and long term forecasts. This model is a simplification of the central nervous system of the human being, being a structure composed of computational units extremely connected to each other (neurons), with capacity of learning, and connections (also called synapses) with the ability to store knowledge in [8]. There are several types of ANN, being classified according to their architecture and learning algorithm. The first concerns how the connections between neurons are organized and the second with the way the neural network adjusts the weights of its connections.

This paper provides a study of energy prices forecasting in futures market, in the Iberian Electricity Market (MIBEL) [9]. Initially, a collection and treatment of historical data of electricity production by technology type, consumption and meteorological data for Portugal and Spain, is carried out. Afterwards, all the data is evaluated through the Spearman correlation coefficient [10], allowing to understand the trend of the price variation of the energy in the MIBEL, compared to the other values. It was also carried out a study of the demonstrated seasonality, as much by the price of the electric power as by the load verified in the Portuguese and Spanish markets, since the load would be one of the data that most influenced the price variation. Therefore, this paper proposes models for forecasting, in order to help reduce the error obtained for the price forecast.

Prediction models are developed using several ANN models. In particular, a monthly forecast model is presented, containing several relevant historical information regarding past consumption, generation, prices, and weather data. The performance evaluation is done through the Mean Absolute Percentage Error (MAPE) and Absolute Percentage Error (APE), and it is concluded that the developed forecast models show a very acceptable accuracy.

2 Proposed Methodology

2.1 Data Correlation

In order to increase the accuracy of the forecast to be made, a correlation analysis of the variables collected with the price of electricity was carried out. In this way, and when evaluating the degree of relationship between variables, it is possible to find out precisely how much one variable interferes with the result of another. The techniques associated with Correlation Analysis represent a fundamental application tool in the Social Sciences and Behavior, Engineering and Natural Sciences. There are several evaluation criteria for this relationship, some for variables that follow a normal distribution and others for variables that do not follow a known theoretical distribution. Several studies are based on Pearson's correlation coefficient [11]. However, there are situations in which the relationship between two variables is not linear, or one of them is not continuous, or the observations are randomly selected. In situations like these, other coefficient alternatives should be applied. The method chosen to study the correlation of the different variables present in the data collected in relation to the price was the Spearman Coefficient [10]. This method, also known as Spearman's Ordinal Correlation Coefficient, measures the binding force between two variables, being as greater as the more monotonous the relative evolution of these variables be, reaching the maximum values of +1 or −1, if they are directly or inversely proportional, respectively. The value indicative of the correlation is represented by ρ, and is presented in (1):

$$\rho = 1 - \frac{6\sum d_i^2}{n(n^2 - 1)} \tag{1}$$

The Spearman coefficient orders the values of the variables x_i and y_i separately, where i represents the value of stations 1, 2, 3, …, n. Then the difference, d_i, of the positions of x_i and y_i, and the square of this value (d_i^2) is calculated. The number of samples is represented by n.

In the analysis of data collected from July 1, 2015 to June 30, 2017, corresponding to the Iberian Electricity Market (MIBEL) [9], seasonal patterns of variability were identified between them, whether daily, weekly or annual, although the selected data to develop the present forecasting model, will only use the second year of information (through July 1, 2016, to June 30, 2017).

Daily patterns are easily identifiable when analysing how the price of electric energy varies throughout the day, being related to the human activity and, of course, to the consumption of this variable. This relationship is copied by load consumption, when comparing both for each of the horizons studied, be it for the relationship of these characteristics throughout the day or the week, mainly. This behaviour, makes that the load represents an important piece of data for the forecast of electricity prices. Figure 1 illustrates: (a) the variation of electricity price with the irradiation; and (b) Load Consumption and Deviation from the Average Value of Irradiation.

From Fig. 1 (a) it is noticeable that there is a negative correlation (inversely proportional evolution) for the data of both variables in the analysed temporal window. This is the data from July 1, 2015 until June 30, 2017, although the data used for the day-ahead forecast model was only July 1, 2016 through June 30, 2017, as they would be relatively short-term forecasts. This option was chosen since no older data would be

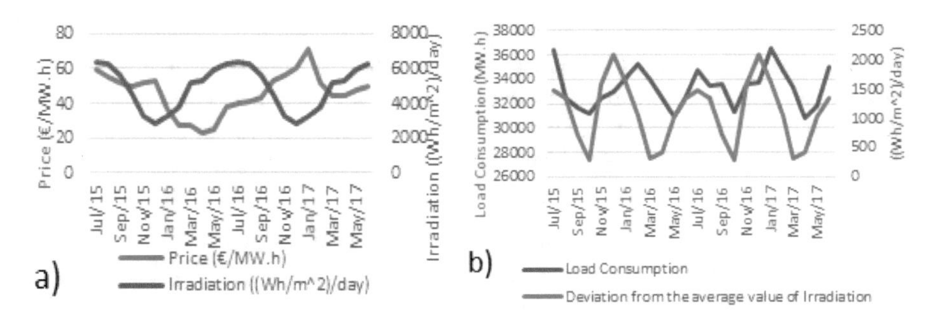

Fig. 1. Variation between: (a) electricity price and Irradiation; (b) load consumption and deviation from the average value of irradiation

required given the apparent correlation between these variables. Although there is some correlation between electricity price and load consumption, and electricity price and irradiation, the correlation between load and irradiation did not have an apparent value. This led to the creation of a new variable: the deviation between irradiation and its own average value. This happens because it is on the seasons with peaks of radiation (be them positive or negative), that the value of load consumption is higher, see Fig. 1 (b).

Using the Spearman Correlation method is possible to analyse the value of correlation between variables, with precision. The analysis is performed in relation to the price and the load consumption. Since in this data set there is a great diversity of variable types, they are separated according to their characteristics. Data refers to production and consumption for the Iberian Peninsula. The notation used as "(D-1)", for example, indicates the value of the referred variable, 24 h before the time for which it is intended to predict (day-1). Likewise, "(D-2)" refers to 48 h before, and "(W-1)" to 168 h before (a week before). In case the reference is "D", it indicates that it is the value for the same period for which it is intended to predict. Initially, in many of the variables tested here, the reference used is "D" as a way to make a previous analysis if it would be worthwhile to test for other references, since the value in this situation will be, in the great majority of variables, higher than previous references, be it 24, 48 or 168 h before. Table 1 presents the Spearman coefficient results obtained for the data related to energy production by technology, and consumption.

Table 1. Price correlations with load consumption

Tested variable	Chronological reference	Correlation value in percentage (ρ)
Load consumption	D	50,30%
	D-1	38,94%
	D-2	28,64%
	W-1	45,11%

The weather data collected for the Iberian Peninsula was also evaluated, and this data were generally more influential in price development. Together with this data, and because they are also meteorological data, although from another source and monthly values instead of day based as in the other cases, solar radiation values will be included. The correlations of these data can be verified in Table 2.

Table 2. Price and load correlations with meteorological data

Tested variable	Chronological reference	Correlation % (ρ) price	Correlation % (ρ) load
Irradiation	D	−83,22%	−4,20%
Average Temperature	D	−57,28%	−4,00%
Wind	D	−34,74%	−15,78%
Precipitation	D-1	−20,96%	−9,33%
Deviation of the average value of Irradiation	D	–	72,73%
Deviation of the average value of Temperature	D	–	51,64%

Analysing the values presented in Table 2, one can see the importance of both solar radiation and temperature in price evolution. Regarding the evolution of the load, the deviation of the average values of solar radiation and of average temperature have a much higher correlation than of the verified values. This is explained by the connection between these, and by the fact that when more extreme temperatures and irradiance values occur, there is also a need for greater consumption, for heating/cooling, or even lighting. Finally, we will analyse the influence that recent historical price values have on this same variable, as well as the history of load correlated to the current load. The values of these Spearman coefficients are shown in Table 3.

Table 3. Price and load correlations with its own historical data

Chronological reference	Correlation % (ρ) Price	Correlation % (ρ) Load
D-1	76,94%	81,74%
W-1	73,79%	90,66%
W-2	71,16%	86,55%

From Table 3 it is noticeable that both the price and the load present a large correlation with their own previous values. These variables will then be the main data to be used in the forecast models to be applied. In addition to the inputs tested in the tables presented in this section, chronological variables will also be used, due to the variability of the load and the price depending on the intensity of the business activities throughout the day, week and, to some extent, the year. With this in mind, the variables of this type to be used will be Hour (H), Day (D), Day of Week (DoW), Month (M), National Holidays (NH) and Seasonality (Summer, Winter or intermediate seasons).

2.2 Forecasting Model

A test ANN is used, based on the nftool and nntool methods of Matlab. The nftool is used to implement, create and train the network, and the nntool is used to simulate the results. Although with a small difference, the method with a smaller error, being the selected method was method 2 (3.52% versus 3.77% of method 1) which used only nntool to create, train and simulate the networks. All simulations and networks were created according to this same method, with the algorithm feedforward backpropagation, in which the information flow occurs unidirectionally, progressing from the input neurons to the output neurons. According to the architecture selected for all the networks developed in this paper, these are multi-layer perceptrons, and in this type of networks where there is a hidden layer that contains a variable number of neurons, this being a fact that makes this type of networks widely used. Two-layer networks were used, and being that there is no universal rule for the selection of the number of neurons N in the hidden layer the rule expressed in (2) has been used, where n is the number of network inputs.

$$N = 3n + 2 \tag{2}$$

In the other parameters to be defined in the networks, Levenberg-Marquardt backpropagation [12] was used as a training function, and as learning and performance adaptation functions the chosen were, learngdm (gradient descent with momentum weight and bias learning function) and mse (mean -squared error), respectively. In layer 1 (hidden layer) the function tansig was used as transfer function, and in layer 2 (output layer) the function purelin.

3 Case Study

Due to the small amount of data collected, compared to those that would be necessary for forecasts of a month or superior time horizon, and also to the greater unpredictability and uncertainty for medium/long term forecasts, it is estimated only one average price of electricity for the following month. For this, the monthly average values of the available variables that showed to be more determinant in the influence shown in the evolution of the price, were used. In order to obtain these same values, new adaptations were made to the data collected, and since it is a forecast that can be considered as being long term, in this case all available data collected were used, that is, from July 1, 2015 until July 30, 2017. However, since one of the entries chosen for the models created was, for example, Price (M-13) (Price 13 months before), data were collected from two new months (July and August 2017) in order to have sufficient input data to obtain previous reference data for the simulation to be carried out. In this case, the month of August 2017 was selected to predict, as a test set, and the remaining months with complete information (from August 2016 to July 2017) for training.

A destructive method is applied in structuring the models, starting with a model with the totality of inputs available, and taking in the following models entries that apparently were not as determinant in the behaviour presented by price. The parameterization used for the simulations to be carried out with regard to this model, after having tested several different alternatives, as well as several training functions of the networks, ended up being the same as that used in the other forecast models (Levenberg- Marquardt), and the aforementioned parameterization, which appears in the nntool tool by default (min_-grad = 1e-07; and max_fails = 6). As for the number of training sessions, it was decided, as in the previous model of price forecast, to take the decision on the values presented in the regression graph presented during the training of the networks, thus ensuring a good adaptation of the networks to the training data.

On the other hand, we did not exaggerate the amount of training data, so that there was no overfitting. This would incur a greater error when given test data he had not experienced before, not knowing how to adapt. Care was taken, not to allow this to occur when stopping the training of the networks when the value *R* (which indicates the correlation of the network with the data) of the validation ceased to increase, stabilizing. Overfitting can also be manifested when too many weights are assigned to some variables by the training function.

The constitution of the models created, as is normal in the long-term forecasting methods, did not contain the same number of input variables as shorter term ones, nor did it have the same ease of obtaining data. The data available for this model are Month (M), Year (Y), Monthly average price of previous months - Price (M-x) for 1, 2, 12 and 13 months - Monthly average load for the month to be forecast (Load Consumption), Solar Radiation (Irradiation), Seasonality (Season) and Average monthly temperature (Aver.Temp.). As previously mentioned, the first model created, contained all variables considered relevant for this forecast. For the second model, it was decided to withdraw the value of the load, and in the third model also the information of the Year and Seasonality. In the fourth, and last, model created, the decision was taken by the withdrawal of these two variables only, maintaining the load.

The simulations of the models in question were followed, with the creation of 5 ANN for each one, in order to obtain 5 different outputs. After calculating the average of the outputs, the value of the absolute percentage error was then evaluated to classify the model error. The values obtained in this process can be seen in Table 4.

Table 4. Obtained results on the evaluation of Monthly Price Forecast

Models	APE (%)
MPM 01 (with Load)	7,79%
MPM 02 (without Load)	5,58%
MPM 03 (without Load)	**0,50%**
MPM 04 (with Load)	5,37%

The model that obtained a smaller error percentage was then the model MPM 03 with an APE value of 0.50%, being also the model that has less variables of the four servants. Since it is a model with no load value included in its inputs, it does not has the need to develop a load forecasting model to obtain this value, unlike the other models, from the other time horizons, daily and weekly, that which results will be compared with the Monthly Price Forecast Model.

Finally, a procedure was applied to handle the error. To achieve this, it were used dispersion measures, as a measure of error management, in order to identify and minimize it. After calculating the mean and standard deviation of the outputs for each period, in the selected final models, a range was constructed by adding and subtracting the resulting standard deviation value to the mean value obtained for each moment. Then, the values of outputs that did not integrate the created interval would not be included in the final calculation of the mean, whose value was then considered for the final predicted value for the moment in question. The objective of this process was to reduce the dispersion of the predicted values, ignoring the values that deviated.

With the application of this method, unlike the other cases, this did not obtain a reduction of the error value of the chosen indicator, the APE. Although the application of the mentioned process reduced the amplitude and the variance of the values obtained in the simulations, the result of the error remained approximately in the same value, worsening slightly, from 0.50% to 0.57%. These values can be somewhat misleading, given the meagre amount of values obtained to train the network in question. However, since it is the prediction of only one value, unlike the other models, it is understandable that a smaller error value is obtained than in the other cases. Figure 2 shows the actual average price value for the month of August 2017, as well as the five predicted values.

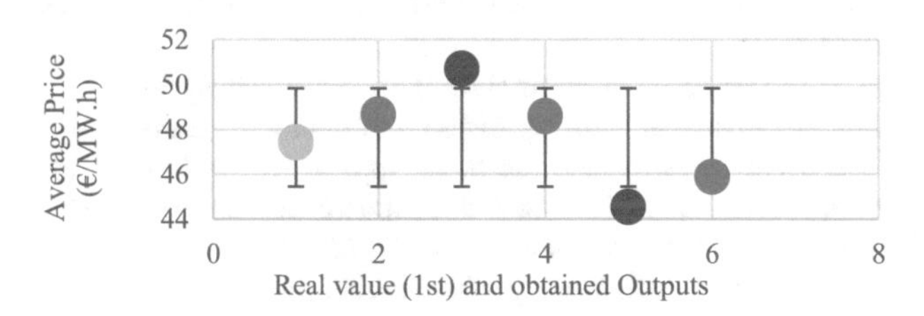

Fig. 2. Actual electricity price value and predicted values

The actual value is in green (1st value) and the following are the ANN forecast. Those whose value is outside the range that translates the standard deviation of the set, are in red, and the remainder in blue. The values of these three blue points were taken into account for the final mean that defined the simulation value, which was used for the evaluation of the error after its management process.

4 Conclusion

Electricity market prices forecast is an active domain of research. Futures contracts prices forecasting, however, is a rather unexplored theme. This paper presents a methodology to forecast futures contracts prices with artificial neural networks and data analysis, using the spearman's rank correlation coefficient. Results using real data from the Iberian electricity market show that the proposed approach is able to achieve very low forecasting errors, through the proper identification of the most relevant variables to be considered in the training process of the artificial neural network.

References

1. Sioshansi, F.P.: Evolution of Global Electricity Markets: New Paradigms, New Challenges, New Approaches (2013)
2. Geng, Z., et al.: Electricity production scheduling under uncertainty: max social welfare vs. min emission vs. max renewable production. Appl. Energy **193**, 540–549 (2017)
3. Mohsenian-Rad, A.H., Leon-Garcia, A.: Optimal residential load control with price prediction in real-time electricity pricing environments. IEEE Trans. Smart Grid. **1**, 120–133 (2010)
4. Nowotarski, J., Weron, R.: Recent advances in electricity price forecasting: a review of probabilistic forecasting. Renew. Sustain. Energy Rev. **81**, 1548–1568 (2018)
5. Al-Musaylh, et al.: Short-term electricity demand forecasting with MARS, SVR and ARIMA models using aggregated demand data in Queensland. Australia. Adv. Eng. Informatics. **35**, 1–16 (2018)
6. Corrêa, J.M., Neto, A.C., Teixeira Júnior, L.A., Franco, E.M.C., Faria, A.E.: Time series forecasting with the WARIMAX-GARCH method. Neurocomputing **216**, 805–815 (2016)
7. Wang, S., et al.: Wind speed forecasting based on the hybrid ensemble empirical mode decomposition and GA-BP neural network method. Renew. Energy. **94**, 629–636 (2016)
8. Pinto, T., Sousa, T.M., Vale, Z.: Dynamic artificial neural network for electricity market prices forecast (2012)
9. MIBEL - Mercado Ibérico de la Electricidad. http://www.mibel.com
10. Zhang, W., et al.: Measuring mixing patterns in complex networks by Spearman rank correlation coefficient. Phys. A Stat. Mech. its Appl. **451**, 440–450 (2016)
11. Mu, Y., Liu, X., Wang, L.: A Pearson's correlation coefficient based decision tree and its parallel implementation. Inf. Sci. (Ny) **435**, 40–58 (2018)
12. Mammadli, S.: Financial time series prediction using artificial neural network based on Levenberg-Marquardt algorithm. Procedia Comput. Sci. **120**, 602–607 (2017)

Demand Response Implementation in an Optimization Based SCADA Model Under Real-Time Pricing Schemes

Mahsa Khorram, Pedro Faria(✉), Omid Abrishambaf, and Zita Vale

GECAD – Research Group on Intelligent Engineering and Computing for Advanced Innovation and Development, Institute of Engineering, Polytechnic of Porto (ISEP/IPP), Porto, Portugal
{makgh, pnfar, ombaf, zav}@isep.ipp.pt

Abstract. Advancement of renewable energy resources, development of smart grids, and the effectiveness of demand response programs, can be considered as solutions to deal with the rising of energy consumption. However, there is no benefit if the consumers do not have enough automation infrastructure to use the facilities. Since the entire kinds of buildings have a massive portion in electricity usage, equipping them with optimization-based systems can be very effective. For this purpose, this paper proposes an optimization-based model implemented in a Supervisory Control and Data Acquisition, and Multi Agent System. This optimization model is based on power reduction of air conditioners and lighting systems of an office building with respect to the price-based demand response programs, such as real-time pricing. The proposed system utilizes several agents associated with the different distributed based controller devices in order to perform decision making locally and communicate with other agents to fulfill the overall system's goal. In the case study of the paper, the proposed system is used in order to show the cost reduction in the energy bill of the building, while it respects the user preferences and comfort level.

Keywords: Optimization · SCADA · Multi-agent system

1 Introduction

Currently, power distribution networks are being updated and move towards the smart grids paradigms [1]. The use of smart gird brings flexibility on the resource management somehow it enables the network operator to have control on electricity consumption and generation [2]. On the other hand, the daily increment of electricity usage forced the network operators to reduce the method of generation by fossil fuels [3], and move towards sustainable and renewable energy resources, especially Photovoltaic

The present work was done and funded in the scope of the following projects: H2020 DREAM-GO Project (Marie Sklodowska-Curie grant agreement No 641794); Project GREEDI (ANI|P2020 17822); and UID/EEA/00760/2013 funded by FEDER Funds through COMPETE program and by National Funds through FCT.

S. Rodríguez et al. (Eds.): DCAI 2018, AISC 801, pp. 21–29, 2019.
https://doi.org/10.1007/978-3-319-99608-0_3

(PV) systems and wind turbines [4]. A significant part of electricity consumption is allocated to commercial buildings, especially office buildings [5]. In this context, Air Conditioners (ACs) have great contribution on the consumption of these kinds of buildings [6]. Demand Response (DR) programs are considered as a solution for managing the consumption of the demand side [7].

DR program is referred to modification of consumption pattern by the end-users in response to the incentives payment by the DR managing entities, which is due to any economic or technical reasons [8]. Real-Time Pricing (RTP) is an example of price-based DR programs, which is applied in day-ahead economic scheduling [9]. In order to implement these programs, the end-users should be equipped with several automation infrastructures in order to be able to perform these programs [10]. Supervisory Control And Data Acquisition (SCADA) system plays a key role in DR implementation, since it offers various advantages in order to have automatic load control in different types of buildings [11].

Multi-Agent System (MAS) is an essential tool for SCADA systems for controlling strategies and exchanging system status [12]. In fact, MAS based SCADA systems would be able to perform complex optimization algorithm, and simultaneously, manage the controllable loads connected to SCADA system [13]. Flexibility and adaptability are two main capabilities that a MAS offers [14].

This paper represents an optimization algorithm implemented in a MAS based SCADA model for an office building. The proposed algorithm manages the consumption of the building under RTP tariff and manages the consumption of the ACs and illumination systems of the building based on defined priorities by the office users. Furthermore, the controlling of the loads is performed through a MAS model, which each agent is associated with particular part of the implemented SCADA system. This enables the model to perform decision making locally and communicate with other agents to fulfill the overall system's goal.

The rest of the paper is organized as follows. The MAS implemented in SCADA model is described on Sect. 2. Section 3 represents the proposed optimization algorithm. A case study is represented in Sect. 4, and its results are illustrated in the same section. Section 5 details the main conclusions of the work.

2 Multi-agent Model Architecture

This section shows the details about the MAS in implemented SCADA system in order to control the consumption of building. The automation infrastructures have been implemented in a part of GECAD research center building, which contains 8 offices, 1 server room, and a corridor. Moreover, there is 7.5 kW PV installation on the building, which supplies a part of total consumption. For managing the consumption of building, three distributed based Programmable Logic Controllers (PLCs) dedicated for a zone including three offices. Therefore, there are three zones somehow each PLC associated with one zone. Moreover, there is a main PLC that is responsible for supervising the other distributed based PLCs. The main controlling panel of the SCADA system including all PLCs is shown on Fig. 1. The controllable loads by the SCADA include lighting systems and ACs, which are controlled by several communication protocols.

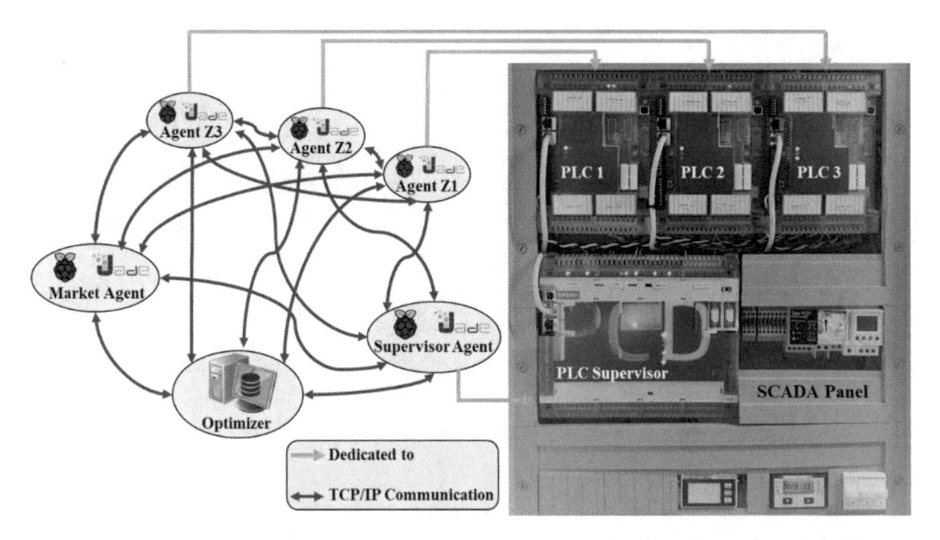

Fig. 1. Multi-agent model of the implemented SCADA system for office building.

Moreover, the real-time consumption of the building is measured through several energy meters. In this model, there are five main agents that each of which is run by a Raspberry Pi (www.raspberrypi.org). As Fig. 1 illustrates, Agent Z1, Z2, and Z3 are devoted for each zone, where these agents are equipped with a PLC for performing controlling decisions locally.

Moreover, Market Agent is responsible to inform the other agents from the real-time electricity price of the market. In this model, the unit called Optimizer is accountable to perform the optimization algorithm, which will be demonstrated in the next section. Additionally, the Supervisor Agent continuously check the status of other agents and if there is any faulty agent, it reconfigures the system. All agents in this system continuously exchanges messages for sharing their latest status through TCP/IP communication. By this way, the response time to any changes will be reduced and adaptability of the system will be increased. Furthermore, flexibility and reconfigurability are two features that will be offered by this MAS.

3 Optimization Algorithm

This section introduces the algorithm implemented in the Optimizer agent, which is responsible for optimizing the power consumption of the building based on RTP. Figure 2 represents the algorithm of this methodology, based on the power reduction of air conditions and lighting system. To achieve the purpose of running this algorithm in the Optimizer agent, all the other agents are obligated to many tasks, such as providing the essential data for the algorithm.

As it is clear in Fig. 2, all data of the building, which are transmitted from other agents, and also the external data received from Market agent, such as electricity price and DR programs, are considered as input data of the algorithm. After definition of the

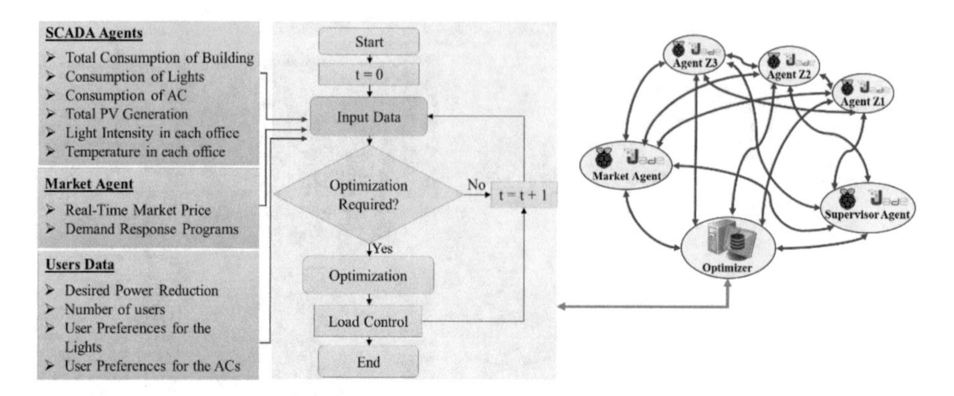

Fig. 2. The procedure of optimization algorithm.

input data, the algorithm performs the decision making for starting optimization. This optimization considered as RTP based, since it will be triggered if the electricity prices goes greater than a specific value. In this optimization, comfort level of user is a critical input. For each AC and light, a specific priority value attributed, which determines contribution of each device in optimization. In the User Data inputs, there is an optional rate for the users, which defines the desired rate of power reduction.

The ACs devices have various priority numbers in each period, since the algorithm should not turn off only specific devices in all periods. Also, in the lights, they are not obligated to participate in the optimization in all periods. This means in certain periods that ACs optimization is not enough to achieve the amount of required reduction of algorithm, lights are as auxiliary part for the ACs. For maintaining the comfort of user, the lights optimization also is based on the priorities defined by the users. Moreover, none of the lights should not be cut completely, and a minimum illumination level is considered for each light.

The proposed methodology is modelled as a linear Programming (LP) optimization problem, which is solved via "Lp_solve" package of Rstudio® (www.rstudio.com). The main objective function of the optimization algorithm is shown by (1), which aims to minimize the Energy Bill (EB).

$$\begin{aligned} Minimize\, EB = \sum_{t=1}^{T}(((\sum_{a=1}^{A} P_{RED.AC(a,t)} \times W_{AC(a,t)}) \\ + (\sum_{l=1}^{L} P_{RED.Light(l,t)} \times W_{Light(l,t)}) + P_{total(t)} - PV_{(t)}) \times COST_{(t)}) \end{aligned} \tag{1}$$

Where $P_{RED.AC}$ is the power that will be reduced from each AC, and $P_{RED.Light}$ is the rate of power that will be reduced from each lights, which are based on the defined input data by users for the rate of desired power reduction. W_{AC} and W_{Light} are the abbreviation of weight of the priority of the ACs, and the lights respectively. P_{total} expresses the total power consumption of the building, PV stands for total PV generation of the building, and *COST* is the electricity price. *T* is the maximum number of periods, and finally, *A* and *L* are maximum number of ACs and lights, respectively.

Equation (2) and (3) shows the limitation of priority number, which should be a value between 0 and 1. Each priority number closer to 0 is the lowest important device for the users and algorithm as well. Equation (4) illustrates the required reduction of the algorithm, which should be decreased to fulfill the goal of the algorithm. Equation (5) shows that the lights will not participate in entire periods of optimization, and they are limited to be reducted only in critical periods.

$$0 \leq W_{AC(a,t)} \leq 1 \quad \forall 1 \leq t \leq T; \forall 1 \leq a \leq A \tag{2}$$

$$0 \leq W_{Light(l,t)} \leq 1 \quad \forall 1 \leq t \leq T; \forall 1 \leq l \leq L \tag{3}$$

$$\sum_{a=1}^{A} P_{RED.AC(a,t)} + \sum_{l=1}^{L} P_{RED.Light(l,t)} = RR_{Total(t)} \quad \forall 1 \leq t \leq T \tag{4}$$

$$RR_{Light(t)} = RR_{Total(t)} - \sum_{l=0}^{L} P^{MAX}_{RED.AC(l,t)} \quad \forall 1 \leq t \leq T \tag{5}$$

RR_{Total} is for total Required Reduction, and $P^{MAX}_{RED.AC}$ is the maximum capacity of ACs for reduction. RR_{Light} stands for required power reduction from lights, which should be a value lower than maximum reduction capacity of all lighting system ($P^{MAX}_{RED.Light(t)}$), as shown on (6). For the limitation of power reduction, (7) shows the technical limitation for each light. and (8) shows the discrete control of ACs, which should be off or on.

$$RR_{Light(t)} \leq \sum_{l=0}^{L} P^{MAX}_{RED.Light(l,t)}; \forall 1 \leq t \leq T \tag{6}$$

$$0 < P_{RED.Light(l,t)} \leq P^{MAX}_{RED.Light(l,t)} \tag{7}$$

$$P_{RED.AC(a,t)} = \left\{0, P^{Max}_{RED.AC(a,t)}\right\} \forall 1 \leq t \leq T; 1 \leq a \leq A \tag{8}$$

4 Case Study and Results

In this section, a case study will test and validate the proposed methodology. As it previously mentioned, main purpose of this paper is to optimize the building consumption with taking advantages of MAS in the implemented SCADA system. There are 8 ACs devices in the building that all are turned on during working hour. Moreover, an AC is located in the server room that is always turned. The lighting system contains 19 fluorescents lamps that are controlled by SCADA via Digital Addressable Lighting Interface (DALI). In this case study, it is considered that SCADA model is configured somehow that if the electricity price is greater than 0.08 EUR/kWh (considered as Base Price), it will perform the optimization algorithm. Also, if the prices increased, the SCADA system specifies more reduction in the optimization algorithm. Figure 3 illustrates obtained results after and before performing the optimization algorithm for 24 h (24 periods).

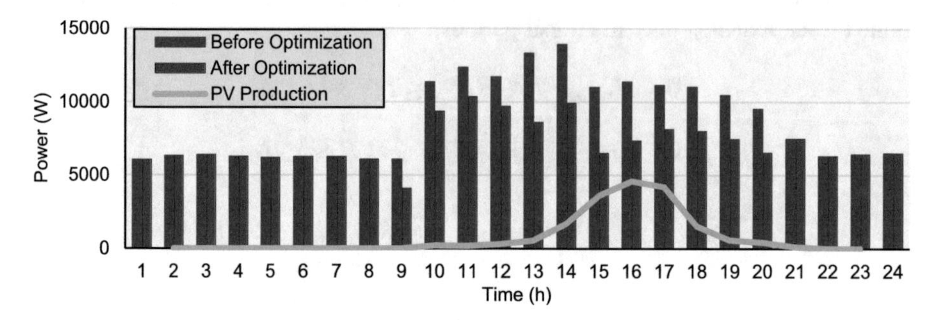

Fig. 3. Total consumption and generation of building.

The consumption and generation curves used in this case study are the real consumption and generation of GECAD building adapted from GECAD database. Moreover, the market prices are for a winter day in 2018 and have been adapted from Portuguese sector of Iberian Electricity Markets (MIBEL – www.omie.es).

As you can see in Fig. 3, the optimization process starts at 9:00 and finishes at 20:00, since the electricity price is higher than Base Price, therefore, the system performs the optimization in order to reduce the energy bill. Furthermore, PV generation profile shown on Fig. 3 is the maximum generation capacity of the system.

In fact, the power reduction shown on Fig. 3, can be related only to the ACs reduction, or in some periods can be the cooperation of ACs and the lights reduction in order to achieve the desired reduction. Figure 4 illustrates the contribution of ACs, and Fig. 5 demonstrates the contribution of lighting system in the optimization process based on RTP for one day.

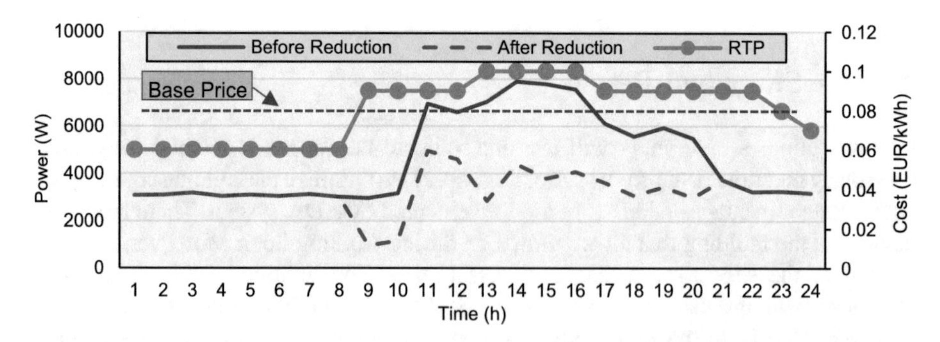

Fig. 4. Consumption reduction in AC devices based on RTP scheme.

The amount of power reduction may be different in each period and depends on the required reduction of the algorithm and the priority of each devices defined by the users. As Figs. 4, and 5 show, whenever the electricity prices increased and goes above the Base Price, the optimization process reduces the consumption of ACs as much as it can, and some periods that ACs reduction would not fulfill the system goal, the optimization reduce the rest of consumption from the lighting system (period #13 to #20 in Fig. 5).

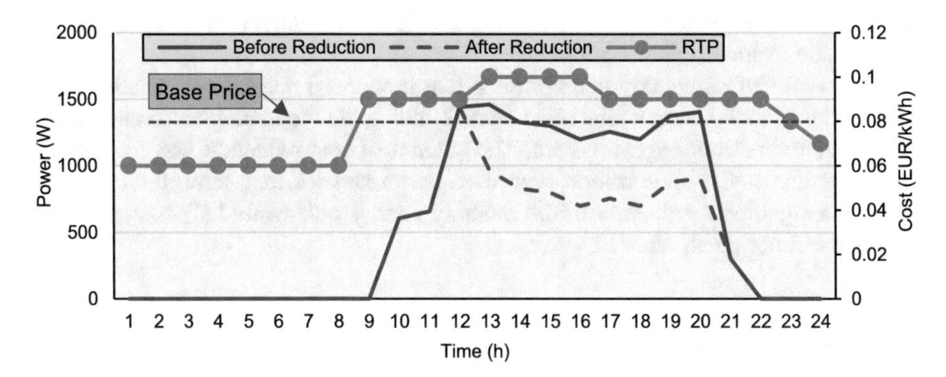

Fig. 5. Consumption reduction in lighting system based on RTP scheme.

Moreover, as Figs. 4 and 5 demonstrates, in period #21 and #22, even though the electricity price is greater than the Base Price, the optimization is not performed, since there was not enough available consumption in order to be reduced. As a last result, Fig. 6 illustrates the effect of optimization in the energy bill of the building for one day.

As it can be seen in Fig. 6, the optimization process leads to reduce the electricity bill of the building from 17.80 EUR to 14.18 EUR, by respecting to the user's preferences.

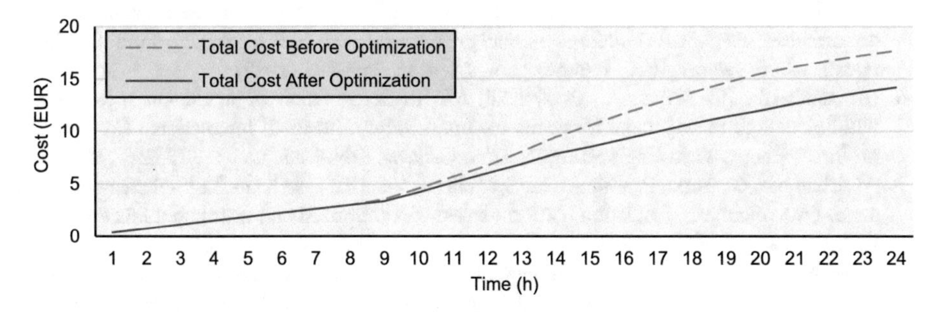

Fig. 6. Accumulated energy cost for 24 h.

5 Conclusions

In this paper, an optimization algorithm has been proposed for a multi-agent based SCADA system. This algorithm considered real-time pricing schemes and optimize the consumption of an office building in the periods that electricity price is greater than a specific value. The main purpose of the paper was to optimize the power consumption and reduce energy bill with take advantages of a multi-agent system. The presented model considered several agents associated with several distributed based controller devices in order to perform decision making locally and communicate with other agents to fulfill the overall system's goal.

The results of case study demonstrated that how the proposed optimization algorithm can reduce the energy bills of an office building via the implemented automation infrastructure and multi-agent system. The amount of cost reduction was for a single day, therefore, if the optimization procedure performed for long-term, the consumer will see a significant reduction in the monthly energy bill, while its preferences and comforts did not much affected.

References

1. Abrishambaf, O., Gomes, L., Faria, P., Vale, Z.: Simulation and control of consumption and generation of hardware resources in microgrid real-time digital simulator. In: IEEE PES Innovative Smart Grid Technologies Latin America (ISGT LATAM), Montevideo, Uruguay, pp. 799–804 (2015)
2. Park, L., Jang, Y., Cho, S., Kim, J.: Residential demand response for renewable energy resources in smart grid systems. IEEE Trans. Industr. Inf. **13**(6), 3165–3173 (2017)
3. Hernandez, L., Baladron, C., Aguiar, J., Carro, B., Sanchez-Esguevillas, A., Lloret, J., Chinarro, D., Gomez-Sanz, J., Cook, D.: A multi-agent system architecture for smart grid management and forecasting of energy demand in virtual power plants. IEEE Commun. Mag. **51**(1), 106–113 (2013)
4. Eddy, Y.F., Gooi, H., Chen, S.: Multi-agent system for distributed management of microgrids. IEEE Trans. Power Syst. **30**(1), 24–34 (2015)
5. Minoli, D., Sohraby, K., Occhiogrosso, B.: IoT considerations, requirements, and architectures for smart buildings – energy optimization and next generation building management systems. IEEE Internet Things J. **4**(1), 269–283 (2017)
6. Esmaeilzadeh, A., Koma, A., Farajollahi, M.: Implementation of intelligent methods of building energy management for economic optimization. In: IEEE International Conference on Smart Energy Grid Engineering (SEGE), Oshawa, ON, Canada, pp. 286–293 (2017)
7. Abrishambaf, O., Faria, P., Gomes, L., Spínola, J., Vale, Z., Corchado, J.: Implementation of a real-time microgrid simulation platform based on centralized and distributed management. Energies **10**(6), 806–820 (2017)
8. Faria, P., Spinola, J., Vale, Z.: Aggregation and remuneration of electricity consumers and producers for the definition of demand-response programs. IEEE Trans. Industr. Inf. **12**(3), 952–961 (2016)
9. Faria, P., Vale, Z.: Demand response in electrical energy supply: an optimal real time pricing approach. Energy **36**(8), 5374–5384 (2011)
10. Tsui, K., Chan, S.: Demand response optimization for smart home scheduling under real-time pricing. IEEE Trans. Smart Grid **3**(4), 1812–1821 (2012)

11. Fernandes, F., Morais, H., Faria, P., Vale, Z., Ramos, C.: SCADA house intelligent management for energy efficiency analysis in domestic consumers. In: IEEE PES Conference on Innovative Smart Grid Technologies (ISGT Latin America), Sao Paulo, Brazil, pp. 1–8 (2013)
12. Manickavasagam, K.: Intelligent energy control center for distributed generators using multi-agent system. IEEE Trans. Power Syst. **30**(5), 2442–2449 (2015)
13. Santos, G., Femandes, F., Pinto, T., Silva, M., Abrishambaf, O., Morais, H., Vale, Z.: House management system with real and virtual resources: energy efficiency in residential microgrid. In: Global Information Infrastructure and Networking Symposium (GIIS), Porto, Portugal, pp. 1–6 (2016)
14. Gazafroudi, A., Pinto, T., Prieto-Castrillo, F., Corchado, J., Abrishambaf, O., Jozi, A., Vale, Z.: Energy flexibility assessment of a multi agent-based smart home energy system. In: IEEE 17th International Conference On Ubiquitous Wireless Broadband (ICUWB), Salamanca, Spain, pp. 1–7 (2017)

Special Session on AI–Driven Methods for Multimodal Networks and Processes Modeling (AIMPM)

Food Supply Chain Optimization – A Hybrid Approach

Paweł Sitek(✉), Jarosław Wikarek, and Tadeusz Stefański

Department of Control and Management Systems, Kielce University of Technology, Kielce, Poland
{sitek, j.wikarek, t.stefanski}@tu.kielce.pl

Abstract. The food sector is a very complex environment influenced by economics, business, industrial, technological, transportation, information, legal and other factors. These factors shape the level of the availability of food, the nature of food products and the delivery method. The efficient and timely distribution of food products is critical for supporting the demands of contemporary consumer market. Without optimal food distribution, modern societies will not survive and will not develop. This paper presents the concepts of hybrid approach to optimization of food supply chain management (FSCM). This approach combines the strengths of constraint logic programming (CLP) and mathematical programming (MP), which leads to a significant reduction in the optimization time and modeling of any type of constraints. Moreover, this paper presents the formal model for optimization of FSCM with different objective functions. Several computational experiments were performed for compare hybrid approach to MP-based approach.

Keywords: Food supply chain management · Hybrid methods
Constraint logic programming · Mathematical programming · Optimization

1 Introduction

Food is a very important part both of everyday life of individuals and of dynamic modern market, as shown by fast changes observed over the last decades. Consumers want high quality, palatable food that looks nice and has a low, competitive price. To address these challenges, the food industry and logistics companies need to ensure that food products reach different customers on time, safely and in the right quality. By identifying and tracing food information along the supply chain in accordance with the "from farm to fork" project can be helpful [1]. Food quality and product waste issues are what make Supply Chain Management (SCM) and Food Supply Chain Management (FSCM) different. Also, constraints such as time constraints, those related to the type of transport, recycling etc. play an important role in FSCM. Given the role and importance of the food sector in the modern market, FSCM optimization becomes a key issue. Application of classical, operations research (OR) [2] based approaches to FSCM optimization have many shortcomings. The most critical of them are: (a) only linear and integer constraints, (b) models completely separated from data, (c) difficulty partial modification of the model and (d) poor efficiency with a large number of integer

S. Rodríguez et al. (Eds.): DCAI 2018, AISC 801, pp. 33–41, 2019.
https://doi.org/10.1007/978-3-319-99608-0_4

decision variables. A hybrid approach [3, 4] has been proposed for modeling and optimization of FSCM that combines OR-based methods with Constraint Logic Programing (CLP) - based methods [5].

2 Problem Description

The FSCM problem being considered involves a supply chain consisting of food processing plants $F = \{f_1, ..., f_i, ..., f_{ZF}\}$, where *ZF* denotes the number of factories, a set of distributors (distribution centers, central warehouses, etc.) $D = \{d_1, ..., d_i, ..., d_{ZD}\}$, where *ZD* denotes the number of centers, a set of customers (stores, retailers, wholesalers) $B = \{b_1, ..., b_i, ..., b_{ZB}\}$, where *ZB* denotes the number of customers, and two sets of vehicles: one set represents the vehicles travelling from food processing plants to distribution centers $A = \{a_1, ..., a_i, ..., a_{ZA}\}$, where *ZA* denotes the number of vehicles, and, the other set of vehicles travelling from distribution centers to customers $Z = \{z_1, ..., z_i, ..., z_{ZZ}\}$, where *ZZ* denotes the number of vehicles. Food processing plants produce/process products $P = \{p_1, ..., p_i, ..., p_{ZP}\}$, where *ZP* denotes the number of product types. Each product type (product) has a specific volume/weight (size) defined by Gz_p and belongs to a given category $G = \{g_1, ..., g_i, ..., g_{ZG}\}$, where *ZG* denotes the number of product categories (e.g. frozen foods, dairy products, bread, fresh vegetables, etc.). Each product can belong to only one category (if product p belongs to category g then $Tz_{p,g} = 1$, otherwise $Tz_{p,g} = 0$). The quantity of product p that can be produced by food processing plant f is defined as $Wz_{f,p}$. The products are ordered by customers. The quantity of product p ordered by customer b is defined as $Oz_{b,p}$. The process of delivery of products to customers takes place through distributors. If product p can be handled by the distributor d, then $Jz_{p,d} = 1$ and if customer b can be supplied by distributor d, then $Uz_{d,b} = 1$. Transport is provided with the use of vehicles of various types $E = \{e_1, ..., e_i, ..., e_{ZE}\}$, where *ZE* denotes the number of vehicle types (refrigerated truck, pickup truck, etc.). If vehicle type e can transport category g product, then $Dz_{e,g} = 1$. On the route between the food processing plant and the distributor, transport is handled by vehicles from set *A*. A maximum volume/tonnage of vehicle a is defined as Hz_a. All the products ordered from a plant by different customers can be transported on the same vehicle (each vehicle can do only one trip from a food processing plant to a distributor). If product p can be transported on vehicle a, then $Nz_{p,a} = 1$. The products in the distribution center are handled (reloaded) and then delivered to customers on vehicles $Z = \{z_1, ..., z_i, ..., z_{ZZ}\}$, where *ZZ* denotes the number of vehicles. If vehicle z is of e type, then $Zz_{z,e} = 1$. The vehicle travels from the distributor to several customers and then returns (each vehicle can only take one such trip in the supply cycle being considered). The maximum volume/tonnage of the vehicle is defined as Wz_z. If vehicle z travels from center d, then $CF_{d,z} = 1$. It is required that each customer receives the ordered products. Products can only be supplied by distributors that are able to provide proper storage and distribution of the products and have appropriate types of vehicles. During transport, the permissible tonnage of the vehicle must not be exceeded. The structure of FSC network has been shown in Fig. 1.

Constraint (1) assures that distributors receive only as many products as the customers have ordered. Constraint (2) ensures not exceeding the production capacity of

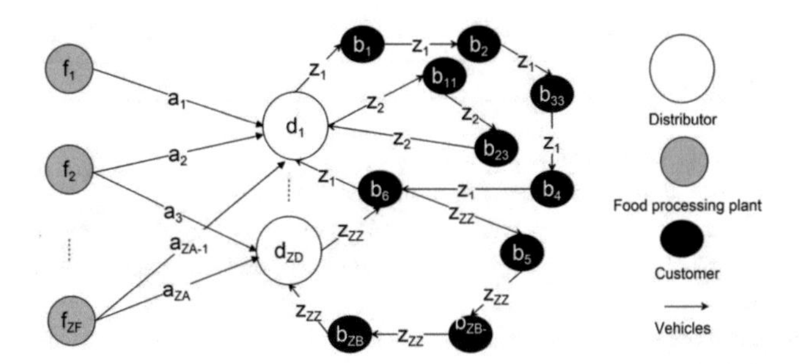

Fig. 1. The structure of FSC network.

the food processing plant (fpp). If the products are delivered from the plant to the distributor, then there must be a vehicle course between the fpp and the distributor (a link between the decision variables $X_{f,p,d,b,a}$ and $Y_{f,d,a}$ - constraints (3) and (4)).

Constraint (5) ensures that vehicle a only performs one trip when executing a given set of orders. The vehicle *a* cannot carry more than its volume/tonnage (6). Constraint (7) determines the total volume/tonnage of the category *g* products delivered to customer *b* by distributor *d*. Constraint (8) ensures the continuity of the route (if the vehicle z arrives at a given point, it has to depart from it). If the category *g* products are delivered to customer *b* by distributor *d*, then the appropriate vehicle trip must be made from the distributor to the customer - linking decision variables $Xh_{z,i,j}$ with variables $Yh_{z,i,j,d,b,g}$ (9, 10). Constraint (11) ensures that the load taken by each vehicle does not exceed the vehicle capacity. Constraint (12) ensures that vehicle z performs at least one trip for a given set of orders. The vehicle trips start and end with the same distributor d (13). Constraints (14) and (16) determine the delivery of products only to those customers who have ordered them. Constraint (15) links decision variables $UO_{d,b,a}$ and $U_{d,b,g}$. Constraint (17) forces the delivery of products of category *g* to customer *b* through distributor *d* on one vehicle. Binarity and integrality are enforced by constraints (18–23). Other parameters and decision variables have been shown in Table 1. Two objective functions FC1 and FC2 were formulated for the FSCM problem. *FC1* defines the minimum cost of delivering customer orders (24). *FC2* minimizes the number of vehicles used to complete a set of customer orders (25). What distinguishes the proposed FSCM problem formalization from the known SCM problem formalization is introducing product categories and linking them to vehicle types.

$$\sum_{f \in F} \sum_{d \in D} \sum_{a \in A} \mathrm{Uz}_{d,b} \cdot \mathrm{Nz}_{p,a} \cdot \mathrm{Jz}_{p,d} \cdot \mathrm{X}_{f,p,d,b,a} \leq \mathrm{Oz}_{b,p} \forall b \in B, p \in P \tag{1}$$

$$\sum_{d \in D} \sum_{b \in B} \sum_{a \in A} \mathrm{X}_{f,p,d,b,a} \leq \mathrm{Wz}_{f,p} \forall f \in F,\ p \in P \tag{2}$$

$$\mathrm{X}_{f,p,d,b,a} \leq \mathrm{ST} \cdot \mathrm{Y}_{f,d,a} \forall f \in F,\ p \in P,\ d \in D,\ b \in B,\ a \in A \tag{3}$$

Table 1. Parameters and decision variables for mathematical model.

Symbol	Description
Parameters	
$Kr_{f,p}$	The cost of product *p* manufactured in plant *f*
$Ks_{f,d,a}$	The cost of transport from plant *f* to distributor *d* by vehicle *a*
$Kz_{z,i,j}$	The cost of transport from point *i* ($i \in D \cup B$) to the point *j* ($j \in D \cup B$) by vehicle *z*
ST	The large fixed value for example the total number of ordered products
Decision variables	
$X_{f,p,d,b,a}$	The quantity of product *p* manufactured in plant *f* upon order placed by customer *b*, delivered to distributor *d* on vehicle *a*
$Y_{f,d,a}$	If vehicle *a* moves from plant *f* to distributor *d*, then $Y_{f,d,a} = 1$, otherwise $Y_{f,d,a} = 0$
$UO_{d,b,g}$	Volume (size) of all products of category *g* delivered to customer *b* from distributor *d*
$Xh_{z,i,j}$	If vehicle *z* moves from a given point *i* (distributor or customer) to point *j* (distributor or customer), then $Xh_{z,i,j} = 1$, otherwise $Xh_{z,i,j} = 0$
$Yh_{z,i,j,d,b,g}$	If vehicle *z* moves from point *i* (distributor or customer) to point *j* (distributor or customer) carrying products of category *g* ordered by customer *b* dispatched by distributor *d*, then $Yh_{z,i,j,d,b,g} = 1$, otherwise $Y_{z,i,j,d,b,g} = 0$
$Fh_{d,b,g,z}$	If products of category *g* ordered by customer *b* are delivered by distributor *d* on vehicle *z*, then $Fh_{d,b,g,z} = 1$, otherwise $Fh_{d,b,g,z} = 0$
$U_{d,b,g}$	If products of category *g* are delivered by distributor *d* to customer *b*, then $U_{d,b,g} = 1$, otherwise $U_{d,b,g} = 0$

$$\sum_{b \in B} \sum_{p \in P} X_{f,p,d,b,a} \geq Y_{f,d,a} \forall f \in F,\ d \in D, a \in A \tag{4}$$

$$\sum_{f \in F} \sum_{d \in D} Y_{f,d,a} \leq 1 \forall a \in A \tag{5}$$

$$\sum_{b \in B} \sum_{p \in P} Gz_p \cdot X_{f,p,d,b,a} \geq Hz_a \cdot Y_{f,d,a} \forall f \in F, d \in D, a \in A \tag{6}$$

$$\sum_{f \in F} \sum_{p \in P} \sum_{a \in A} Gz_p \cdot Tz_{p,g} \cdot Dz_{e,g} \cdot Zz_{a,e} \cdot X_{f,p,d,b,a} = UO_{d,b,g} \forall d \in D, b \in B, g \in G \tag{7}$$

$$\sum_{j \in B \cup D} Xh_{z,i,j} = \sum_{j \in B \cup D} Xh_{z,j,i} \forall z \in Z, i \in B \cup D \tag{8}$$

$$Xh_{z,i,j} \leq \sum_{d \in D} \sum_{b \in B} \sum_{g \in G} Yz_{z,i,j,d,b,g} \forall z \in Z, i \in B \cup D, j \in B \tag{9}$$

$$Yh_{z,i,j,d,b,g} \leq Xh_{z,i,j} \forall z \in Z, i \in B \cup D, j \in B \cup D, d \in D, b \in B, g \in G \quad (10)$$

$$\sum_{d \in D} \sum_{b \in B} \sum_{g \in G} UO_{d,b,g} \leq (\sum_{d \in D} \sum_{b \in B} \sum_{g \in G} Yz_{z,i,j,d,b,g}) \cdot Vz_z \forall z \in z, i \in B \cup D, j \in B \cup D \quad (11)$$

$$\sum_{j \in B} Xh_{z,d,j} \leq 1 \forall z \in Z, d \in D \quad (12)$$

$$\sum_{j \in B} Xh_{z,d,j} \leq CF_{z,d} \forall z \in Z, d \in D \quad (13)$$

$$\sum_{j \in B} \sum_{z \in Z} Yh_{z,j,i,d,b,g} - \sum_{j \in B} \sum_{z \in Z} Yh_{z,i,j,d,b,g} = U_{d,b,g} \forall i \in O, d \in D, b \in B, g \in G \quad (14)$$

$$UO_{d,b,a} \leq ST \cdot U_{d,b,g} \forall d \in D, b \in B, g \in G \quad (15)$$

$$Yh_{z,i,j,d,b,g} = Fh_{d,b,g,z} \forall z \in Z, i \in B, j \in B, d \in D, b \in B, g \in G \quad (16)$$

$$\sum_{z \in Z} Fh_{z,d,b,g} = 1 \forall h \in H, d \in D, b \in B, g \in G \quad (17)$$

$$X_{f,p,d,b,a} \in Z^+ \forall f \in F, p \in P, d \in D, b \in B, a \in A \quad (18)$$

$$Y_{f,d,a} \in \{0,1\} \forall f \in F, d \in D, a \in A \quad (19)$$

$$Yh_{z,i,j,d,b,g} = \{0,1\} \forall z \in Z, i \in B \cup D, j \in B \cup D, d \in D, b \in B, g \in G \quad (20)$$

$$Xh_{z,i,j} = \{0,1\} \forall z \in Z, i \in B \cup D, j \in B \cup D \quad (21)$$

$$Fh_{d,b,g,z} = \{0,1\} \forall z \in Z, b \in B, g \in G, d \in D \quad (22)$$

$$U_{d,b,g} = \{0,1\} \forall b \in B, g \in G, d \in D \quad (23)$$

$$\begin{aligned} &\min \sum_{f \in F} \sum_{d \in D} \sum_{a \in A} Ks_{f,d,a} \cdot Y_{f,d,a} \\ &+ \sum_{f \in F} \sum_{p \in P} \sum_{d \in D} \sum_{b \in B} \sum_{a \in A} Kr_{f,p} \cdot X_{f,p,d,b,a} + \sum_{z \in Z} \sum_{i \in D \cup B} \sum_{j \in D \cup B} Kz_{z,i,j} \cdot Xh_{z,i,j} \end{aligned} \quad (24)$$

$$\min \sum_{f \in F} \sum_{d \in D} \sum_{a \in A} Y_{f,d,a} + \sum_{z \in Z} \sum_{i \in D} \sum_{j \in B} \cdot Xh_{z,i,j} \quad (25)$$

3 Methodology – A Hybrid Approach

Based on numerous studies and our earlier gained experience, it shows that the CLP environment [5, 6], offers a very good environment for representing the knowledge, information and methods needed for modeling complex problems such as FSCM problem. Effective search for the solution in the CLP depends considerably on the effective constraint propagation, which makes it a key method of the constraint-based approach. Constraint propagation embeds any reasoning that consists in explicitly forbidding values or combinations of values for some variables of a problem because a given subset of its constraints cannot be satisfied otherwise [5].

Based on our previous work [3, 4, 7], we observed some advantages and disadvantages of MP and CLP environments. An integrated approach of CLP and MP can help to solve optimization problems that are intractable with either of the two methods alone [5, 8–10]. Our approach differs from the known integration of CLP/MP [8–10]. This approach called hybridization consists in the combination of both environments and the transformation of the modeled problem. The transformation is seen as a presolving method. In general, it involves elimination from the model solutions space of those points which are unacceptable. It is determined on the basis of data instances and model constraints. For example in FSCM problem, based on analysis of instances of data, you can specify that some products cannot be manufactured in some plants and cannot be delivered by some vehicles, etc. The general concept of the hybrid approach is shown in Fig. 2.

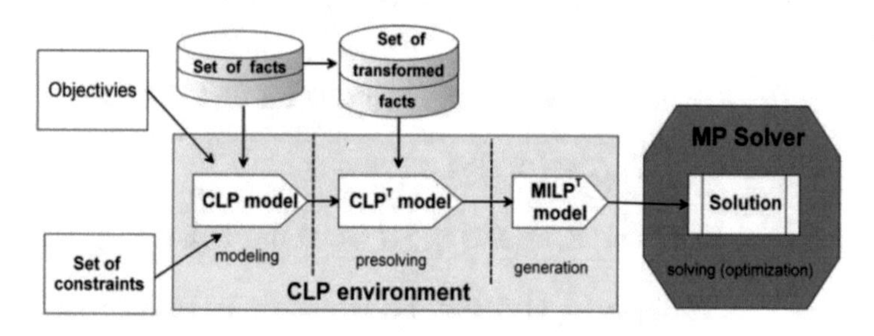

Fig. 2. The concept of the hybrid approach.

This hybrid approach consists of several phases: modeling, presolving, generation phase and solving. It has two inputs and uses the set of facts. Inputs are the set of objectives and the set of constraints to the model of a given problem. Based on them, the primary model of the problem is generated as a CLP model, which is then presolved. The built-in CLP method (constraint propagation [5]) and the method of problem transformation designed by the authors [3, 4] are used for this purpose. Presolving procedure results on the transformed model CLPT. This model is the basis for the automatic generation of the MILP (Mixed Integer Linear Programming) model, which is solved in MP (with the use of an external solver or as a library of CLP).

4 Computational Experiments

All the experiments relate to the FSCM problem with three plants ($f = 1..3$), fifteen customers ($o = 1..15$), eight vehicles ($a = 1..8$), nine vehicles ($z = 1..9$), ten types of products ($p = 1..10$), three categories of products ($g = 1..3$), three distributors (*d* = *1..3*), and six sets of orders (from *E(50)* to *E(175)*). Computational experiments were carried out for the model (Sect. 2) implemented using mathematical programming (two environments Lingo and Gurobi) and the hybrid approach (Lingo/Gurobi and ECLiPSe). Experiments were made for different sets of orders. The results are shown in Table 2 and Figs. 3 and 4.

Table 2. Results of computational examples.

Orders E(n)	V	C	Mathematical Programming							
			Lingo				Gurobi			
			FC1		FC2		FC1		FC2	
			Value	T	Value	T	Value	T	Value	T
E(50)	20289	5345	5646	23	5	9	5646	6	5	4
E(75)	20289	5345	6756	345	5	61	6756	12	5	45
E(100)	20289	5345	7856	567	7	103	7856	78	7	62
E(125)	20289	5345	10123**	900*	9	234	10123	234	9	121
E(150)	20289	5345	14756**	900*	9	678	14756	567	9	356
E(175)	20289	5345	24672**	900*	12	845	23567	876	12	434
Orders E(n)	V	C	Hybrid							
			Lingo				Gurobi			
			FC1		FC2		FC1		FC2	
			Value	T	Value	T	Value	T	Value	T
E(50)	3223	3745	5646	5	5	7	5646	3	5	3
E(75)	4789	3934	6756	10	5	18	6756	8	5	8
E(100)	5435	4129	7856	56	7	45	7856	18	7	11
E(125)	6234	4227	10123	123	9	98	10123	34	9	36
E(150)	6975	4434	14756	456	9	145	14756	45	9	45
E(175)	7435	4634	23567	567	12	356	23567	56	12	67

T - calculation time; V - the number of decision variables; C - the number of constraints; * - Interruption of the calculation after 900 s; ** - feasible solution

The results (Table 2) clearly indicate the superiority of the hybrid approach over mathematical programming. The time to solve the FSCM problem for *FC1* and *FC2* using the hybrid approach is up to 5 times shorter than using the same mathematical programming. The use of the hybrid approach also significantly reduces the size of the model, e.g., the number of decision variables is reduced by a factor of 6.

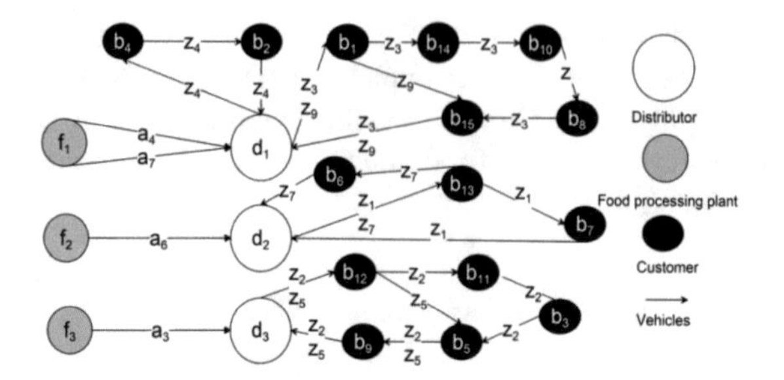

Fig. 3. Optimal delivery network for *E(125) Fc1.*

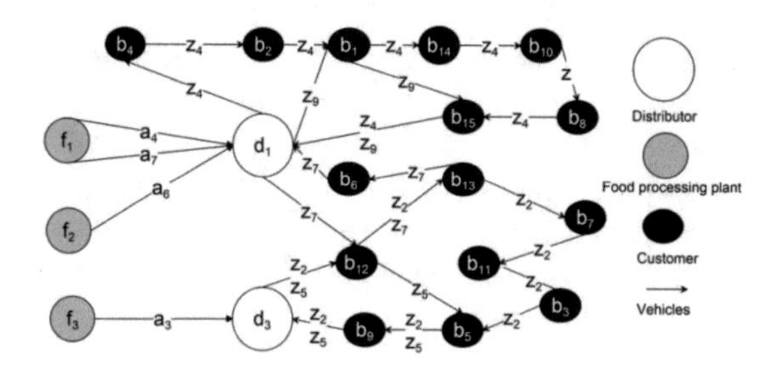

Fig. 4. Optimal delivery network for *E(125) Fc2.*

A detailed analysis of the values of decision variables allows determining optimal delivery networks (routes, number of vehicles, etc.) (Figs. 3 and 4).

5 Conclusion

This paper provides an effective and robust method (hybrid approach) for modeling and optimization of a FSCM problem. The method incorporates two environments: (i) mathematical programming and (ii) constraint logic programming. The application of the hybrid approach allows the optimization of larger size problems within a shorter time compared to the use of mathematical programming (Table 2). The data storage in the form of a set of facts allows easy integration of the proposed hybrid approach with database systems, data warehouses, and even flat files, e.g. XML, which provides high flexibility to the hybrid approach.

The integration of CLP with metaheuristics such as GA (Genetic Algorithms) or ACO (Ant Colony Optimization), capable of solving industrial size problems is planned, as is complementing the hybrid approach with fuzzy logic [11]. In the next stage of research, it is planned to apply the hybrid approach to production problems such as [12].

References

1. Banaeian, N., Mobli, H., Nielsen, I.E., Omid, M.: Criteria definition and approaches in green supplier selection – a case study for raw material and packaging of food industry. Prod. Manuf. Res. **3**, 149–168 (2015)
2. Schrijver, A.: Theory of Linear and Integer Programming. Wiley, New York (1998)
3. Sitek, P., Wikarek, J.: A hybrid programming framework for modeling and solving constraint satisfaction and optimization problems. Sci. Program. **2016** (2016). Article ID 5102616, https://doi.org/10.1155/2016/5102616
4. Sitek, P., Wikarek, J., Nielsen, P.: A constraint-driven approach to food supply chain management. Ind. Manag. Data Syst. **117**(9) (2017). Article number 600090, https://doi.org/10.1108/IMDS-10-2016-0465
5. Rossi, F., Van Beek, P., Walsh, T.: Handbook of Constraint Programming (Foundations of Artificial Intelligence). Elsevier Science Inc., New York (2006)
6. Bocewicz, G., Nielsen, I., Banaszak, Z.: Iterative multimodal processes scheduling. Annu. Rev. Control **38**(1), 113–132 (2014)
7. Sitek, P., Wikarek J., Grzybowska, K.: A multi-agent approach to the multi-echelon capacitated vehicle routing problem. In: Corchado J.M. et al. (eds.) Highlights of Practical Applications of Heterogeneous Multi-Agent Systems. The PAAMS Collection. PAAMS 2014. Communications in Computer and Information Science, vol 430. Springer, Cham (2014). https://doi.org/10.1007/978-3-319-07767-3_12
8. Hooker, J.N.: Logic, optimization, and constraint programming. INFORMS J. Comput. **14** (4), 295–321 (2002)
9. Bockmayr, A., Kasper, T.: A framework for combining CP and IP, branch-and-infer, constraint and integer programming. In: Toward a Unified Methodology Operations Research/Computer Science Interfaces, pp. 59–87, 27 (2014)
10. Milano, M., Wallace, M.: Integrating operations research in constraint programming. Ann. Oper. Res. **175**(1), 37–76 (2010)
11. Kłosowski, G., Gola, A., Świć, A.: Application of fuzzy logic in assigning workers to production tasks. In: 13th International Conference Distributed Computing and Artificial Intelligence, AISC, vol. 474, pp. 505–513 (2016). https://doi.org/10.1007/978-3-319-40162-1_54
12. Nielsen, I., Dang, Q., Nielsen, P., Pawlewski, P.: Scheduling of mobile robots with preemptive tasks. In: Advances in Intelligent Systems and Computing, vol. 290, pp. 19–27 (2014). https://doi.org/10.1007/978-3-319-07593-8_3

The Use of the Simulation Method in Analysing the Performance of a Predictive Maintenance System

Sławomir Kłos(✉) and Justyna Patalas-Maliszewska

Faculty of Mechanical Engineering, University of Zielona Góra,
Licealna 9, 65-417 Zielona Góra, Poland
{s.klos, j.patalas}@iizp.uz.zgora.pl

Abstract. The progressive automation of manufacturing systems results, on the one hand, in a reduction in the number of production workers but necessitates, on the other hand, the development of maintenance systems.

The concept of Industry 4.0 (I4.0) includes implementation of predictive/preventive maintenance as an integral part of manufacturing systems. In this paper, an analysis of the different structures of manufacturing systems, using the simulation method is proposed, in order to evaluate the resistance of a system to change in the availability of manufacturing resources. The parallel-serial manufacturing system is considered where the availability of resources and the capacity of the buffers are input values and the throughput and average product lifespan, that is, the particular detail relating to the time remaining within a system, are output values. The simulation model of the system is created using Tecnomatix Plant Simulation.

Keywords: Predictive maintenance · Computer simulation
Parallel-serial manufacturing system · Throughput · Average product lifespan

1 Introduction

1.1 Predictive Maintenance

Predictive maintenance entails monitoring the condition and prognosis of the future behaviour of a manufacturing system, where decision-making, *vis-à-vis* maintenance, is based on the results of prediction [1]. The concept of I.40, assumed integration and the close co-operation of the technical resources of manufacturing systems with human resources. Simple operations will be undertaken with the co-operation of robots and machines which will interact with products in order to optimise production flow.

Maintenance will play an ever increasing *rôle* but maintenance workers will be expensive and quotas of them will be hard to fill. In the paper, a predictive maintenance system is implemented as a set of buffer capacities and connections between the buffers and manufacturing resources, in a parallel-serial manufacturing system. The simulation model of the system is created using Tecnomatix Plant Simulation Software. The effectiveness of the system, that is, the throughput and average product lifespan, is analysed for the different buffer capacities and availabilities of manufacturing resources

S. Rodríguez et al. (Eds.): DCAI 2018, AISC 801, pp. 42–49, 2019.
https://doi.org/10.1007/978-3-319-99608-0_5

defining the simulation experiments. The predictive maintenance system is implemented using a control system, based on dispatching rules. Dispatching rules are allocated within the manufacturing resources, with the buffers deciding the destination of the details, *that is, the next operation or buffer*, using the *Round Robin* dispatching rule. The research problem can be formulated as follows: *Given, is a parallel-serial manufacturing system. How may the effectiveness of the predictive maintenance system, using computer simulation, be analysed?*

1.2 Literature Overview

Predictive and preventive maintenance is the objective for many research studies. *Mori and Fujishima* introduce a remote monitoring and maintenance system for machine tool manufacturers [2]. *Dong et al.* present the monitoring and maintenance of equipment systems for mine safety. They establish a predictive maintenance system which is based on the technology of the *Internet of Things* in order to change the existing method for the maintenance of coal mining equipment [3]. *Susto et al.* present the methodology of multiple classifier, machine learning (ML) for predictive maintenance (PdM). The PdM methodology proposed, allows dynamic decision rules to be adopted for maintenance management and can be used with high-dimensional data problems. *Almeida* presents a multi-criteria decision model to support decision makers in choosing the best maintenance interval, based on the combination of conflicting criteria, such as reliability and cost. He proposes a procedure for using the model, which is based on the Multi-Attribute Utility Theory (MAUT). *Nguyen et al.* proposed a novel, predictive maintenance policy with multi-level decision-making for a multi-component system of complex structures. The main idea was to propose a decision-making process considered on two levels, *viz.*, the system level and component one. The main objective of the decision rules at the system level was to address whether preventive maintenance actions were needed regarding the predictive reliability of the system. Computer simulation can be used to build complex manufacturing systems, analyse the effectiveness of preventive or corrective maintenance and improve overall, operational reliability, the utilisation of manufacturing resources and the productivity of manufacturing systems [7]. In the literature, many decision-support tools have been proposed, based on the computer-simulation method for effective maintenance operations. For example, *Ni and Jin* propose mathematical algorithms and simulation tools in order to identify data-driven, short-term, throughput bottlenecks, the prediction of *windows of opportunity* for maintenance, the prioritisation of maintenance tasks, the joint production-scheduling and maintenance-scheduling of systems and the management of maintenance staff [8]. *Roux et al.* recommend tools be combined in order to collaborate, optimising multi-component preventive-maintenance problems. The structure of the maintenance-production system is modelled using a combination of timed, petri-nets and Parallel Discrete-Event System-Specification models and is implemented in a virtual-laboratory environment [9]. Simulation software, such as, Arena Simulation Software, Tecnomatix Plant Simulation, is often used in the implementation of maintenance system cycle times, in the reduction of discrete manufacturing systems and in the modelling of network and transport processes [10–12].

2 Simulation Model of a Manufacturing System

2.1 Manufacturing System Structures

The parallel serial manufacturing system includes three, production lines with each line performing the same three, technological operations (see Fig. 1). The proposed structures guarantee that the system is deadlock-free and starvation-free.

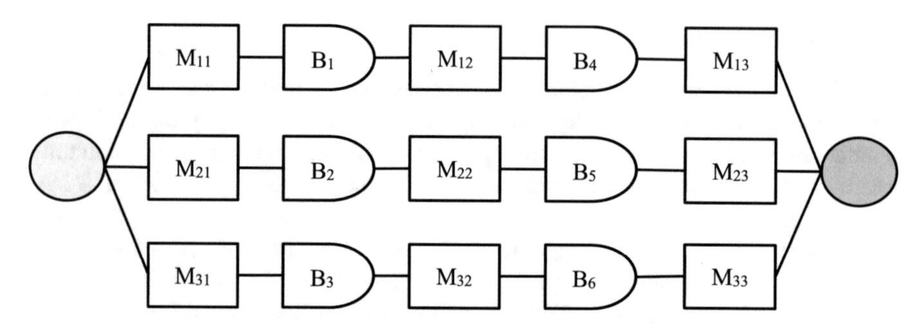

Fig. 1. The structure of a parallel-serial manufacturing system – variant V1

The groups of machines (M_{11}, M_{21}, M_{31}), (M_{21}, M_{22}, M_{32}) and (M_{13}, M_{23}, M_{33}) are identical and perform the same technological operations; however, the availability of each machine may vary.

Variant V1 of the manufacturing system (Fig. 1) includes three, independent manufacturing lines without any interconnection between manufacturing resources and buffers.

Variant V2 (see Fig. 2) includes the same manufacturing resources and buffers but each machine and each buffer is connected to two other machines or buffers.

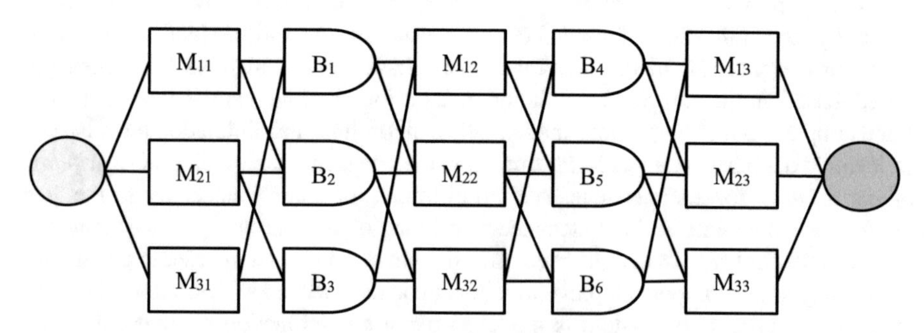

Fig. 2. The structure of a parallel-serial manufacturing system – variant V2

For example, machine M_{11} is connected to buffers B_1, B_2 while buffer B_2 is connected with machines M_{22}, M_{32}. The buffers and machines send details, evenly, to connected resources, on the basis of the *Round Robin* dispatching rule.

The last variant (V3) of the structure of the parallel-serial manufacturing system is presented in Fig. 3.

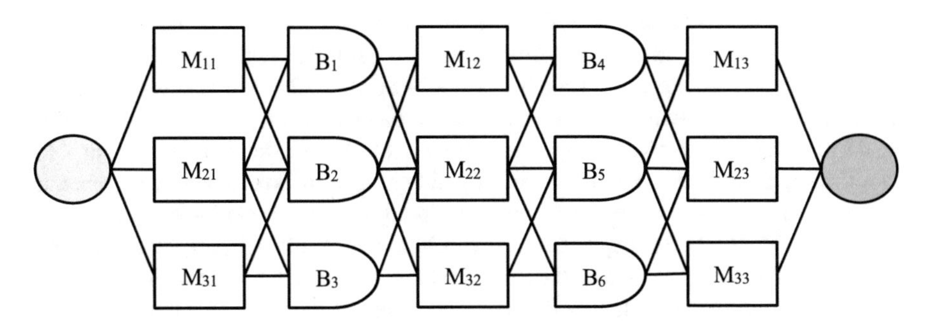

Fig. 3. The structure of a parallel-serial manufacturing system – variant V3

In the model of manufacturing systems, the external buffers (B_1, B_3, B_4, B_6) and the machines (M_{11}, M_{31}, M_{12}, M_{32}, M_{13}, M_{33}) are connected with two others buffers or machines and internal buffers (B_2, B_5) while machines (M_{21}, M_{22}, M_{23}) are connected with three other buffers or machines. The distribution of details is effected using the same *Round Robin* dispatching rule. The dispatching rules and buffers allocated in the system, enable the effectiveness of the system to be increased during any periods of low availability of manufacturing resources.

2.2 The Simulation Model

Manufacturing resources are denoted as M_{11}, M_{21}, M_{12}, M_{21}, M_{32}, M_{13}, M_{23} and the buffers as B_1, B_2, B_3, B_4, B_5, B_6. The processing and set-up times are based on a lognormal distribution. Lognormal distribution is a continuous distribution in which a random number has a natural logarithm, corresponding to a normal distribution. The realisations are non-negative, real numbers. The density of the lognormal distribution Lognor (σ, μ) is calculated as follows:

$$f(x) = \frac{1}{\sigma_0 x \sqrt{2\pi}} \cdot \exp\left[\frac{-\ln(x-\mu_0)^2}{2\sigma_0^2}\right] \tag{1}$$

where σ and μ are respectively the mean and standard deviations and are defined as follows:

$$\mu = \exp\left[\mu_0 + \frac{\sigma_0}{2}\right] \tag{2}$$

$$\sigma^2 = \exp(2\mu_0 + \sigma_0^2) \cdot (\exp(\sigma_0^2) - 1) \tag{3}$$

The maximum density function is defined as:

$$\exp(\mu_0 - \sigma_0^2) \tag{4}$$

The values of operation times for all machines are defined by $\sigma^2 = 120$ and $\mu = 10$. The values of setup times are defined as $\sigma^2 = 480$ and $\mu = 20$. In the system, four products (A, B, C, D) are manufactured, based on the following sequence of the sizes of the production batches: A - 100, B - 300, C - 80, D - 120. It was assumed that the availability of manufacturing resources could be changed from 70%–90% and that the capacity of the buffers could be changed from 1 to 10. For the availability ranges of manufacturing resources and buffer capacity, 50 simulation experiments was generated randomly. In Table 1, the first 10 examples of input data for simulation experiments are presented, with the buffer capacities and machine availabilities being given as a percentage and the output values of the throughput of the system, *along with the average product lifespan*, being evaluated hourly. The results of the experiments are presented in the next chapter.

Table 1. First 10 examples of input data for the simulation experiments

Exp	B_1	B_2	B_3	B_4	B_5	B_6	M_{11}	M_{21}	M_{31}	M_{12}	M_{22}	M_{32}	M_{13}	M_{23}	M_{33}
1	5	9	1	4	1	7	75	71	71	80	88	89	82	83	86
2	3	5	2	3	3	10	89	74	77	85	83	84	78	80	76
3	7	10	4	6	5	2	77	74	76	89	89	79	73	82	71
4	1	8	4	2	4	6	80	85	74	88	81	87	80	77	86
5	8	4	10	7	5	7	72	75	84	71	73	90	89	77	71
6	8	3	1	7	6	3	78	86	86	85	84	84	72	77	86
7	3	9	1	4	2	2	85	82	86	75	88	78	81	81	89
8	2	2	5	6	8	1	84	72	72	79	88	81	81	74	81
9	5	10	4	9	3	8	80	72	84	82	74	89	79	87	83
10	10	2	9	6	4	3	76	82	73	82	72	81	80	88	85

3 The Results of the Simulation Experiments

The simulation experiments are performed using Tecnomatix Plant Simulation Software. For each experiment, 3 observation are made. The time taken up by each experiment is 80 real hours (10 days). The results of simulation experiments are presented in Figs. 4 and 5. The results that generally illustrate the best effectiveness of the system are obtained for variant V3 of the structure of the manufacturing system and may be seen in Fig. 4. For example, the difference between the throughput of the V1 and V3 variants is 5,52 details per hour. The results of simulation experiments for variants V2 and V3 are similar for the throughput of the system. The opposite is the case for product lifespan. For example, for experiment 14, the details remain, on average, 10 min longer in the V2 system than in the V1 system. Generally speaking- and in order to increase efficiency within the system- variants V2 or V3 should be taken into account, while in order to reduce the average product lifespan of work-in-progress, variant V1 should be selected. Based of the results of simulation experiments, assumptions for the predictive maintenance system may now be determined. For defined availabilities of manufacturing resources, that is, for the reliability of machines, an allocation of buffer capacity and a variant of the structure of the system may now be proposed.

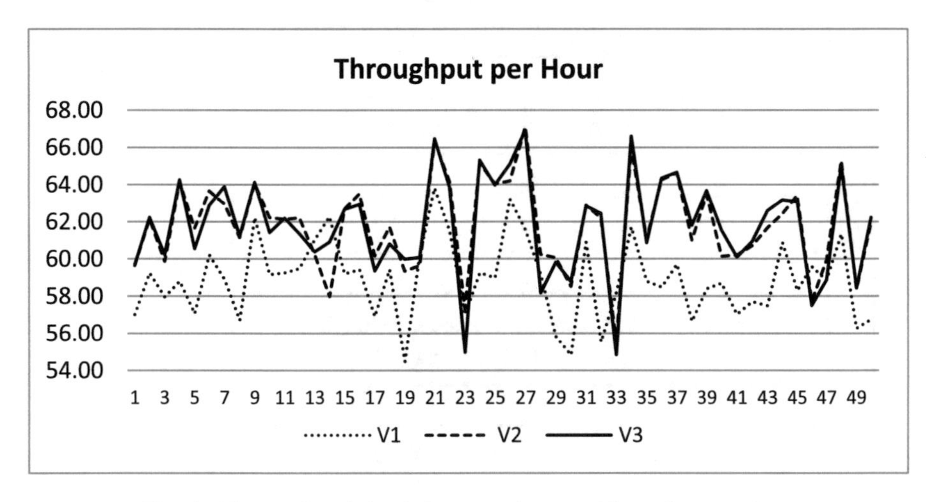

Fig. 4. The results of simulation experiments - Throughput per hour

In the next chapter, a methodology for the implementation of a predictive maintenance system is proposed, the end conclusions are formulated and the direction of further research is presented.

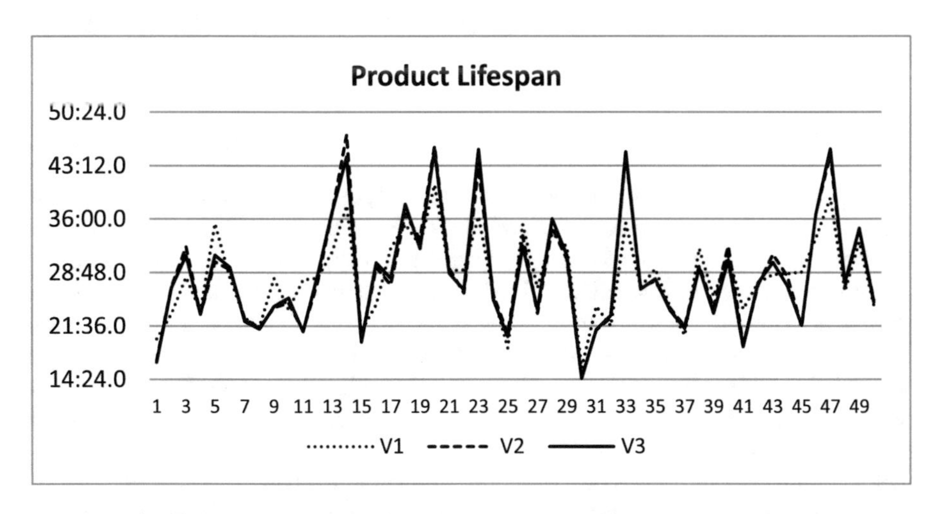

Fig. 5. The results of simulation experiments – Average product lifespan

The results present the relationships between the allocation of buffer capacity, the availability of manufacturing resources, as well as the throughput and lifespan of the system for the defined structure of the parallel-serial manufacturing system.

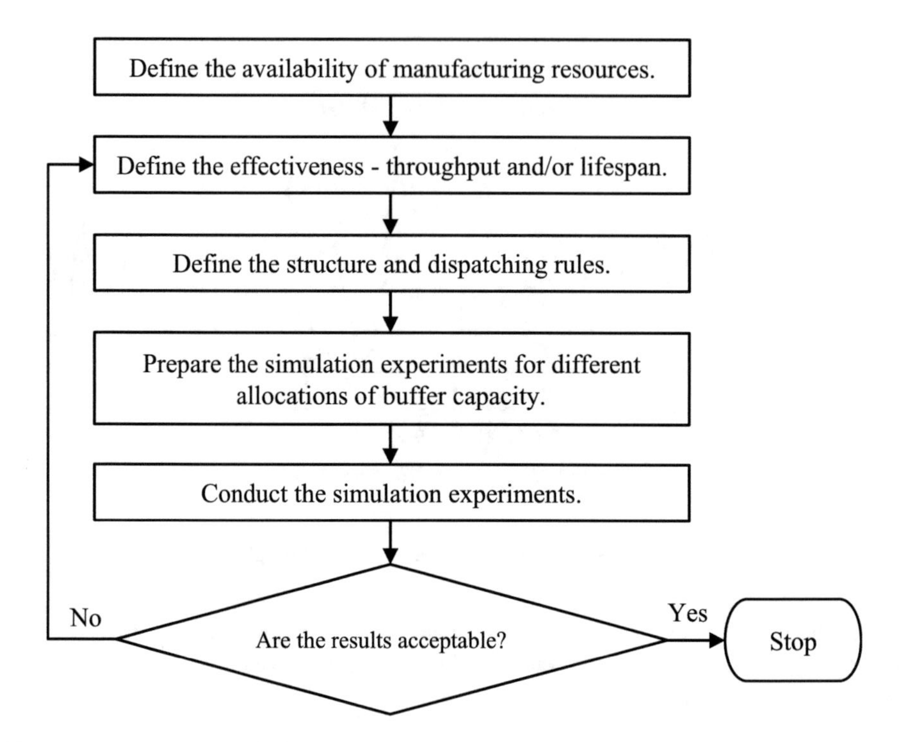

Fig. 6. Methodology for the introduction of a predictive, maintenance procedure into a parallel serial manufacturing system

4 Conclusions

In the paper, the simulation method is used to analyse the effectiveness of a discrete manufacturing system. Simulation experiments are conducted for three variants of the structure of a manufacturing system. The structure includes the availability of manufacturing resources, the allocation of buffer capacities and the interaction between the objects, determining the different the flow of the material. The flow of the material is controlled by dispatching rules which are based on the *Round Robin* protocol. The predictive maintenance system requires that the allocation of buffer capacity be determined and also that the structure of the manufacturing system be selected. In Fig. 6, the methodology for the implementation of predictive maintenance, for a parallel serial manufacturing system, using the simulation method, is presented.

In the first step, the availability and the reliability of manufacturing resources should be estimated. Next, the effectiveness level, regarding throughput and work-in-progress, should be determined. In the next step, the interaction between the manufacturing resources and buffers should be defined and the dispatching rules should be determined. If the structure of the parallel-serial manufacturing system is ready, an allocation of buffer capacity should be proposed; the simulation model of the system is now ready. Based on the results of the simulation experiments, the correct choice of buffer allocation, in order for an acceptable effectiveness to be achieved within the system, can be done. In further research, the model of a production system, with manufacturing and service workers, will be taken into account.

In addition to the availability of manufacturing resources, the availability of logistics systems could be considered as transportation and storage resources. The results of the work will be verified on the basis of a real, manufacturing system in an automotive branch.

References

1. Raza, A., Ulansky, V.: Modelling of predictive maintenance for a periodically inspected system. In: The 5th International Conference on Through-life Engineering Services (TESConf 2016), Procedia CIRP 59, pp. 95–101 (2017)
2. Mori, M., Fujishima, M.: Remote monitoring and maintenance system for CNC machine tools. Procedia CIRP **12**, 7–12 (2013)
3. Dong, L., Mingyue, R., Guoying, M.: Application of internet of things technology on predictive maintenance system of coal equipment. In: 13th Global Congress on Manufacturing and Management, GCMM, Procedia Engineering, vol. 174, pp. 885–889 (2017)
4. Susto, G.A., Schirru, A., Pampuri, S., McLoone, S.: Machine learning for predictive maintenance: a multiple classifier approach. IEEE Trans. Ind. Inf. **11**(3), 812–820 (2014)
5. Almeida, A.T.: Multicriteria model for selection of preventive maintenance intervals. Qual. Reliab. Eng. Int. **28**(6), 585–593 (2012)
6. Nguyen, K.-A., Do, P., Grall, A.: Multi-level predictive maintenance for multi-component systemsm. Reliab. Eng. Syst. Saf. **144**, 83–94 (2015)
7. Negahban, A., Smith, J.S.: Simulation for manufacturing system design and operation: literature review and analysis. J. Manuf. Syst. **33**, 241–261 (2014)
8. Ni, J., Jin, X.: Decision support systems for effective maintenance operations. CIRP Ann. Manuf. Technol. **61**, 411–414 (2012)
9. Roux, O., Duvivier, D., Quesnel, G., Ramat, E.: Optimization of preventive maintenance through a combined maintenance–production simulation model. Int. J. Prod. Econ. **143**(1), 3–12 (2013)
10. Gangala, C., Modi, M., Manupati, V.K., Varela, M.L.R., Machado, J., Trojanowska, J.: Cycle time reduction in deck roller assembly production unit with value stream mapping analysis. In: Recent Advances in Information Systems and Technologies. WorldCIST 2017, Advances in Intelligent Systems and Computing, vol. 571, pp. 509–518 (2017)
11. Klos, S., Trebuna, P.: The impact of the availability of resources, the allocation of buffers and number of workers on the effectiveness of an assembly manufacturing system. Manag. Prod. Eng. Rev. **8**(3), 40–49 (2017)
12. Bocewicz, G., Muszyński, W., Banaszak, Z.: Models of multimodal networks and transport processes. Bull. Pol. Acad. Sci. Tech. Sci. **63**(3), 635–650 (2015)

Modelling of Knowledge Resources for Preventive Maintenance

Justyna Patalas-Maliszewska(✉) and Sławomir Kłos

Institute of Computer Science and Production Management,
University of Zielona Góra, Zielona Góra, Poland
{J.Patalas,S.Klos}@iizp.uz.zgora.pl

Abstract. This paper focusses on modelling expert knowledge resources for preventive maintenance. Expert knowledge resources are defined as those employees engaged in a company's external maintenance who are important in the realisation of maintenance activities in the supported industries as well as those employees whose work is focussed on the application of knowledge. The aim of this work is to elaborate methods for acquiring and formalising this expert knowledge, in order to improve the manner of giving instructions *via* manuals, currently in use in maintenance areas. The approach presented in the form an IT tool, dedicated to the automotive industry, is implemented.

Keywords: Expert knowledge · Expert knowledge resources
Preventive maintenance

1 Introduction

Preventive maintenance (PM) is that part of the maintenance strategies within an industry [1], which requires to be formulated and implemented, in order to reduce the likelihood of the occurrence of breakdown. For manufacturing enterprises, one of the main expenditure items is the cost of maintenance [2, 12]. Employee availability, to render PM service, implies long-term contracts with supported companies [8]. Currently, we can distinguish PM, both as design-based preventive maintenance- which includes an estimation of the time taken up by such service- and as condition-based maintenance, this being the monitoring of the deterioration of any equipment) [11, 13].

We can define a company as a set of resources and a set of processes, as well as the connections linking these sets [3]. Masood stated, that knowledge is an important resource of maintenance, as well as an essential factor of the individual [10]. Bouzidi-Hassini et al. [4] and Wang and Liu [15] considered human resources an essential factor in the optimisation model. We, therefore, seek to provide a model for the acquisition and formalisation of expert knowledge, in order to ensure the expert knowledge of those employees engaged in PM service, within the context of PM.

In this paper, special attention is paid to presenting our approach to the modelling of expert knowledge resources for preventive maintenance. In the first stage of our approach, the experts in the maintenance company are identified, according their competences, using our expert knowledge questionnaires. According to Fan [6] expert

S. Rodríguez et al. (Eds.): DCAI 2018, AISC 801, pp. 50–57, 2019.
https://doi.org/10.1007/978-3-319-99608-0_6

knowledge is the expression of the core competences of the workers in a company. Next, we propose a model for the acquisition and formalisation of expert knowledge.

Finally, using the Algorithm - k-means Clustering and Distance Method and the Euclidean Distances Method, it is possible to assign the expert knowledge, thus acquired, to defined, manual instructions for new maintenance workers and prepare new instructions, based on the experience of expert knowledge resources, in order to provide continuous, experimental support for manufacturing companies.

2 A Model for the Representation of Knowledge Resources for Preventive Maintenance

The economic importance of responding quickly to customer requirements, in supported industries, has fuelled interest in external maintenance companies; such companies are expanding their rôle in the value chain by offering advanced services or complete solutions to other companies [10].

We assume that there are n, nϵN, manufacturing companies being supported by external maintenance staff. The problem in this paper is how to acquire and formalise expert knowledge in an external maintenance company, in order to improve the safety of the stability of the system in a supported industry. The proposed approach, in presenting methods for the acquisition, formalisation and clustering of expert knowledge, provides an opportunity to identify knowledge resources for preventive maintenance in industry. Therefore, the following stages are involved in the model for the identification of knowledge resources (see Fig. 1).

Stages one and two are based on the results of the research literature; consequently, the five, web-based questionnaires create each of the competences, for each maintenance worker viz. Technical Competences (TC), Organisational Competences (OC), Behavioural Competences (BC), Social Competences (SC) and Personal Competences (PC). The workers had to complete each web-based questionnaire by selecting one answer for each question. Based on the results, and on the solutions to the algorithms, values for each competence TC, OC, BC, SC and PC for the each maintenance worker were formulated and based on those results, the manager of a maintenance company can determine the importance of each competence for each supported industry.

In Stage 3, we identify the set of expert knowledge resources (EKR set), based on the answers to the questionnaires by the maintenance workers and the importance of each of the competences to the supported company. This means that a different EKR set will be defined for each supported industry.

In Stage 4 we propose using the verbal analysis method for solving the problem of acquiring expert knowledge [9]. It is assumed that a sound recording may be obtained, using a smartphone device, in order to carry out a maintenance order in an industry. The records obtained by observing maintenance workers is then converted from the verbal form to the written form using Automatic Speech Recognition technology (ASR) and the tool: Google Speech Api [5, 7]. This is a highly efficient method; the Commercial API programming interface or the Google Web Speech Api, free solution may also be used to great advantage.

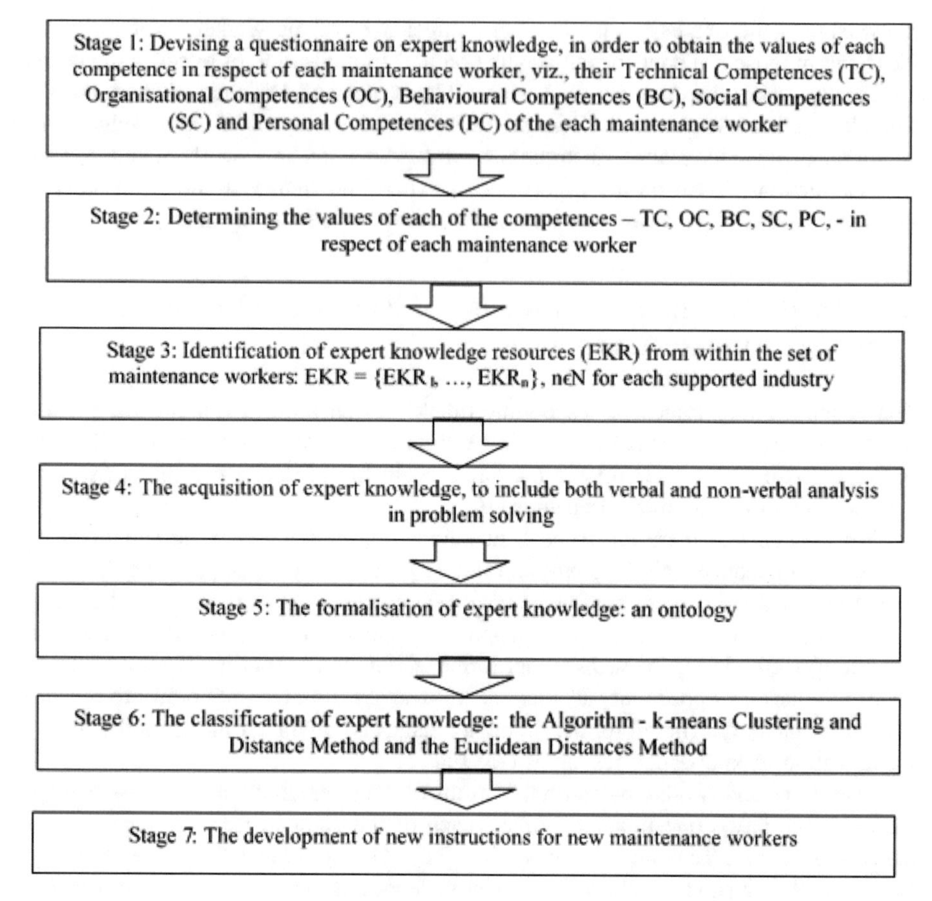

Stage 1: Devising a questionnaire on expert knowledge, in order to obtain the values of each competence in respect of each maintenance worker, viz., their Technical Competences (TC), Organisational Competences (OC), Behavioural Competences (BC), Social Competences (SC) and Personal Competences (PC) of the each maintenance worker
Stage 2: Determining the values of each of the competences – TC, OC, BC, SC, PC, - in respect of each maintenance worker
Stage 3: Identification of expert knowledge resources (EKR) from within the set of maintenance workers: EKR = {EKR₁, ..., EKRₙ}, n∈N for each supported industry
Stage 4: The acquisition of expert knowledge, to include both verbal and non-verbal analysis in problem solving
Stage 5: The formalisation of expert knowledge: an ontology
Stage 6: The classification of expert knowledge: the Algorithm - k-means Clustering and Distance Method and the Euclidean Distances Method
Stage 7: The development of new instructions for new maintenance workers

Fig. 1. An approach to the modelling of knowledge resources for preventive maintenance

We know that the received record can be presented by using formulated ontology. This ontology is created based on the instruction manual available for supported machines. Thanks to the adopted formalism, it is possible to recognise the records received and assign them to the current instruction manual and also to distinguish new elements of expert knowledge, which will be distinguished by the application of the adopted ontology.

Next, we use the Algorithm - k-means Clustering and Distance Method and the Euclidean Distances to classify new elements of expert knowledge within the current instruction manual.

The concept of a representation for the modelling of knowledge resources for preventive maintenance (see Fig. 1) in the form an IT tool, dedicated to the automotive industry is implemented and is presented in the next section.

3 The Applications Within the Automotive Industry, of a Knowledge Resources Model for Preventive Maintenance

In order to illustrate the possibility of answers to our research questions, let us consider the situation. The problem under consideration is that of defining improved instructions, for new maintenance workers, based on expert knowledge, in order to improve the stability of safety support within the automotive industry. Below, is an extract from the application, based on the proposed concept (see Fig. 1).

According to Stage 1 (see Fig. 1) for each competence, a knowledge web-questionnaire is defined.

Figure 2 presents an extract from the employees' web-questionnaire, in order to make the obtaining of the values for Personal Competences (PC) easier, viz.

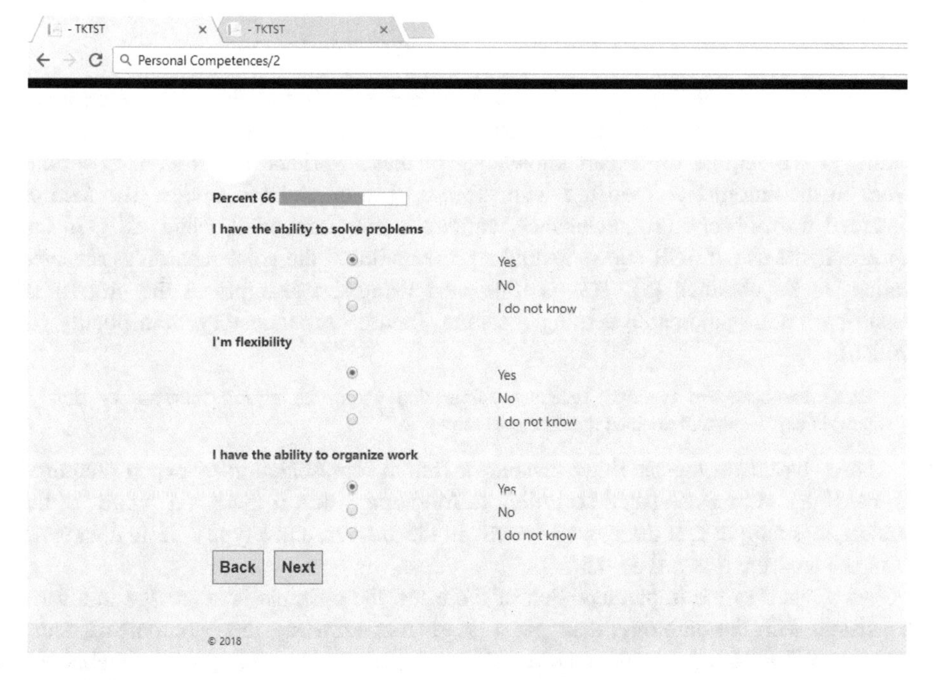

Fig. 2. An extract from the web questionnaire for maintenance workers, detailing their Personal Competences

Each maintenance worker completes a defined survey form for each competence: TC, OC, BC, SC and PC. Using an algorithm to evaluate each worker's responses, it is possible to determine the value for each worker's competences [14].

Next, the manager in the maintenance company describes the importance of each worker's competences to each supported company, in our case-study, describing them to the automotive industry (Fig. 3).

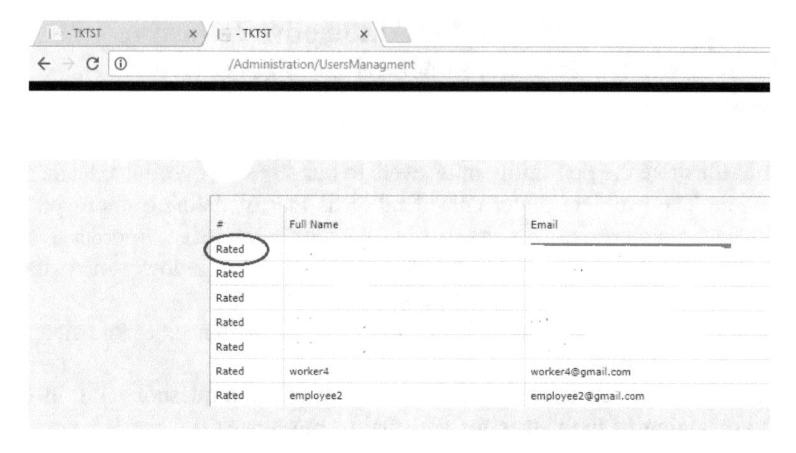

Fig. 3. An extract from the determination of the importance of each competence of a maintenance worker

Finally, based on the survey results, we receive the set of the expert knowledge resources (EKR) from the maintenance workers – in our case for the automotive industry. We acquire the expert knowledge of these workers by undertaking service work in the automotive industry, with the use of a smartphone device. The records obtained from observing maintenance workers is then converted, using ASR. In our approach, the use of ASR allows textual representation of the voice recordings received earlier, to be obtained [5]. This is illustrated using the example of the process of repairing a semi-automatic welding machine, failure: unsatisfactory weld quality (in Polish):

> "Przed rozpoczęciem naprawy ustalono potencjalną przyczynę awarii: niewłaściwy drut spawalniczy. Zabrałem ze sobą drut samoosłonowy …"

Next, based on the ontology created, a formal representation of expert, acquired knowledge can be established according to following rules: if there is a "word" in the answer, it is a type: 1, if there is no "word" in the answer, it is a type 0, if no data word is in the answer, it is a type 0.5.

As a result of the implementation of the rules, the example is expressed in a form consistent with the ontology; example = [start (rozpoczynać), repair (naprawa), cause (przyczyna), failure (awaria), none, welding wire (drut spawalniczy), take (zabierać), self-shield wire (drut samoosłonowy)].

Based on the ontology and the defined rules, formal representation of expert acquired knowledge can be established for this example: [1, 1, 1, 1, 0.5, 1, 1, 1].

Currently, 20 different activities in the maintenance area in the automotive industry were undertaken by maintenance workers and were recorded. With regard to defining the ontology, each activity was classified using the Algorithm - k-means Clustering and Distance Method and the Euclidean Distances Method with Statistica, ver. 13.3 to the two, standard instructions chosen, it being assumed that two instructions, with 70 defined words, would be assembled.

Figure 4 presents the results of the research.

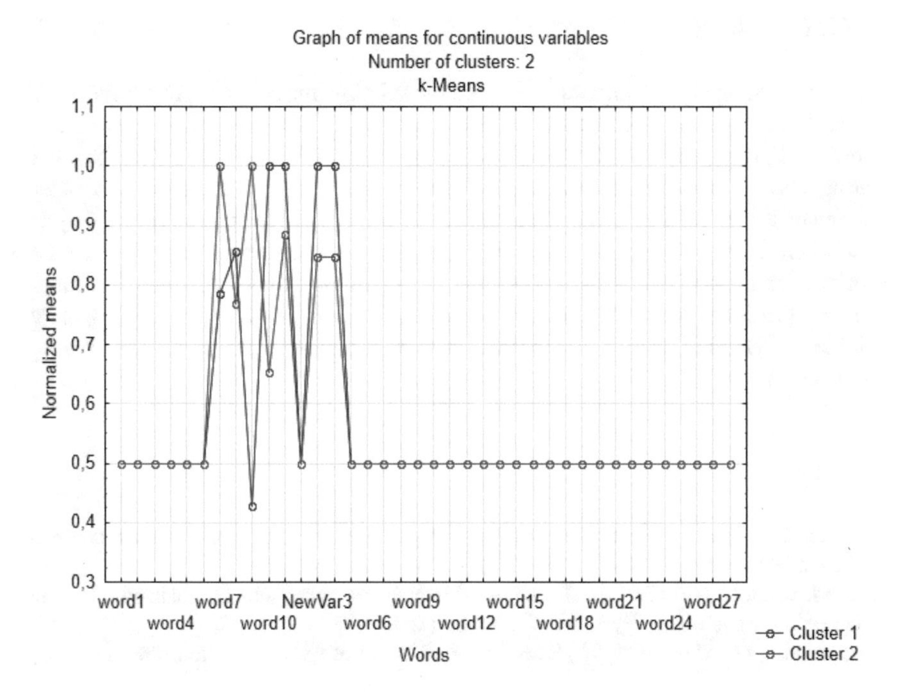

Fig. 4. Groups of activities using the Algorithm - k-means Clustering and Distance Method and the Euclidean Distances Method

Seven activities for the first instruction were classified, while for the second instruction, thirteen were classified. In our case study, therefore, we observe that we can develop a new form for instruction No. 2, based on the thirteen activities taken by expert knowledge resources (EKR) from the maintenance workers.

This manual instruction will be useful to new employees, because, in addition to the standard instruction, it will include expert knowledge based on real problem cases, resolved in the area of maintenance. The proposed solution is significant in preventive maintenance, where there is a need to send a new employee to a malfunction and the so-called source of expert knowledge is absent or is no longer working or has retired. The new instruction prepared, based on the experience of expert knowledge resources, will allow the new employee to carry out repairs reliably. Currently, the authors' work includes verification of the proposed solution in a selected automotive company, in particular in the area of building new instructions based on collected data from expert knowledge resources. Unfortunately, the difficulties encountered in implementing maintenance models is an issue scarcely discussed in the literature.

4 Conclusions

Expert knowledge is an important factor for building models for preventive maintenance. The approach presented is able to manage expert knowledge resources in a maintenance company with the web-based application making it easier for knowledge sharing among employees. Implementation of this model will allow the external maintenance company to provide constant, expert support for production enterprises, even where there is high employee turnover. Furthermore, this application could be developed in the context of an expert blog, which would provide discussion forums, in order to facilitate comment on and discussion of the problems of related, service activities. It seems that this approach is a good example for a model of preventive maintenance.

References

1. Ahmad, R., Kamaruddin, S.: An overview of time-based and condition-based maintenance in industrial application. Comput. Ind. Eng. **63**, 135–149 (2012)
2. Bevilacqua, M., Braglia, M.: The analytic hierarchy process applied to maintenance strategy selection. Reliab. Eng. Syst. Saf. **70**, 71–83 (2000)
3. Bocewicz, G., Muszyński, W., Banaszak, Z.: Models of multimodal networks and transport processes. Bull. Pol. Acad. Sci. Tech. Sci. **63**(3), 635–650 (2015)
4. Bouzidi-Hassini, S., Tayeb, F.B., Marmier, F., Rabahi, M.: Considering human resource constraints for real joint production and maintenance schedules. Comput. Ind. Eng. **90**, 197–211 (2015)
5. Dudek, A., Patalas-Maliszewska, J.: A model of a tacit knowledge transformation for the service department in a manufacturing company: a case study. Found. Manag. **8**(1), 176–177 (2016)
6. Fan, H.: Business Strategic Management: Knowledge Capital and Integration of Core Competence. Shanghai Foreign Language Education Press, Shanghai (2004)
7. Google Cloud Platform. https://cloud.google.com/video-intelligence/
8. Johnsen, T., Howard, M., Miemczyk, J.: UK defence change and the impact on supply relationships. Supply Chain Manag. Int. J. **14**(4), 270–279 (2009)
9. Lemke, J.: Analyzing Verbal Data: Principles, Methods, and Problems. Springer International Handbooks of Education, vol. 24, pp. 1471–1484 (2012)
10. Masood, T., Royand, R., Harrison, A., Xu, Y., Gregson, S., Reeve, C.: Integrating through-life engineering service knowledge with product design and manufacture. Int. J. Comput. Integrated Manuf. **28**(1), 59–74 (2015)
11. Murthy, D.N.P., Rausand, M., Østerås, T.: Product Reliability Specification and Performance. Springer, London (2008)
12. Nielsen, P., Michna, Z., Do, N.A.D., Sørensen, B.B.: Lead times and order sizes-A not so simple relationship. Adv. Intell. Syst. Comput. **429**, 65–75 (2016)
13. Öhmana, M., Finneb, M., Holmströma, J.: Measuring service outcomes for adaptive preventive maintenance. Int. J. Prod. Econ. **170**(B), 457–467 (2015)
14. Patalas-Maliszewska, J., Krebs, I: Decision model for the use of the application for knowledge transfer support in manufacturing enterprises. In: Abramowicz, W. (ed.) Business Information Systems Workshops 2015, LNBIP. Springer International Publishing, Heidelberg (2015)

15. Wang, S., Liu, M.: Multi-objective optimization of parallel machine scheduling integrated with multi-resources preventive maintenance planning. J. Manufacturing Syst. **37**, 182–192 (2015)

Practical Application of a Multimodal Approach in Simulation Modeling of Production and Assembly Systems

Pawel Pawlewski(✉)

Poznan University of Technology, ul. Strzelecka 11, 60-965 Poznań, Poland
pawel.pawlewski@put.poznan.pl

Abstract. This paper presents the results of research carried out in recent years in the area of modeling and simulation of production and assembly systems. The main goal of the article is to show the practical application of the multimodal approach in the simulation modeling of production and assembly systems and the benefits resulting from it. The article describes the issues of simulation modeling and presents a multi-level model of behaviors of the cyclic processes system, which is the backbone of the multimodal approach. The main part of the article is the concept of using a multimodal approach in simulation modeling and an example of its application. The main contribution of the paper is to define the methodology of simulation modeling based on cyclical processes - a multimodal approach and an approach to modeling from a high level where the basic object is a workstation.

Keywords: Multimodal approach · Cyclic processes · Simulation modeling

1 Introduction

The article presents the results of research carried out in recent years in the area of modeling and simulation of production and assembly systems. In these studies, simulation software based on discrete event simulation (DES) was used. The collected experience and observations allowed to identify the research gap in simulation modeling. This gap has already been noticed by the author [1], which resulted on the one hand with the concept of description of processes implemented in the production system using a multimodal approach, on the other hand prompted the author to further work. The effect of these works is described in this article.

The main goal of the article is to show the practical application of the multimodal approach in the simulation modeling of production and assembly systems and the benefits resulting from it.

The most important points of the article are the presentation of the originality of this approach in the context of traditional methodologies for building simulation models, proposing a mechanism for describing cyclical processes and showing a model built in the described way.

S. Rodríguez et al. (Eds.): DCAI 2018, AISC 801, pp. 58–66, 2019.
https://doi.org/10.1007/978-3-319-99608-0_7

The main contribution of the author is to define the methodology of simulation modeling based on cyclical processes - a multimodal approach and an approach to modeling from a high level where the basic object is a workstation.

The article is divided into 4 sections. The first section introduces the issues of simulation modeling. The second section describes a multi-level model of behavior of the cyclic processes system, which is the backbone of the multimodal approach. The third section describes the concept of using a multimodal approach in simulation modeling, and the fourth section presents the conclusions and outline of further research.

2 Simulation Modeling of Production and Assembly Systems

Simulation is the imitation of a real-world process or system. Simulation is used in many context, such as modeling, i.e. mapping of the real system, understanding of system behavior, virtual (and visual) assessment of possible consequences of actions. Simulation is also used for analyzing, experimenting and testing ideas and alternatives before making decisions on actions and resource involvement [2]. Simulation is a collection of methods and techniques to which we include discrete simulation, continuous simulations (including systems dynamics), Monte Carlo method (including static simulations in a spreadsheet), managerial games, qualitative simulation, agent simulation and others. In [3], the application ranges are presented according to the levels of abstraction (Fig. 1) and the corresponding simulation modeling methods (Fig. 2).

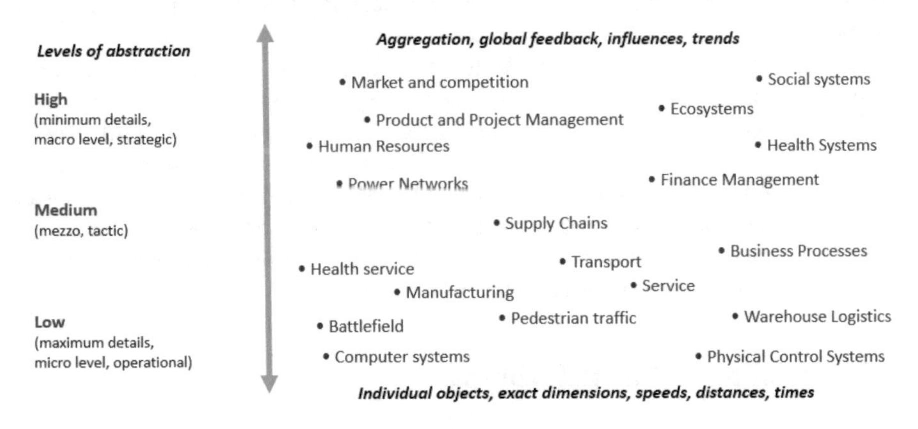

Fig. 1. Application ranges for abstraction levels - based on [3]

The essence of system dynamics is thinking in terms of feedback. This approach was developed in the 1950s by Forrester [4]. They are included in continuous simulation, for applications with a high level of abstraction - strategic. There is software available on the market that supports system dynamics modeling, e.g. Vensim (www.vensim.com), PowerSim (www.powersim.com).

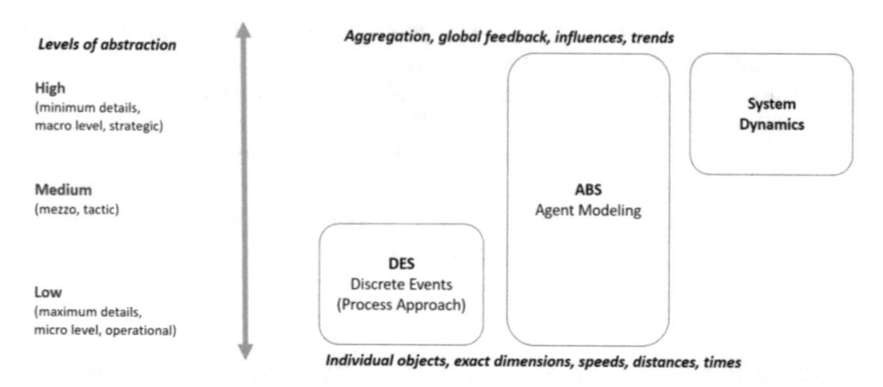

Fig. 2. Methods of simulation modeling in relation to levels of abstraction - based on Borshchev [3]

From the point of view of the research area described in this article, methods based on discrete events and agent modeling are of interest because they are used at the tactical level and above all at the operational level. Generally, it can be concluded that the discrete-based approach is a process-based approach and agent-based modeling can be compared to a task-driven approach [5]. However, it should be borne in mind that some authors believe that the task approach cannot be identified with agent modeling [6].

The software based on agent modeling available on the market (e.g. Swarm, NetLogo, Repast, ASCAPE) is primarily software focused on modeling social behavior - interactions between objects. Their application in the modeling of production systems is not important.

Many of simulation software available on the market are dedicated to simulation modeling of production and assembly systems (e.g. FlexSim, Arena, Tecnomatix, Anylogic, Witness, etc.). These systems are based on discrete events but also provide the ability to control tasks. However, their main task is to model the so-called "Flow process" (or process flow) focusing on what is flowing through the process. A typical, simplest process model is a model consisting of a source (generator) of flow elements, a buffer (in which a queue of flow elements may be created), a processor (performing an operation) and runoff or ending place of the process.

3 Basics of a Multimodal Approach

The origins of multimodal thinking can be found in the definition of a multimodal transport system. In accordance with the Convention on International Multimodal Transport of Goods [7], a multimodal transport system is defined as an internally integrated system of transporting goods along (supply chain) with associated services using at least two modes of transport based on a multimodal transport contract. The multimodal transport contract is carried out by one multimodal carrier, which assumes full overall responsibility for the performance of the contract. In the case of intermodal transport, at least two types of transport also participate, but its special feature is that only one unit load is used on the entire freight belt. In short, we can assume that the intermodal supply chain is a specific type of multimodal logistics, which is characterized by a unitary load unit, fixed throughout the entire freight line. In logistic practice, the most commonly used multimodal solutions are sea and air transport as well as rail and air transport. Multimodality in transport results from the development of containerization: different modes of transport have become more connected due to the fact that modes of transport, storage and loading of unified loading units had to become similar. The main multimodal transport units are: JTI container, UTI swap body, ITU trailer. In summary, logistic multimodality is characterized by the following features:

1. use of at least two types of transport,
2. single charge agreement,
3. one contractor responsible for the delivery of goods,
4. total price for the goods delivery service,
5. loading and handling of the entire loading device (e.g. container, transporter or means of transport).

In the context of transport networks, a multimodal network is one in which two or more types of transport are modeled (such as walking, riding a train or driving a car). Alternatively, in utility networks, a multimodal network may consist of various transmission and distribution systems [8].

By expanding this thinking, it is possible to define multimodal processes as processes implemented along roads consisting of roads of local processes [9, 10]. A multi-layered model of behavior of the system of concurrent cyclic processes was developed - Fig. 3.

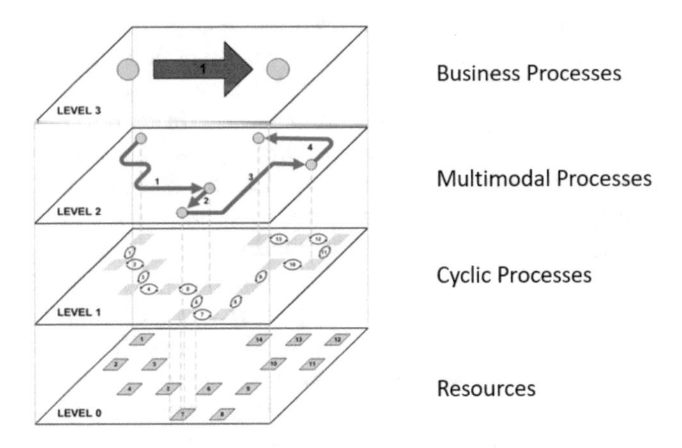

Fig. 3. Multilayered model of behaviors of the system of concurrent cyclic processes [10]

4 The Use of a Multimodal Approach in Simulation Modeling

The previous parts of the article describe the simulation modeling of production systems and the multimodal approach. What is the relationship between them? Traditionally, a process approach is used to model the production flow, creating a model according to the "top-down" rule from top to bottom, bringing the simulation model to life at every step. Of course, simulation models are built to answer the question asked by the engineer to solve a specific problem. In many cases, this approach is appropriate and sufficient. In the literature on the subject, recommendations can be found to start the simulation project from building a map of the value stream [3]. In such cases a process approach is used in which the part flow through the production system is presented [11]. This approach is also described in the literature on simulation modeling. This approach is also recommended by simulation software producers, e.g. Simio, Anylogic, Arena, Tecnomatix, FlexSim. However, you can take a deeper look, because it can be assumed that this approach is approximated due to the "top-down" rule, i.e. modeling at a high level of abstraction and going down - taking into account more details depending on the needs. This is in accordance with the classic method of Artur Hall [12].

However, when building a simulation model, we build it with elements. These elements (also called objects) reflect or approximate the real system. These elements are abstract objects (e.g. graphic icons) or objects to which certain behaviors are assigned. Graphic icons are elements of graphical modeling languages, e.g. VSM (Value Stream Mapping) [13], IDEF0 [14] or OFD (Object Flow Diagram) [15]. In turn objects that are assigned certain behaviors form the basis for building models in simulation programs - e.g. source, buffer, processor, operator, etc. These elements are used in modeling at different levels of abstraction depending on the needs [2, 3].

It seems interesting to try to change this way of thinking. You can make an attempt to analyze the modeled system in such a way that going down and down as far as possible and back again. This article presents observations and conclusions from such a "journey". The manufacturing system in the automotive industry has been analyzed and modeled. The process is performed on several welding stations, where welding operations are carried out successively and then the assembly and packaging operation of the product is carried out. It is a pipeline production. An attempt was made to model the flow of parts. Parts flow in containers. The conclusion that appeared was that you should focus on the flow of containers. In the analyzed system, the flow area was defined from the so-called where there are already prepared containers with parts for the buffer with containers of finished products, which is located in front of the finished products warehouse. The warehouse flows were not analyzed and modeled before the supermarket and after the container buffer with ready products.

The containers "flowed" from the supermarket to the fields of workstations, where the welding operation was carried out, then the assembly (in containers or on the logistic trolley) "floated" to the next workstation, up to the buffer in front of the finished product warehouse. Assembly operations have been modeled taking into account the installation time (described with the appropriate statistical distribution), disruptions (failures) and planned breaks. The focus was on the flow. Containers themselves do not flow through the system - there is always a mechanism that causes this flow. Containers as found in the supermarket are static - they have no initiative to flow (this is not in line with the process approach). As a result, the focus was on how the flow is implemented and provoked. The mechanism that realizes the flow is well known to all - in the case of non-automatic flow, operators who use various means of transport such as pallets, logistic trucks, forklifts, logistic trains and in the case of automatic flow are robots, conveyors, agv trucks. Generally, in the subject literature this is considered separately as internal transport. Production flow management is the management of mechanisms that realize this flow. The described mechanisms usually work in a cycle: approach the storage area (it can be a highlighted place on the floor, shelf of the rack), get the container, move to the target storage field, put away the container, repeat the cycle (or change the cycle). There was an idea to focus on these mechanisms and try to model them in such a way that they would be useful. Taking such an attempt, it was associated that a very good theoretical and practical model is the multimodal approach described in the previous section of this article. Cyclic processes describe the mechanisms that create the flow while the multimodal process describes the flow process itself.

Figure 4 shows the situation in described assembly system (using automatic welding stations) where the containers flow using the milk run system (Fig. 4 - green arrows – CYCLE MR) – logistics train. When the train stops at the "bus stop" near the assembling workstation, the driver performs the cycle between logistics trolley and storage location (Fig. 4 - blue arrows – CYCLE DR). The last cycle is performed by operator at the workstation – he loads the part, travels to welding station and put it in welding (Fig. 4 - yellow arrows – CYCLE OP). All these activities are repeated.

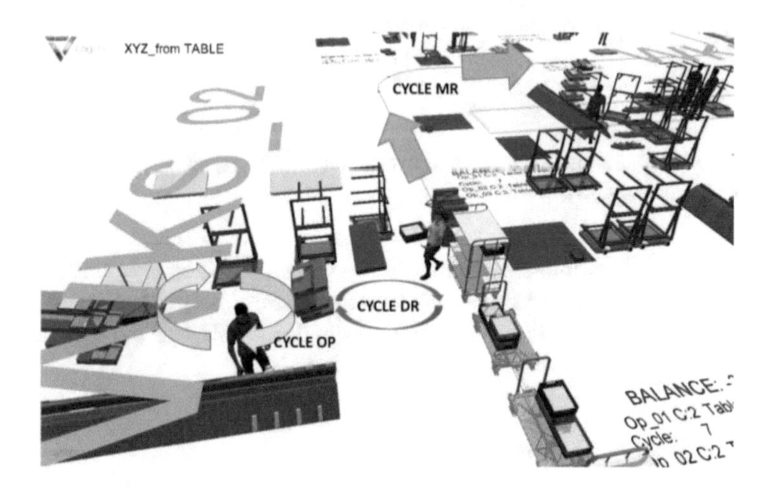

Fig. 4. Simulation model with marked locations for cyclic processes - own study

The question arose whether it is possible to build a simulation model of the production system, in which only the flow mechanisms are modeled using primarily cyclic processes. During the simulation project for the client from the automotive industry, a successful attempt to build a simulation model in the described way was made. The main flow was identified based on the BOM (Bill of Materials), an array of the so-called PFEP (Plan for Every Part) assigning a part, a team, a container product, routes were built (cyclic processes) for each operator, robot - contractor, flow producer. Routes have been described in a special language for which an interpreter has been prepared in a simulation program. This language consists of 57 instructions. One instruction - for example "Load a Part from a Container" that performs the removal of a part from a container located in a specific storage field has the form:

P_08 LoadFromTote 1

P_08 is the identifier of the reloading place where the container with the parts is located, LoadFromTote is the name of the instruction and 1 is the number defining how many parts should be taken (in this case one part). This simple structure enable to describe behaviours of operators. Figure 5 shows the example of two parts of list of tasks stored in tables which describe behaviours of operator and graphic presentation of relations between cycles performed by one operator.

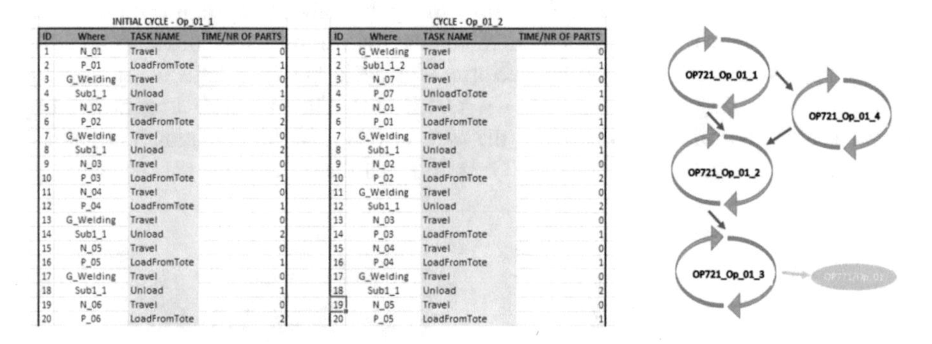
INITIAL CYCLE - Op_01_1

ID	Where	TASK NAME	TIME/NR OF PARTS
1	N_01	Travel	0
2	P_01	LoadFromTote	1
3	G_Welding	Travel	0
4	Sub1_1	Unload	1
5	N_02	Travel	0
6	P_02	LoadFromTote	2
7	G_Welding	Travel	0
8	Sub1_1	Unload	2
9	N_03	Travel	0
10	P_03	LoadFromTote	1
11	N_04	Travel	0
12	P_04	LoadFromTote	1
13	G_Welding	Travel	0
14	Sub1_1	Unload	2
15	N_05	Travel	0
16	P_05	LoadFromTote	1
17	G_Welding	Travel	0
18	Sub1_1	Unload	1
19	N_06	Travel	0
20	P_06	LoadFromTote	2

CYCLE - Op_01_2

ID	Where	TASK NAME	TIME/NR OF PARTS
1	G_Welding	Travel	0
2	Sub1_1_2	Load	1
3	N_07	Travel	0
4	P_07	UnloadToTote	1
5	N_01	Travel	0
6	P_01	LoadFromTote	1
7	G_Welding	Travel	0
8	Sub1_1	Unload	1
9	N_02	Travel	0
10	P_02	LoadFromTote	2
11	G_Welding	Travel	0
12	Sub1_1	Unload	2
13	N_03	Travel	0
14	P_03	LoadFromTote	1
15	N_04	Travel	0
16	P_04	LoadFromTote	1
17	G_Welding	Travel	0
18	Sub1_1	Unload	2
19	N_05	Travel	0
20	P_05	LoadFromTote	1

Fig. 5. Example of sets of instructions stored in tables and relations between cycles which describe behaviours of operator – own study

The whole has been implemented in the FlexSim simulation program environment treating it as so-called SOS (Simulation Operating System). The FlexSim program was chosen because of the possibility of working directly in 3D and the openness of the system.

5 Conclusions and Future Research

The article presents the practical application of a multimodal approach in simulation modeling of manufacturing systems. Changing the approach to modeling resulted in shortening the time of building the simulation model, shifting most of the work to the data preparation stage. The main reason was using the workstation template and the developed scripting language. The simulation model built on the basis of this approach, or rather the method of its construction, prompts the author to continue this work. The reaction of people from the industry, which is presented in the way of thinking described in the article, is promising.

Directions for further work are:

1. completion of work on the language describing the activities carried out by cyclic processes, especially since the instructions of the language so far developed are understood and accepted by production and logistics engineers,
2. preparation of the environment for simulation modeling based on cyclic processes (using available simulation software as the equivalent of the operating system),
3. formal description of the whole - using the formal apparatus offered by the authors of works from the area of multimodal cyclic processes models,
4. preparation of technologies for collecting and preparing data in the enterprise for simulation modeling.

The main purpose of this work is to prepare a technology that answers the question how to prepare data in an enterprise to build a simulation model quickly and without having deep knowledge about simulation modeling in a given tool offered on the market. The purpose of using this technology is to solve problems in the area of production and assembly regarding: design and modification of the layout of the production plant/production space, design and redesign (improvement) of intralogistics (Milk Run), work balancing and Yamazumi analyzes.

Acknowledgements. The work was carried out as part of the POIR.01.01.01-00-0485/17 project, "Development of a new type of logistic trolley and methods of collision-free and deadlock-free implementation of intralogistics processes", financed by NCBiR.

References

1. Pawlewski, P.. Multimodal approach to modeling of manufacturing processes. Procedia CIRP **17**, 716–720 (2014). Variety Management in Manufacturing — Proceedings of the 47th CIRP Conference on Manufacturing Systems (2014)
2. Beaverstock, M., Greenwood, A., Nordgren, W.: Applied Simulation. Modeling and Analysis using Flexsim, Flexsim Software Products, Inc., Canyon Park Technology Center, Orem, USA (2017)
3. Borshchev, A.: The Big Book of Simulation Modeling. Anylogic North America, Lisle (2013)
4. Forrester, J.W.: Industrial Dynamics. Pegasus Communications, Waltham (1961)
5. Siebers, P.O., Macal C.M., Garnett, J., Buxton, D., Pidd, M.: Discrete-event simulation is dead, long live agent-based simulation! J. Simul. **4**(3), 204–210 (2010)
6. Weimer, Ch.W., Miller, J.O., Hill, R.R.: Introduction to agent based modeling. In: Roeder, T.M.K., Frazier, P.I., Szechtman, R., Zhou, E., Huschka, T., Chick, S.E. (eds.) Proceedings of the 2016 Winter Simulation Conference (2016)
7. United Nations Convention on International Multimodal Transport of Goods (Genewa, 24 mai 1980 r.) (1980)

8. Hoel, E.G., Heng, W.L., Honeycutt, D.: High performance multimodal networks. In: Bauzer, M.C., et al. (eds.) SSTD 2005. LNCS, vol. 3633, pp. 308–327. Springer, Heidelberg (2005)
9. Bocewicz, G., Nielsen, P., Banaszak, Z., Quang, V.Q.: Cyclic steady state refinement: multimodal processes perspective. In: Bjørge, J.F., et al. (eds.) Advances in Production Management Systems, series: IFIP Advances in Information and Communication Technology, vol. 384, pp. 18–27. Springer, Heidelberg (2012)
10. Pawlewski, P.: Multimodal approach to model and design supply chain. In: Proceedings of IFAC MIM Conference, St. Petersburg (2013)
11. Nyemba, W.R., Mbohwa, C.: Modelling, simulation and optimization of the materials flow of a multi-product assembling plant. Procedia Manuf. **8**, 59–66 (2017)
12. Hall, A.D.: A Methodology for Systems Engineering. Princeton University Press, Princeton (1962)
13. Rother, M., Shook, J.: Learning to see. Value-Stream Mapping to Create Value and Eliminate Muda, The lean Enterprise Institute, Inc. (2003)
14. Santarek, K.: Organisational problems and issues of CIM systems design. J. Mater. Process. Technol. **76**, 219–226 (1998)
15. Greenwood, A., Pawlewski, P., Bocewicz, G.: A conceptual design tool to facilitate simulation model development: object flow diagram. In: Pasupathy, R., Kim, S.-H., Tolk, A., Hill, R., Kuhl, M.E. (eds.) Proceedings of the 2013 Winter Simulation Conference (2013)

Method of Quality Assessment of the Implementation of Design Patterns Used in Production

Rafał Wojszczyk[1(✉)] and Piotr Stola[2]

[1] Faculty of Electronics and Computer Science,
Koszalin University of Technology, Koszalin, Poland
rafal.wojszczyk@tu.koszalin.pl
[2] Quick-Solution, Koszalin, Poland
piotr.stola@quick-solution.net

Abstract. Programmers in agile production teams very willingly are implementing design patterns during their work. They often do based on own experience, which does not include additional profits that can be expected after implementing the patterns.

The aim of the article is to show the process and results of the verification method of quality assessment of the implementation of design patterns used in production. The experiment conducted together with a real, working by the agile production methods, company.

Keywords: Design patterns · Software quality assessment
Practical verification

1 Introduction

A programmer implementing design patterns does so on the sample templates from [2, 4] and his own knowledge, during which he usually focuses on achieving the purpose of the pattern (solving the programming problem). The implementation of the pattern goal in accordance with the template from [2] does not mean a beneficial implementation, because each computer program is different. The preferred implementation of the template is a fragment of the source code that meets additional expectations, otherwise it provides benefits in selected criteria. Assuming the low development and integration cost criterion, this means that the template code will not require additional modifications when expanding and integrating with this code. Therefore the cost of the development will consist of the cost of adding new parts of the code that use existing pattern implementation. In this context, a programmer working in an agile team after doing his job (writing the source code, usually without complete documentation) is looking for the answer to the question: *will the implementation of a given design pattern provide the benefits expected from this pattern?* The answer to this question enables the author's method of assessing the quality of implementation of design patterns. The aim of the article is to present the results of the use of the method in the production environment.

S. Rodríguez et al. (Eds.): DCAI 2018, AISC 801, pp. 67–74, 2019.
https://doi.org/10.1007/978-3-319-99608-0_8

2 Quality of Implementation of Design Patterns and Alternative Methods

2.1 Quality of Pattern Implementation

The criteria of the assessment of quality in terms of the cost of development and software integration are one of the most important for the vendor. The vendor, who constantly keeps and develops his product, even for many years, should take care of the fact that the cost of running and development are as low as possible. For that purpose design patterns are used. It has been widely accepted, that programmers are implementing patterns on the second level of quality, i.e. so that the implementation meets only the presented aim of the pattern, e.g. one instance of the object in the Singleton pattern. First level of implementation quality is undesirable, such an implementation contains errors, e.g. the public constructor of the class of Singleton pattern. Both 1st and 2nd level of implementation quality does not provide the benefits, that were explained in the introduction, this is only ensured by implementation on the third level of quality. Level 0th is a special case when there is no fragment in the code that matches the pattern. A comparison of all quality levels is shown in Fig. 1. Leaving the implementation on the 1st and the 2nd level in the production software will cause additional costs in the future.

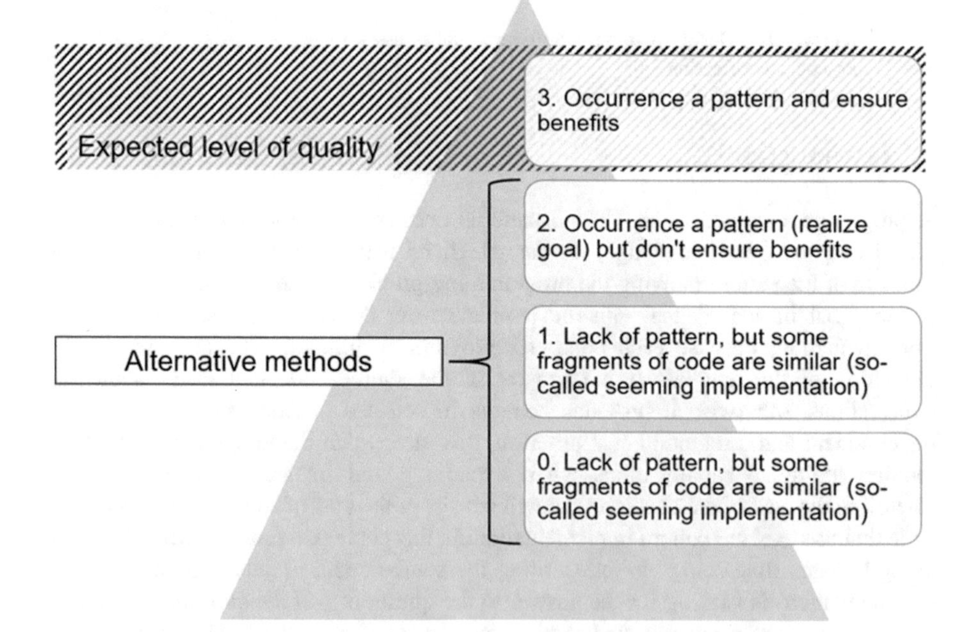

Fig. 1. A comparison of the levels of implementation of design patterns quality.

2.2 Alternative Methods

The quality of the source code is commonly associated with object-oriented software metrics. Unfortunately, popular metrics do not apply to the implementation of design patterns, despite the cost of use acceptable in agile vendor teams. Amongst the scientific research related to this issue, the dominating problem is the search for occurrence design patterns [6, 7]. The result of the method of finding the occurrence is the number of occurrences of patterns in examined part of the program code or the equivalent of the code. One occurrence of the pattern in most methods is only an information about a compatibility of a part of the code with the template describing reference pattern, on the basis of this part of the code is classified as the occurrence of the pattern. Most methods searching for an occurrence of patterns works in binary, i.e. indicates an occurrence of the pattern or no pattern, which corresponds to an estimation of the assessment of 2nd or 0th level of the quality of implementation. Chosen methods additionally enable to show an incomplete occurrence of the pattern (e.g. it contains errors or deficiency in implementation), which corresponds to 1st level of implementation quality. The cost of using methods searching for occurrence of the patterns is in most cases accepted agile vendor teams. Other research concern methods of verification of pattern implementations, which once again rely on showing the compliance of the tested part of the code with design pattern template [3, 5]. The result of the implementation verification method is the indication of a part of the code, that is compatible with the pattern template. Full compliance with the template corresponds to the 2nd level of quality of implementation, while the exceptions from this correspond to 1st and 0th level. Cost of using methods verifying the implementation of patterns is bigger than possibilities of the agile team, since detailed documentation is required. To sum up, alternative methods do not allow to distinguish implementation compatible with the 3rd level from the 2nd level of quality, i.e. it is not possible to assess whether the implementation of a given pattern provides the expected benefits, including lower costs of development and integration.

3 Experiment Environment

3.1 Method of Quality Assessment of Design Patterns

The used method, called Danyko (abbreviation from polish words) was described in [8]. The action of the method starts with the model building: program code, pattern and metrics. Then, the program code equivalent is evaluated in regard to the template pattern. Pattern template is described by a set of characteristics and metrics that specify them, which is an analogy to known quality models, e.g. ISO 9126. In other words, the assessment consists in measuring the distance of the program code from the pattern template, then the distance is interpreted in regard to the chosen assessment criteria. Result of interpretation is the appropriate level of implementation quality.

3.2 Design Patterns and Software Used

Verification of the method carried out in cooperation with the company Quick-Solution from Koszalin, which provided the source code. The code consists of 11 projects (packages), 52 unique classes that contain 3374 lines of the code. The company specializes in custom-made software in Microsoft .NET technology. The basic area of software produced is web, mobile and desktop applications. The team works according to the Scrum methodology, which is supervised by the team leader, who is also a Scrum Master. Scrum belongs to the group of agile methodologies, therefore, according to Agile Manifesto, the amount of project documentation is limited to the minimum, because the program is more important than the extensive documentation.

Experiment was carried out using the Command and Factory patterns, which belong to one of the most popular patterns.

The aim of the Command pattern is [2]: *encapsulation of requests in the form of an object*. This allows the client to be parameterized using different requests, and putting requests in queues and logs, as well as provide and undo operation support. Implementation of the pattern is useful when many different operations can be performed on one object (e.g. a bank account). Figure 2 shows a class diagram with an example pattern implementation, on the basis of [2]. The diagram from the Fig. 2 shows a structural variant, the modification of this variant is a variant with dynamic mapping (connections in Client class are created dynamically, e.g. by reflection mechanism or injection of dependencies).

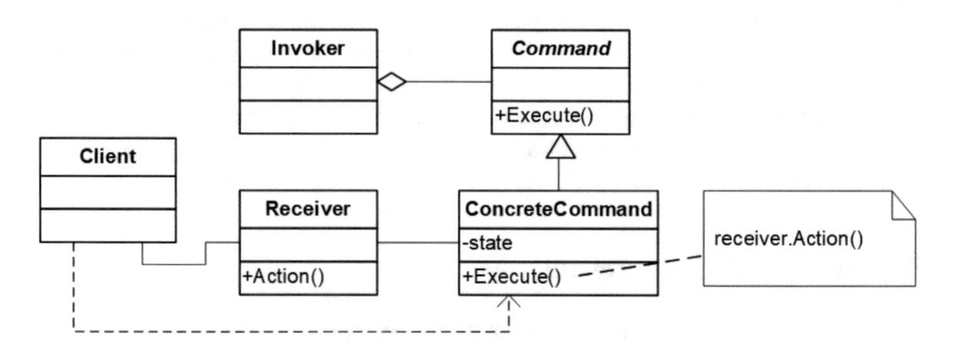

Fig. 2. Class diagram of command pattern.

Elements, of which the Command pattern is made of [2]:

- *AbstractCommand* class – declares a common point to perform operations, other names: parent class, parent type, general command,
- *ConcreteCommand* – includes the implementation of the *Execute* operation in the form of calling appropriate operation of the *Receiver* object, other names: Concrete command, subclass,
- *Receiver* – executes a specific command (algorithm), other names: recipient,
- *Client* – creates objects of specific commands and determines connections (maps) with recipients, other names: map, connection mapping,

- *Invoker* – request servicing of the command, other names: sender.

The purpose of the Factory pattern is [2] *to define the interface for creating objects, while the act allows subclasses to determine the class of a given object the creation process is passed to the subclasses*. The implementation of the pattern is useful when different objects carrying information can be created from one operation. Figure 3 shows the class diagram of the sample implementation pattern, on the basis of [2, 4]. In [2] Factory patterns, i.e. Abstract Factory and Factory Method are described separately, although they are included in one group. In practice, however, programmers unify these patterns and define them as two variants of the Factory pattern.

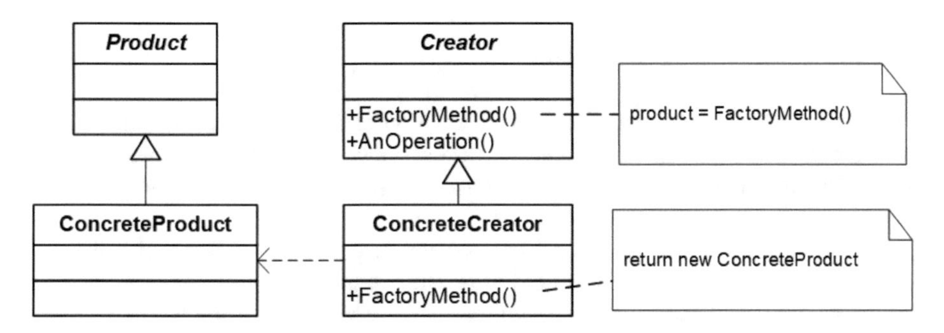

Fig. 3. Class diagram of Factory pattern.

Elements that Factory patterns is made of [2]:

- *Product* – declare the interface of objects generated by the factory, other names: product,
- *ConcreteProduct* – includes the implementation of the Product class, other names: a specific product,
- *Creator* – contains a declaration of the vendor methods that returns *Product* objects, other names: vendor,
- *ConcreteCreator* – override the method from *Creator* to return a copy of the *ConcreteProduct* class, other names: concrete vendor.

The use of the Factory pattern was planned and designed by programmers developing the application. The pattern was described in the project documentation, which contained only selected parts of the application. Then the pattern was implemented in one of the first iterations in accordance with the knowledge of the team. However, using the Command pattern was not initially planned, which is why the pattern was implemented as a part of refactoring. Changes resulting from refactoring were not applied to the project documentation, which caused discrepancies between the source code and the original documentation. The experiment was carried out in the iteration, in which refactorization was carried out to the Command pattern. In the release plan (backlog) of later iterations (regarding to iteration in which the experiment was carried out), the development of the software is planned, which requires the use of implementation of mentioned design patterns.

The experiment consisted of 3 stages. In the first stage, a "brainstorm" was carried out together with the Quick-Solution team, whose aim was to develop reference implementations of both used patterns. The developed implementations meet the third level of quality, i.e. they meet the expected goal of the standard and provide a low development cost. In the case of a Command pattern, the cost of adding a new type of command should be lowered. In the case of Factory pattern, the cost of adding new types of objects to existing operations and creating new operations that can generate different objects should be lowered. Then, in the second stage, the reference implementations were mapped to the required models used by the method. The third stage is the assessment process itself [9].

4 Results of the Experiment

The results of the assessment of the implementation of design patterns obtained during the experiment are presented in Table 1, these are only parts of the code below the 3rd level of implementation (for the Command pattern from C1 to C4, the Factory pattern - F1).

In the case of the Command pattern, several errors occurred. Probably many of them would not be improved as part of further work, which in the case of expanding the software with new features in the future means more work. By using the method, the errors can be corrected even during the iteration of the experiment. In total, about 10% of the pattern code is below the 3rd level of implementation quality.

In the case of the Factory pattern (which was previously designed) there was only one error related to access modifiers. These types of errors are often the result of oversight of the implementers and, presumably, they would be successively repaired as part of other code work, however, that they are significantly extended in time. By assessing the quality of implementation of this pattern, the detected errors can be repaired earlier. In total, only 2% of the pattern code is below the 3rd level of implementation quality.

The cost of changes to be made resulting from the detected defects is small in the case of the Factory pattern. The programmers decided that the error can be noticed and corrected "by the way" during other works. Much more interest was raised by the results of the quality assessment of the Command pattern. In the Quick-Solution team's opinion, "locating these errors at the beginning of the implantation phase is extremely effective and saves much more time than if the code was released in the production version". The cost of work without the use of the Quick-Solution team method Danyko was estimated at 20 man-hour, which contains time devoted to finding faults, testing, implementation work. A mixed method was used for estimation, based on the expert's assessment and estimation by analogy. After hearing the results of the quality assessment, the team estimated the cost of work for 6 man-hour (implementation work), therefore the estimated savings in Software development is 14 man-hour. This time-consuming time does not take into account the time needed to develop reference implementations, however it is a one-time cost. Once developed models will be used many times in the production application of the method. The described time consumption does not take into account the time needed to develop reference

Table 1. Results of experiment

Error number	Element in the source code	Quality level	Error and suggestion of changes	Consequences
C1	Command	2	Type is a class, the class should be replaced with an interface	Limited ability to inherit in specific commands (some commands may require inheriting from the selected class, e.g. related to the ORM context)
C2	Concrete-Command. Execute (int pam)	2	Occurrence the method with a similar signature, the method name should be changed	Possible a risk that the programmer may use a different Execute method than expected
C3	Client	2	Not all specific commands are included in the connection map, add the missing commands to the map	Unused specific commands can be deleted or re-implanted
C4	Invoker1, Invoker2,	1	There is a single call to the so-called in-line (the override type declaration was omitted), the call should be preceded by ICommand declarations	It disrupts the use of the pattern Command, limits the flexibility of the code, the execution of the selected commands is beyond the control of the Command pattern
F1	Serialize Factory. Get Serializer	2	Internal modifier (limited availability of the method), changes the access modifier to public	Limited availability of the factory, will not be available outside the package (risk of reimplementation)

implementations, however it is a one-time cost. Once developed models will be used many times in the production application of the method.

5 Summary

The work briefly discusses the concept of quality implementation of design patterns and three quality levels to which pattern occurrence can be classified. It also presents the problem of evaluating the quality of pattern implementation from the point of view of a programmer in an agile production team. This problem required the development of an author's method Danyko. The work presents the results of method verification in the production application. During verification, it was shown that the use of the method saves 14 man-hour. Without using the method, the vendor team would have to recon with 20 man-hour of hidden costs. The obtained result confirms that the developed method has potential in practical application, which will help in reducing selected costs in software development.

Further work involves refining the tool supporting the method and integration with IDE environments. There is also a plan to extend the method with new types of patterns and adaptation to other popular solutions belonging to the so-called good practices.

References

1. Czyczyn-Egird D., Wojszczyk R.: Determining the popularity of design patterns used by programmers based on the analysis of questions and answers on Stackoverow.com social network. In: Communications in Computer and Information Science, vol. 608, pp. 421–433. Springer, Cham (2016)
2. Gamma, E., et al.: Design Patterns: Elements of Reusable Object-Oriented Software. Addison-Wesley Professional, Boston (1994)
3. Mehlitz, P.C., Penix, J.: Design for verification using design patterns to build reliable systems. In: Proceedings of Working on Component-Based Software Engineering (2003)
4. Metsker, S.J.: Design Patterns in C#, 1st edn. Addison-Wesley Professional, Boston (2004)
5. Nicholson, J., et al.: Automated verification of design patterns: a case study. Sci. Comput. Program. **80**, 211–222 (2014)
6. Singh, R.R., Gupta, M.: Design pattern detection by Greedy algorithm using inexact graph matching. Int. J. Eng. Comput. Sci. **2**(10), 3658–3664 (2013)
7. Tsantalis, N., et al.: Design pattern detection using similarity scoring. IEEE Trans. Softw. Eng. **32**(11), 896–908 (2006)
8. Wojszczyk, R., Wójcik, R.: The model of quality assessment of implementation of design patterns. In: Advances in Intelligent Systems and Computing, vol. 474, pp. 515–524. Springer, Cham (2016)
9. Wojszczyk, R.: The experiment with quality assessment method based on strategy design pattern example. In: Advances in Intelligent Systems and Computing, vol. 656, pp. 103–112. Springer, Cham (2018)

Routing and Scheduling of Unmanned Aerial Vehicles Subject to Cyclic Production Flow Constraints

G. Bocewicz[1(✉)], P. Nielsen[2], Z. Banaszak[1], and A. Thibbotuwawa[2]

[1] Faculty of Electronics and Computer Science, Koszalin University of Technology, Koszalin, Poland
bocewicz@ie.tu.koszalin.pl
[2] Department of Materials and Production, Aalborg University, Aalborg, Denmark
{peter, amila}@mp.aau.dk

Abstract. The focus is on a production system in which material handling operations are carried out by a fleet of UAVs. The problem formulated for the considered case of cyclic multi-product batch production flow is a material handling cost problem. To solve this problem, it is necessary to designate the routes and the corresponding schedules for vehicles that make up the given UAV fleet. The aim is to find solutions that minimizes both the UAV downtime and the takt time of the cyclic production flow in which operations are performed by the UAVs. A declarative model of the analyzed case was used. This approach allows us to view the problem as a constraint satisfaction problem and to solve it in the OzMozart constraint programming environment.

Keywords: UAV · Production flow · Vehicle routing problem
Cyclic scheduling

1 Introduction

Material handling operations in discrete production systems can be carried out using various modes of internal transport, such as industrial robots and manipulators, cranes, conveyors, self-propelled trucks, palletizers, automated guided vehicles, single-track cranes and overhead cranes [1, 6, 7, 12]. In addition to the aforementioned machines, unmanned aerial vehicles (UAVs) increasingly considered for material and product movements between and within departments and even between workstations [7, 15].

UAVs and UAV technologies, on one hand, enables more flexible transfer of materials and goods between workstations (by exploiting the possibilities of operating in 3D space), but, on the other hand, they generate new problems related to the organization and maintenance of the planned production flow. Typical limitations on the implementation of UAVs in material handling tasks include the need for periodic battery replacement/charging, overlapping air corridors designated for UAV movement as well as selected technical parameters such as payload capacity and speed.

S. Rodríguez et al. (Eds.): DCAI 2018, AISC 801, pp. 75–86, 2019.
https://doi.org/10.1007/978-3-319-99608-0_9

The present research addresses the problems of routing and scheduling of a UAV fleet, taking into account the limitations associated with the cyclic nature of flow in the production system in which the fleet is deployed. The focus of the study are solutions that minimizes both the UAV downtime and the takt time of the cyclic production flow in which operations are performed by the UAVs. The results fall within the scope of research, reported in previous papers, on the levelling of multi-product batch production flows [2] and repetitive production flow balancing [1].

The remainder of the paper is structured as follows. Section 2 provides an overview of the most important areas of application of UAVs, with a focus on selected material handling problems, which determine effective production flow. An example of a UAV Fleet Routing and Scheduling problem is discussed in Sect. 3. Section 4 presents a declarative model of the organization of production flow that allows to consider the problem of UAV routing as a constraint satisfaction problem and, consequently, to solve it in the OzMozart programming environment. The key conclusions are formulated and the main directions of future research are suggested in Sect. 5.

2 Related Work

The most commonly used method of ensuring the feasibility of UAV routes is to solve a Path Planning problem [13] consisting of finding a feasible trajectory of a UAV visiting a given sequence of waypoints in a three-dimensional space, without considering the vehicle's dynamics (e..g. deviations caused by disturbances in the environment). Finding feasible trajectories in UAV routing problems is a complex task, which is why very few studies have attempted to simultaneously solve the routing and the trajectory optimization problems for a fleet of UAVs. In this kind of joint-optimization problem, in which a fleet of UAVs has to visit a set of waypoints assuming generic kinematics and dynamics constraints, the solutions have to satisfy constraints imposed by the operating environment and those related to collision avoidance between UAVs and obstacles [4, 10].

The Vehicle Routing Problem (VRP) approaches used for routing UAVs can be grouped into several classes following the general classifications of VRPs [3]. However, the majority of VRP approaches that are specifically applied in the routing and scheduling of UAVs belong to the following categories: periodic VRP (the deliveries are done in a cyclic manner), dynamic VRP, multiple depot VRP (the vendor uses many depots to supply the customers), and capacitated VRP [14].

A special place in path planning studies is occupied by the Cyclic-Routing UAV Problem (CRUAVP). The UAV path-planning decision version of the CRUAVP can be seen as boiling down to a simple recurrent UAV path-planning problem in which each target must be visited not only once but repeatedly, i.e., at intervals of prescribed maximal duration. Problems of this type have long been considered in many other fields such as logistics, robotics, and supply chain management [9], where a common scenario concerns missions in which a set of targets have to be visited by a limited number of UAVs [8]. Among the studies that take into account the kinematic and dynamic constraints of UAVs, the majority are aimed at developing models that allow one to calculate the total power consumption of UAVs in different flight scenarios including horizontal movement, vertical movement, and hovering.

In the case of CRUAVP, which has the same constraints as the Multi-objective Periodic VRP, a mathematical formulation of the problem as a multi-criteria optimization model can be considered. In such a model, the total distances travelled by the UAVs (to be minimized), customer satisfaction (to be maximized) and the number of UAVs used (to be minimized) are considered simultaneously [5, 8].

To overcome the limitations connected with the fact that UAVs have to land frequently to change or re-charge their batteries. To reduce the costs of returning to the charging station, battery charging depots are either allocated at points along the trajectory which UAVs follow in repeating cycles, or located on vehicles of a ground fleet in the solutions proposed here. These two types of solutions generate different problems; in the first case, the problem of allocating a recharging/battery replacement station boils down to maximizing battery life (so that batteries that are still in working order should not be replaced). In the second case, the problem in question is a two-echelon [11] cooperated routing problem for a ground vehicle (GV) traveling on a road network and its UAV traveling in areas beyond the road to visit a number of targets unreached or even unreachable by the GV [16].

3 UAV Fleet Routing and Scheduling

3.1 Illustrative Example

In a flow production system consisting of five workstations, whose structure is shown in Fig. 1, two different products are manufactured at the same time.

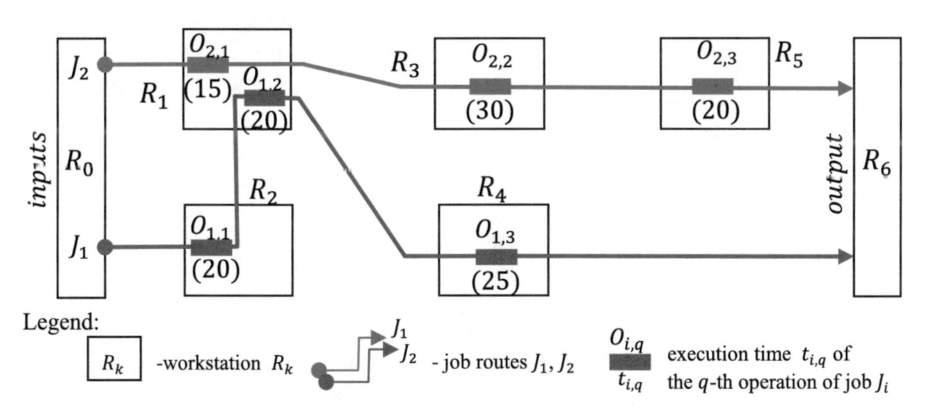

Fig. 1. Structure of a multi-item batch flow production system

The technological route of product J_1, marked in red, runs through workstations R_2, R_1, and R_4 whose respective technological operation times are 20 u.t. (units of time) for $O_{1,1}$, 20 u.t. for $O_{1,2}$ and 25 u.t. for $O_{1,3}$. In turn, the technological route of product J_2, marked in blue, runs through workstations R_1, R_3 and R_5, whose respective technological operation times are 15 u.t. for $O_{2,1}$, 30 u.t. for $O_{2,2}$, and 20 u.t. for $O_{2,3}$.

When travel times between workstations, changeover times, and loading/unloading times are omitted, it is easy to notice that workstation R_1, which is the bottleneck in the production flow, determines production takt time $TP = 35$, see Fig. 2. Let us assume that transport operations $o_1, \ldots, o_8$ are carried out by a fleet of three UAVs: U_1, U_2, and U_3. An example of a flight trajectory of U_1 is shown in Fig. 3.

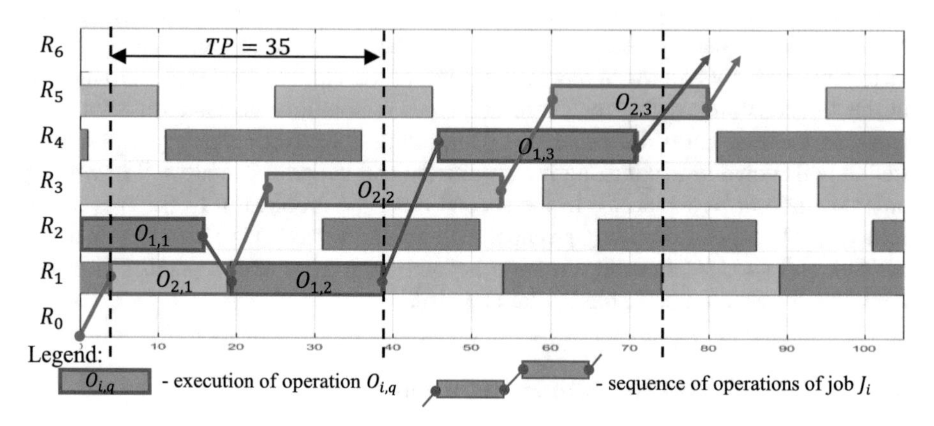

Fig. 2. Gantt's chart of production flow for zero values of transport time between workstations, changeover time, and loading/unloading time

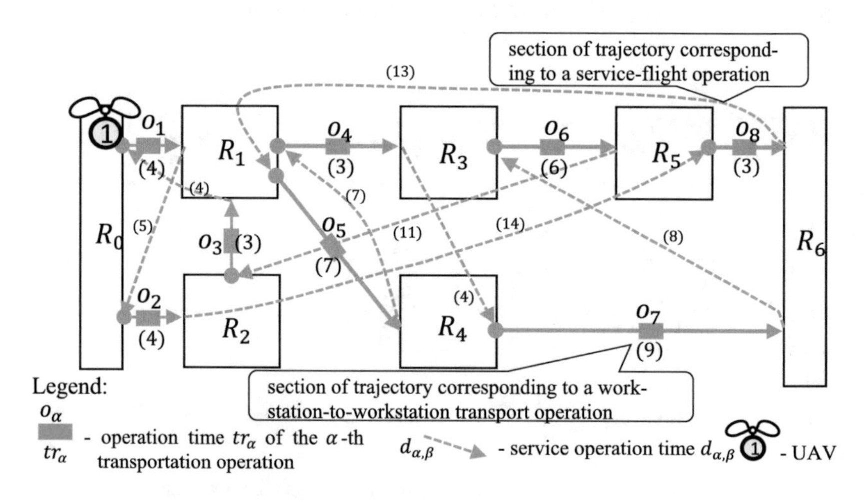

Fig. 3. An example of a trajectory of transport route U_1 corresponding to takt time $TP = 53$

Assuming that each of the UAVs moves (at appropriate intervals) along the same trajectory (running along selected corridors Fig. 4), that the time of each operation o_α and the times of service flights (between successive operations o_k, o_l of the trajectory) are selected depending on the current needs of the production process, e.g. as in Fig. 3, a Gantt's chart of the production flow has the form shown in Fig. 5. Figure 4 shows a map

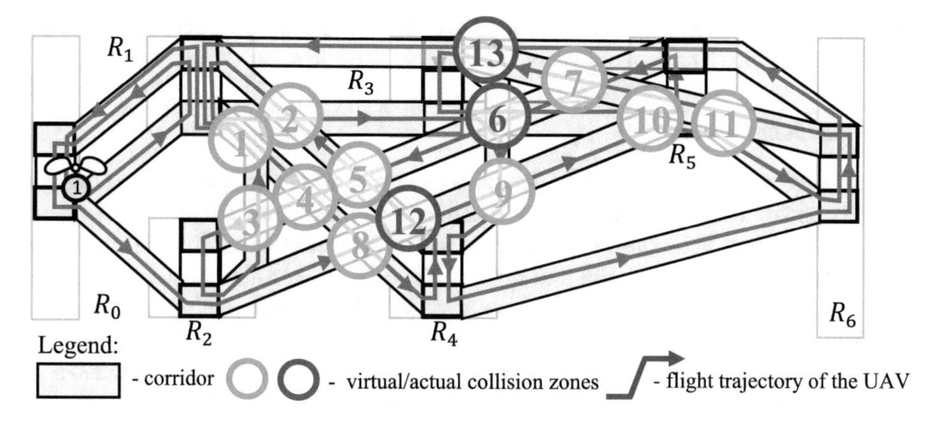

Fig. 4. Map of air corridors used by the UAVs following the same trajectory as in Fig. 3

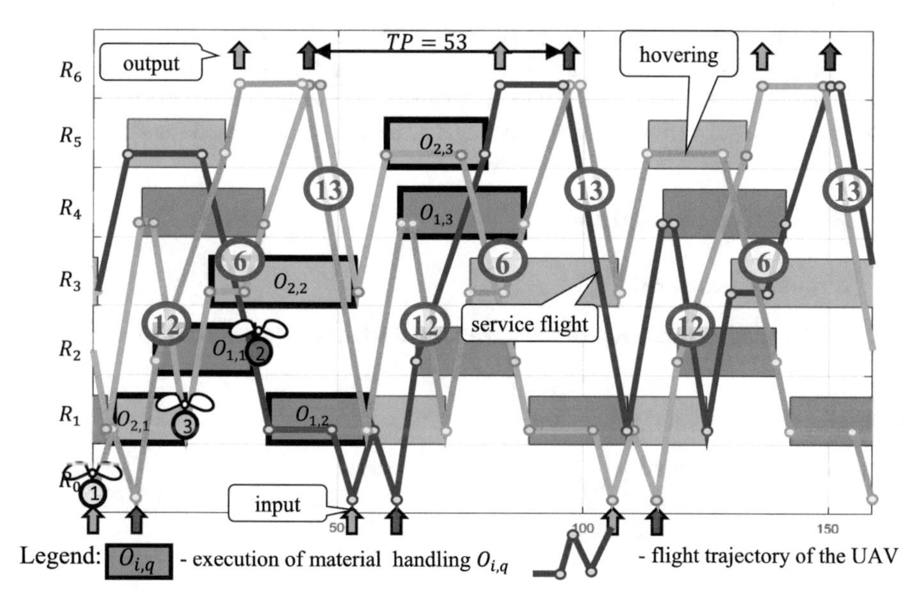

Fig. 5. Gantt chart of production flow taking into account material handling operations (including pick/place and set-up activities) and service flight operations, UAVs hovering

of air corridors used by the UAVs traveling along the trajectory of Fig. 3, with virtual and actual collision zones marked on it. The first type of collision zones are formed by intersecting parts of the corridors. In these zones, there are no actual collisions between UAVs, which travel at intervals according to the schedule shown in Fig. 5. In the actual collision zones, UAVs may collide as they move simultaneously along different, intersecting corridors. In the case under consideration, when transport times for workpiece batches have non-zero values, production takt time increases by 18 u.t., (*TP* = 53 u.t.), see Fig. 5. Also, the Gantt chart of production flow indicates that there are zones in the production flow where the UAVs are likely to collide. These observations mean that the schedule from Fig. 5 cannot be accepted.

3.2 Problem Description

The example analyzed in Sect. 3.1. demonstrates that the value of production takt time depends on the times of technological operations and the times related to the movement, loading and unloading of products, as well as workstation changeover times. This means that if the originally takt time of $TP = 36$ u.t. is to be maintained at non-zero handling operation times, other ways of scheduling and routing the available UAVs must be found.

The considered problem assumes that a given set of workstations allocated at different points is to be served periodically (within a time window, e.g. equal to production takt time) by several capacitated UAVs charged from a set of spatially distributed battery charging depots. The goal is to minimize the number of UAVs needed, such that each customer is reached with the right delivery at the right moment of time, while obeying the capacity and the battery change or re-charge constraints. A secondary objective is to minimize the total distance travelled. Since UAV routing and scheduling problems are known to be NP-hard, the literature related to this problem primarily consist of various heuristic methods or meta-heuristic based search strategies.

4 Declarative Modelling

4.1 Problem Statement

The mathematical formulation of the model considered is determined by the following parameters:

Symbols:

R_k:	resource k;	J_i:	job i;
$O_{i,q}$:	operation q of J_i;	U_v:	transport process v (v-th UAV);
o_α:	transport operation α;		
b_α:	index of transport operation which precedes o_α;		
f_α:	index of transport operation which follows o_α;		
$K_{\mu,\varepsilon}$:	corridor linking resources R_μ and R_ε.		

Sets and sequences:

R: the set of resources R_k (workstations);

K: the set of links $K_{\mu,\varepsilon}$ (corridors between workstations);

J: the set of jobs J_i, (production processes);

O_i: sequence of operations for J_i: $O_i = \left(O_{i,1}, \ldots, O_{i,q}, \ldots, O_{i,lm_i}\right)$;

p_i: route of J_i, sequence of resources on which operations $O_{i,q}$ are executed: $p_i = \left(p_{i,1}, \ldots, p_{i,q}, \ldots, p_{i,lm_i}\right)$, $p_{i,q} \in R$;

Q_k: the set of operations executed on R_k;

$\mathcal{O}$: the set of transport operations o_α;
S_k: the set of transport operations started from R_k, $S_k \subseteq \mathcal{O}$;
E_k: the set of transport operations ending on R_k, $E_k \subseteq \mathcal{O}$;
U: the set of transport means U_v (transport processes);
B: sequence of predecessor indices of transport operations, $B = (b_1, \ldots, b_\alpha, \ldots, b_\omega), b_\alpha \in \{0, \ldots \omega\}$;
F: sequence of successor indices of transport operations, $F = (f_1, \ldots, f_\alpha, \ldots, f_\omega), f_\alpha \in \{1, \ldots \omega\}$.

Parameters:

m: number of resources; h: number of links;
n: number of jobs, l: number of transport means;
lm_i: number of operations of J_i; ω: number of transport operations,
$t_{i,q}$: operation time of $O_{i,q}$; tr_α: operation time of o_α,
$d_{\alpha,\beta}$: travel time between resource at which operation o_α ends and resource at which operation o_β begins;
TP^*: maximum value of production takt time TP.

Variables:

TP: production takt time; $x_{i,q}$: start time of operation $O_{i,q}$,
$y_{i,q}$: end time of operation $O_{i,q}$; xt_α: start time of operation o_α,
yt_α: end time of operation o_α;
xs_α: the moment the resource occupied by UAV is released after completion of operation o_α;
b_α: index of the transport operation preceding operation o_α (operations o_{b_α} and o_α are executed by the same UAV); $b_\alpha = 0$ means that o_α is the first operation of the system cycle;
f_α: index of the transport operation following o_α, (operations o_α and o_{f_α} are executed by the same UAV).

Constraints:

I. For job operations (production processes):

$$y_{i,q} = x_{i,q} + t_{i,q}, \quad q = 1 \ldots lm_i, \quad \forall J_i \in J, \tag{1}$$

$$y_{i,q} \leq x_{i,q+1}, \quad q = 1 \ldots (lm_i - 1), \quad \forall J_i \in J, \tag{2}$$

$$y_{i,q} \leq x_{i,q} + TP, \quad q = 1 \ldots lm_i, \quad \forall J_i \in J, \tag{3}$$

$$(y_{i,a} \leq x_{j,b}) \vee (y_{j,b} \leq x_{i,a}), \text{ when } O_{i,a}, O_{j,b} \in Q_k, \forall R_k \in R, \tag{4}$$

$$TP \leq TP^*. \tag{5}$$

II. For UAVs (transport process operations):

$$yt_\alpha = xt_\alpha + tr_\alpha, \quad \alpha = 1,2,\ldots,\omega, \tag{6}$$

$$b_\alpha = 0, \quad \forall\alpha \in BS, BS \subseteq BI = \{1,2,\ldots,\omega\}, |BS| = l \tag{7}$$

$$b_\alpha \neq b_\beta \quad \forall\alpha,\beta \in BI \backslash BS, \alpha \neq \beta, \tag{8}$$

$$f_\alpha \neq f_\beta \quad \forall\alpha,\beta \in BI, \alpha \neq \beta, \tag{9}$$

$$(b_\alpha = \beta) \Rightarrow (f_\beta = \alpha), \forall b_\alpha \neq 0, \tag{10}$$

$$[(b_\alpha = \beta) \wedge (b_\beta \neq 0)] \Rightarrow (yt_\beta + d_{\beta,\alpha} \leq xt_\alpha), \alpha,\beta = 1,2,\ldots,\omega, \tag{11}$$

$$[(f_\alpha = \beta) \wedge (b_\beta = 0)] \Rightarrow (yt_\alpha + d_{\alpha,\beta} \leq xt_\beta + TP), \alpha,\beta = 1,2,\ldots,\omega, \tag{12}$$

$$xs_\alpha \geq yt_\alpha, \quad \alpha = 1,2,\ldots,\omega, \tag{13}$$

$$[(f_\alpha = \beta) \wedge (b_\beta \neq 0)] \Rightarrow (xs_\alpha = xt_\beta - d_{\alpha,\beta}), \ \alpha,\beta = 1,2,\ldots,\omega, \tag{14}$$

$$[(f_\alpha = \beta) \wedge (b_\beta = 0)] \Rightarrow (xs_\alpha = xt_\beta - d_{\alpha,\beta} + TP), \alpha,\beta = 1,2,\ldots,\omega, \tag{15}$$

$$\begin{gathered}[(xs_\alpha < yt_\beta) \wedge (xs_\beta - TP < yt_\alpha)] \vee [(xs_\beta < yt_\alpha) \wedge (xs_\alpha - TP < yt_\beta)], \\ \forall o_\alpha, o_\beta \in S_k, \quad k = 1,\ldots,m,\end{gathered} \tag{16}$$

$$\begin{gathered}[(xs_\alpha < yt_\beta) \wedge (xs_\beta - TP < yt_\alpha)] \vee [(xs_\beta < yt_\alpha) \wedge (xs_\alpha - TP < yt_\beta)], \\ \forall o_\alpha, o_\beta \in E_k, k = 1,\ldots,m,\end{gathered} \tag{17}$$

$$\begin{gathered}[(xs_\alpha < xt_\beta) \wedge (xt_\beta - TP < yt_\alpha)] \vee [(xt_\beta < yt_\alpha) \wedge (xs_\alpha - TP < yt_\beta)] \\ \forall o_\alpha \in E_k, \forall o_\beta \in S_k, k = 1,\ldots,m.\end{gathered} \tag{18}$$

III. For transport and production processes (linking UAVs with jobs)

$$x_{i,q} = yt_\alpha + c \times TP, c \in \mathbb{N}, \forall o_\alpha \in E_k, \forall O_{i,q} \in Q_k, k = 1,\ldots,m, \tag{19}$$

$$y_{i,q} = xt_\alpha + c \times TP, c \in \mathbb{N}, \forall o_\alpha \in S_k, \forall O_{i,q} \in Q_k, k = 1,\ldots,m. \tag{20}$$

Question: Do there exist routes (represented by sequences B, F) for the given UAV fleet (set U), which guarantee the existence of a production schedule $(x_{i,q}, xt_\alpha)$ that allows the achievement of takt time TP which does not exceed the given takt time TP^*?

The above decidability problem, can be seen as Constraint Satisfaction Problem:

$$CS = (\mathcal{V}, \mathcal{D}, \mathcal{C}) \tag{21}$$

where: $\mathcal{V} = \{B, F, X, XT\}$ is a set of decision variables, where $X = \{x_{i,q} | i = 1\ldots n, q = 1\ldots lm_i\}$, $XT = \{xt_\alpha | \alpha = 1,2,\ldots,\omega\}$; $\mathcal{D}$ is a discrete finite set of domains of variables $\mathcal{V}$; $\mathcal{C}$ is a set of constraints describing the following relations: the execution order of job operations (1)–(3) and UAV operations (19), (20); exclusion of job operations (4) and UAV operations performed on shared resources (16)–(18). These constraints ensure cyclic routes (7)–(10), and determine the execution order of transport operations (6), (11)–(15) and production takt time requests (5).

To solve the problem formulated in this way *CS* (21), one must determine such values (determined by $\mathcal{D}$) of decision variables *B*, *F* (UAV routes) and *X*, *XT* (production schedules and transport operation schedules), for which all the constraints $\mathcal{C}$ (including the mutual exclusion constraint, the cyclic operation execution constraint, etc.) will be met. These types of problems are typical solved using constraint programming CP/CLP environments, such as OzMozart, IBM ILOG, ECL^{i}PS^{E}.

4.2 Computational Results

A solution to the organization of workstation-to-workstation transport should ensure, given the number of UAVs used, a required production takt time *TP*. The solution shown in Fig. 5 is an example of optimal organization of workstation-to-workstation transport (that guarantees the shortest production takt time $TP = 53$) for a fleet consisting of three UAVs: $U = \{U_1, U_2, U_3\}$. This solution, however, does not allow simultaneous use of intersecting flight corridors (collision zones in Figs. 4 and 5) which, in practice, may lead to collisions of UAVs moving along these corridors. Two approaches are proposed here to avoid this complication.

In the first solution, UAVs can move on two levels (at two heights), see Fig. 6. This solution assumes that in zones of actual collision, UAVs travel at different heights. The UAV which moves along corridors: $K_{6,1}$. $K_{5,2}$. $K_{2,5}$, (marked orange in Fig. 6) travels at the higher level in a collision-free mode. The placement of flight channels at different heights makes it possible for production flow operations to be executed in accordance with the schedule presented in Fig. 5 ($TP = 53$) (collisions no longer occur in zones 6, 12, and 13). The schedule have been obtained by solving a CS problem (21); the calculations were performed in the OzMozart environment (calculation time: 86 s, Intel Core i5-3470 3.2 GHz, 8 GB RAM). A second solution, in which all UAVs fly at the same height, is shown in Figs. 7 and 8. This solution has been obtained by solving *CS* (21), taking into account the additional mutual exclusion constraints on UAVs in collision zones (calculation time: 120 s). The alternative UAV trajectories shown in Fig. 7, which use a different subset of corridors, include U_2 and U_3 which move along the same route: $po_{2,3}$ and U_1 which moves along route po_1 (see Fig. 7). The job execution schedule resulting from such an organization of transport operations is shown in Fig. 8. UAVs no longer collide in the intersecting corridors, but the *TP* is higher $TP = 57$. It is easy to see that, with the same fleet of UAVs, the second solution (in which vehicles move at the same height) is associated with a longer total flight route and, thus, a longer production takt time $TP = 57$ u.t. The experiments carried out in this study indicate that, it is possible to achieve a production takt time of $TP = 38$ u.t.. However, this requires the use of a fleet of 4 UAVs working at three different heights.

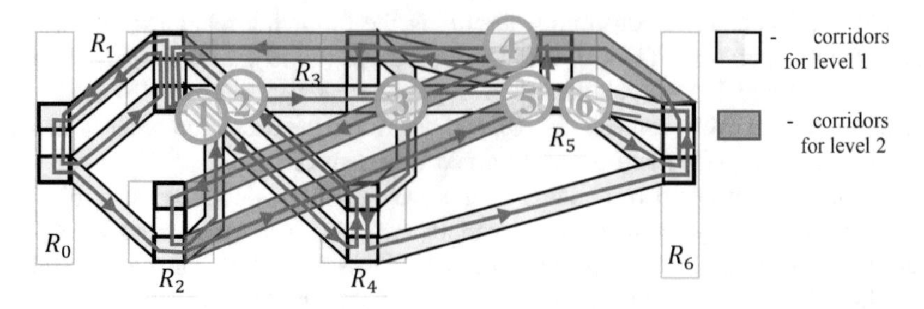

Fig. 6. Map of air corridors used by UAVs following the same flight trajectory as in Fig. 3 – the air corridors highlighted in orange are at a higher level

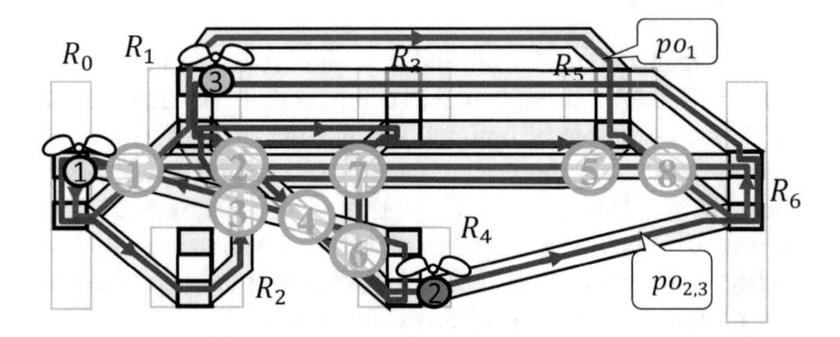

Fig. 7. Alternative map of air corridors for UAVs which guarantees a collision free schedule

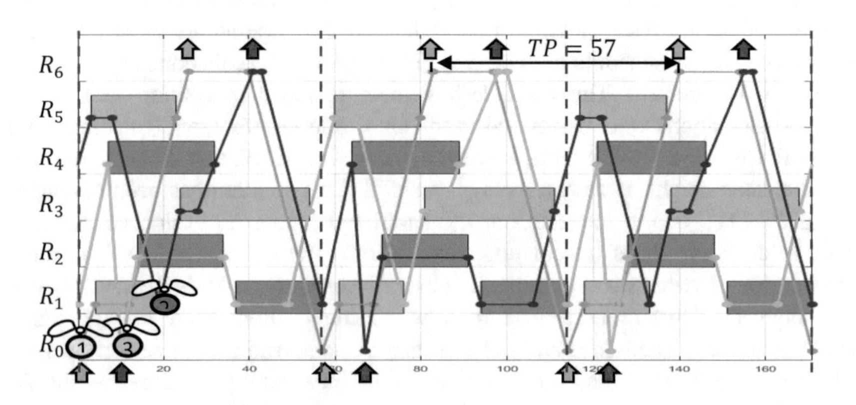

Fig. 8. Gantt chart of production flow for the UAV trajectories from Fig. 7

5 Conclusions

Opportunities for new applications of UAVs for material delivery tasks are especially related to the possibility of flexible arrangement of collision-free flight/transport routes in 3-D space. The flexibility of solutions, which make use of UAVs, is particularly welcome in situations requiring frequent changes in periodic service to-and-from delivery and collection depots (e.g. when supply chains are restructured due to changes in the organization of sorting, storage, and supply replenishment activities, etc.). In our future research, we would like to explore the issues related to the allocation of battery charging/replacement stations. In particular, in view of the need to take into account the restructuring of supply chains, the research will be focused on mobile battery exchange points.

Acknowledgements. The work was carried out as part of the POIR.01.01.01-00-0485/17 project, "Development of a new type of logistic trolley and methods of collision-free and deadlock-free implementation of intralogistics processes", financed by NCBiR.

References

1. Bocewicz, G., Nielsen, I., Banaszak, Z.: A diophantine set-driven approach to part sets cycle time scheduling and repetitive flow balancing. In: Advances in Intelligent Systems and Computing. Springer (in print)
2. Bocewicz, G., Nielsen, I., Banaszak, Z.: Towards leveling of multi-product batch production flows. A multimodal networks perspective. In: Proceedings of the 2018 IFAC Symposium on Information Control Problems in Manufacturing (INCOM 2018) (in print)
3. Braekers, K., Ramaekers, K., Nieuwenhuys, I.V.: The vehicle routing problem: state of the art classification and review. Comput. Ind. Eng. **99**, 300–303 (2016)
4. Coutinho, W.P., Fliege, J., Battarra M.: The Unmanned Aerial Vehicle Routing and Trajectory Optimisation Problem, Work in Progress, University of Southampton
5. Drucker, N.: Cyclic Routing of Unmanned Aerial Vehicles, Master of Science Research Thesis, Technion, Israel Institute of Technology, Haifa (2014)
6. Gola, A., Kłosowski, G.: Application of fuzzy logic and genetic algorithms in automated works transport organization. Adv. Intell. Syst. Comput. **620**, 29–36 (2018). https://doi.org/10.1007/978-3-319-62410-5_4
7. Hayat, S., Yanmaz, E., Muzaar, R.: Survey on unmanned aerial vehicle networks for civil applications: a communications viewpoint. IEEE Commun. Surv. Tutor. **18** (2016). https://doi.org/10.1109/COMST.2016.2560343
8. Ho, H.M., Ouaknine, J.: The cyclic-routing UAV problem is PSPACE-complete. In: Pitts, A. (ed.) Foundations of Software Science and Computation Structures, FoSSaCS 2015. Lecture Notes in Computer Science, vol. 9034, pp. 328–342 (2015)
9. Jin, Z., Shima, T., Schumacher, C.J.: Optimal scheduling for refueling multiple autonomous aerial vehicles. IEEE Trans. Rob. **22**(4), 682–693 (2006)
10. Myers, D., Batta, R., Karwan, M.: A real-time network approach for including obstacles and flight dynamics in UAV route planning. J. Def. Model. Simul. Appl. Methodol. Technol. **13**, 291 (2016)
11. Sitek, P., Wikarek, J.: A hybrid approach to the optimization of multiechelon systems. In: Mathematical Problems in Engineering, vol. 2015. https://doi.org/10.1155/2015/925675
12. Sobaszek, Ł., Gola, A., Kozłowski, E.: Application of survival function in robust scheduling of production jobs. In: Proceedings of the 2017 Federated Conference on Computer Science and Information Systems (FEDCSIS), pp. 575–578. IEEE, New York (2017)
13. Song, B.D., Kim, J., Morrison, J.R.: Rolling horizon path planning of an autonomous system of UAVs for persistent cooperative service: MILP formulation and efficient heuristics. J. Intell. Rob. Syst. **84**, 241 (2016). https://doi.org/10.1007/s10846-015-0280-5

14. Thibbotuwawa, A., Nielsen, P.: Unmanned Aerial Vehicle Routing Problems: A literature review (in print)
15. Khosiawan, Y., Nielsen, I.: A system of UAV application in indoor environment. Prod. Manuf. Res. **4**(1), 2–22 (2016). https://doi.org/10.1080/21693277.2016.1195304
16. Zhihao, L., Zhong, L., Jianmai, S.: A two-echelon cooperated routing problem for a ground vehicle and its carried UAV. Sensors **17**(5), 1144 (2017). https://doi.org/10.3390/s17051144

World Wide Web CBIR Searching Using Query by Approximate Shapes

Roman Stanisław Deniziak and Tomasz Michno(✉)

Kielce University of Technology,
al. Tysiaclecia Panstwa Polskiego 7, 25-314 Kielce, Poland
{s.deniziak,t.michno}@tu.kielce.pl

Abstract. Nowadays more and more images are stored in the World Wide Web. There are a lot of photo galleries, media portals and social media portals where users add their own content, but also they would like to find the proper ones. The problem of searching for an image is not trivial. Objects present on images may have e.g. different colors, backgrounds or orientations. Moreover, the image may contain many other details which may be hard to be described by words. This paper presents a new system which may be used to query for images from the internet which is based on our Query by Approximate Shapes algorithm. The main idea of the proposed approach is to gather images from the internet. Next, all images are processed using our algorithm which is based on decomposing objects into a set of simple shapes. During the query, depending on its type, an example image or a sketch is used. For both types a graph is constructed which is compared with graphs in the database.

Keywords: CBIR · Multimedia databases · Query by sketch

1 Introduction

Nowadays more and more images are stored in the World Wide Web. There are a lot of photo galleries, media portals and social media portals where users add their own content, but also they would like to find the proper ones. The problem of searching for an image is not trivial. Objects present on images may have e.g. different colors, backgrounds or orientations. Moreover, the image may contain many other details which may be hard to be described by words.

The searching for images in the world wide web may be reduced to searching for images in the multimedia database. The algorithms which are used to query multimedia databases may be divided into three categories: KBIR, SBIR and CBIR algorithms. The KBIR stands for Keywords Based Image Retrieval algorithms and they are based on adding annotations which describes all details present in the image stored in the database. Most often a set of keywords is used, added manually by humans which may cause some mistakes in case of insufficient knowledge about objects [3]. The KBIR algorithms provide very precise

S. Rodríguez et al. (Eds.): DCAI 2018, AISC 801, pp. 87–95, 2019.
https://doi.org/10.1007/978-3-319-99608-0_10

results for objects which are easy to describe by words, e.g. the brand or colors. In other situations, the results may be imprecise [9,15].

The SBIR stands for Semantic Based Image Retrieval. The algorithms are partially similar to KBIR algorithms but are easier to use by users because they allow writing queries as phrases which are more natural. The phrases are mapped onto so called semantic features which are correlated with image details [9,15]. After mapping, the features are used to perform query. Some SBIR algorithms use graphical queries which are transformed into semantic features. An example is a Li et al. research [10] which uses sketches. After drawing, they are mapped onto textual annotations which are compared with textual annotations of the 3D models in the database.

The CBIR stands for Content Based Image Retrieval which uses images as a query [3]. There could be distinguished two types of algorithms: low level and high level [1]. The low level algorithms use global image features e.g. a normalized color histogram [11], a spatial domain image representation [12], a bag of words histogram [14], a difference moment and entropy [7] or MPEG-7 descriptors [8]. The low level algorithms provide precise results when a query is used for searching for similar images (e.g. paintings). When an image with the same object but different other details or background is needed they are insufficient and the high level algorithms are more suitable. The main idea of the high level algorithms is to decompose image into smaller parts. Most often regions which gathers pixels with similar colors [1] and creates uniform areas are used. All regions are transformed into a graph in order to store mutual positions between each of them. During the query, the graph is compared with graphs in the database. The region-based high level algorithms provide precise results but they need example image which may be problematic to obtain for some situations. In case of no example image, there are also CBIR algorithms which uses sketches drawn by users. One of the example may be [6] where the sketch is transformed into lower resolution image and compared with others using edges detection algorithms. The results are quite precise, but as other low level algorithms only for searching for paintings or whole images.

This paper presents a new system which may be used to query for images from the World Wide Web which is based on our Query by Approximate Shapes algorithm [1–5]. The main idea of the proposed approach is to gather images from the internet with all information about their urls, urls of websites and if available some descriptions. Next, all images are processed using our Query by Approximate Shapes algorithm which is based on decomposing an object present in the image into a set of simple shapes like line segments, arches and more complex shapes based on them. After that a graph is constructed for each image which contains all information about shapes and, as a metadata, all urls and description. All graphs are stored in the tree-based database. During the query, depending on its type, an example image or a sketch is used. For both types a graph is constructed which is compared with graphs in the database.

The motivation for us was to provide system which allows users to search for images in the World Wide Web using an image example or a hand drawn sketch. The assumptions were as following:

- two types of queries: using an image as a query and using a sketch
- easy addition of new entries (images and connected with them urls) without need of training the whole structure
- fast queries of huge database.

The paper is organized as follows: the first section describes different approaches to obtaining images from multimedia databases and our motivation to providing a new web image searching system. In the third section, we present the architecture of the proposed system. The next section presents the experimental results. The fourth section is the conclusion and future works description. The last one, is a list of references used in the paper.

2 The System Architecture

The proposed system consists of three modules: the web crawler, the image indexing database and the user interface. The web crawler is responsible for gathering images from the world wide web. The image indexing database stores the graphs of images and urls to websites using query by approximate shapes algorithm. The user interface allows queering the indexing database both using an example image and a user-drawn sketch. The overview of the system is shown in the Fig. 1.

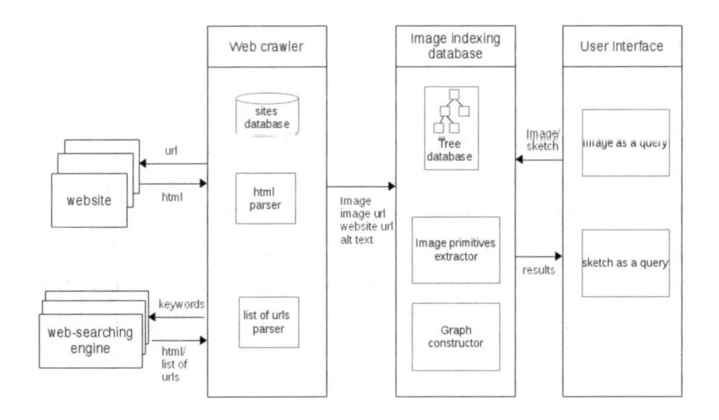

Fig. 1. The system overview

2.1 Web Crawler

The web crawler module is the part of the system which is used to gather images from the world wide web and to store them in the database. It is able to process websites and web-searching engines. The web crawler stores addresses to visit in a database. The processing of websites consists of getting the first not visited address from the database and analyzing its HTML code. When an <img> html tag is encountered, the src url is stored in connection with the website url and everything is send to the image indexing database. Moreover, when an <a> is encountered, also the source url is retrieved. After that it is added to the database when not visited. The processing of web-searching engines is different. Firstly the set of keywords are retrieved from the database. Then, according to the specific engine, an api or a http request is used. When using an api, a list of images and urls connected with them is retrieved. When a http request is used, the html is retrieved and it is processed as normal websites. After that, the results are send to the image indexing database.

2.2 Image Indexing Database

The image indexing database is used to process and store images and theirs metadata. The image url, the website url and alternative text are stored in order to be shown to users as a part of the query result.

The image indexing database is based on our Query by Approximate Shapes algorithm. The main idea is to represent an image as a set of simpler shapes called primitives. In [1] we propose to use line segments and arches as a first detected shapes and then to construct more complex ones: poly lines, polygons, polyarches and arc-sided polygons (Fig. 2). Because detecting arches is problematic, in practical implementations we used Circular Hough Transform to detect circles and processing of detected connected line segments which are suspected to be arches. After primitives detection, they are transformed into a graph. The graph stores for each primitive its description (called attribute) which is independent from the image resolution. In [2] for line segments we propose to use the angle of line slope, for arches the angle of the arc. Optionally for the line segment its length and for the arc its radius or diameter may be used, both in relation to the whole object size. As the attributes for the more complex primitives the number of segments is used and the attributes of each segment. Moreover, the graph stores the mutual relationships between primitives - their positions between each others and connections [2]. As an improvement of the algorithm we propose to use additionally the information about positions between primitives in the object graph's. As a position information the 4 geographical directions (with 4 additional intermediate) are used. Moreover, the distances may be also stored, e.g. using relations to the whole object bounding box or using 3 compartments: near, medium, far. With such addition, the algorithm is able to localize primitives and check if they are matched properly with the graphs stored in the database.

In order to store graphs, the database structure is also proposed. The database is based on a tree which stores the common parts of graphs in parent nodes (called common nodes) and graphs in leaves (data nodes). Moreover,

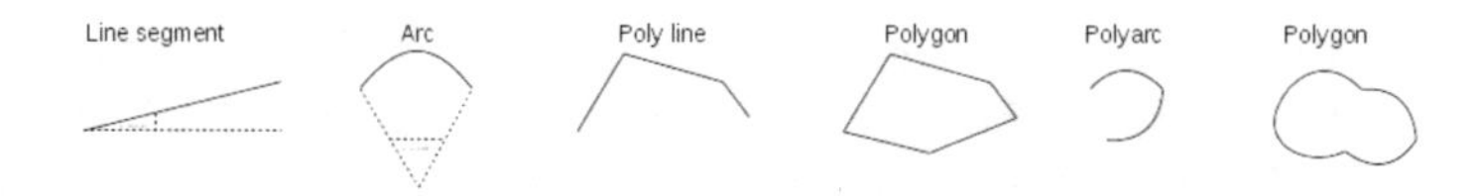

Fig. 2. The primitives used to describe objects

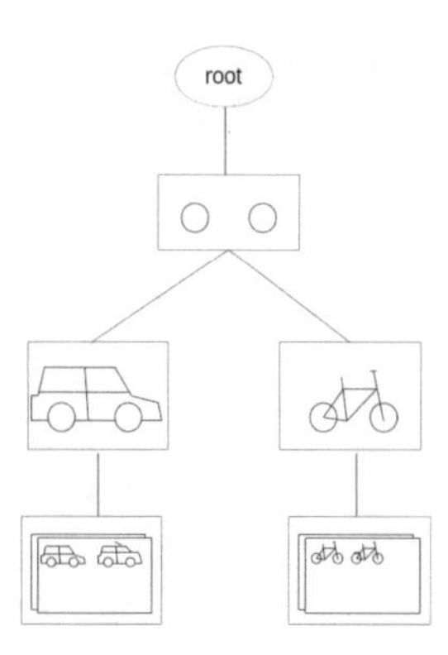

Fig. 3. The example tree structure

if graphs are very similar, they are stored in the same data node, which reduces the height of a tree. Due to the database structure, all similar graphs are grouped in the same subtree which highly decreases the time of processing a query. The example tree is shown in the Fig. 3.

The data nodes store graphs in vectors called slices. The most similar graph to the parent common node is added into the first slice, in the first position. The least similar is stored as the last element. In order to provide ability to process queries in parallel there may be more than one slice [2].

In order to organize the image indexing database structure it is divided into a three parts: image primitives extractor, graph constructor and tree database. The image primitives extractor is responsible for extracting primitives from an image. The graph constructor processes all extracted primitives in order to calculate all theirs attributes and to create a graph. The tree database is responsible for adding new graphs to the database and processing queries.

2.3 User Interface

The user interface for the system is based on two interfaces: using an image as a query and using a sketch as a query. The user is able to select which one would be more suitable for the specific requirements. Using an image as a query is more suitable for situations where an user has the image of an object which has to be used as a query. The sketch as a query is more suitable for users who does not have a proper image of a desired object or does not have the full knowledge about the object (e.g. its full shape).

The image as a query interface consists of retrieving from an user the example image and sending it to the Image Indexing Database module's Image Primitives extractor part. Then a graph is constructed and compared with other graphs in the database.

The sketch as a query interface provides users a set of predefined shapes which may be used for drawing a desired object. Due to the predefined shapes, drawing for users without high drawing skills is much easier. After drawing primitives, the sketch is sent to the Image Indexing Database module's Graph Constructor part and then the query is executed. The prototype interface of our system is shown in the Fig. 4(a).

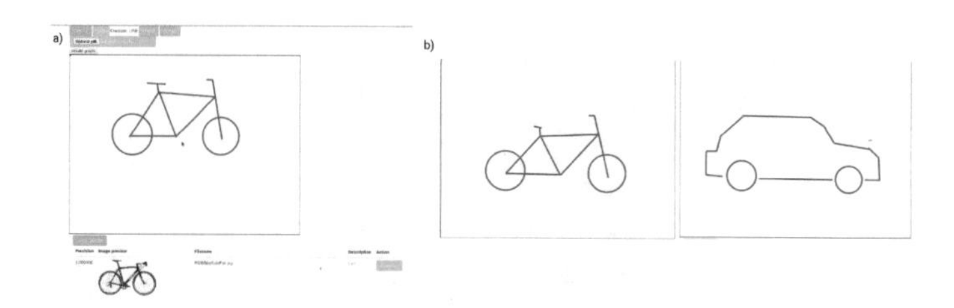

Fig. 4. (a) The GUI sketch interface of our prototype application. Under the sketch, the first result of a query is visible. (b) The sketches used for queries using a sketch

2.4 Other Applications

The proposed system is not only able to be used as a web-searching CBIR engine, but also as e.g. smart e-commerce search.

The smart e-commerce search may be very useful for users who know how the product looks, but does not know its name. In this application, the web-crawler module may be replaced by the module which provides product image, name, page and description.

3 Experimental Results

The proposed system was evaluated using an experimental application. As described in Sect. 2, the application was divided into a three modules: the web crawler, the image indexing database and the user interface. The web crawler was written in python because of its very good http request processing. The image indexing database was written in C++ and OpenCV library for image processing. The user interface was developed using python, Django framework and HTML5. The web crawler was evaluated in order to check if it gather all images from the website and it was successful with all tested websites. Another tests were done in order to evaluate the image indexing database. As a measurements the commonly used *precision* and *recall* coefficients were used, defined as follows:

$$precision = \frac{the\ number\ of\ images\ from\ the\ searched\ class\ in\ results}{the\ number\ of\ results} \quad (1)$$

$$recall = \frac{the\ number\ of\ images\ from\ the\ searched\ class\ in\ results}{the\ number\ of\ images\ from\ the\ searched\ class\ in\ the\ database} \quad (2)$$

As images used for tests we used images gathered by the web-crawler which consisted of a car, a bicycle and a motorbike images - all real life images from the world wide web. We tested both image as a query and sketch as a query interfaces. The results are presented in the Table 1, the sketches used for queries are shown in Fig. 4(b). As can be seen, the query using images has given higher precision values than the sketches. This is caused by using as a sketch the very schematic queries (Fig. 4(b)) and comparing them with automatically created graphs from images gathered by web crawler, without any further manual modifications. The recall value was higher for the bicycle object classes because of higher similarity of all images. The cars images contained different brands, models and types of cars which caused some problems for classification.

Table 1. The experimental results

Query object	Top 10 precision		Recall	
	Image	Sketch	Image	Sketch
Bicycle	0.8	0.7	0.86	0.78
Car	0.7	0.6	0.5	0.5

4 Conclusion and Future Works

In this paper we presented a new Content Based Image Retrieval from the World Wide Web system. The system consists of three modules: The Web crawler, The Image Indexing Database and The Graphical User Interface. The Web crawler retrieves images and information about them from the internet and then send to the Image Indexing Database which creates a graph from detected shapes and stores it with the metadata. The GUI is used for preparing queries. The experimental results prove that our solution provides good results when querying both for images and sketches.

As the future works we would like to execute tests on higher number of images and classes of objects. Moreover, other graphs comparisons algorithms should be tested, e.g. based on optimization methods with constraints [13]. Some works on improving the recall parameter also should be done.

References

1. Deniziak, R.S., Michno, T.: Content based image retrieval using query by approximate shape. In: 2016 Federated Conference on Computer Science and Information Systems (FedCSIS), pp. 807–816. IEEE, Gdańsk (2016). https://doi.org/10.15439/2016f233
2. Deniziak, R.S., Michno, T.: New content based image retrieval database structure using query by approximate shapes. In: 2017 Federated Conference on Computer Science and Information Systems (FedCSIS), pp. 613–621. IEEE, Prague (2017). https://doi.org/10.15439/2017F457
3. Deniziak, R.S., Michno, T.: Query by shape for image retrieval from multimedia databases. In: Kozielski, S., Mrozek, D., Kasprowski, P., Małysiak-Mrozek, B., Kostrzewa, D. (eds.) Beyond Databases, Architectures and Structures. CCIS, vol. 521, pp. 377–386. Springer, Ustroń (2015). https://doi.org/10.1007/978-3-319-18422-7_33
4. Deniziak, R.S., Michno, T.: Query-by-shape interface for content based image retrieval. In: 2015 8th International Conference on Human System Interaction (HSI), pp. 108–114. IEEE, Warsaw, June 2015. https://doi.org/10.1109/HSI.2015.7170652
5. Deniziak, R.S., Michno, T., Krechowicz, A.: The scalable distributed two-layer content based image retrieval data store. In: 2015 Federated Conference on Computer Science and Information Systems (FedCSIS), pp. 827–832. IEEE, Łódź (2015). https://doi.org/10.15439/2015F272
6. Kato, T., Kurita, T., Otsu, N., Hirata, K.: A sketch retrieval method for full color image database-query by visual example. In: [1992] Proceedings. 11th IAPR International Conference on Pattern Recognition, pp. 530–533. IEEE, The Hague (1992). https://doi.org/10.1109/ICPR.1992.201616
7. Kriegel, H.P., Kroger, P., Kunath, P., Pryakhin, A.: Effective similarity search in multimedia databases using multiple representations. In: 2006 12th International Multi-Media Modelling Conference. IEEE, Beijing (2006). https://doi.org/10.1109/MMMC.2006.1651355
8. Lalos, C., Doulamis, A., Konstanteli, K., Dellias, P., Varvarigou, T.: An innovative content-based indexing technique with linear response suitable for pervasive environments. In: 2008 International Workshop on Content-Based Multimedia Indexing, pp. 462–469. IEEE, London (2008). https://doi.org/10.1109/CBMI.2008.4564983
9. Li, C.-Y., Hsu, C.-T.: Image retrieval with relevance feedback based on graph-theoretic region correspondence estimation. IEEE Trans. Multimedia **10**(3), 447–456 (2008). https://doi.org/10.1109/tmm.2008.917421
10. Li, B., Lu, Y., Shen, J.: A semantic tree-based approach for sketch-based 3d model retrieval. In: 2016 23rd International Conference on Pattern Recognition (ICPR), pp. 3880–3885. IEEE, Cancun (2016). https://doi.org/10.1109/ICPR.2016.7900240
11. Mocofan, M., Ermalai, I., Bucos, M., Onita, M., Dragulescu, B.: Supervised tree content based search algorithm for multimedia image databases. In: 2011 6th IEEE International Symposium on Applied Computational Intelligence and Informatics (SACI), pp. 469–472. IEEE, Timisoara (2011). https://doi.org/10.1109/SACI.2011.5873049
12. Shih, T.K.: Distributed Multimedia Databases. IGI Global, Hershey (2002)

13. Sitek, P., Wikarek, J.: A hybrid programming framework for modeling and solving constraint satisfaction and optimization problems. Sci. Programm. **2016**, Article ID 5102616 (2016). https://doi.org/10.1155/2016/5102616
14. Śluzek, A.: Machine vision in food recognition: attempts to enhance CBVIR tools. In: Ganzha, M., Maciaszek, L.A., Paprzycki, M. (eds.) Position Papers of the 2016 Federated Conference on Computer Science and Information Systems, FedCSIS 2016. PTI, Gdańsk (2016). https://doi.org/10.15439/2016f579
15. Wang, H.H., Mohamad, D., Ismail, N.A.: Approaches, challenges and future direction of image retrieval. J. Comput. **2**(6) (2010)

Defect Prediction in Software Using Predictive Models Based on Historical Data

Daniel Czyczyn-Egird(✉) and Adam Slowik

Department of Electronics and Computer Science, Koszalin University of Technology, Sniadeckich 2 Street, 75-453 Koszalin, Poland
daniel.czyczyn-egird@cicomputer.pl,
aslowik@ie.tu.koszalin.pl

Abstract. Nowadays, there are many methods and good practices in software engineering that aim to provide high quality software. However, despite the efforts of software developers, there are often defects in projects, the removal of which is often associated with a large financial effort and time. The article presents an example approach to defect prediction in IT projects based on prediction models built on historical information and product metrics collected from various data repositories.

Keywords: Data mining in software · Defect prediction models
Software metrics

1 Introduction

In the era of the constant development of computers and software, there is great need for new and increasingly advanced IT systems, which require, in addition, certain functionalities, as well as the highest level of reliability. Unfortunately, the vast majority of software is burdened with defects causing unstable operation of certain functionalities or can cause a malfunction of the entire system. The defect appears in the software when the person creating the system makes a mistake that may occur at various stages of software development, such as requirements analysis, system documentation design (general/detailed design), test plan, inappropriate test scripts and source code, etc.

Therefore, one important aspect is the testing process, during which the tester performs specific test cases and can observe whether the results of these tests coincide with expectations. Any deviations from expectations are treated as incidents that need to be checked and explained. All defects and problems detected should be saved to issue tracking systems (ITS) and/or version control systems (VCS) for further analysis and finding an attempt to solve the problem.

Data repositories in which the above information is stored can be an interesting source of knowledge for researchers and developers, who deal with the issues of software improvement processes (SPI) [1] or quality assurance (QA).

The article aims to present a general approach to the problem of predicting software defects. It is based on models operating on historical data from repositories and analysing data based on predictive classifiers.

S. Rodríguez et al. (Eds.): DCAI 2018, AISC 801, pp. 96–103, 2019.
https://doi.org/10.1007/978-3-319-99608-0_11

2 Related Works

There are many tools supporting the work of programmers and testers in their daily work on information systems. Among these tools are those that allow for convenient and quick testing of solutions created both at source code and post-compilation levels. Finding errors in the software is extremely important to ensure the final delivery of products without defects. However, continuous testing and debugging of systems involves spending on the use of human resources (programmers, testers) as well as financial resources [2]. Research on the prediction of defects has thus generated more and more interest on the part of both practitioners and researchers.

Ramler and Himmelbauer [3] have proposed prediction of defects using predictive models associated with software systems at the level of their modules. The modules can be files, classes and components, as well as subsystems of a given system. These modules are described by sets of attributes (e.g. by code metrics or number of changes in a given iteration), which are available by extracting them from various data sources such as databases or source code repositories.

There are also many studies in which the authors devote a lot of time to the issues of data acquisition from repositories, in addition to focusing only on models. Several tools have been developed for tasks related to data acquisition and supporting predictive modelling (e.g. defect prediction).

Jureczko and Magot [4] have prepared a *QualitySpy* [5] open-source framework (Apache 2.0 license [6]) whose task is to read and collect raw data from source code and event repositories, as well as user-defined metrics. Their project has focused on two modules for data acquisition and reporting. In the last released version, the framework allowed metrics to be read in classes for Java technology and reading JIRA [7] system events, as well as entries from the Subversion (SVN) [8] – version control system.

D'Ambros and Lanza [9] have suggested a tool to support collaborative software evolution analysis through a web interface going by the name of – *Churrasco*. It is an open source tool that retrieves and processes data from the Bugzilla and SVN systems, based on the FAMIX meta-model, which is independent of the programming technology used. In addition, the object relational mapping module (GLORP), the fact extraction module (MOOSE) and the SVG module for visualisation were used.

Madeyski and Majchrzak [10] have developed *Defect Prediction for software systems (DePress)*, a special framework that aims to extend the measurement of software and data integration for predictive purposes (defect prediction, cost/effort prediction). The *DePress* framework is based on the KNIME [11] project and extends it through a set of plugins, enabling models of data flow to be built in a graphical, simple and transparent way for the user. The main assumption of the project was prediction of software defects based on historical data. For this purpose, a set of plugins was prepared responsible for collecting, transforming and analysing data. The authors focused on data acquisition, data transformation and reporting operations, while at the same time leaving statistical operations and data mining [12] to a proven KNIME environment that has appropriate built-in mechanisms.

One important aspect upon which the authors placed emphasis is archiving and sharing data sets outside (e.g. for other researchers who wanted to test their own solutions). In this type of research, it is vital that repositories of historical data and prepared predictive models are as public as possible (after an earlier process of anonymization of content protected by commercial copyright).

3 Defect Prediction Using Predictive Models Based on Historical Data

3.1 Defect Prediction and Software Development Life Cycle

The defect can be introduced at any stage of the process called SLDC (Software Development Life Cycle) [13]. Therefore, it is very important that the testers are involved from the beginning of the life cycle of the software to detect and remove defects.

The sooner the defect is located and repaired, the cost of maintaining the quality will be lower. For example, if the defect is identified in the requirement analysis phase, then the cost of repair will be reduced to modify the requirements on the appropriate document. However, if the requirements are incorrectly described and implemented, and the defect is detected only during the testing phase, then the cost of the repair will be very high and will involve the improvement of the requirements and specifications as well as the change in the implementation.

It will also require a further testing process. In this article, the authors focus on the implementation and testing phase because, in relation to the level of these phases, it is possible to obtain relevant historical data from version control systems (VCS) and issue tracking systems (ITS), as long as these are stored and maintained.

3.2 General Assumptions

One of the main assumptions of the defect prediction operation is to determine the sources of historical data on the basis of which the entire prediction process will take place. The data in which we are interested can be obtained from software configuration management systems and error tracking systems.

There are currently many types of these systems on the market. Those proving to be most popular and most frequently used in professional programming teams are, respectively, the top five version control systems: Subversion, Git, CVS, Mercurial, Perforce, and the top five for bug tracking systems: JIRA, Github, Redmine, Bugzilla, BitBucket (study conducted by RebelLabs based on surveys in over 100 IT companies) [14].

Acquisition of data from the aforementioned systems can take place in two ways: directly or indirectly. The direct approach allows access to system repositories most often by connecting to their database or by means of appropriate mechanisms that allow reading information from these databases. In the case of the JIRA system, in addition to parametrized SQL queries directly executed on the appropriate tables in the database (the database engine supporting the JIRA system is MSSQL), there is also an API that returns the expected results using appropriate requests.

JIRA also offers its own micro-language JIRA Query Language (JQL). It's the most flexible way to search for issues in JIRA and can be used by everyone: developers, testers, project managers, and even non-technical business users. This method can be dedicated to those who have no experience with database queries, as well as those who want faster access to information in JIRA. The second way to access data is based on exchange files, which are exported manually from systems through appropriate interfaces, e.g. to CSV or XML format.

However, analysis of exported files may be the easiest and most frequently used way to access data if, for example, due to the lack of rights, we do not have direct access to the system and/or its database. In the case of version control systems, the same two options are available for selection. For example, for Git or Subversion, we can also try to connect to their (file) databases and search for artifacts that interest us or use data to export data to swap files.

The downloaded historical data should be unified in some way and cleared of unnecessary fields (e.g. from the information about the authors of a given entry) before being transformed into a common format, e.g. a tabular form, which will be explored in subsequent stages.

In addition to the historical data downloaded, source code metrics should also be used. These metrics should be appropriately linked to historical data to enable the construction of prediction classifiers. Without including the relevant metrics (e.g. number of lines of code in the file, the number of classes in the file and the degree of class nesting), it would not be possible to apply the predictive model to newly created systems where historical data does not exist.

3.3 An Exemplary Approach to Creating a Prediction Model

One of the methods of building a predictive model is the two-stage approach. In the first stage, we build and trial our model, while in the second stage it is used to predict the new input data. In each of the two stages, we can distinguish several of the following phases (Fig. 1):

1. Data acquisition phase – data collection processes.
2. Data transformation phase – processes of unification and matching of input data.
3. Data mining phase – processes related to the discovery of new knowledge.
4. Reporting phase – processes of presenting results and their archiving.

From the research point of view, a vital element for further research, if the most difficult one, is the selection of appropriate classifiers in the third phase. For this purpose, different techniques are used from areas related, for example, to evolutionary algorithms, swarm intelligence algorithms [15], artificial neural networks or fuzzy logic.

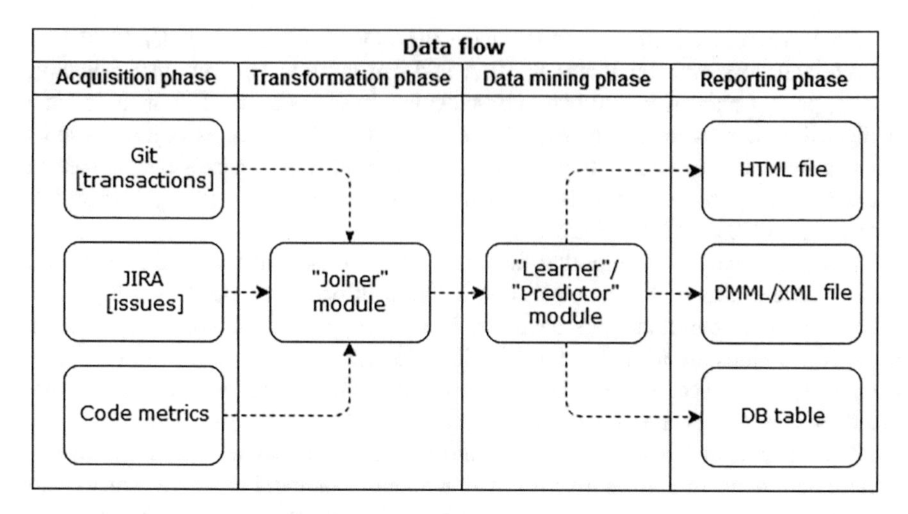

Fig. 1. Data flow in defect prediction process - diagram

4 Example of Defect Predictions

4.1 Selection of the Project for Analysis

For the following study, one of the main criteria for the selection of analysis software was free access to archival data contained in version control systems and error tracking systems. This criterion is met by open source projects; here, it was decided to choose the Apache Maven [16] system. It is a tool that automates software development for the Java platform. The various functions of Maven are realised through plug-ins, which are automatically downloaded at their first use. The file defining the method of building the application is called POM (Project Object Model). Just like other Apache products, Maven is distributed under the Apache License. This tool was first released in July 2004 and is constantly being developed; in fact, its latest release dates back to October 2017 (version 3.5.2 for Java 7). Maven has been developed throughout its history by a large group of programmers - open source software supporters (about 100 people involved in the project over a dozen years, recent releases supported by 10 constantly working programmers). The fact of such an active development of the chosen tool is undoubtedly caused by the lack of limitations in the development of this software, along with unqualified access to source code repositories, in accordance with the Apache Foundation license policy.

Source code repositories are stored in Subversion and Git [17] systems, while the defects detected are reported in the JIRA system.

4.2 Data Acquisition

Maven source codes are stored both in the Subversion system and the Git system. However, recent times have shown the growing popularity of software developers affiliated with the Apache foundation to migrate their projects to the Git system. Thus, the choice was made to download data from the Git system. For this purpose, appropriate tools were used, among others, GraphQL API v4 [18] for Github (data query language) and a set of bash scripts operating on the Git system. The range of dates for these activities is January - December 2017. In Table 1, the general data information from Git are presented:

Table 1. General data information from Git

Time period	01.01.2017–31.12.2017
Release	3.5.0–3.5.2
Commits (Transactions)	752
Files changed	5053
No. of active developers	10

The next step was to acquire data from the JIRA system for a given period of time. From the JIRA repository, all the events were described as Bug with status Closed/Resolved, and final result Fixed/Duplicated. These were taken into account, although the priority parameter for the need to repair the defect was omitted to obtain a wider range of information about events. In Table 2, the general data information from JIRA are presented:

Table 2. General data information from JIRA

Time period	01.01.2017–31.12.2017
Release	3.5.0–3.5.2
No. of issues	175
No. of issues (type: Bug)	111
Improvement tasks	40

In the next stage, an attempt is made to pair artifacts taken from the Git system with artifacts taken from the JIRA system. Such pairing is possible if, for example, in transaction descriptions (Git) there are descriptions of solved problems reported in the JIRA system. For example, you should search (e.g. with the use of regular expressions) string with the number of a given event, e.g. “Resolved issue MNG-6221”, where the JIRA side describes the event: “(MNG-6221) Maven cannot build with Java 1.8.0_131”. The last stage of data acquisition is downloading the source code metrics. This can be performed using tools dedicated to the given programming technology. In the case of Maven for Java, you can use the JavaNCSS library.

4.3 Data Unification Process

During the transformation and normalization process, the data (columns) unnecessary from the prediction view point are deleted or the assigned values are set to zero. Each artifact that was downloaded in the previous step and related to each other receives a description of the properties *HasDefects* and *NoOfDefects*, with the appropriate values describing whether for a given class/method from the source code, there are any defects that were reported in the JIRA system and repaired in the next transaction to the version control system (Git).

4.4 Prediction Process and Results

The prediction methodology should be based on a minimum of three stages: selection of attributes for prediction, model construction and its validation. In the first stage, we select the attributes on which the model will be based. The use of all attributes (historical data and metrics) may lead to the re-loading of the model; therefore, it is recommended that only those that stand out be selected (you can use, for example, backwards elimination).

The next stage is the construction of a model that can be based on one classifier or several linked together, creating the so-called hybrid classifier. One popular classifier used for all predictions are decision trees. For a given case, the prediction attribute is *HasDefects*, which tells you if a defect has been found in the given input set. The input variables selected in the first stage are independent variables in this case.

As the last stage, validation of the model may be based, for example, on cross-validation (K-fold cross validation), where the input data is divided into two segments (learning and validating) and the entire process is repeated several times with random division (different each time).

While the results of the model work will be satisfactory, it can be applied to subsequent project releases, especially during the code creation and testing phase. All operations related to modeling and testing can be performed, for example, in the KNIME research environment, which has a set of modules supporting the techniques of creating predictive models. It also allows you to save results to a database, or export to XML as a PMML model (useful for further research).

5 Summary and Future Works

The article aims to present a general approach to the problem of predicting defects in software, based on prediction models operating on historical attributes obtained from version control systems, issue tracking systems and metrics obtained from the source code level.

A general mechanism of the defect prediction process based on the analysis of selected Apache Maven data repositories has been presented. This tool is characterized by a long presence on the IT market (since about 2004), so the version control systems (Git) as well as issue tracking systems (JIRA) contain a lot of interesting data for exploration. Based on them, a predictive model can be built, which after proper validation will be able to predict defects in the future in a given project. What will translate into measurable benefits in the form of time savings (testing) and financial resources (corrections).

The subject on effective defect prediction is a very developmental subject and raises the real, modern problems of the IT market, therefore further work is foreseen on prediction of defects. Future work plans should focus on own capabilities, competitive implementation of tools supporting acquisition, standardization and predictive modeling. In addition, there is a need to build more effective predictive models, for this purpose further interest may be focused, for example, on hybrid prediction classifiers.

References

1. Petersen, K., Wohlin, C.: Software process improvement through the Lean Measurement (LEAM) method. J. Syst. Softw. **83**(7), 1275–1287 (2010)
2. Wojszczyk, R.: Quality assessment of implementation of strategy design pattern. In: Advances in Intelligent Systems and Computing, vol. 620, pp. 37–44. Springer (2018)
3. Ramler, R., Himmelbauer, J.: Building defect prediction models in practice. In: Handbook of Research on Emerging Advancements in Software Engineering, pp. 540–565 (2014)
4. Jureczko, M., Magott, J.: QualitySpy: a framework for monitoring software development processes. J. Theor. Appl. Comput. Sci. **6**(1), 35–45 (2012)
5. Jureczko, M. and Contributors: Quality Spy. http://java.net/projects/qualityspy
6. The Apache Software Foundation, Apache License, Version 2.0. http://www.apache.org/licenses/LICENSE-2.0.html
7. JIRA. Atlassian. https://www.atlassian.com/software/jira
8. SVN. Enterprise-class centralized version control. https://subversion.apache.org/
9. D'Ambros, M., Lanza, M.: Distributed and collaborative software evolution analysis with churrasco. Sci. Comput. Program. **75**, 276–287 (2010)
10. Madeyski, L., Majchrzak, M.: Software Measurement and defect prediction with depress extensible framework. Found. Comput. Decis. Sci. **39**(4), 249–270 (2014)
11. Berthold, M.R., Cebron, N., Dill, F., Gabriel, T.R., Meinl, T., Ohl, P., Sieb, C., Thiel, K., Wiswedel, B.: KNIME: the konstanz information miner. In: Studies in Classification, Data Analysis, and Knowledge Organization (GfKL 2007). Springer (2007)
12. Czyczyn-Egird, D., Wojszczyk, R.: The effectiveness of data mining techniques in the detection of DDoS attacks. In: 14th International Conference on Distributed Computing and Artificial Intelligence, vol. 620, pp. 53–60. Springer (2018)
13. Kazim, A.: A study of software development life cycle process models. Int. J. Adv. Res. Comput. Sci. **8**(1) (2017)
14. RebelLabs. Developer Productivity Report 2013 – How Engineering Tools & Practices Impact Software Quality & Delivery. http://bit.ly/2nQVSFh. Accessed Feb 2018
15. Slowik, A., Kwasnicka, H.: Nature inspired methods and their industry applications - swarm intelligence algorithms. IEEE Trans. Ind. Inform. **14**(3), 1004–1015 (2018)
16. Apache Maven. https://maven.apache.org/
17. GitHub Inc. http://www.github.com
18. GraphQL API v4. https://developer.github.com/v4/

The Use of Artificial Neural Networks in Tomographic Reconstruction of Soil Embankments

Tomasz Rymarczyk[1], Grzegorz Kłosowski[2], and Arkadiusz Gola[2(✉)]

[1] Research and Development Centre, Netrix S.A., Lublin, Poland
tomasz@rymarczyk.com
[2] Lublin University of Technology, Lublin, Poland
{g.klosowski,a.gola}@pollub.pl

Abstract. This paper deals with the problem of tomographic reconstruction of objects inside embankments. The article presents a new method of tomographic reconstruction of images of such technical objects as flood banks and dams. The concept is based on a neural controller that converts electrical signals into individual pixels of the image. The proposed solution provided very good quality of mappings for both small and large objects hidden inside flood embankments and dams.

Keywords: Neural networks · Multilayer perceptron · Imaging tomography

1 Introduction

The article presents a new concept of tomographic imaging based on artificial neural networks. The primary reason why the existing systems for monitoring technical facilities like dams and embankments do not employ tomography is low resolution of the tomographic image. There are various methods and techniques for tomographic imaging, however, none provides sufficient quality to find a wide range of practical applications [1–3]. Many concepts assume in advance that objects inside the tested dam or flood embankment are of specific shapes or their number is predetermined. Such models are ineffective due to the unpredictability of real situations. Bearing in mind that the reliability of technical facilities of the dam or levee type is of critical importance there is no place for pre-assumed uncertainty of forecasts.

Monitoring is understood as a regular and systematic collection and analysis of quantitative and qualitative data obtained from measurements and observations of phenomena. It is carried out over a specified time, at regular intervals, in a continuous and long-term process [4, 5].

Geotechnical monitoring consists in observing the behaviour of the surface and the geotechnical structure prior to, in the process of and upon its erection (PN-EN ISO 18674-1:2015-07E), which means that it constitutes an integral part of designing stage based on the observational method (PN-EN 1997-1:2008).

Flood embankment is a highly important and current issue, which is also emphasised by the fact that at present a modern system for river embankment monitoring is

S. Rodríguez et al. (Eds.): DCAI 2018, AISC 801, pp. 104–112, 2019.
https://doi.org/10.1007/978-3-319-99608-0_12

being developed in Poland. SAFEDAM is an effective flood prevention system, which also aids crisis management actions when the flood does occur. The aim of the SAFEDAM project is to create a levee monitoring system employing a non-invasive, unmanned aerial platform, as well as satellite and aerial imagery. The system integrates the resources of IMGW (Institute of Meteorology and Water Management) and CODGiK (Head Office of Geodesy and Cartography), with the data obtained from the UAV, optic and radar satellite imagery. Resources obtained from Sentinel satellites will be implemented to present the reach of flood water. SAFEDAM system is currently under development and is expected to be completed by the end of 2018 [6].

Present methods of flood bank monitoring predominantly consist in observation, which may be conducted underground or on the surface. A typical surface observation method involves making standard rounds of the flood banks.

The procedure for periodical and emergency monitoring of flood embankment condition precisely specifies the equipment used by the controllers during inspection, which includes: a control form, a notepad, a camera, binoculars, a metre stick, a water container, a flashlight and a filter for water material analysis. The schematic representation of embankment inspection is shown in Fig. 1.

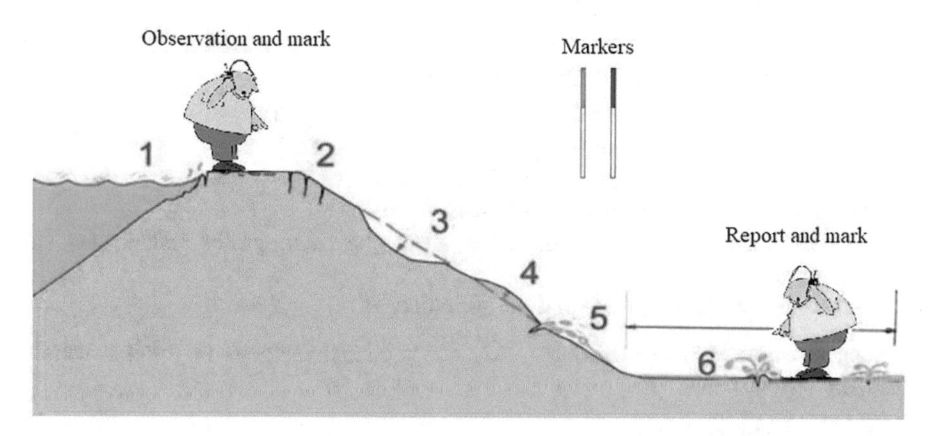

Fig. 1. Scheme of the technical embankment assessment: 1 – the riverside escarp; 2–4 – settling, cracks and landslides of the crest and the landside escarp; 5–6 – seepage through embankments body and substratum [7]

The control team should consist of at least two inspectors. The person standing at the top of the technical embankment is responsible for controlling the area including the riverside escarp and the crest of the flood bank, whereas the other one assesses the condition of the body of the embankment and the substratum. The damage in the flood bank is marked and noted in the protocol, while the inspectors are to inform appropriate service of the detected damage.

The observational method still enjoys a wide appeal. Figure 2 shows major advantages of the method.

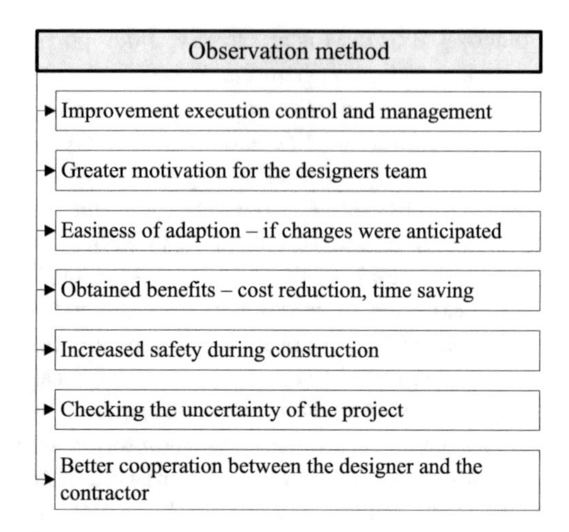

Fig. 2. Main advantages of the observational method [8]

Offering a wide range of advantages, the observational method does have its limitations and disadvantages, especially in comparison with the computer-aided technical embankment monitoring methods, which include [9]:

- inadequate time between measurements and their interpretation
- measurements cause failure (semi-destructive testing methods)
- failure mechanism/parameter cannot be measured
- change of failure mechanism during construction
- costs for changes during construction are higher than profits minus costs of monitoring
- communication between site and design office.

The disadvantages of employing only observational methods may entail certain risks, *e.g.*:

- quickly changing loads
- unwillingness of authorities
- time restrictions
- calculation methods and tools do not always allow for proper use of observational methods.

Many attempts are being made to compensate for the deficiencies of the observational method, usually by implementation of innovative IT solutions. Numerous research centres are conducting studies into the development of specific methods which will increase the effectiveness of the observational method through appropriate metering hydrotechnical structures and application of computer algorithms. Figure 3 shows a model algorithm for the process carried out with the observational method.

In the last decade several of Poland's significant structures for raising water levels have been equipped in fully automated metering systems for real-time monitoring of their condition. The systems in questions implement several types of sensors:

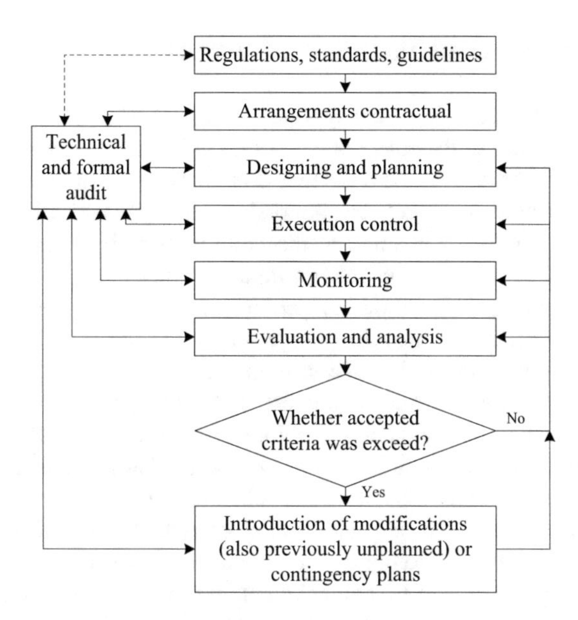

Fig. 3. Scheme of the observational method [8]

tensometric, electro-resistant, string, induction, ultrasonic, hydraulic, optical *etc*. The sensor-management instrumentation is modern electronic equipment and signals are mainly transmitted via optical-fibre cables.

Regardless of what type of converter is used in a sensor, they can be grouped according to the measured quantity. Therefore, we may distinguish sensors and instrumentation for water-level measurement in piezometers and water tanks, quantity of filtered water, seepage volume, linear and angular displacement, internal and external deformation of concrete structures and surface of metal structures, as well as forces, loads and pressures exerted by the technical structure on the ground surface meteorological quantities. The data collected by sensors is implemented in *e.g.* determining pressure and temperature corrections for other sensors in the Automatic Systems of Technical Control of Dams (ASTKZ).

The existing and currently used methods are known to be considerably erroneous, which hinders proper assessment of the condition of such technical objects as dams or flood banks. Periodical inspection and site inspections do not allow for regular technical data collection in a format which enables long-term data comparison or trend analysis. The mechanisms of potential flood bank failure remain undetermined and it is therefore difficult to predict it in time. No inspection and classical geotechnical and geophysical examination can guarantee detection of damage in advance [10].

This paper presents a study developing a new concept for monitoring river embankments and other hydrotechnical structures. The concept employs an improved computer tomography method with the application of custom neural networks.

2 Modelling the Artificial Neural Network

Electric tomography consists in conversion of electric signals into images. In the analysed case study the study object is the flood bank, inside which 16 bar electrodes were placed. The electrodes are 1.5 m long and are introduced in line into the section of the body of embankment at regular intervals. Subsequently, after connecting the electric current of specified parameters, the voltage disparity between pairs of electrodes is measured to produce 208 electric input parameters.

The values of voltage depend on the electric conductivity of the river embankment material. The parameter is subject to change relative to the amount of water in soil, which is a critical factor in the inspection in question.

The basic reason for the limited interest in tomography methods application in embankment inspection is low quality (resolution) of produced images. This defect of the method hampers correct identification of dangerous changes in hydrotechnical earth structures. In order to tackle this specific problem concerning the method, a series of experiments implementing the physical model of river embankment structures was conducted. Reconstruction of tomographic images, *i.e.* converting the electric signals into a coloured section image of embankment, was carried out with the system of artificial neural networks. The section of the structure was divided into 2012 pixels, and the neural network model converted 208 electric input signals into 2012 pixels of the image after reconstruction. The scheme for converting electric impulses into pixels of the image is presented in Figs. 4 and 5.

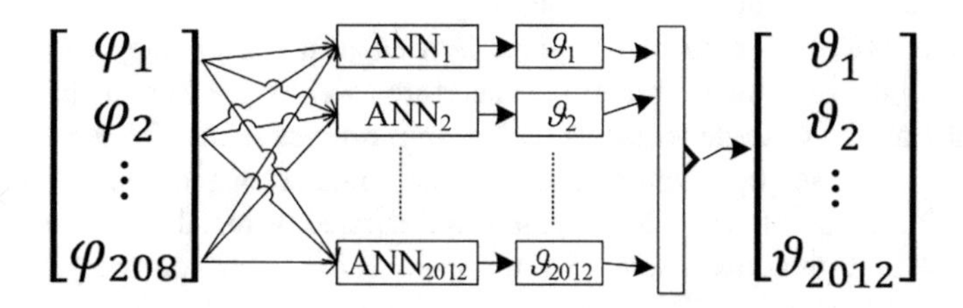

Fig. 4. Conversion of electric signals into pixels of the image

Figure 4 shows a vector consisting of 208 input quantities. Each input sends signals to one of 2012 separately trained artificial neural networks (ANN). Each network generates one signal ϑ_i ascribed to a specific pixel of the image. Figure 5 shows conversion of the input vector into the section image of river embankment.

3 Test Methodology and Results

During the research it was proven that the best solution in the field of tomographic imaging is the concept that allows the image to be reproduced from individual pixels. To this end, simulation tests and experiments were carried out to develop a neural driver processing input electrical signals from electrodes into images.

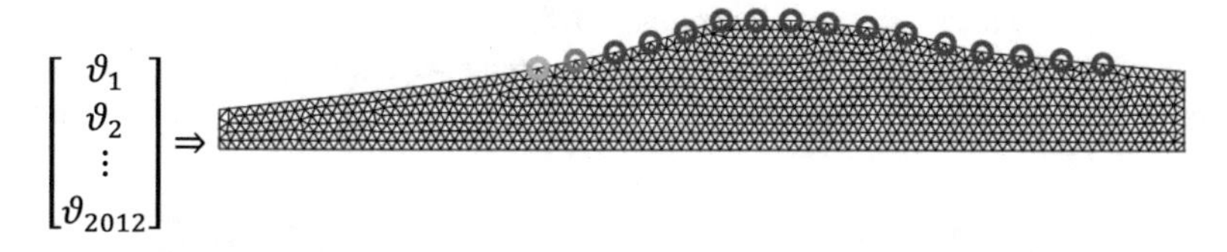

Fig. 5. Conversion of input vector of real numbers into the image of flood bank section

Common neural models that convert electrical signals into images use a variety of input and output variables. The output is a set of pixels which ordered and properly arranged on the grid, make up the output image.

During the research, artificial neural networks like the multilayer perceptron and convolutional neural networks were used. The decision of using neural networks for this purpose was taken because of their value when forecasting and optimizing many technical problems [11, 12]. During the research various variants of neural networks models were tested. In particular, such parameters of multi-layer perceptron as: number of hidden layers, number of neurons in hidden layer and the type of net training algorithm were differentiated. Finally from all analyzed variants the model of neural network including one hidden layer with 10 neurons was selected. To train the net the Levenberg-Marquardt algorithm was used.

The neural network was trained on 10442 pattern cases. The entire data set was divided into 3 sets: training set, validation set and testing set. The network training was carried out by means of the Levenberg-Marquardt algorithm. Training automatically stops when generalization stops improving, as indicated by an increase in the mean square error of the validation samples. Training results of the best developed network are shown in Fig. 6.

	Samples	MSE	R
Training:	7310	4.87890e-5	9.99907e-1
Validation:	1566	1.19163e-3	9.97999e-1
Testing:	1566	1.02137e-3	9.98134e-1

Fig. 6. Network training results

The highest Mean Squared Error (MSE) was recorded for the validation and testing sets, whereas significantly lower MSE was observed for the training set, where it amounted to 4.87890e−5. Mean Squared Error is the average squared difference between outputs and target values. Lower values are better. Zero means no error. The training set showed the lowest learning error value, which is typical and regular for network training. Another measured indicator was regression R. An R value of 1 means a close relationship, 0 a random relationship. As it is shown in Fig. 6 in all three cases R is close to 1. This particularly concerns the testing and validation sets, which is highly desired. Values close to 1 indicate good fit of input (results of the network) and target data (of training, validation and testing sets).

Figure 7 shows correlation diagrams for the analysed network. The scatter of results beyond the target is visible but the correlation coefficient is high, which is emphasised by correlation lines for both analysed cases: the testing set and overall.

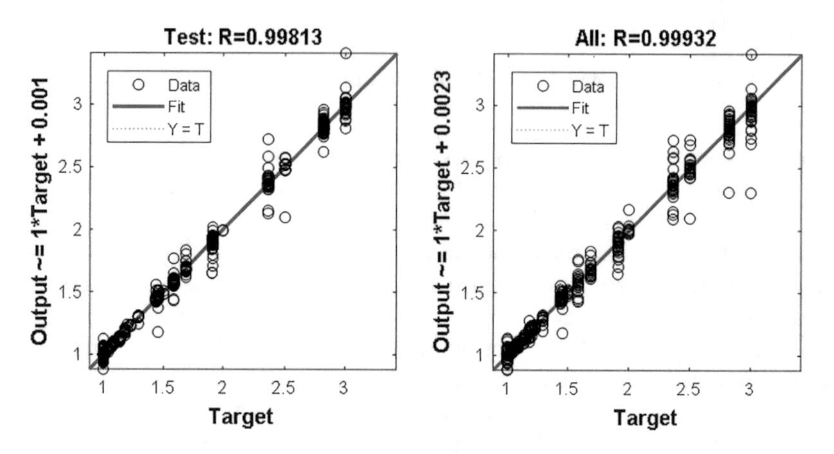

Fig. 7. Correlation diagrams for network learning

Figure 8 shows a randomly selected example of the analysed flood bank. The top image shows the target image of the flood bank with visible seepage. The middle image was generated by the neural network based on the electric signals obtained from the embankment. The differences are notably small, hence the accuracy of the output is considerably high.

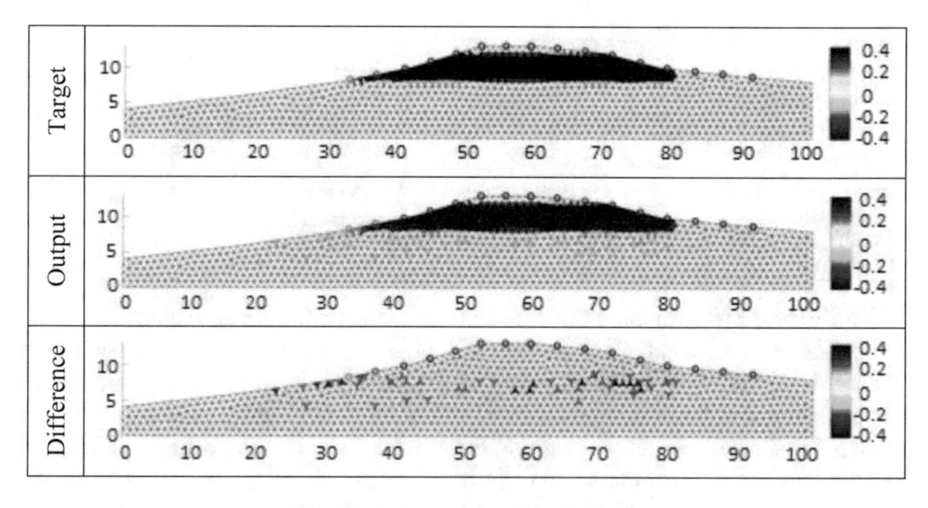

Fig. 8. Output against target values

4 Conclusion

The results showed good quality, especially when detecting medium and large objects. In the case of very small inclusions (disturbances in the internal structure of the embankment), the network generated erroneous, ambiguous images. It was noted that images for generated small objects - although erroneous – show certain similarity.

Two sets of neural networks were developed. The first set of networks was trained for tasks related to detection of medium and large objects. The second set was expected to correctly detect small objects and was trained for such objects. The third neural network as the classifier was also developed. Its task was to classify the image into the appropriate controller - the first (dedicated for medium and large objects) or the second (for small objects). This solution provided very good quality of mappings for both small and large objects hidden inside flood embankments and dams.

References

1. Rymarczyk, T.: New methods to determine moisture areas by electrical impedance tomography. Int. J. Appl. Electromagn. Mech. **37**(1–2), 79–87 (2016)
2. Banasiak, R., Wajman, R., Sankowski, D., Soleimani, M.: Three-dimensional nonlinear inversion of electrical capacitance tomography data using a complete sensor model. Prog. Electromagn. Res. (PIER) **100**, 219–234 (2010)
3. Grudzien, K., Chaniecki, Z., Romanowski, A., Niedostatkiewicz, M., Sankowski, D.: ECT image analysis methods for shear zone measurements during silo discharging process. Chin. J. Chem. Eng. **20**(2), 337–345 (2012)
4. Borecka, A., Stopkowicz, A., Sekuła, K.: Metoda obserwacyjna i monitoring geotechniczny w świetle przepisów prawa do oceny zachowania podłoża i konstrukcji inżynierskich. Przegląd Geologiczny **65**(10), 685–691 (2017)
5. Jasiulewicz-Kaczmarek M.: Practical aspects of the application of RCM to select optimal maintenance policy of the production line. In: Nowakowski, T., et al.: Safety and Reliability: Methodology and Applications - Proceedings of the European Safety and Reliability Conference, ESREL 2014, Taylor & Francis Group, London, pp. 1187–1195 (2015)
6. http://www.space24.pl/safedam-system-monitoringu-walow-przeciwpowodziowych-nie-tylko-z-satelitow. Last Accessed 23 Jan 2018
7. Ślizewski, B.: Wybrane zagadnienia bezpieczeństwa wałów przeciwpowodziowych w Niemczech. Woda-Środowisko-Obszary Wiejskie **7**, 45–57 (2007)
8. Nicholson, D., Tse, C., Penny, C.: The observational method in ground engineering: principles and applications. Report 185. Construction Industry Research and Information Association, London (1999)
9. Korff, M., De Jong, E., Bles T.J.: SWOT analysis observational method applications. In: Delange, P., Desrues, J., et al. (eds.) Proceedings of the 18th International Conference on Soil Mechanics and Geotechnical Engineering, pp. 1883–1888 (2013)
10. Borecka, A., Sekuła, K., Kessler, D., Majerski, P.: Zastosowanie testowych czujników pomiaru temperatury w quasi-przestrzennych (3D) sieciach pomiarowych w hydrotechnicznych budowlach ziemnych – wyniki wstępne. Przegląd Geologiczny **65**(10), 748–755 (2017)

11. Kowalski, A., Rosienkiewicz, M.: ANN-Based hybrid algorithm supporting composition control of casting slip in manufacture of ceramic insulators. In: Advances in Intelligent Systems and Computing, vol. 527, pp. 357–365 (2017)
12. Sitek, P., Wikarek, J.: A hybrid programming framework for modeling and solving constraint satisfaction and optimization problems. Sci. Program. **2016**, Article No. 5102616 (2016)

Occurrences Management in a Smart-City Context

Mário Ferreira[(✉)], João Ramos, and Paulo Novais

Algoritmi Centre, Department of Informatics, University of Minho, Braga, Portugal
a70441@alunos.uminho.pt, {jramos,pjon}@di.uminho.pt

Abstract. With the expected increase of the population that lives in the urban cities, the concept of a smart city is becoming more concrete due to the necessity of adaptation, considering its sustainability and competitiveness. The hope is that modern society will be able to deal with the multitude of issues that urban inhabitants are already facing, which will doubtless be further exacerbated as cities continue to expand. Therefore, instead of traveling to the city hall to register an occurrence, the user may have a 24-h service that is portable and easier to create events. Its proficiency and facility to get into a smartphone is a plus point to encourage the citizens to use it, thus using existing resources to make the city smarter. This service will be the application (mobile and web) built within the scope of this thesis, which allows to record the occurrences efficiently. In this way, Artificial Intelligence will perform an important role in this field, since it recognizes patterns and, in an efficient way, find and suggest better solutions, which will be applied with a machine learning method in order to anticipate occurrences to make the city proactive, trying to anticipate and solve their problems in advance.

Keywords: Smart city · Occurrences · Machine learning

1 Introduction

The world is expanding, as is the population that inhabits it. In these days, there is a term that is being more and more used: overcrowding. In recent decades there has been a substantial increase in population growth and this provokes a necessary growth in the cities. Figure 1 depicts the exponential growth of population.

Europe has around three-quarter of population in urban areas and this number tends to increase. Although it is in the cities that is the majority of the population and the engines of the European economy, it depends a lot on more distant regions to stock up on resources such as food, water, energy and others [2].

Environmental challenges and city evolution are connected, although many cities fail to combine a sustainable environment with social and economic problems resulting in pollution, traffic, among others. However, a more developed city does not necessarily mean a less sustainable and energy efficient one. Population

S. Rodríguez et al. (Eds.): DCAI 2018, AISC 801, pp. 113–120, 2019.
https://doi.org/10.1007/978-3-319-99608-0_13

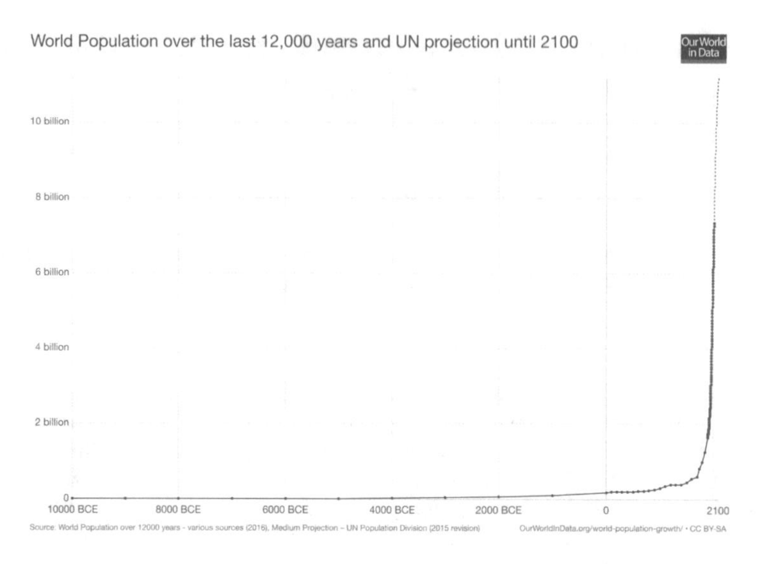

Fig. 1. World population over the last 12,000 years and UN projection until 2100, retrieved from [6]

density in cities means, in reality, shorter journeys to work and services, and more walking, cycling or greater use of the public transportation systems, while apartments in multi-family houses or blocks require less heating and less ground space per person. As a consequence, urban residents consumes less energy and land per capita than rural residents [2].

With such an increase in density population and technology evolution, cities are obligated to grow, mainly applying technology, in order to become a smart-city.

The concept of smart city is being more used all over the world with different names and in different circumstances, but in short a 'smart city' is an urban region that is highly advanced in terms of overall infrastructure, sustainable real estate, communications and market viability. It is a city where information technology is the principal infrastructure and the basis for providing essential services to residents. There are many technological platforms involved including, but not limited to, automated sensor networks and data centers.

In this work the focus is in management of occurrences in a city, that is, how to make this process more efficient and faster, which will turn the city smarter.

The main goal is to develop a software which effectively and efficiently manages occurrences in a city using artificial intelligence, in order to improve and speed up the resolution of these problems, using feedback from the entire population.

This paper is organized as follows: Sect. 2 describes the current state of the art considering the occurrences management topic. In this section it is also highlighted the gaps that need to be fulfilled. Section 3 presents current approaches, which includes their features, functionalities and advantages of their use.

In Sect. 4, a description of the application that will be developed is provided. This also includes the planned architecture and the description of some machine learning methods that may be used. Finally, in Sect. 5, the main conclusions are drawn.

2 Occurrences Management

Cities are seeking a way to be smarter, taking concrete actions in order to achieve their determined success through specific strategic plans [5].

An occurrence can be any incident related to the city and the citizens in general, which usually does not influence just one individual. Thus, an occurrence may be due to lack of lighting, flooding, illegal parking, road in deteriorated conditions, that would normally be solved by the municipality. This process may take or less time according to the necessities and to the number of affected citizens.

In this way, there are some projects in order to fund smart city initiatives in the European Union (EU). It is shown that, not only the EU, but also various private-sector businesses are interested and connected to the smart city concept, which can help and encourage in the development of some initiative in this field. Besides this, the EU Smart Cities and Communities platform also offers tools for all phases of a smart city development process [9].

Having a platform that helps the city to manage this type of incident, would make the city more reactive to its problems. Since any citizen can report an occurrence, this is detected much faster than it actually is and it will be more likely to be solved quicker.

There are several cases that indicate the mismanagement of occurrences. For example, in Silves, Portugal, one kilometer of road took 16 years to build. The same happened in Vila Real, taking about 20 years. In addition, transparency in these resolutions is not always the best, as there is often controversy over the values of some incidents.

In this moment, it is possible, for example, to report a failure with street lighting or report missed, uncollected or lost bins in Blackburn government website. It is also possible to report a problem with a traffic light in United Kingdom government website. In other words, it is also achievable to observe that there is a possibility of reporting specific occurrences by some governments. However, in addition to being often unnoticed, they are too specific, or rather do not guarantee the possibility of reporting in the same way any other type of occurrence.

Also, in Portugal, there are projects to start manage occurrences between the police, firemen and civil protection, which shows there are an increased concern in the management of current or emergency occurrences [1].

There is a clear need to learn how to manage occurrences well in a city since the old methods, compared to the current population, are not effective. It is necessary to take into account the existing technology and to take advantage of it in order to genuinely improve the life of the human beings and the place where they live, which are mostly cities.

Making the city smarter may be an answer to this problem. In Santander, Spain, the installation of some GPRS and RFID sensors around the city have cut energy costs by as much as 25% and waste management costs by 20% [7].

3 Current Approaches

Countries are being increasingly worried about the fact of having an efficient occurrence management and, therefore, it have appeared in the market several apps or systems that may do this management and receive information by the citizens. Certain systems arise from the city councils, but there are also other systems that are more embracing and could be adapted to more than one city.

The main goal is to enable citizens to report problems in real-time and have access to the state of the process and its resolution. Thus, occurrences in cities are evaluated and resolved more quickly and efficiently.

This approach to managing occurrences is not revolutionary and there are already some cities trying to apply it, as is the case of Sintra that created the "Sintra Project" [3] or the case of Lisbon [1].

An example of an application trying to make occurrences management easier is CitiAct (available on the Play Store). Figure 2 demonstrates its use and operation. In this mobile application it is possible to register occurrences and to specify the type of it, choosing one of the options for predefined occurrence types. However, it is not possible to observe the statistics, such as number of occurrences per day/total, place with more occurrences, among others.

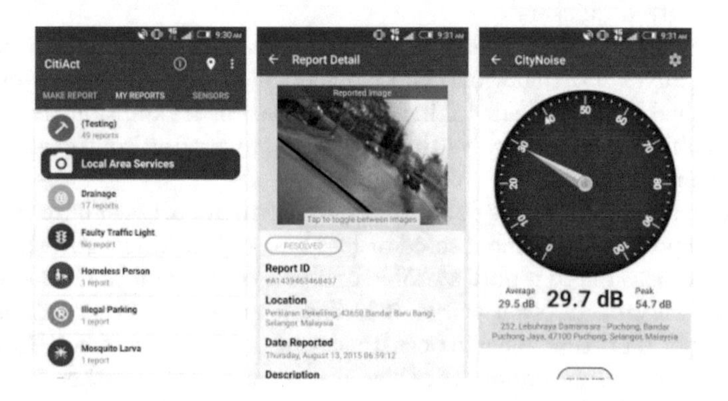

Fig. 2. Screenshots from "citiAct" mobile app

Despite not having this possible statistical observation, it has very interesting sensors such as the noise in the city, among others. This application doesn't have any machine learning applied, so it does not learn with the data that is stored and do not use it for anything.

Applications like the one previously described make the city more reactive, that is, since the detection is faster, the reaction to this occurrence is also faster and greatly simplify this whole process of occurrence management. In order to achieve this reaction, it is necessary to detect these problems as soon as possible. Thus, is important to enable citizens to register the occurrence in a mobile device, which facilitates the identification of the location and time of the occurrence. Using a mobile device the creation of a occurrence is easier, allowing to add pictures and to define the type of it.

4 System Description

In previous sections it was exposed the occurrence management problem and the advantages that a system may bring to the city, turning it into a smart city. There are few platforms that manage this situations in a simple and efficient way and, at the same time, spending few resources. Thus, the goal is to make a mobile application that runs on Android and iOS (which covers about 99.6% of the smartphones in the world [10]) and also a web application. In this way, any citizen, whether in New York (USA) or Braga (Portugal), may virtually report incidents wherever they are.

Although both will presumably have the same functionalities, the mobile application is more suitable to quickly report an occurrence and the web version is more appropriate to verify statistics and analyze existing occurrences.

Using a device, a smartphone or computer, the citizen may report any occurrence or even make a suggestion, attaching to the incident a photo for better observation, a description, title and type of occurrences within existing ones (for example, lighting). The user must be able to observe all occurrences on the map or in a list, filtering by proximity or type, and propose a resolution, being able to submit a budget.

In order to build an application capable to efficiently gather the occurrences, it will be separate in some categories. In this context, a category should be define as a class of incidents that a citizens find in the city, this incident can also be define as occurrence. This division will help the citizens to improve the city and not confusing them when they need to choose the category to submit the occurrence. It will be more easy and quick to resolve them, since there are some structure. For example, the categories should be divided in: Water & Sewers, Pets & Animals, Roads & Road signs, Lighting & Energy, Garden & Environment, Forest Cleaning, Cleaning & Conservation, Pavement & Sidewalks, Garbage Collection, between others. Besides this, the users of the app can add a suggestion or select "other" and, manually, choose another category.

An important feature is to receive feedback from other citizens, who can upvote or downvote an occurrence, suggestion, and even proposed resolutions. In this way, it is easier to obtain proposals for an occurrence on the part of the municipality, as this whole process is more transparent and takes into account the opinion of society.

Finally, and the most important and differentiating aspect of this approach, is the implementation of a machine learning method to this application that allows, in view of the data, to be able to anticipate and predict future occurrences, and automatically propose the resolution to the citizens.

4.1 System Framework

In Fig. 3 the architecture of the application is schematized. The goal is to make a mobile and web application. For the mobile, the framework to use is React-Native, a framework that uses the same fundamental UI building blocks as regular iOS and Android apps. For the web version, the technology to use is React JS (React and javascript). This allows us to reuse some code, since both versions are based on the same technology, which is React, in addition to the experience of those involved in the project.

For the backend, the technologies to use will be Java and Spring. Using them is possible to build a RESTful API to communicate with the applications, using HTTP methods, like GET, POST, PUT, DELETE, since it is a simple web service and that ensures consistency, since the same service is used in both applications. For the persistence layer hibernate, an object-relational mapping tool, will be used with MySQL as a database. For the webcontainer, apache tomcat will be considered. In addition to having experience with these technologies, Java allows an accessible integration with machine learning methods to be applied, using existing libraries such as Weka.

4.2 Reasoning Methods

At the base of a real smart-city is often artificial intelligence. In the form of water management, public transport, etc., and there are several examples that demonstrate this use of AI.

The power of using large volumes of data and learning from them is huge and can be used in many ways and will try to be adapted to this specific case: occurrences.

In this work, the goal is to implement a machine learning method in the framework, using all the data provided by the citizens, such as problems or feedback about a resolution. Furthermore, we will apply a machine learning method to learn through the data, with all the occurrences and proposed solutions, its location, time interval from its identification to its resolution, among others and use this to improve the application.

So far, it has not been possible to verify any application that uses artificial intelligence to improve the performance of this detection or resolution of occurrences, which is why it is the core part of this application.

However, there are various methods and it is necessary to investigate which best applies to this case. The following are some of the methods that could be appropriate for the case study:

Deep Learning [8] can be used to anticipate occurrences. Most deep learning methods use neural network architectures, usually with a lot of hidden layers. We need a method that can learn with the data, without being supervised (after

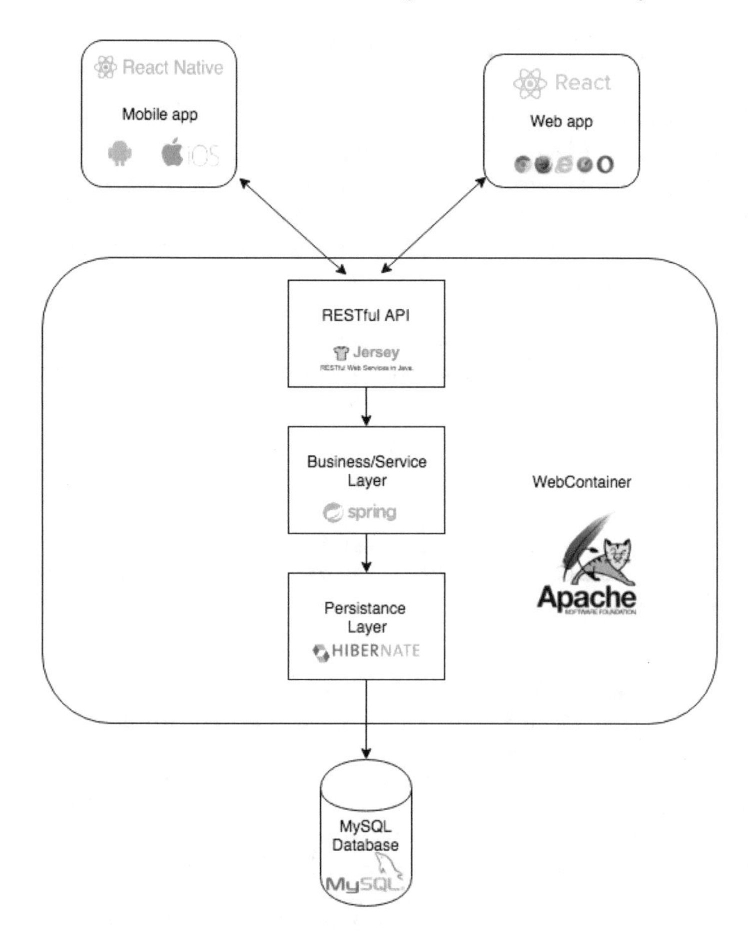

Fig. 3. Architecture of the application

the learning supervised phase), based on the existing and already resolved occurrences, and taking into account the number of occurrences in a given location, it should be possible through this method to attempt to predict occurrences, their location and type.

Case Based Reasoning [4] is indicated to choose citizens to solve a problem based on previous occurrences. This method is the process of solving new problems based on solutions of similar past problems. Like an human, based on previous experience, there may be a set of possible actions to retrieve the best solution to a problem. This method use mainly 4 phases, the four R's, retrieve, reuse, revise and retain. Retrieve from memory cases that are relevant to a problem, reuse solutions to the target problem, revise if necessary to adapt and improve, and finally retain the new problem and its solution.

5 Conclusions and Future Work

After the application is developed, we proceed to the choice and implementation of a suitable machine learning method to learn from the collected information. It is necessary to take into account the amount of information that will be dealt with at an early stage, among other factors. It is expected that more such platforms will be developed and event management relentlessly improved to benefit both citizen and city.

Acknowledgements. This work has been supported by COMPETE: POCI-01-0145-FEDER-007043 and FCT - Fundação para a Ciência e a Tecnologia within the Project Scope: UID/CEC/00319/2013.

References

1. Lisboa vai ter plataforma para gestão de ocorrências (2017). https://www.sapo.pt/noticias/economia/lisboa-vai-ter-plataforma-para-gestao-de_596874968acee80f0d638ea2
2. European Environment Agency: Urban environment (2017). https://www.eea.europa.eu/themes/urban/intro
3. Genésio, F., Nascimento, E.: Sintra cria aplicação de denúncia de problemas (2016). http://www.cmjornal.pt/portugal/cidades/detalhe/sintra-cria-aplicacao-de-denuncia-de-problemas
4. Lopez, R., Plaza, E.: Case-based reasoning: an overview. AI Commun. **10**, 21–29 (1997)
5. Nam, T., Pardo, T.A.: Conceptualizing smart city with dimensions of technology, people, and institutions. In: Proceedings of the 12th Annual International Digital Government Research Conference on Digital Government Innovation in Challenging Times - dg.o 2011 (2011). https://doi.org/10.1145/2037556.2037602
6. Roser, M., Ortiz-Ospina, E.: World Population Growth (2018). https://ourworldindata.org/world-population-growth
7. SCC Europe Staff: How Santander, Spain is using sensors to tackle waste (2017). https://eu.smartcitiescouncil.com/article/how-santander-spain-using-sensors-tackle-waste
8. Schmidhuber, J.: Deep learning in neural networks: an overview. Neural Netw. **61**, 85–117 (2015)
9. Toppeta, D.: How Innovation and ICT The Smart City vision: How innovation and ICT can build smart, liveable, sustainable cities. Think report (2010)
10. Vincent, J.: 99.6 percent of new smartphones run Android or iOS (2016). https://www.theverge.com/2017/2/16/14634656/android-ios-market-share-blackberry-2016

Special Session on Social Modelling of Ambient Intelligence in Large Facilities (SMAILF)

Exploiting User Movements to Derive Recommendations in Large Facilities

Jürgen Dunkel[1(✉)], Ramón Hermoso[2], and Florian Rückauf[1]

[1] Hannover University of Applied Sciences and Arts, Hannover, Germany
juergen.dunkel@hs-hannover.de, florian.rueckauf@stud.hs-hannover.de
[2] University of Zaragoza, Zaragoza, Spain
rhermoso@unizar.es

Abstract. This paper provides an innovative approach for taking advantage of user's movement data as implicit user feedback for deriving recommendations in large facilities. By means of a real-world museum scenario a beacon infrastructure for tracking sojourn times is presented. Then we show how sojourn times can be integrated in a collaborative filtering algorithm approach in order to outcome accurate recommendations.

Keywords: Context-aware recommender systems
Collaborative filtering · Beacon technology

1 Introduction

During the last two decades, Recommender Systems have become a major field of research, aiming to design and develop systems that suggest accurate help in decision-making processes in which the user does not have sufficient experience [12]. In the last years, with the boost of smart mobile devices and network technologies, Context-Aware Recommeder Systems (CARS) have emerged as an approach to allow users to obtain recommendations by using contextual information, such as location, time, weather, etc. [1].

Large facilities presents itself as a very promising area in which advances in CARS may be applied. There exist some works on evacuation guidance on large smart buildings [4,10] in which authors take into account the context around users to suggest best routes for evacuation in case of emergencies. Visitors in large facilities barely want to continuously disclose explicit feedback about their movements during their stay. Thus, this type of explicit feedback is often not available. A possible solution to cope with this issue is to infer user preferences from the so-called implicit feedback, in which opinions are extracted through user behavior observation [8]. In this paper, we put forward a method to obtain implicit user feedback by observing the users trajectories. The system collects sojourn times to estimate the user actual interest in an item and integrates these values into a collaborative filtering approach. We present a case study in which we propose a CARS for museum visitors, in which sojourn times are monitored by Bluetooth Low Energy (BLE) sensors.

S. Rodríguez et al. (Eds.): DCAI 2018, AISC 801, pp. 123–131, 2019.
https://doi.org/10.1007/978-3-319-99608-0_14

The paper is organized as follows: In Sect. 2 we present a real-world museum scenario that is used to explain our approach. Then, Sect. 3 describes the conceptual framework the paper is based on. We present the mechanism for recommendation based on sojourn times in Sect. 4. In Sect. 5 we refer to the related work existing in the literature. Finally, we summarize the paper and sketch some future avenues in Sect. 6.

2 Application Scenario

To convey the idea of our approach, we will use a recommender system for museum visitors. Recently, we implemented the Samba (Situation-aware Museum Recommender System with Beacons) system: a recommender system for the Landesmuseum Hannover, a museum of natural history in Hannover, Germany[1]. A rough overview of our system architecture is depicted in Fig. 1.

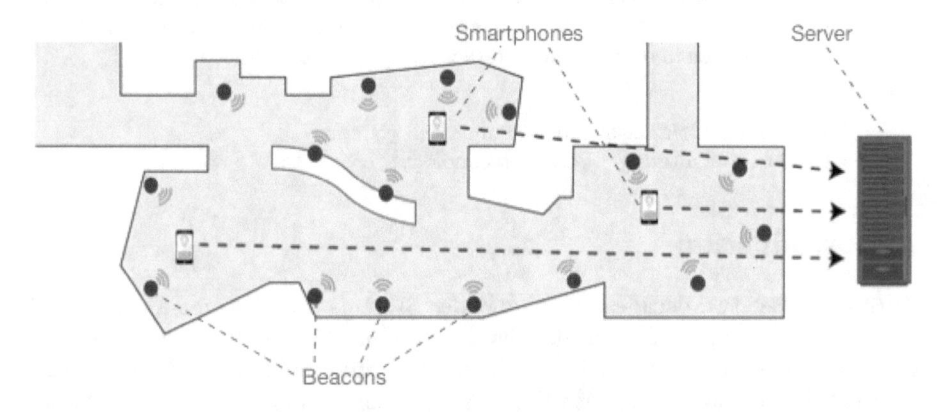

Fig. 1. System architecture of the recommender system

For deriving the users positions and sojourn times in front of certain exhibits, the museum is equipped with a beacon infrastructure. Beacon technology has recently been introduced for supporting indoor navigation[2,3]. A beacon device is located beneath a certain exhibit and sends continuously a unique ID that can be read by any smartphone. In our system, each museum visitor uses the Samba app running on her smartphone, which infers her positions from the received beacon signals and derives the sojourn times at particular positions. All smartphones are transmitting continuously position data and sojourn times to a central server that executes the recommendation algorithm and replies personalized recommendations to the visitors. The first version of Samba implemented

[1] https://www.landesmuseum-hannover.niedersachsen.de/.
[2] https://developer.apple.com/ibeacon/.
[3] https://developers.google.com/beacons/.

a recommendation approach was based on Semantic Web technologies [7]. Currently, we are extending the system with a recommender module that exploits the movement data of the museum visitors as described in this paper.

3 Conceptual Framework

Our recommendation approach makes use of the user's movement data. How movement data is transformed into a recommendation is shown by Fig. 2. Shadowed blocks represent phases taking place in the user's mobile device. The sojourn time s_{ue} of user u in front of certain exhibit e serves as an implicit rating: the longer a user stays the more she likes the exhibit. Interpreting sojourn times as user specific ratings requires several data preparation steps: collection, sojourn time calculation, data cleaning and normalization. In this section, we first will present how the beacon infrastructure could provide the required data. Then, we will discuss how to prepare movement data for using it in recommender system.

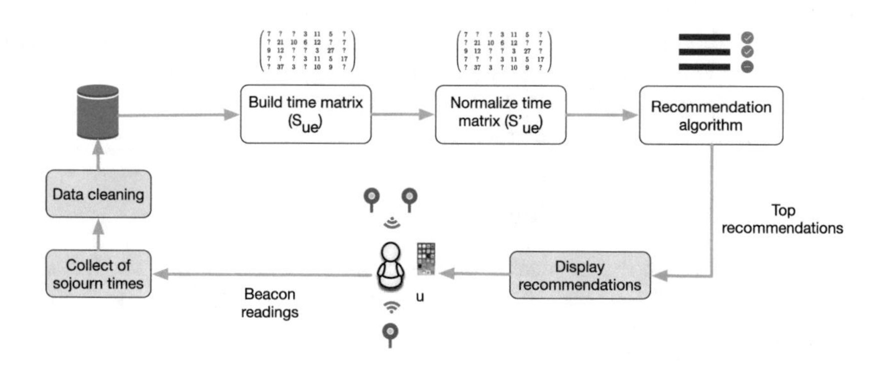

Fig. 2. Schema of the recommendation mechanism

3.1 Collecting Beacon Data

There are several technologies to support the localization of users [4]. In our Samba approach, we use beacon technology. A beacon is a device with a well-know position, which uses Bluetooth LE to send a unique ID in a certain frequency (e.g. 5 times per second). This beacon ID can be read by any smartphone within the range of that beacon - the beacon range could be adjusted to a few meters. Smartphones scan their environment for beacon signals in a predefined frequency, e.g. every second.

Figure 3 shows a situation where a user is in the range of three different beacons. Her smartphone reads the three beacon signals in her proximity. Then, for each beacon it knows about the beacon ID and the corresponding signal strength,

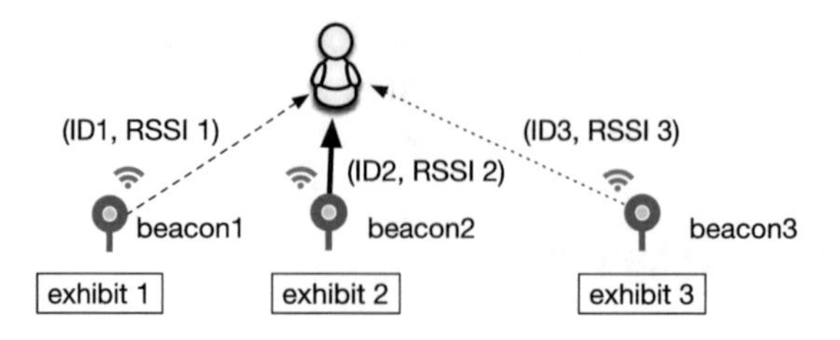

Fig. 3. Using beacons for user localization

the so-called RSSI (received signal strength indicator). The beacon with the highest RSSI value (here: beacon2) is assumed to be the nearest one. Usually the smartphone collects beacon signals over a longer time interval (between 1 and 2 s) to compensate signal jitter.

The nearest beacon determines the position of the user, which can be assigned to the related exhibit. When moving around the scanned beacon signals will change and the beacon with the strongest signal will define the new position of the user. Thus, for each user u we get beacon data that describes her path in the building. In particular, her phone provides the sequence of IDs from the beacons she passed and the corresponding arrival times in $path(u) = \{(ID_1^u, t_1^u), (ID_2^u, t_2^u), ..., (ID_n^u, t_n^u)\}$, where ID_i^u is the ID of the i-th beacon read by user u, and t_i^u denotes the user's arrival time at this beacon.

3.2 Measuring Sojourn Times

As we mentioned above the user's phone is able to collect a path based on the beacons detected. To infer sojourn times we need to calculate the temporal distance $\delta_{i,i+1}^u$ the user needed to go from one beacon i to the following $i+1$ in the sequence $path(u)$: $\delta_{i,i+1}^u = t_{i+1}^u - t_i^u$. Thus we measure the time elapsed between the arrival times of two consecutive beacons in the user's path.

3.3 Data Cleaning

It is important to remark that not all $\delta_{i,i+1}^u$ values can be considered as relevant sojourn times to build the matrix S. The main problem is to identify when a user has actually been staring at an exhibit or if the calculated δ time is due to any other reason (e.g. the regular time a user needs to go from i to $i+1$ without stops on her way). Sojourn times below a certain threshold could be considered confusing too. Short times mean that the user just passed by without paying particular attention to an exhibit, so they do not represent a conscious interest in the exhibit. Therefore, we clean the data by using Eq. 1.

$$s_{ue} = \begin{cases} \delta_{i,i+1}^u - \xi(u,i,i+1) & \texttt{if} \quad \delta_{i,i+1}^u - \xi(u,i,i+1) \geq \Theta \\ ? & \texttt{otherwise} \end{cases} \tag{1}$$

in which $\xi(u, i, +1)$ stands for a function that approximates the time for user u to go from the area of beacon i (exhibit e) until reaching the area of beacon $i+1$, and Θ is a threshold time to consider the sojourn time as relevant. Although out of the scope of this paper, note that ξ function might take into account different factors, such as the distance between beacons, the regular walking pace of user u or the occupancy of the area.

3.4 Building the Sojourn Time Matrix

Once the data collected from beacons has been cleaned we need to build the sojourn time matrix. Figure 4 shows the sojourn time matrix $S = [s_{ue}]$. Usually, the matrix S is *sparse*, i.e. most of its elements are zero, since users typically visit only a fraction of the exhibits in a museum. We represent this value with the symbol ?. Note that after data cleaning $\forall s_{ue} \in S \neq ?$, $s_{ue} \geq \Theta$ must hold.

exhibit:	1	2	3	4	5	6	avg
user1	167	152	?	?	45	52	104
user2	?	?	183	212	85	76	139
user3	58	45	162	181	?	?	111,5
user4	118	111	26	39	135	?	85,8

Fig. 4. Example of the sojourn times matrix S

3.5 Sojourn Time Matrix Normalization

There are users who tend to contemplate at exhibits for a long time, while others tend to jump quickly from one exhibit to another (which does not necessarily entail a worse satisfaction of the user). For that reason, we apply a normalization step to smooth those potential different types of behaviour. We use a *z-score normalization* approach, calculating matrix $Z = [z_{ue}]$ as shown in Eq. 2.

$$z_{ue} = \frac{s_{ue} - \mu_u}{\sigma_u} \tag{2}$$

where μ_u is the mean of all non-zero sojourn times of a user u, σ_u stands for the corresponding standard deviation, and s_{ue} is the sojourn time of user u at exhibit e. Note that this type of normalization ensures z_{ue} is derived from a normal distribution $N(\mu_u = 0, \sigma_u = 1)$.

4 Recommending Approach

In this section we present our recommending approach by integrating sojourn times into classic collaborative filtering (CF) algorithms. CF methods produce user specific recommendations of items based on patterns of ratings or usage without need for exogenous information about either items or users.

User-Based CF (UBCF). To implement this technique we use a k-Nearest Neighbour (k-NN) approach, in which the user is recommended with items that top-k similar users liked in the past. This mechanism can be described as follows:

1. Using matrix S the system calculates similarities between any pair of users. We use the Pearson's correlation method as presented in Eq. 3, in which $z_{ue}, z_{ve} \in S$ and $e \in E_u \cap E_v$ where E_x is the set of exhibits visited by user x.
$$w_{u,v} = \frac{\sum_{e \in E_u \cap E_v} z_{ue} \cdot z_{ve}}{\sqrt{\sum_{e \in E_u \cap E_v} z_{ue}^2} \sqrt{\sum_{e \in E_u \cap E_v} z_{ve}^2}} \tag{3}$$
2. Then the system ranks – for the active user u – the values $w_{u,v}$ and chooses only the top-k users most similar to u.
3. Let us suppose we want to calculate the prediction $\hat{z}_{u,e'}$ (for user u) of the normalized sojourn time for an unvisited exhibit e'. Then Eq. 4 is used,
$$\hat{z}_{u,e'} = \frac{\sum_{v \in K} z_{ve'} \cdot w_{u,v}}{\sum_{v \in K} w_{u,v}} \tag{4}$$

where K stands for the set of top-k users most similar to u. Even though the z-values estimate the expected normalized sojourn time for a user visiting an exhibit we need to unwind this value to the obtain the expected sojourn time. We can use Eq. 2 to calculate the estimated $\hat{s}_{u,e'}$ as follows:

$$\hat{s}_{ue'} = \hat{z}_{ue'} \cdot \sigma_u + \mu_u \tag{5}$$

Item-Based CF (IBCF). This method is similar to UBCF but exploiting similarity between exhibits instead of between users. This method first analyzes the user-item matrix to identify relationships between different items, and then uses these relationships to indirectly compute recommendations for users. This method attempts to solve some issues related to: the scalability of the system (systems containing more users than items), efficiency (only calculates similarities between items instead of between users such as in user-based approaches) and tends to be more stable, since changes in user profiles are supposed to occur in a more frequent fashion than in item profiles. The method works as follows:

1. The system computes similarities between items for any user by using z-score values. Note that in UBCF the algorithm works on rows of matrix Z while ITCF works on columns instead. Similarities can be computed by using many different approaches. In this paper we use the Adjusted Cosine function, which takes different types of user behavior into account. For instance, a *slow* visitor could spend higher sojourn times even in front of exhibit she does not like, in comparison to fast visitors. Adjusted Cosine function attempts to alleviate this issue as shown in Eq. 6 (with the set of all users U).
$$w_{e,e'} = \frac{\sum_{u \in U} z_{ue} \cdot z_{ue'}}{\sqrt{\sum_{u \in U} z_{ue}^2} \sqrt{\sum_{u \in U} z_{ue'}^2}} \tag{6}$$

2. To calculate the prediction $\hat{z}_{u,e'}$ (for user u) of the normalized sojourn time for an unvisited exhibit e' Eq. 7 is used.

$$\hat{z}_{u,e'} = \frac{\sum_{e \in Q} w(e,e') \cdot z_{ue}}{\sum_{e \in Q} w(e,e')} \tag{7}$$

where $e \in Q$ are the most similar exhibits to e the user u has already visited. Finally, as we presented for the CF approach we would calculate the sojourn time as explained in Eq. 5.

5 Related Work

As we pointed out in previous sections, implicit acquisition of user movement could be used to model user context [11,14]. It allows representing relevant contextual features by observing user's behaviour in an unobtrusive way. Time has been traditionally used to model context in recommender systems. For instance in [2], time, among other parameters, defines the context of the user when recommending the type of music the user prefers in different day or week periods. However, in this paper we have focused on using time as a reliable implicit indicator to estimate user's ratings. In this line we find in [9] a CF approach for e-commerce websites, in which the time when a product in an e-commerce platform was launched and the time when it was purchased by the user are stored. However, it deals with virtual environments (e-commerce websites), which tend to be easier to monitor, while our approach tackles the use of sojourn times in real-world environments in which dynamics may entail more complexity.

To our knowledge there does not exist any approach using sojourn times to infer visitor preferences in museums. Nevertheless, there are some relevant contributions related to our work. For example, MusA [13] is a generic framework that allows building multimedia guides for mobile devices, including a vision-based indoor positioning system. Moreover, it provides thematic tours designed by professional curators or museum staff. Another interesting work is SmARTweet [5], a location-based approach that detects exhibits near visitors and then displays their corresponding information (history, art period, etc.) using multimedia resources. Furthermore, the application incorporates a collaborative filtering mechanism to recommend personalized exhibits based on the visitors' interests (collected from questionnaires) and visitors' behaviour collected by tracking the users throughout the museum. In the same line, UbiCicero [6] is a system that suggests recommendations based on the detection of nearby exhibits by using RFID readers assembled in mobile devices. A more complex recommender system for mobile devices for museum visitors can be found in [3]. This system takes into account different contextual features such as the visitor location, expertise and time to adapt the recommendations to her/his interests. It uses a hybrid recommender mechanism based on collaborative filtering combined with a post-filtering semantics-based approach.

6 Conclusions

In this paper, we have presented a recommendation approach that is adapted to the situation in large facilities, where we interpret the user's sojourn times as their implicit feedback. We introduced a beacon infrastructure for tracking user movements, which we have already implemented in a museum of natural history. We showed how movement data must be preprocessed for recommendation algorithms, especially how different user behavior can be taken into account. We adapted both User-based and Item-based CF so that they can deal with sojourn times. The main advantage of out approach is that it can provide qualified recommendations without any explicit user feedback. As future lines of research we want to transfer our approach to other classic recommendation algorithms such as Content-based Filtering. Another idea is to apply machine learning approaches on user's movement data for deriving semantic recommendation rules.

Acknowledgments. This work has been supported by the projects DGA-FSE, TIN2015-65515-C4-4-R, TIN2016-78011-C4-3-R and Universidad de Zaragoza - Ibercaja-CAI fellowship IT 9/17.

References

1. Adomavicius, G., Tuzhilin, A.: Context-aware recommender systems. In: Ricci, F., Rokach, L., Shapira, B., Kantor, P.B. (eds.) Recommender Systems Handbook, pp. 217–253. Springer (2011)
2. Baltrunas, L., Amatriain, X.: Towards time-dependant recommendation based on implicit feedback. In: Workshop on Context-Aware Recommender Systems (CARS 2009) (2009)
3. Benouaret, I., Lenne, D.: Personalizing the museum experience through context-aware recommendations. In: 2015 IEEE International Conference on Systems, Man, and Cybernetics, pp. 743–748 (2015)
4. Billhardt, H., Dunkel, J., Fernandez, A., Lujak, M., Hermoso, R., Ossowski, S.: A proposal for situation-aware evacuation guidance based on semantic technologies. In: 4th International Conference on Agreement Technologies (AT2016), Valencia, vol. 2016, pp. 493–508 (2016)
5. Chianese, A., Marulli, F., Moscato, V., Piccialli, F.: SmARTweet: a location-based smart application for exhibits and museums. In: International Conference on Signal-Image Technology and Internet-Based Systems (SITIS), pp. 408–415. IEEE Computer Society (2013)
6. Ghiani, G., Paternò, F., Santoro, C., Spano, L.D.: Ubicicero: a location-aware, multi-device museum guide. Interact. Comput. **21**(4), 288–303 (2009)
7. Hermoso, R., Dunkel, J., Krause, J.: Situation awareness for push-based recommendations in mobile devices. In: 19th International Conference on Business Information Systems (BIS), Leibzig, vol. 2016, pp. 117–129 (2016)
8. Hu, Y., Koren, Y., Volinsky, C.: Collaborative filtering for implicit feedback datasets. In: 2008 Eighth IEEE International Conference on Data Mining, pp. 263–272, December 2008
9. Lee, T.Q., Park, Y., Park, Y.-T.: A time-based approach to effective recommender systems using implicit feedback. Expert Syst. Appl. **34**(4), 3055–3062 (2008)

10. Lujak, M., Billhardt, H., Dunkel, J., Fernández, A., Hermoso, R., Ossowski, S.: A distributed architecture for real-time evacuation guidance in large smart buildings. Comput. Sci. Inf. Syst. **14**(1), 257–282 (2017)
11. Perera, C., Zaslavsky, A., Christen, P., Georgakopoulos, D.: Context aware computing for the internet of things: a survey. IEEE Commun. Surv. Tutorials **16**(1), 414–454 (2014)
12. Resnick, P., Varian, H.R.: Recommender systems. Commun. ACM **40**(3), 56–58 (1997)
13. Rubino, I., Xhembulla, J., Martina, A., Bottino, A., Malnati, G.: MusA: using indoor positioning and navigation to enhance cultural experiences in a museum. Sensors **13**(12), 17445–17471 (2013)
14. Yürür, O., Liu, C.H., Sheng, Z., Leung, V.C.M., Moreno, W., Leung, K.K.: Context-awareness for mobile sensing: a survey and future directions. IEEE Commun. Surv. Tutorials **18**(1), 68–93 (2016)

Using Queueing Networks to Approximate Pedestrian Simulations

Ismael Sagredo-Olivenza(✉), Marlon Cárdenas-Bonett, and Jorge J. Gómez-Sanz

Department of Software Engineering and Artificial Intelligence, Research Group on Agent-Based, Social and Interdisciplinary Applications, Complutense University of Madrid, Madrid, Spain
{isagredo,marlonca,jjgomez}@ucm.es

Abstract. Pedestrian or crowds simulation is a complex and expensive task that involves a plethora of technologies. In certain exigent environments, for example, whether we want to use a complex simulation in a machine learning system, in real-time decision making or when the user does not need the details of the simulation, this computational cost may not be desirable. Having simpler models is useful if you want to use these simulations in those exigent environments or we just want to obtain approximate calculations of these simulations quickly. In this paper, we propose a simplified model of simulation based on a network of configurable queues that helps us to approximate the results of a complex simulation in a very short time, while maintaining a high representativeness of the real simulation.

Keywords: Crowd simulation · Queueing theory
Simulation as a service

1 Introduction

Crowd or pedestrian simulation is a challenging application area for agent-based modeling and simulation [4,11]. Exist an active field of research about this topic thanks to its utility in several applications such as evacuations planning, designing and planning pedestrian areas, subway, designing complex sport-stadiums and entertainment among others. In the past to extract information from a simulation, the user needed a plethora of evaluation methods to apply, for example, the direct observation, photographs, and time-lapse films. All of these methods needed the human intervention. Nowadays, it is usually used multi agent-system models as a way to face this task [8,9] and although several systems have even been commercialized, for example: Vadere[1], Pedestrian Dynamics[2] or PEDSIM[3]. All of them spend a lot of time carrying out the simulation.

[1] http://www.vadere.org/.
[2] https://www.incontrolsim.com/product/pedestrian-dynamics/.
[3] http://pedsim.silmaril.org/.

S. Rodríguez et al. (Eds.): DCAI 2018, AISC 801, pp. 132–139, 2019.
https://doi.org/10.1007/978-3-319-99608-0_15

Consequently, the use of them becomes prohibitive in certain environments. For example, on the one hand, whether we use a complex simulation as the function in a reinforcement learning [6] or in the fitness function in a genetic algorithm, the time expended in both examples whether we used a complex simulation can be excessive. On the other hand, it is not always necessary to have markedly reliable simulations. Sometimes you look for speed, even if the information provided is not so accurate.

The aim of this work is proposing a fast model of crowd simulation that allows, with little computational effort, to create low-cost simulations that allow the use of these models in very demanding computing environments, such as the methods mentioned above. Another reason to use this simpler model is whether the user wants to make a quick evaluation of the principal characteristics of the simulation roughly, without going into detail of what is the individual behavior of each agent that belong the simulation.

To accomplish this goal, we have used the fundamentals of the queueing theory [1], to modeling the environment and the movements of the pedestrians in a simulated environment. The queueing theory is widely used in multiples dominoes because is one of the most typical phenomenon which is observed in the cities, buildings, supermarkets, banks, a concert hall and many other human activities [12].

The remainder of this paper is organized as follows. Next, we show our approach proposed in Sect. 3. In the Sect. 4 we show the experimental results achieved and finally in Sect. 5, we present some conclusions and future works.

2 Background

The queueing network models occur throughout many parts of the world. These networks must be designed and controlled in order to simulate the flow pedestrian traffic in many human processes like super-market, bank or any other producer-consumer system. But can modeling more complex environment like traffic control for signalized intersections, vehicular and pedestrian network evacuation, and many other situations [3].

A simple queue system can be defined as a Fig. 1. In this figure, it can see a queering buffer of finite or infinite size. This queue it is filled with the customers that are arriving. One o more server are serving customers, extracting them from the queue. Each server may expend a specific time to serve the customer.

The queueing network model is widely used in the literature to simulate a lot of human processes. For example, it has been used in [2] that using models M/G/C/C for the analysis of pedestrian traffic flows or in [10] where the authors propose networks of queues based on the routes established by pedestrians or vehicular circulation. And finally in [7] where the authors using queueing network in combination with a social force algorithm to study the dynamics of crowded movement.

Keeping these in mind We propose, based on these previous works, to use queueing networks to simulate the behavior of the pedestrians that walking

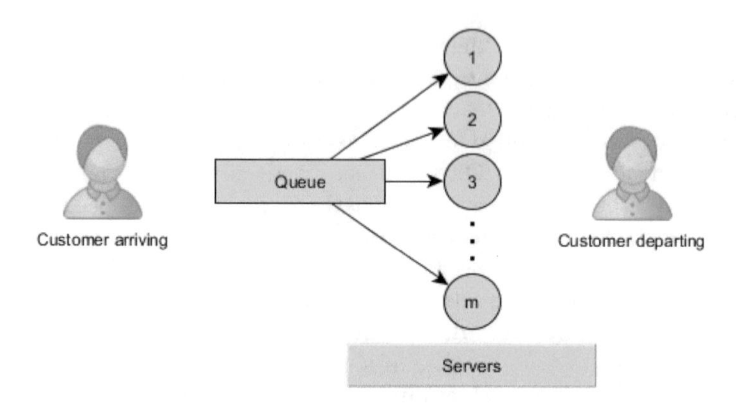

Fig. 1. Sketch of a elementary queue system

through different places, but instead of using probabilistic models of client arrival, we simulate these arrivals with the aim to improving fidelity from the simulation. The particularities of the model used are detailed in the next section.

3 Proposed Approach

As we have explained before, the proposal of this work is to get a simplified model that abstracts the main aspects of the pedestrian simulation with the aim to improve the performance of this simulation when the execution time is critical, keeping the simulation fidelity high.

This simplified model abstracts the spaces of a building using a undirected graph where on one hand, the nodes are the spaces that the pedestrians can transit and on the other hand, the edges of the graph are doors and others structures that communicate the spaces among them. We assume, for simplicity, that these structures can be traversed by the pedestrian in zero units of time. Each node internally is modelled with a queue of a specific size. Although the queuing theory is very successful and has been used to study queuing systems and simulations, it does not consider the effect of walking distance between two nodes [12]. The time that pedestrians expend walking between two points is modelled in our system using a variable that establishes the time that an individual takes to cross the node. In order to simplify the definitions, we assume that all individuals walk with the same constant velocity. But this system can works properly with a multiplier of this velocity for each individual, to model different capacities of the pedestrians.

The pedestrians will be enqueue when they arrive at a node. Furthermore, each edge in the graph can extract a number of pedestrians buffered in a node for each time interval. Each pedestrian has a path to follow defined in the simulation and will select the edge by which they will leave a node. A pedestrian can abandon its current node when it stays in this node enough time to traverse it. Among the candidates to leave the node in a certain time interval by a particular edge, the edge allows the passage to the next node only a number of pedestrians equal to its capacity at most. Figure 2 shows a schema of our queueing network system.

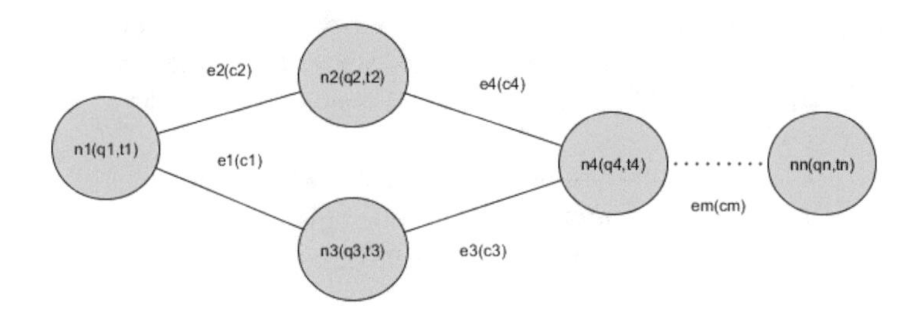

Fig. 2. Example of a queueing network

Taking it into a count, we define a queueing network $G(N, E)$ as an undirected graph composed of a set of nodes $N = \{n_1, n_2, ...n_n\}$ that represent the different partitions of the space to be simulated. These partitions can correspond to rooms, corridors or other instances of the surface where pedestrians walk or be created by the designer of the simulation based on other criteria. And a set of edges $E = \{e_1, e_2...e_m\}$ that represent the interconnection between two nodes $n_i, n_j \in N$ such that:

$$e_k : n_i \leftrightarrow n_j \quad \text{where} \quad e_k \in E \tag{1}$$

Each node has a pair of values q_i, t_i that represents the queue size (q) and the average time to cross the node (t) take into account a default pedestrian speed 1.25 m/s [5].

$$n_i(q_i, t_i) \quad \forall \quad n_i \in N \tag{2}$$

Each edge has a parameter (c) that represents the number of pedestrians that can move from the current node to the next node in each time interval.

$$e_k(c_k) \quad \forall \quad e_k \in E \tag{3}$$

One pedestrian $p_k \in P = \{p_i, p_2...p_p\}$ can be defined as a tuple of consecutive movements $p_k = \{n_1, n_2, ...n_k\}$. Each movement move the pedestrian front the current node n_c to the next n_i when the pedestrian has waited a certain time $\overline{W_k}$.

Then, a movement of a certain pedestrian p_k that want move from n_c to n_i having stayed at the node n_c during $\overline{W_k}$ units of time can be defined as $(p_k \rightarrow M(n_c(q_c, t_c), n_i(q_i, t_i), \overline{W_k}))$ and this movement is possible if:

$$M(n_c, n_i, \overline{W_k}) \rightarrow \begin{cases} \exists e_k(c_k) \in E | e_k : n_c \leftrightarrow n_i \quad \text{and} \\ \overline{W_k} \geqslant t_c \quad \text{and} \\ |n_i| < q_i \quad \text{and} \\ |e_k| < c_k \end{cases} \tag{4}$$

Where $|n_i|$ is the current size of the queue in the node n_i and $|e_k|$ is the number of pedestrian transit by the edge e_k before does it p_k in this time interval.

This model permits us, configuring properly the parameters q and t for each node and c for each edge, modeling several environments to simulate the pedestrian flow without going into details like a real simulation would do. Since the movement of individuals is a simulated movement, we can use different speeds for each one and create different waves of pedestrians without the need to use probabilistic models.

But is important to measure whether this model can modeling with enough precision the behavior of a more complex simulation and how much deviates from the simulation. In the next Section, we describe an experiment who tries to clarify this.

4 Experiments

The aim of the experiment is to test whether our approximation using a queueing network allows increasing the performance of the simulation, keeping high the representativeness of the real simulation.

The experiment measured the number of pedestrians detected in the different partitions realized on the first floor of the informatics faculty at the Complutense University of Madrid. We have carried out two simulations. First, with a complex graphics simulator and next with our simplified model. The simulation consisted of the entrance to the faculty of 125 students at the first hour of the morning. All students enter through the main door. There are 25 students per classroom, five classes and all of them are filled, when the simulation has finished.

Figure 3 shows the structure of the first floor of the faculty and the space partitioning that has been made of it. We have delimited the space trying that all the exits of a node have a similar distance among them. That it is important because the time of a pedestrian takes to cross a node in our model is the same regardless by which edge has entered to the node or by which edge will exit.

With the aim to tunning the queueing network parameters, we measure in the graphical simulation the different parameters of the node. For example, the time that expended a pedestrian to cross a node, the size of the doors or the capacity of the different partitions realized.

When the model was configured, we executed the same simulation in both simulators (the graphical simulator and our queueing network model). The aim of this comparison is measure whether the number of pedestrians detected in both simulators are similar and to determinate whether if these differences are significant.

To carry out the experiment, the graphical simulation was run during 50 s. On the contrary, the queueing network model took only a few seconds to complete the same simulation time as the graphical simulator.

The results of the experiment are shown in the Fig. 4

As it can observe in the experiment. In the most of the nodes, the results are similar between the simulator and the queueing network. But exist on certain

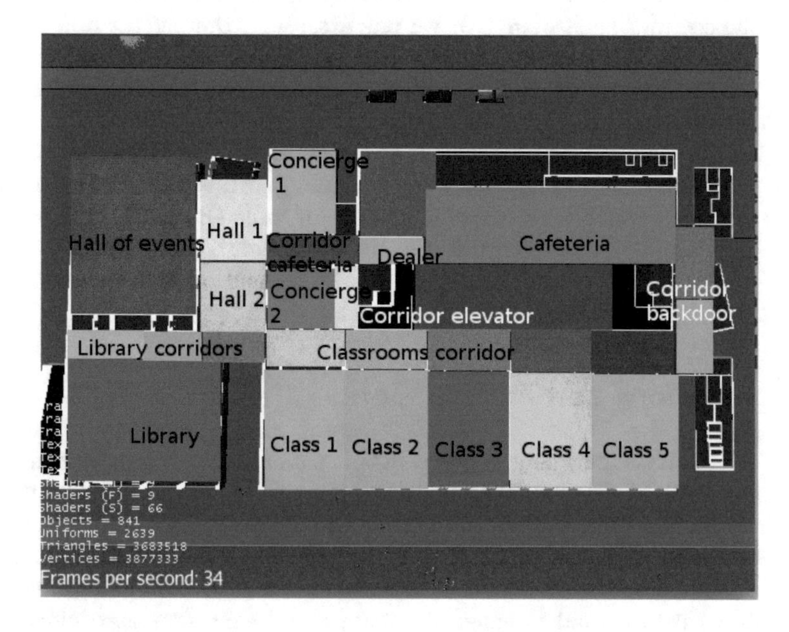

Fig. 3. The space partition of the faculty floor one using queueing network

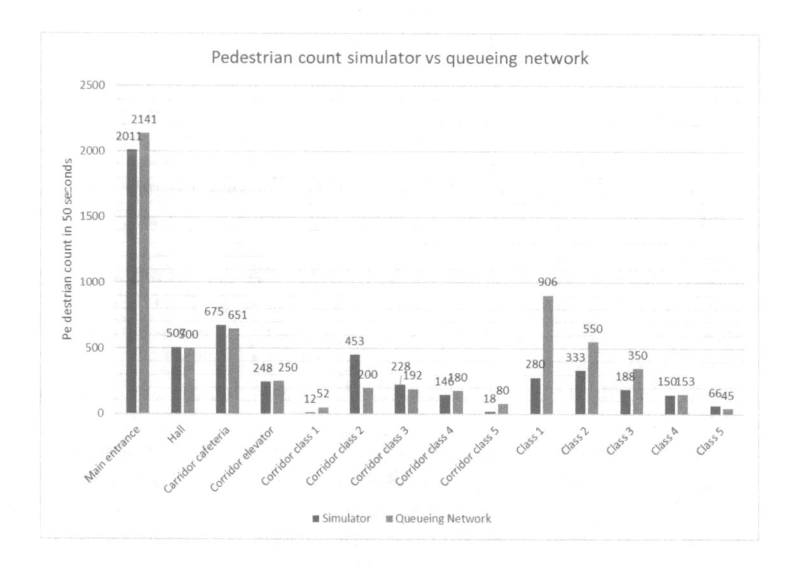

Fig. 4. Experiment results comparing the simulation and the queueing network

nodes where the differences are important. For example in classroom 1, classroom 2 or in the corridor classroom 1. If we pay attention, the biggest differences in the number of pedestrians are concentrated in the final nodes of the pedestrian routes, for example in the classrooms. This is because the differences between the real simulation and our model are accumulated and magnified over time.

It seems that the configuration of the parameters by hand based on a simulation does not allow sufficiently to specify the configuration values of the network in all nodes. So it would be interesting to look for some automatic system to calculate these weights of the network. On the other hand, the graphics simulator that we have used has some problems with the navigation of the characters, so it also generates errors derived from this fact.

5 Conclusions and Future Works

In the preliminary experiment exposed in this paper, we have a promising result about the utility of this model to approximate a simulation with our queueing network. However, some nodes have been approximated better than others. The hand-made tunning of the parameters is not enough to achieve the best configuration to modeling the system. But the results are good enough to ensure the model can be used to make a good approximation to the simulated model.

As a future work, we want to calculate an optimal configuration of this network model using search techniques. For example, genetic algorithms.

It seems probable that a genetic algorithm may be able to find the configuration of the network parameters that match the results obtained in the counting of the individuals in each room. With this technique, we could allow building a model much closer to the simulator that we intend to approximate.

References

1. Bolch, G., Greiner, S., de Meer, H., Trivedi, K.S.: Queueing Networks and Markov Chains: Modeling and Performance Evaluation with Computer Science Applications. Wiley, New York (2006)
2. Cheah, J.Y., Smith, J.M.: Generalized M/G/C/C state dependent queueing models and pedestrian traffic flows. Queueing Syst. **15**, 365–386 (1994)
3. Cruz, F.R., Smith, J.M., Medeiros, R.: An M/G/C/C state-dependent network simulation model. Comput. Oper. Res. **32**(4), 919–941 (2005)
4. Helbing, D., Molnar, P.: Social force model for pedestrian dynamics. Phys. Rev. E **51**(5), 4282 (1995)
5. Knoblauch, R., Pietrucha, M., Nitzburg, M.: Field studies of pedestrian walking speed and start-up time. Transp. Res. Rec. **1538**(1538), 27–38 (1996)
6. Martinez-Gil, F., Lozano, M., Fernández, F.: Multi-agent reinforcement learning for simulating pedestrian navigation. In: International Workshop on Adaptive and Learning Agents, pp. 54–69. Springer (2011)
7. Mu, H.: A study of left-right preference of pedestrian counter flow in passageways. In: Transportation Research Board 95th Annual Meeting Transportation Research Board (2016)

8. Pan, X., Han, C.S., Dauber, K., Law, K.H.: A multi-agent based framework for the simulation of human and social behaviors during emergency evacuations. Ai Soc. **22**(2), 113–132 (2007)
9. Pelechano, N., Allbeck, J.M., Badler, N.I.: Controlling individual agents in high-density crowd simulation. In: Proceedings of the 2007 ACM SIGGRAPH/Eurographics Symposium on Computer Animation, Eurographics Association, pp. 99–108 (2007)
10. Smith, J.M.: Application of state-dependent queues to pedestrian/vehicular network design. Oper. Res. **42**(3), 414–427 (1994)
11. Ulicny, B., Thalmann, D.: Towards interactive real-time crowd behavior simulation. In: Computer Graphics Forum, vol. 21, pp. 767–775. Wiley Online Library (2002)
12. Yanagisawa, D., Suma, Y., Tanaka, Y., Tomoeda, A., Ohtsuka, K., Nishinari, K.: Methods for improving efficiency of queuing systems. In: Peacock, R.D., Kuligowski, E.D., Averill, J.D. (eds.) Pedestrian and Evacuation Dynamics, pp. 297–306. Springer, Boston (2011)

An Agent-Based Simulation Model for Emergency Egress

Álvaro Carrera, Eduardo Merino, Pablo Aznar, Guillermo Fernández, and Carlos A. Iglesias(✉)

Intelligent Systems Group, Universidad Politécnica de Madrid,
Av. Complutense, 30, 28040 Madrid, Spain
{a.carrera,eduardo.merinom13,pablo.aznar,g.fernandeze,
carlosangel.iglesias}@upm.es
http://www.gsi.dit.upm.es/

Abstract. Unfortunately, news regarding tragedies involving crowd evacuations are becoming more and more common. Understanding disasters and crowd emergency evacuation behaviour is essential to define effective evacuation protocols. This paper proposes an agent-based model of egress behaviour consisting of three complementary models: (i) model of people moving in a building in normal circumstances, (ii) policies of egress evacuation, and (iii) social models for integrating models (e.g. affiliation) that explain the social behaviour and help in mass evacuations. The proposed egress model has been evaluated in a university building and the results show how these models can help to better understand egress behaviour and apply this knowledge for improving the design and execution evacuation plans.

Keywords: Agent-based social simulation · Evacuation protocol
Emergency egress · Affiliation model

1 Introduction

Unfortunately, tragedies involving crowd evacuation appear frequently in the news, including natural and man-made disasters, such as massive parties, terrorist attacks and sports events. In these cases, many people become injured by the chaos produced because they did not know how to manage these situations. Thus, it is important to progress on the study of these situations learning from the past experiences and improving emergency plan strategies.

Emergency evacuation, known as egress, is a critical component of emergency response and requires developing in advance evacuation preparation activities ensuring people can get to safety in case of emergency. In order to define effective evacuation protocols, understanding disasters and crowd emergency evacuation behaviour is essential [1].

S. Rodríguez et al. (Eds.): DCAI 2018, AISC 801, pp. 140–148, 2019.
https://doi.org/10.1007/978-3-319-99608-0_16

There are many research theories about mass psychology with the objective of understanding crowd behaviours in emergencies from several points of view: decision making, exit times, clinical issues and crowd behaviour [1]. For example, one of the most influential is exit selection and the time it takes to evacuate. Aspects, such as their familiarity with exits or their visibility are very important to choose the exit way. People personality has become another relevant aspect because it also influences to follow or cooperate with other users when they have to select the exit [6].

From another point of view, there are a lot of models and theories. Some previously mentioned studies [1,6] have analysed clinical issues, such as freezing or becoming disassociated from reality, which are also potentially dangerous. However, those researchers found other interesting findings; around 50% of emergency survivors referred unambiguously to a sense of unity or togetherness with the rest of the crowd during the emergency. Also, there is one model which explains why family groups often escape or die together, this is the affiliation model [1].

A full-scale evacuation demonstration is not viable because of ethical, practical and financial issues [8]. Therefore, models and simulations of crowd behaviour are widely used to analyse the effectiveness of evacuation preparation activities. Different computer-based simulation approaches are used in the literature for evacuation, such as flow dynamics [7] or cellular automate [5]. However, agent-based simulation (ABS) has been used as the preferred method to simulate crowd behaviour [14], because agents are particularly suitable for modelling human behaviour.

This paper proposes the use of agent-based social simulation for modelling an egress evacuation through three complementary models: (i) model of people moving in a building in normal circumstances, (ii) policies of egress evacuation, and (iii) social models for integrating models (e.g. affiliation) that explain the social behaviour and help in mass evacuations. The proposed egress model has been evaluated in a university building.

The remainder of this paper is structured as follows. In Sect. 2 the different simulation models analysed are presented. Whereas Sect. 3 shows the scenario used in the experimentation, the evaluation methodology and the results. Finally, a general discussion of findings and future research is presented in Sect. 4.

2 Simulation Model

The design of the simulation model is based on three complementary models. Firstly, we model the behaviour and daily routines of the occupants. Second, we model the behaviour and actions of the occupants in response to the threat and during the evacuation. This model is organised in two models, a model of the different policies for leaving the building and a model of the different social behaviours relevant for emergencies such as the affiliation model [4] or mobility disabilities [2]. The different models that are studied are implemented in the simulation by defining some parameters, such as new states and positions or different movement speeds.

Crowd Modelling in Buildings. The behaviour and activity of the occupants in the building are represented through occupancy agents, which can be defined using real patterns and characteristics. Agents have been modelled using probabilistic state machines. The states are the main engine to model the actions of people in the building. The occupancy agents performance in the building is controlled by the simultaneous action of schedules and Markov chains, in contrast with conventional static schedules.

Emergency Policy Models. When an emergency situation occurs, the occupants who are performing normal daily actions have to face it. According to [13], during the impact period, from 10–25% of the people remain united and calm, 75% manifest disorderly behaviour, bewilderment and 10–25% show confusion and panic. So we have modelled the agents in the worst case, they take sometime to react. When they notice that the emergency happens they try to go out. For this goal we have implemented three ways or policies for leaving the building.

1. *Nearest gate.* By means of this policy, the occupants leaving the building use the nearest gate when the emergency occurs. This technique could be the fastest way to evacuate the building, but it is not the safest considering that the fire could be in the way to this exit. If the fire progresses quickly and it is near from one of the exits, this fact could be dangerous if the agent tries to evacuate in that direction.
2. *Safest gate.* In this case, the occupants leave the building by the furthest exit in relation to the initial position of the fire. The path to this safest exit could be dangerous if this path contains any fire position.
3. *Less crowded gate.* The occupants, either quickly evaluating the decisions of the rest of the occupants or using information provided by a system of the building, knows an approximate value of the agglomeration at the exits and decides to go to the less crowded gate.

Social Models in Emergencies. On the other hand, two social behaviours are highlighted. Firstly, affiliation model has been chosen according to [14] because it is one of the most important crowd behaviour together with different kinds of reactions when an emergency happens. Other works [14] have observed that when there is an evacuation, families try to exit together by the same exit, meet at a point or someone take care of the youngest member and the rest leave the building. So in this project, the last behaviour is modelled. When occupants are defined they could belong to a family, each family have one child and one parent. Not every agent belongs to a family only a 20% have been modelled in this way to analyse the different results between agents with family and individual agents. When the emergency happens, each member of a family try to leave the building using the same way as his family whereas the child of each family stay stopped in his position waiting to his parent. This parent mentioned previously look for the child and go to the child's position. When they are together they try to leave the building in the same way as their family did. On the other hand, individual agents leave the building in their own way. An example of the different Markov states, the decisions and the probabilities assigned to them can be seen in Fig. 1.

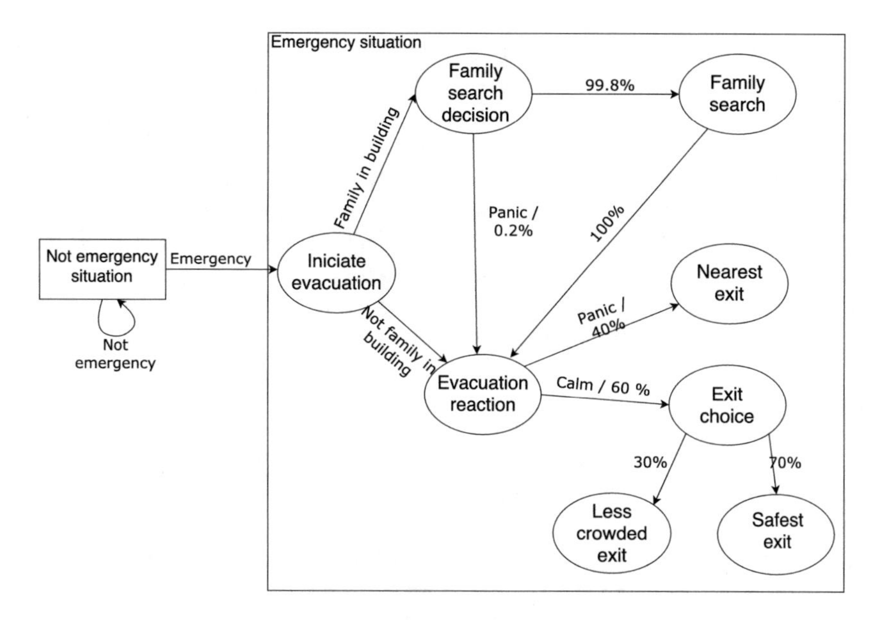

Fig. 1. Example of an emergency behaviour of the occupation

Secondly, modelling people with mobility disabilities, such as old people, is considered. These disabilities allow us to treat this kind of agents in a different way because their velocity is slower than normal agents and they could take more time to react when there is an emergency.

3 Experimentation

3.1 Implementation and Scenario

Recently a large number of Agent Based Social Simulation (ABSS) platforms have been made available, as recently surveyed in [11]. The most popular ones, MASON, NetLogo [12] and Repast [3] are developed in Java and provide general facilities for developing agent systems and its visualisation. In a previous work regarding evacuation, we selected UbikSim [10], which extends MASON and adds facilities for indoor visualization as well as an agent behaviour based on probabilistic state machines. Given the popularity of the Python ecosystem and its growing use for machine learning, we have opted for Mesa [9] in this work, which is developed in Python, includes a browser-based visualization module and allows using IPython notebooks for running the simulations and analyzing the simulation data. Since Mesa does not provide specific facilities for indoor modeling and visualization, we have extended Mesa as explained in this paper.

The system has been evaluated in a real-life scenario, the ground floor of Building B of the School of Telecommunication Engineers (ETSIT) of the Universidad Politécnica de Madrid (UPM). This floor has a rectangular shape and an area of 1600 m^2. The building is formed by four types of room: offices, laboratories, classes and transition spaces, such as halls, corridors or resting areas. The floor is divided into 41 rooms distributed as 14 offices, 20 laboratories and 4 classes and 1 hall and 2 corridors. There are two available exits, one to exit from building and another to enter to other connected building.

In order to model crowd and people in a normal situation at the university, firstly some data was collected. The collection was performed by a survey answered by 25 people, which had the aim to find the main habits of the people at the university. Some questions were about work time entry, stops for lunch and time in each dependency (classes, labs...).

3.2 Evaluation

The aim is to find the best policy of evacuation in each situation and type of occupant. This project is focus on:

1. *Exit time*: This time covers from the beginning of the emergency situation until the occupant has left the building, if he has succeeded.
2. *Number of dead agents*: This metric inform us about the number of agents that have died during the simulated evacuation.

Furthermore, five research questions are suggested, which are answered with averaged data obtained by successive simulations, executed with 50 agents. The questions raised are presented below together with the experimentation results.

RQ1: Does emergency time affect the evacuation? To answer this question, the simulation is made in three different hours. The obtained results are: at 10:20, 22 agents were in the building and 1 dead; at 12:00, 41 agents in the building and 2 dead, and at 3:00 p.m., 10 agents in the building and none dead. Depending on the hour there will be more or less agents in the building.

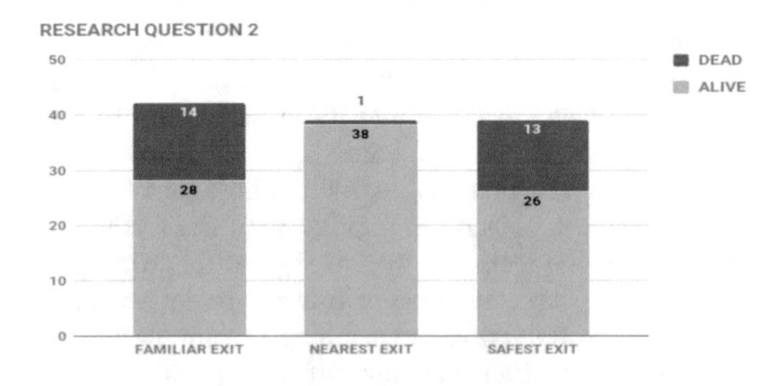

Fig. 2. RQ2 results

RQ2: What policy is more effective saving people? In this case, three simulations are made because there are three different policies: familiar exit, nearest exit and safest exit. The following chart shows the number of agents alive and dead in each situation. The total number differs due to the probabilistic model based on stochastic Markov chains. As can be seen in Fig. 2, it can be concluded that the best policy to save people is 'Nearest exit'.

RQ3: Which policy gets the best evacuation time? In this case study, an evaluation of which policy leads to the best evacuation time is made. With this aim, the simulation features are the same than in RQ2. The obtained results, which are measured as the average time of evacuation in seconds, are 40,14 s using the *Familiar exit* policy; 42,77 s using the *Safest exit* and 27,65 s using the *Closest exit.*

Firstly, the best policy to get the best exit time is the 'Nearest exit' policy. On the other hand, the other policies have a similar exit time considering that can be the same exit in many cases.

RQ4: How do mobility problems impact on the evacuation? The simulations are made in this study including agents representing elderly people (50%), which walk slower than the general ones. A comparison between young agents and elderly agents using the nearest exit policy is presented in Figs. 3 and 4.

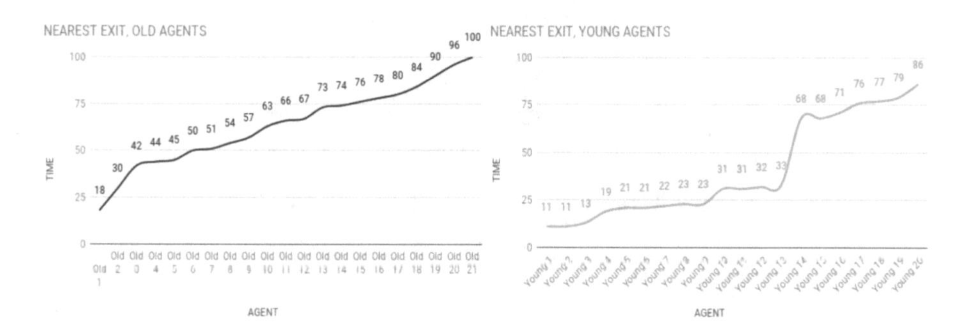

Fig. 3. RQ4 results: elderly agents

Fig. 4. RQ4 results: young agents

As it is obvious, elderly agents need more time to leave the building. However, some young agents are delayed by the elderly agents due to help them and follow them to the exit. This fact made young people more vulnerable to be dead as we can see in the chart in Fig. 5. In this chart can be also seen that elderly agents have more have more probability of dying. Specifically, the average time to leave the building is 63,71 s (elderly agents) and 40,80 s (general agents).

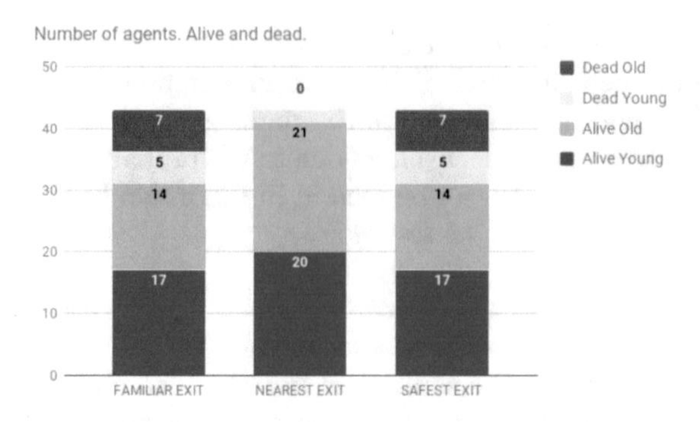

Fig. 5. RQ4 results: number of dead and alive agents by type

RQ5: How do family ties affect in the evacuation? In this case, two families, which are formed by one child, one parent and another three agents, are included in the simulation. The results are evaluated comparing familiar exit and nearest exit policies. The data corresponding to both cases is presented in Table 1.

Table 1. RQ5 results

	Exit time (seconds)			
	Familiar exit		Nearest exit	
Member	Family 1	Family 2	Family 1	Family 2
Parent	41	75	43	Dead
Child	44	74	45	Dead
Member 3	Dead	77	41	30
Member 4	46	Dead	43	36
Member 5	51	Dead	52	Dead

With the information of both cases, it can be affirmed that family ties affect the evacuation time, and also increase the probability of death. This fact is due to when the parent goes to look for the child they delay the exit time and the threat could be increased.

4 Conclusions and Future Work

In this study, an evacuation simulator has been developed in order to study effective evacuation protocols. The aim is to analyse the best policy to leave the building defining and modelling agents with different features. Specifically, two social behaviours are used: affiliation model based on family relationships and modelling of people with mobility disabilities. In addition, three different strategies or policies to abandon the building have been studied.

First, Nearest gate, the occupants leave the building using the nearest gate when the emergency occurs. Second, Safest gate, the occupants leave the building by the farthest exit in relation to the initial position of the fire. Finally, less crowded gate, the occupants, either quickly evaluating the decisions of the rest of the occupants or using information provided by a system of the building about the less crowded gate.

In this paper, five research questions are suggested, which provide five main conclusions. First, the number of dead agents depends on the time the emergency occurs, being the worse scenario in hours of a greater concurrence of people in the building. Second, the best policy or strategy to leave the building is the 'Nearest exit' policy. Third, elderly occupants need more time to leave the building, and also some young occupants are delayed by the elderly agents. Finally, family ties affect in the evacuation time, and also increase the probability of dead.

The models considered, the policies proposed, and the results obtained are a useful aid to define new evacuation strategies and highlight the most relevant aspects that should be considered during their planning, determining the effective ways of setting appropriate evacuation policies.

Acknowledgements. This work is supported by the Spanish Ministry of Economy and Competitiveness under the R&D projects SEMOLA, EmoSpaces (RTC-2016-5053-7) and ITEA3 Citisim (ITEA3 15018, funded by CDTI), by the Regional Government of Madrid through the project MOSI-AGIL-CM (grant P2013/ICE-3019, co-funded by EU Structural Funds FSE and FEDER); and by the Ministry of Education, Culture and Sport through the mobility research stay grant PRX16/00515.

References

1. Challenger, W., Clegg, W.C., Robinson, A.M.: Understanding crowd behaviours: guidance and lessons identified. UK Cabinet Office, pp. 11–13 (2009)
2. Christensen, K., Sasaki, Y.: Agent-based emergency evacuation simulation with individuals with disabilities in the population. J. Artif. Soc. Soc. Simul. **11**(3), 9 (2008)
3. Collier, N.: Repast: an extensible framework for agent simulation. Univ. Chicago's Soc. Sci. Res. **36**, 2003 (2003)
4. Drury, J., Cocking, C., Reicher, S.: Everyone for themselves? A comparative study of crowd solidarity among emergency survivors. Brit. J. Soc. Psychol. **48**(3), 487–506 (2009)
5. Fromm, J.: The Emergence of Complexity. Kassel University Press, Kassel (2004)
6. Guy, S.J., Kim, S., Lin, M.C., Manocha, D.: Simulating heterogeneous crowd behaviors using personality trait theory. In: Proceedings of the 2011 ACM SIGGRAPH/Eurographics Symposium on Computer Animation, pp. 43–52. ACM (2011)
7. Gwynne, S., Galea, E.R., Owen, M., Lawrence, P.J., Filippidis, L.: A review of the methodologies used in the computer simulation of evacuation from the built environment. Build. Environ. **34**(6), 741–749 (1999)

8. Liu, Y., Sun, C., Wang, X., Malkawi, A.: The influence of environmental performance on way-finding behavior in evacuation simulation. In: Proceedings of the BS2013: 13th Conference of International Building Performance Simulation Association, Le Bourget Du Lac, France, pp. 25–30 (2013)
9. Masad, D., Kazil, J.: Mesa: an agent-based modeling framework. In: 14th PYTHON in Science Conference, pp. 53–60 (2015)
10. Sánchez, J.M., Carrera, Á., Iglesias, C.Á., Serrano, E.: A participatory agent-based simulation for indoor evacuation supported by google glass. Sensors **16**(9), 1360 (2016)
11. Sánchez, J.M., Iglesias, C.A., Sánchez-Rada, J.F.: Soil: an agent-based social simulator in python for modelling and simulation of social networks. In: LNAI, vol. 10349, pp. 234–245. Springer, June 2017
12. Tisue, S., Wilensky, U.: Netlogo: a simple environment for modeling complexity. In: International Conference on Complex Systems, Boston, MA, vol. 21, pp. 16–21 (2004)
13. Vega, M.F.: Ntp 390: La conducta humana ante situaciones de emergencia: análisis de proceso en la conducta individual. Instituto Nacional de Seguridad e Higiene en el, Trabajo, pp. 1–11 (1993)
14. Wolfram, S.: A New Kind of Science, vol. 5. Wolfram Media, Champaign (2002)

On the Use of Elevators During Emergency Evacuation

Qasim Khalid[1(✉)], Marin Lujak[2], Alberto Fernández[1], and Arnaud Doniec[2]

[1] CETINIA, University Rey Juan Carlos, Madrid, Spain
{qasim.khalid,alberto.fernandez}@urjc.es
[2] IMT Lille Douai, Lille, France
{marin.lujak,arnaud.doniec}@imt-lille-douai.fr

Abstract. In this paper, we study the use of elevators during the time of emergency evacuation. We review the State-of-the-Art and analyze the risks, benefits, and safety measures associated with the use of elevators during hazardous situations. In the end, we discuss some open issues related with the use of elevators in emergency evacuation and propose a smart agent-based elevator system (SABES) along with some research directions towards safer, more responsive, and reliable usage of elevators based on the emergency context.

Keywords: Elevators · Building evacuation · Lifts
High-rise buildings · Human behavior

1 Introduction

Imminent hazards that cannot be eliminated without endangering occupants of a building such as the acts of terrorism and natural disasters, e.g., earthquakes, tsunamis or floods, pose elevated risk to life hazard or risk of injury especially in high-rise buildings where the occupants cannot evacuate fast enough. Since the severity, location, and dynamics of these events cannot be predicted with any degree of certainty, an open issue in high-rise buildings is how to facilitate safe and timely evacuation in the case of imminent danger.

During evacuation in highly crowded spaces, an evacuation approach has to consider building safety, crowd movements, congestion, and human factor (see, e.g., [11,23,24]). Lately, many architectural advancements have been made in high rise buildings for evacuation, as, e.g., the installation of escape chutes (special kinds of emergency exits), external escape rescue systems, and suspended rescue platforms [32], but still the dynamic coordination of the use of the various evacuation means in a specific evacuation context remains an open topic.

Additionally, in high rise buildings, elevators are the main means for reaching higher floors as they are easier to use and considerably faster than stairs. Unfortunately, during an evacuation, their usage is not considered safe in fire or earthquake. Moreover, they increase the time of evacuation as they work in

S. Rodríguez et al. (Eds.): DCAI 2018, AISC 801, pp. 149–156, 2019.
https://doi.org/10.1007/978-3-319-99608-0_17

batches with limited loading capacity [31]. Still, elevators remain the main means of evacuation for disabled people as they cannot use stairs autonomously. This is the reason why, in this paper, we study the use of elevators during evacuation and the agent-modeling solutions that improve their evacuation efficiency.

This paper is organized as follows. In Sect. 2, we give the motivation for this work and review the State-of-the-Art. In Sect. 3, we propose a model for efficient coordination of elevators in an evacuation. We conclude our study in Sect. 4 with discussion and future work.

2 Motivation and Related Work

In this section, we review the state-of-the-art related to the modeling and simulations of the use of elevators in an evacuation. We also debate over the importance of elevators for people with physical and sensory disabilities.

There is a vast literature about the use of elevators in an evacuation, e.g., [31]. The elevators have become the essential part of high-rise building evacuation because it is quite difficult to evacuate a multi-story building using stairs only. However, the use of elevators in fire and earthquake is dangerous. This is why there have been multiple approaches to design fire elevators that resist smoke and fire to a certain level [11].

Ma et al. in [26] concluded that in the case of only a few people to evacuate, elevators are a better option to evacuate a high-rise building regardless of their high loading/unloading times. Min and Yu in [28] proposed a hybrid strategy for evacuation in which both stairs and elevators can be used efficiently. However, if the number of evacuees is high, they conclude that stairs are preferable. Additionally, Harding et al. in [12] considered that elevators are dangerous during evacuation due to their limited occupancy which results in long waiting queues that cause congestion.

Elevator aided evacuation efficiency can be increased up to a certain limit [26], however, Heyes and Spearpoint in [13] concluded that elevators take more time compared to stairs because of long waiting times and congestion in the lobby area. For this purpose, Liao et al. in [20] conducted some interviews with occupants to understand the risks and complexities associated with elevators. People expressed fear concerning safety in using elevators in evacuation due to any electrical/mechanical fault, fire or smoke. Ding et al. in [7] discussed a fire scenario, where fire and smoke can quickly spread all over the buildings and enter into the elevator shaft, known as piston effect [15]. Groner [11] reveals that though the use of elevators is risky and not allowed during an evacuation, they can be used by firefighters. Consequently, Klote et al. suggest that it is challenging to design such an elevator for emergency evacuation because of their complexity (see, e.g., [15]) and a cost factor as they require extra floor area for the lobby and proper electrification (see, e.g., [7])

Chertkoff and Kushigian [5] explain how poor decisions and no-support from the management of building staff can lead to more casualties during an evacuation. Furthermore, Kinsey et al. [14] highlight two underlying factors that affect the evacuation in a high-rise building: the height of the building and human behavior. To analyze the behavior of evacuees in a detailed manner, Koo et al. [17] created a panic model and concluded that the reason of panic-related behaviors could be the sudden announcement of hazardous situation and increased evacuation velocities.

The behavior of occupants may be unpredictable in case of sudden emergencies and hazards due to stress and nervousness which increases in the case of high evacuation speed, congestion, and when the safety of the evacuation path is unsure [13,23]. As a result, situations like herding and stampeding may occur that might lead to crushing fatalities. In this regard, Lujak and Ossowski in [24] proposed an evacuation route architecture for smart spaces which optimizes the travel time and route safety by considering the panic-induced human factors of stampeding and herding. Billhard et al. in [2] also proposed intelligent evacuation guidance for smart buildings. Their system provides an individual evacuation route recommendation to every occupant considering real-time congestion and other evacuation dynamics.

Ding et al. and Liao et al. [8,21] analyzed the behavior of evacuees by changing the physical characteristics of elevators such as sizes of doors and connected lobby areas. The evacuation time can be reduced by improving these characteristics. According to Heyes and Spearpoint [13], the decision of selecting stairs or elevators is entirely dependent on evacuees and situation. However, typically, as the floor height increases, the preference is towards elevators. Therefore, Chow [6] suggested that the situation awareness and mental stability of evacuees during emergency evacuation are necessary.

Kuligowski et al. in [19] discuss 30 evacuation models and only 3 support elevators in an evacuation. Furthermore, Koo et al. in [18] experimented with a 24-story building using an agent-based simulation model to analyze the behavior of evacuees due to congestion, blocking and capacity drop effects which result in larger evacuation time. Ma et al. in [25,26] designed a network model using simulation to explain a configuration of elevators for evacuation in a building. They also purposed a concept of special refuge floor which has special fire elevators.

On the other hand, Ding et al. in [9] proposed the shortest elevator evacuation route method by using Pathfinder [1]. They concluded that the evacuation time matters on the number of elevators, age group of occupants and pedestrian flow. Due to this, Noh et al. [29] proposed an efficient evacuation model to handle pedestrian flow problem with the help of time-expanded network flow model. Also, Groner in [11] proposed a decision model that relocates evacuees to a safer location to avoid congestion and other behavioral issues.

Research on safe evacuation for the people with disabilities in high rise buildings started in early 1970's by proposing a special type of elevators (e.g., [3,4,16]). Therefore, American Society of Mechanical Engineers has taken steps to update the concept of using an elevator during emergencies especially for the people with disabilities [31]. In this regard, Proulx in [30] proposed a buddy system in which the people with disabilities will be assisted by the able-bodied people with the help of some special devices and sensors. Also, Manley and

Kim in [27] proposed a method to classify people according to their mobility characteristics with respect to danger at the time of a sudden hazard.

3 A Proposal for an Intelligent Elevator System

There is a need for a decision making support for the elevator usage in the evacuation that should satisfy the following requirements:

(i) Robustness and reliability. (ii) Dynamic and autonomous adaptation, scheduling, and control according to the emergency situation. (iii) Prioritization of evacuees based on their level of mobility and sensory disabilities. (iv) Avoidance of panic-induced situations by a priori assignment of evacuees to use it. (v) Integration with existing State-of-the-art intelligent evacuation architectures.

To this end, we propose the concept of a Smart Agent-Based Elevator System (SABES) inspired by [10,22]. The proposed system will be able to work in a real-time manner as an additional layer on the existing evacuation architectures as well as individually. The functioning of our proposed system is as follow.

The block diagram of the proposed system is shown in Fig. 1. This agent-based system consists of a central coordinator agent and three major modules, i.e., sensors, optimization and a real-world operation module. Also, these modules have been further divided into sub-modules based on the functionality and connected to a central coordinator agent which is the system's decision-making module, as seen in Fig. 1. Further details of every module and sub-modules are discussed below.

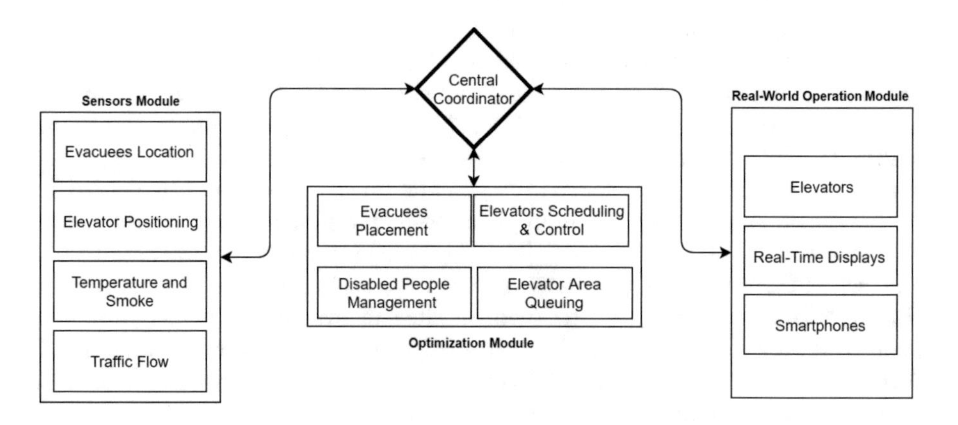

Fig. 1. Block diagram of proposed intelligent elevator system

Sensors Module. This module consists of multiple sensors that receive the information about the evacuees and the building in real-time. This information is fused and sent to the central coordinator.

Evacuees Location Sub-module. To sense the location of all evacuees either individually or as a group, a location sensor is required. Therefore, we use iBeacons and smartphones of evacuees to find their location.

Elevator Positioning Sub-module. These sensors determine the position and status of the elevators, e.g., the floor origin and destination, state (empty/occupied) and direction of movement together with the number of people assigned to the elevator in the next service and using it momentarily.

Temperature and Smoke Sub-module. Temperature and smoke sensors will be installed in such a way that they could keep measuring the temperature and smoke of every part of the building. These sensors will be installed in the elevators and lobby area of each floor. If there is a rise in temperature or smoke, the elevators will be stopped at a nearest safe floor to avoid any severe accident.

Traffic Flow Sub-module. Traffic flow sensors should be installed on the stairs, lobby areas and corridors to detect the congestion and other panic-induced situations.

Optimization Module. This module consists of four sub-modules which perform different tasks, and these sub-modules are also connected to central coordinator as the transceivers of relevant information.

Evacuees Placement Sub-module. It assigns evacuees the safest place close by. This sub-module works by managing evacuees with respect to their position and location. After this, it continuously communicates with the central coordinator to get the latest information of evacuees and their locations. Then it processes the information on a real-time basis and gives safe place recommendations near to them.

Elevator Scheduling and Control Sub-module. It is dedicated to the management of elevators. This sub-module performs several tasks related to elevators such as it detects the current status of elevators concerning their floor positions. It also calculates the number of occupants of elevators with the help of a sensor. The most important function of this sub-module is to detect the smoke and temperature in a real-time manner so that it could stop the operation of elevators before any major risk. By doing this, risks associated with elevators and severe accidents will be avoided.

Elevator Area Management Sub-module. It manages and controls the traffic flow near the elevators and exit, i.e., lobby area and stairs. The purpose of this sub-module is to avoid the congestion and other panic-induced issues of evacuees during the time of emergency.

Disabled Persons Management Sub-module. It prioritizes and manages the evacuation of people with disabilities and impairments. It also allocates the nearest place to elevators in case of waiting queues or if the elevators stop suddenly.

Real-World Operation Module. All the outputs and other operations that occur in or having any connection with real world have been included in this module. The purpose of this module is to help out the evacuees using displays and conveying messages with the help of the sub-modules of this module. Therefore the smartphones, displays, signs, and elevators are the sub-modules of this module.

All sub-modules of sensors, optimization and real-world operation modules coordinate with the central coordinator by sending information to it in a real-time manner. An algorithm is deployed in the central coordinator by which it controls every module. Also, it manages all outputs of the system, i.e., real-world operation module. The proposed elevator system keeps on sensing the instantaneous locations of each evacuee as well as the current status of elevators and the traffic flow of selected areas in the buildings.

In the case of emergency evacuation, our proposed model is activated as follows. First of all, the sensors module activates all of its sub-modules and start sending the real-time data to central coordinator agent. For example, the instantaneous location of each occupant in the building will be detected by his/her smartphone and sent to the central coordinator. Then the central coordinator computes the temperature and smoke status with the help of the dedicated sensors. Meanwhile, the status of the elevators is also checked by the central coordinator agent. After getting all required data from all sensors on the real-time basis, the algorithm of central coordinator will then proceed to the optimization module to compute the safest location and routes for the occupants of the building according to their priorities.

The central coordinator agent passes the instantaneous information of occupants, elevators and the situation of every part of the building to the optimization module where every sub-module works according to its relevant functionality. Eventually, after getting all the information and processing it in a real-time manner, the algorithm will suggest to the occupants whether to use the elevators or it is not safe to use them. If the system suggests using the elevators after checking the safety status, the system also gives route guidance to all evacuees to find the elevators. However, to avoid the congestion and panic-induced situation of evacuees, the system makes a queue or waiting line in which the people with physical and sensory disabilities will be on the highest priority on using elevators because it will be most difficult for them to use stairs if the operation of elevators is stopped afterward. The people who will be waiting for their turn will be suggested to wait in the waiting rooms which have been specially designed for emergencies as those rooms will be fire and smoke proof. Meanwhile, the system will also recommend the safest evacuation route to everyone using stairs. In this way, it will increase the rate of evacuation to some extent.

4 Conclusion and Future Work

In this study, we gave a brief review of the state-of-the-art literature related to the use of elevators during an emergency evacuation. Several techniques were discussed regarding the efficient and safe use of elevators. However, we noticed that the usage of elevators in evacuation is still an open issue for engineers and researchers.

Hence, we identified some challenges that can be resolved to improve the functioning of elevators during an evacuation. We hypothesize that if we design a real-time intelligent control system for evacuation and integrate the elevators with that system, the evacuation process will be more secure, rapid and trustworthy. This is why we proposed the concept of a Smart Agent-Based Elevator System (SABES) and described its main modules and submodules.

The people having mobility and sensory disabilities are the most critical evacuees for our evacuation system as they are not able to use stairs independently. For this purpose, we have to propose and design some effective techniques for them as well so that they could be evacuated safely.

As a future work, we plan to develop and further simulate the proposed SABES system to schedule and control the elevators as well as to manage the long waiting queue problems in the lobby and nearby areas to avoid the congestion and panic-related behaviors.

Acknowledgments. Work partially supported by the Autonomous Region of Madrid (grant "MOSI-AGIL-CM" (S2013/ICE-3019) co-funded by EU Structural Funds FSE and FEDER), project "SURF" (TIN2015-65515-C4-4-R (MINECO/FEDER)) funded by the Spanish Ministry of Economy and Competitiveness.

References

1. Pathfinder: Agent based evacuation simulation, thunderhead engineering consultants, inc. https://www.thunderheadeng.com/pathfinder/. Accessed 04 Aug 2018
2. Billhardt, H., Dunkel, J., Fernández, A., Lujak, M., Hermoso, R., Ossowski, S.: A proposal for situation-aware evacuation guidance based on semantic technologies. In: Multi-agent Systems and Agreement Technologies, pp. 493–508. Springer (2016)
3. Boyce, K.E., Shields, T.J., Silcock, G.W.H.: Toward the characterization of building occupancies for fire safety engineering: capabilities of disabled people moving horizontally and on an incline. Fire Technol. **35**(1), 51–67 (1999)
4. BS BSI: 9999 code of practice for fire safety in the design, management and use of buildings. BSI Global (2008)
5. Chertkoff, J.M., Kushigian, R.H.: Don't Panic: The Psychology of Emergency Egress and Ingress. Praeger Publishers, Westport (1999)
6. Chow, W.K.: Evacuation in a supertall residential complex. J. Appl. Fire Sci. **13**(4), 291–300 (2005)
7. Ding, N., Chen, T., Zhang, H.: Experimental study of elevator loading and unloading time during evacuation in high-rise buildings. Fire Technol. **53**(1), 29–42 (2017)
8. Ding, N., Zhang, H., Chen, T., Luh, P.B.: Evacuees behaviors of using elevators during evacuation based on experiments. Transp. Res. Procedia **2**, 594–602 (2014)
9. Ding, Y., Yang, L., Weng, F., Fu, Z., Rao, P.: Investigation of combined stairs elevators evacuation strategies for high rise buildings based on simulation. Simul. Model. Pract. Theory **53**, 60–73 (2015)
10. Dou, L., Zong, Q., Ji, Y.: A mixed robust optimization and multi-agent coordination method for elevator group control scheduling. In: 2010 International Conference on Logistics Systems and Intelligent Management (ICLSIM), vol. 2, pp. 1034–1038, January 2010
11. Groner, N.E.: A decision model for recommending which building occupants should move where during fire emergencies. Fire Saf. J. **80**, 20–29 (2016)
12. Harding, P.J., Amos, M., Gwynne, S.: Prediction and mitigation of crush conditions in emergency evacuations. In: Pedestrian and Evacuation Dynamics 2008, pp. 233–246. Springer, Heidelberg (2010)

13. Heyes, E., Spearpoint, M.: Lifts for evacuation - human behaviour considerations. Fire Mater. **36**(4), 297–308 (2012)
14. Kinsey, M.J., Galea, E.R., Lawrence, P.J.: Investigating evacuation lift dispatch strategies using computer modelling. Fire Mater. **36**(5–6), 399–415 (2012)
15. Klote, J.H., Levin, B.M., Groner, N.E.: Emergency elevator evacuation systems. In: Proceedings of the 2nd Symposium on Elevators, Fire, and Accessibility, pp. 131–149 (1995)
16. Kobes, M., Helsloot, I., de Vries, B., Post, J.G.: Building safety and human behaviour in fire: a literature review. Fire Saf. J. **45**(1), 1–11 (2010)
17. Koo, J., Kim, B.I., Kim, Y.S.: Estimating the effects of mental disorientation and physical fatigue in a semi-panic evacuation. Expert Syst. Appl. **41**(5), 2379–2390 (2014)
18. Koo, J., Kim, Y.S., Kim, B.I.: Estimating the impact of residents with disabilities on the evacuation in a high-rise building: a simulation study. Simul. Model. Pract. Theory **24**, 71–83 (2012)
19. Kuligowski, E.D., Peacock, R.D., Hoskins, B.L.: A review of building evacuation models. US Department of Commerce, NIST (2005)
20. Liao, Y.J., Lo, S.M., Ma, J., Liu, S.B., Liao, G.X.: A study on people's attitude to the use of elevators for fire escape. Fire Technol. **50**(2), 363–378 (2014)
21. Liao, Y.J., Liao, G.X., Lo, S.M.: Influencing factor analysis of ultra-tall building elevator evacuation. Procedia Eng. **71**, 583–590 (2014)
22. Lujak, M., Billhardt, H., Dunkel, J., Fernández, A., Hermoso, R., Ossowski, S.: A distributed architecture for real-time evacuation guidance in large smart buildings. Comput. Sci. Inf. Syst. **14**(1), 257–282 (2017)
23. Lujak, M., Ossowski, S.: On avoiding panic by pedestrian route recommendation in smart spaces. In: 2016 IEEE International Black Sea Conference on Communications and Networking (BlackSeaCom), pp. 1–5. IEEE (2016)
24. Lujak, M., Ossowski, S.: Evacuation route optimization architecture considering human factor. AI Commun. **30**(1), 53–66 (2017)
25. Ma, J., Lo, S., Song, W.: Cellular automaton modeling approach for optimum ultra high-rise building evacuation design. Fire Saf. J. **54**, 57–66 (2012)
26. Ma, J., Chen, J., Liao, Y.J., Siuming, L.: Efficiency analysis of elevator aided building evacuation using network model. Procedia Eng. **52**, 259–266 (2013)
27. Manley, M., Kim, Y.S.: Modeling emergency evacuation of individuals with disabilities (exitus): an agent-based public decision support system. Expert Syst. Appl. **39**(9), 8300–8311 (2012)
28. Min, Y., Yu, Y.: Calculation of mixed evacuation of stair and elevator using EVACNET4. Procedia Eng. **62**, 478–482 (2013). 9th Asia-Oceania Symposium on Fire Science and Technology
29. Noh, D., Koo, J., Kim, B.I.: An efficient partially dedicated strategy for evacuation of a heterogeneous population. Simul. Model. Pract. Theory **62**, 157–165 (2016)
30. Proulx, G.: Evacuation planning for occupants with disability. Fire Risk Management Program, NRC-IRC, Canada (2002)
31. Ronchi, E., Nilsson, D.: Fire evacuation in high-rise buildings: a review of human behaviour and modelling research. Fire Sci. Rev. **2**(1), 7 (2013)
32. Ronchi, E., Nilsson, D.: Modelling total evacuation strategies for high-rise buildings. In: Building Simulation, vol. 7, pp. 73–87. Springer (2014)

Time Analysis of the Integration of Simulators for an AmI Environment

Álvaro Sánchez-Picot[1(✉)], Diego Sánchez-de-Rivera[1], Tomás Robles[1], and Jaime Jiménez[2]

[1] Department of Telematics Systems Engineering, Universidad Politécnica de Madrid, Avenida Complutense no. 30, 28040 Madrid, Spain
alvaro.spicot@gmail.com
[2] Ericsson, Hirsalantie 11, 02420 Kirkkonummi, Finland

Abstract. There is a trend nowadays where each time there are more and more devices around us and all of them need to be connected between them and to the Internet so that the data they provide or the service they offer can be accessed anywhere anytime and the can coordinate between them to achieve a greater goal. Aside from the devices, the other important agents in this environment are the people that move around while the devices monitor their activities. With all this complex environment in mind it becomes clear that the use of simulators to improve it is necessary. In this paper we analyze the communication necessary in our AmI environment simulator composed of an engine, an existing social simulator and an existing network simulator. We also propose a mathematical model for the times of the different messages sent between the simulators.

Keywords: Simulation · Social simulation · Network simulation
Ambient intelligence

1 Introduction

We are in a technologically world where every day more devices surround us. Some of these devices are just newer versions of what has been existing for a long time such as computers, televisions, air conditioning units or printers but others are relatively newer such as smart speakers, smartwatches or smart thermostats. This devices have invaded our homes to give us as much information as possible of what is happening in it and to react with external data to what might happen in the near future. We want to know how well or bad do we sleep, we want our house to start heating up or cooling up before we get home and we want our fridge to inform us of what we should buy in the supermarket.

We want our buildings to be sensitive [9] and responsive to the presence of people and other environmental factors such as the weather. This is what we know as Ambient Intelligence (AmI), an environment where all the devices that exist in it cooperate together to achieve a common goal. All these devices are orchestrated by an intelligence behind, usually a central system, that manages

S. Rodríguez et al. (Eds.): DCAI 2018, AISC 801, pp. 157–164, 2019.
https://doi.org/10.1007/978-3-319-99608-0_18

the data generated in the sensors in the building such as temperature, humidity or luminosity and following certain rules control the different actuators so that the people inside the building benefit from the result. A light might be turned off when nobody is present or the air conditioning might start working when there is a lot of people inside all without the need of a human interaction and if possible so that the people inside do not notice the changes.

All this devices are expensive and deploying the infrastructure to create an AmI environment is a complex task. There is a lot involved aside from the devices such as the communication between them the central unit that coordinates everything or the security required so that nobody can use the system as it is not intended. In this paper we propose a tool that helps with the deployment of these environments simulating different compositions of the elements present in them to get valuable information before even deploying the environment. This tool we call Hydra, is a simulator composed of a Network simulator, a Social simulator and an engine that coordinates both. This way we have an AmI environment simulator.

This paper is the continuation of our previous work [6], where we analyze the different phases present in a simulation using Hydra and the times and delays required to execute them. In Sect. 2 we present the explain the key terms used in this work. In Sect. 3 we show the architecture used in the tool. In Sect. 4 we analyze the times necessary in each step of the simulation. Finally in Sects. 5 and 6 we present the conclusions and discuss some possible future work.

2 Related Work

This section describes the related work with the tool that we present in the paper. These includes AmI environments and also the simulations, specifically both the social simulation and the network simulation.

2.1 Ambient Intelligence

AmI corresponds to a discipline where our everyday environments are sensitive to the events that happen in their surrounding. In order to achieve this goal we require sensors to recollect information about what is happening in the environment, actuators that enables an interaction with the environment influencing on it, a network that enables communication between all the entities in the AmI environment and a way to orchestrate all the entities in the environment [3].

An environment needs to be sensitive, adaptive, transparent, ubiquitous and intelligent to be considered AmI [4].

2.2 Simulation

A simulation is a simplification of the real model in order to work easily with it with the idea of experimenting with it in order to achieve a particular goal. A simple mode can be simulated using a mathematical model but once it gets more complex the simulation needs to be done in a computer [7]. In this work we focus on two computer simulations, social simulation and network simulation.

Social simulation studies the interaction among social entities taking into account their psychology and their behavior, both between people and with the people and the environment. There are two main types of social simulation: system level simulation that analyzes all the scene as a whole and agent-based simulation where we model the different elements in the simulation as agents with its own behavior being an agent a person or a device. In this work we will be using agent-based simulation as it is more adapted to AmI environments [5].

There are different agent based Social Simulators (SS) such as MASON, Repast or Swarm [1], each with its own characteristics and usually particularized for a certain case study. Some of them work with a 2D environment while others have a 3D one. All of them include some kind of physical engine to calculate the collisions between the agents and the environment. These simulators work synchronously, so that all the elements inside the simulation are updated every certain time called step.

The SS specializes in the behavior of the human and it can simulate other elements in an AmI environment such as sensors or actuators but it is not designed to simulate details of these elements such as packages sent over the network with the information.

A network simulator (NS), models the behavior of a network and the different elements that compose it, as well as the packages sent between them [2]. These elements include routers, computers as well as sensors and actuators.

There are several NS nowadays both open-source such as NS or OMNet++ and proprietary such as OPNET or NETSIM [8]. These simulators are asynchronous, and some examples of the events that can happen are sending packets, or a device that connects to the network. After the simulation ends they generate a log that contains all these events, useful for a future analysis of the network.

NS are very good at simulating the network and the actuators and sensors in an AmI environment and can simulate the other elements in this environment, such as the people moving around, using specific algorithms for their movement but they simulation will not be very realistic as their behavior would be limited.

3 Architecture

In this chapter we present the architecture of our tool called Hydra. There are three main components in the architecture of Hydra as can be seen in Fig. 1: the engine, the NS and the SS, all briefly explained below. More information about the architecture can be found in our previous work [6].

The engine has several key tasks in this architecture. One of them is the initialization of the simulation using the parameters defined by the user and the stored data of the scenario that is going to be simulated. This data includes the models of the different objects in the simulation classified in environment such as walls, doors or furniture and agents such as persons and cybernetic devices. Another task of the engine is the coordination of both the SS and the NS during

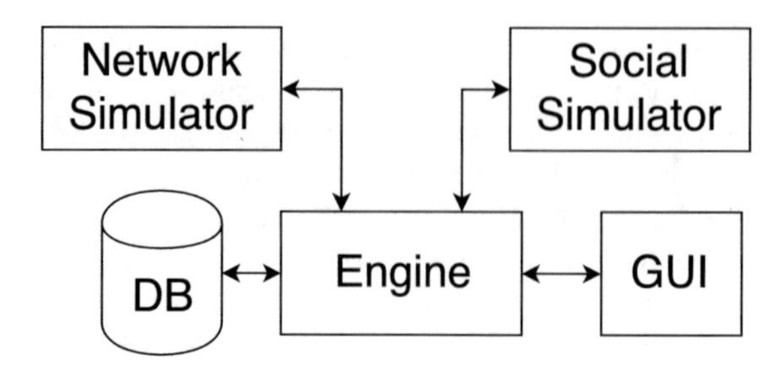

Fig. 1. General architecture

the execution of the simulation alternating between them as the different events are processed until an ending event is reached. Finally another engine task is to decompose the different information sent by the simulations so that each simulator receives only the data relevant to it. The engine also has access to a database to load the models of the simulation and store the data once the simulation is finished and also has access to a GUI where the user will configure the parameters of the simulation.

The Network Simulator will run a network simulation using the data received from the engine. It contains all the models related to network elements such as protocols and wireless technologies. It will start the simulation and the pause it and resume as requested by the engine until the ending of the simulation is reached. The NS will store all the data generated during the simulation and once it finishes it will be sent to the engine.

The Social Simulator will run a social simulation with the data received from the engine. It contains all the different models of people behavior and collisions. It will start the simulation when the engine sends an initialization message and then will update the agents in the simulation each time the engine requests it until the simulation ends. The SS will store all the data generated during the simulation until it ends when this information will be sent to the engine.

4 Simulation Phases Analysis

As seen in the previous chapter there are four different communication phases between the simulators and the engine and these are: Initialization, SS event, NS event and ending. In this chapter we will analyze in detail those messages and the time needed since the process is started until it ends.

4.1 Initialization of the Simulation

During the process of the initialization of the simulation there are different steps necessary to complete.

First the engine needs to access the database in order to retrieve the data related to the simulation scenario and the models of the different agents involved in the simulation (people, cybernetic devices and their behavior) as well as the models of the environment objects (walls, doors, windows, furniture and communication lines). This different models are explained in more detail in our previous work [6]. We name this time t_{idb}.

Second, the data obtained from the database is then decomposed for each simulator (t_{ien}). The SS requires data related to the agents and their behavior but not in deep details of the communication used in the scenario. Otherwise the NS requires all the data relevant in the communication such as wireless technologies, protocols or propagation delays but only requires the position of the agents, not their behavior. All this information is particularized with the parameters introduced by the user. More information about the configuration screens used in Hydra can be found in our previous work [6].

Finally this data is sent to the SS (t_{tss}) and the NS (t_{tns}). The SS will process the data and start its own simulation with the data received (t_{iss}). Once it has started the simulation is paused and the Engine is informed about it (t_{tss}). Similarly the NS will start a simulation with the data received and pause the simulation (t_{ins}). Then the engine is informed that the simulation has started (t_{tns}).

After all this steps are completed the simulation is ready to begin. The engine will process the events in the queue and it can be either one corresponding to the SS or the NS.

The total time necessary to complete the initialization phase is as follows:

$$T_{init} = t_{idb} + t_{ien} + max(2 * t_{tss} + t_{iss}, 2 * t_{tns} + t_{ins})$$

4.2 Social Simulation Event

As commented in Sect. 2 Mason, the SS we our using in our tool Hydra, and usually most SS works in steps, updating the agents status every step that corresponds to a time period inside the simulation, in our case, a parameter defined by the user during the configuration of the simulation. This is what we have named SS event. The process of this phase is completed during different steps.

When the engine extracts a SS event it creates a message with all the information required by the SS to update its status (t_{ssen1}). This information includes the actual time in the simulation as well as some event predefined by the user that might have triggered.

The message created by the engine is then sent to the SS (t_{tss}) that uses this data to update the status of its own simulation what usually translates in the update of the position of the agents (t_{uss}). This changes in the simulation are then stored in a message that is sent back to the engine (t_{tss}).

The message created by the SS informs the engine that the SS has finished updating its status but some of the changes in the state of the simulation can be necessary in the NS. The engine creates a message with the information relevant to the NS (t_{ssen2}) and then sends it to the NS (t_{tns}).

Then the NS uses the data in the message to update the state of its own simulation (t_{uns}). This data can for example contain the new position of the agents that have moved since the previous event in the simulation. The NS then sends a message to the engine to inform it that it has finished updated but no more information is included as the NS has not executed new events in its simulation (t_{tns}), it just updated the information of some elements.

Finally when the engine receives the update from the NS the execution of the SS event is finished and it can proceed with the next event in the queue.

The total time necessary to complete the SS event phase is as follows:

$$T_{sse} = t_{ssen1} + 2 * t_{tss} + t_{uss} + t_{ssen2} + 2 * t_{tns} + t_{uns}$$

4.3 Network Simulation Event

Differently to how the SS works, the NS events are asynchronous as explained in Sect. 2. When the next event in the queue corresponds to the NS there are different steps necessary to complete until the event is processed.

First the engine generates a message with the information necessary to the NS (t_{nsen1}). This information includes the actual time in the simulation and the time when this event ends (this corresponds with the time of the next SS event). This message is then sent to the NS (t_{tns}).

Then the NS extracts the data in the message and proceeds to advance its own simulation during the time specified in the message (t_{uns}). During this simulation time new events will be generated inside the simulation and all the events will be processed as long as their execution time is less than the current event finishing time. An example of events executed during this event could be a router sending a IP packet to a computer that will generate an event when the packet arrives to the computer that once processed might generate new packages sent over the network. When the NS stops updating the simulation (based on the event time define in the message) it then generates a new message with data from the updates and sends it to the engine (t_{tns}).

The engine then processes this message and generates a new message with information from the updates of the event relevant to the SS (t_{nsen2}). An example of this information could be a notification sent to a mobile phone, so its owner might react to this event. The message is then sent to the SS (t_{tss}).

Then the SS updates its agents with the information received so that it influences future updates of the simulation (t_{uss}). It the generates a message to the engine just to inform that it has finished updating the state (t_{tss}).

Finally, once the engine receives the message from the SS the processing of the NS event finishes and it can proceed with the next event in the queue.

The total time necessary to complete the NS event phase is as follows:

$$T_{nse} = t_{nsen1} + 2 * t_{tns} + t_{uns} + t_{nsen2} + 2 * t_{tss} + t_{uss}$$

4.4 Ending the Simulation

There are different conditions that can make the simulation end. The user can stop the simulation manually or he may have defined a time or another condition to end. After the ending condition is triggered different steps are necessary to save all the generated data until the simulation can arrive to an end.

Firstly the engine sends a message to the SS (t_{tss}) and the NS (t_{tns}) informing about the termination of the simulation.

When the SS receives the message it will start to end its own version of the simulation saving the data that has been generated in a message and then safely stopping the simulation liberating all the data from memory (t_{ess}). The generated data is then sent to the engine (t_{tss}). A process in the SS will still keep running in case a new simulation is later started.

When the NS receives the message from the engine it will also start to end its own version of the simulation. Similarly to what the SS has done, the NS will save all the data generated during the simulation in a file and proceed to safely stop the simulation liberating all the data from memory (t_{ens}). The stored data includes information about all the packages sent during the simulation as well as information from the different devices. A process in the NS will still keep running waiting for a new simulation to start. The data is then sent to the engine (t_{tns}).

When the engine has received the information from both the simulation it will analyze the data (t_{een}) and proceed to store it in the database (t_{edb}) so that the simulation can be analyzed later in time. After saving the data, the simulation is complete.

The total time necessary to complete the ending phase of the simulation is as follows:

$$T_{end} = max(2 * t_{tss} + t_{ess}, 2 * t_{tns} + t_{ens}) + t_{een} + t_{edb}$$

5 Conclusions

In this paper we present a tool called Hydra that is composed of two simulators, a SS and a NS in order to get an AmI simulator that can help in the deployment and update of AmI environments. It enables the simulation of different scenarios with different use cases that can detect possible errors and reduce costs of a real deployment by processing the data generated in the simulations and observing certain patterns.

In this paper we analyze the different phases of the simulation where the engine coordinates both the SS and the NS during the initialization, updates of the different simulators and the ending of the simulation. We also calculate the different times and delays required in each phase useful for a future inclusion of the simulator in a real time environment.

6 Future Work

The tool presented in this paper can be greatly improved. We comment some of them.

The time analysis done requires real data obtained from the simulator that enables the comparison of the different delays and times to get an idea of where improvements could be made.

The time analysis done would be very useful with the inclusion of the simulation in a real time environment where the data can be obtained from real devices. This time analysis could help us to know the limits of the tool such as the maximum number of devices where the simulation can keep up with the real time data.

Acknowledgements. These results were supported by UPM's "Programa Propio", the Autonomous Region of Madrid through program MOSI-AGIL-CM (grant P2013/ICE-3019, co-funded by EU Structural Funds FSE and FEDER) and has also received funding from the Ministry of Economy and Competitiveness through SEMOLA project (TEC2015-68284- R).

References

1. Allan, R.J.: Survey of agent based modelling and simulation tools. Science and Technology Facilities Council, ISSN 1362-0207 (2010)
2. Breslau, L., Estrin, D., Fall, K., Floyd, S., Heidemann, J., Helmy, A., Yu, H.: Advances in network simulation. Computer **33**(5), 59–67 (2000)
3. Cabitza, F., Fogli, D., Lanzilotti, R., Piccinno, A.: Rule-based tools for the configuration of ambient intelligence systems: a comparative user study. Multimedia Tools Appl. **76**(4), 5221–5241 (2017)
4. Cook, D.J., Augusto, J.C., Jakkula, V.R.: Ambient intelligence: technologies, applications, and opportunities. Pervasive Mob. Comput. **5**(4), 277–298 (2009)
5. Macal, C., North, M.: Introductory tutorial: agent-based modeling and simulation. In: Proceedings of the 2014 Winter Simulation Conference, pp. 6–20. IEEE Press, December 2014
6. Sánchez-Picot, Á., de Andrés, D.M., Sánchez, B.B., Garrido, R.P.A., de Rivera, D.S., Robles, T.: Towards a simulation of AmI environments integrating social and network simulations. In: AmILP@ ECAI (2016)
7. Sokolowski, J.A., Banks, C.M. (eds.): Principles of Modeling and Simulation: A Multidisciplinary Approach. Wiley, New York (2011)
8. Yuan, X., Cai, Z.P., Liu, S.H., Yu, Y.: Large-scale network emulation software and its key technologies. Comput. Technol. Dev. **7**, 003 (2014)
9. Bordel Sánchez, B., Alcarria, R., Martín, D., Robles, T.: TF4SM: a framework for developing traceability solutions in small manufacturing companies. Sensors **15**(11), 29478–29510 (2015)

Special Session on Communications, Electronics and Signal Processing (CESP)

Blur Restoration of Confocal Microscopy with Depth and Horizontal Dependent PSF

Yuichi Morioka[1(✉)], Katsufumi Inoue[1], Michifumi Yoshioka[1], Masaru Teranishi[2], and Takashi Murayama[3]

[1] Osaka Prefecture University, Sakai, Japan
morioka@sig.cs.osakafu-u.ac.jp
[2] Hiroshima Institute of Technology, Hiroshima, Japan
[3] Juntendo University, Tokyo, Japan

Abstract. Confocal microscopy is a popular technique for 3D imaging of biological specimens. Confocal microscopy images are degraded by residual out of focus light. Several restoration methods have been proposed to reduce these degradations. The major one is Richardson Lucy based deconvolution (RL). Even when employing this method images are still blurry. This is mainly caused due to spherical aberration that depends on the distance from lens. Hence, in the previous study, the restoration method is taking only the depth direction into account. In this paper, predicting PSF more correctly, an image restoring method using RL method and Point Spread Function that is considered based on the depth and horizontal effect of direction, is proposed.

Keywords: Depth and horizontal dependant point spread function Richardson Lucy method · Confocal microscope

1 Introduction

In recent years, chemical and medical fields disclosed a lot of facts regarding cell components and their functions. Accordingly, there is an increasing demand to further deepen the knowledge of the cells. In particular, the confocal microscopy has been used as a technique for observing alive cells [1]. Although, due to the refractive index difference between the cells and surrounding materials, spherical aberration and so on, the observed image is defocused. The observed images become degraded due to the blur effect. To examine the living cells, the development of the method to restore degraded images is important. For restoration of the defocused images, Point-Spread-Function (PSF) estimation is required. To predict PSF, beads were scattered and pictures whose pixel size are as same as confocal microscopy were taken. Beads images are used as PSF. It is known that PSF are mainly depended on the distance from the lens [2]. Therefore, in the previous study, the restored method is taking only the depth direction

S. Rodríguez et al. (Eds.): DCAI 2018, AISC 801, pp. 167–174, 2019.
https://doi.org/10.1007/978-3-319-99608-0_19

into account. Due to only considering the depth direction, PSF prediction is not correct.

Confocal microscopy has a geometrically operating imaging scanning device named nipkow [3]. A disk that horizontally rotates with respect to the imaged object. We thought that this rotation effects horizontal blur. After some experiment, we suppose that PSF also depends on the horizontal direction. In this paper, from this assumption, an image restoring method using RL method and Point Spread Function that is considered based on the depth and horizontal effect of direction are proposed.

1.1 Mathematical Model of Image Restoration

The image restoration can be modeled with the PSF. Such a general linear model can be written as

$$I(x, y, z) = f(x, y, z) * g(x, y, z) \tag{1}$$

where I is the observed (measured) image, f is the object that we want to obtain (corresponding to the distribution of fluorescent markers inside the specimen), g is the PSF, and $*$ is a three-dimensional convolution. It is required to deconvolve I using g to obtain f.

1.2 Point Spread Function for Confocal Microscopy

From Eq. (1), I is known value, though g (PSF) is unknown value, therefore we need to predict PSF.

To predict PSF, pictures were taken where beads are scattered. The size of the beads are under 1 pixel, and the pixel value is 255 (pixel value range of microscopy are $0 \sim 255$). Since Eq. (1) the size of beads are under 1 pixel, f of beads image become 1 pixel, the observed beads image become PSF at the point.

Confocal microscope scans objects fastened by cover glass with varying the depth of focus and it observes excitation of fluorescent material at each focus. It then obtains 3-dimensional images from that information. It is a problematic but to the refraction that is caused by the difference of refractive index between cover glass and objection. It is also known that when the depth of the focus become deeper, the refraction increases. Therefore, PSF is changed mainly depending on the depth [2]. PSF can be known by taking a beads picture. However, it is difficult to take beads images at all coordinates. This is due to the difficulty in distributing the beads in the desired way. Therefore the next point to be considered is to interpolate PSF to restore cell images.

As it also can be seen from Figs. 1(a) and 2(a), their vertical blur are larger than horizontal one. From the previous research, it is known that images are not restored completely varying the PSF based only on the depth [2]. From Figs. 1(b)

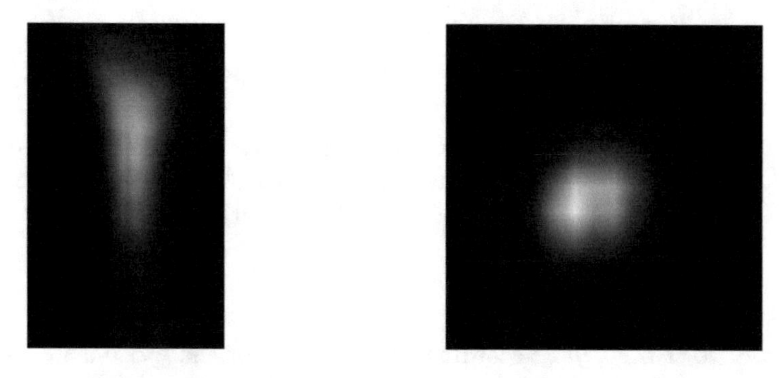

(a) *Vertically sliced beads image* (b) *Horizontally sliced beads image*

Fig. 1. Sliced beads images where $(x, y, z) = (300, 407, 27)$

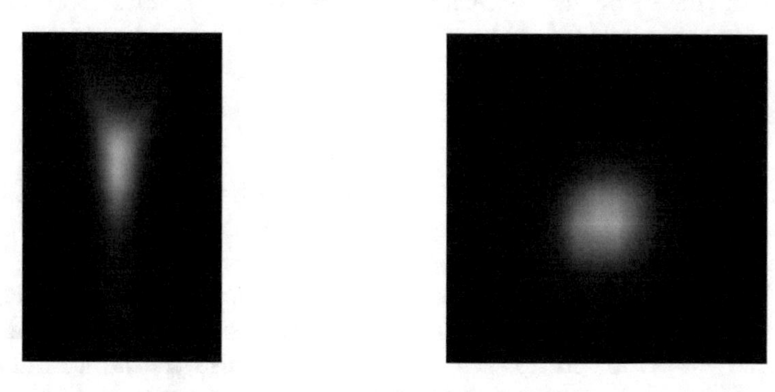

(a) *Vertically sliced beads image* (b) *Horizontally sliced beads image*

Fig. 2. Sliced beads images where $(x, y, z) = (467, 84, 344)$

and 2(b), it can be seen that their blur also varied horizontally. It means that varying the PSF based only on depth does not restore the images completely. We think that PSF varies mainly depended based on depth, though it also depended based on horizontal direction. Therefore this paper purposes a method to predict a PSF considering not only depth but also horizontal direction.

2 Conventional Method

2.1 Restoring Method (Richardson Lucy)

Richardson Lucy (RL) method [4,5] is adopted for restoring images in conventional method. We also adopted this method in proposed method. The algorithm is written as below equations.

$$f_{r+1}(\tau) = f_r(\tau) \int \frac{I_0}{I_r} g(x - \tau) dx \tag{2}$$

$$I_r(\tau) = \int f_r(x) g(x - \tau) dx \tag{3}$$

where r is number of iteration, τ is place of pixel, and x is integration variable. I_r, f_r, and g are respectively the rth measured observed image, object image, and PSF. Observed image is used as I_0. In this paper, g is fixed.

2.2 PSF Selection

In order to select PSF, according to the depth of the cell image, beads where they are up to 9.6 μm are observed. A three-dimensional cell image is divided into 4 partitions in the depth direction. PSF where they are the most close to the center of the each partition in depth direction are selected. RL method is used with these PSF in the each divided cell image. After restoration, divided cell images are combined.

3 Proposed Method

To verify which horizontal direction does the beads blur, Principal Component Analysis (PCA) to each beads image of with two kinds datasets were performed. As remarked before, observed image blur mainly in the depth direction [2]. Therefore, the first principal component of the beads image becomes the depth direction. Second principal component was considered since it is supposed that PSF has some anisotropy in not only the depth but also the horizontal direction. As images are 3-dimensional, second principal component and third principal component are on the horizontal direction. We chose the beads whose second principal component are 1.1 times larger than the third. This is because, if the value is less than 1.1 times, the beads form appear to be circular. Therefore the direction of second principle component may become greatly changed by a sight noise and error.

From Fig. 3, the points in both (a) and (b) is divided into 2 parts, though the difference of angle of second principal component between them is π. Therefore the angles of two parts of components can be regarded as the same. It is possible to see that the angle of second principal component of the beads images are almost the same at any points. Therefore, it is supposed that the PSF of confocal microscopy horizontally spread in a certain direction. We propose three-dimensional linear regression of the PSF based on this assumption. The formula is written as

$$G_{xyz}(x, y, z) = P_x x + P_y y + P_z z + P_d \tag{4}$$

where G_{xyz} is the PSF, P_x, P_y, and P_z are the coefficient of x, y, and $z(x, y, z$ are coordinates). P_d is a constant.

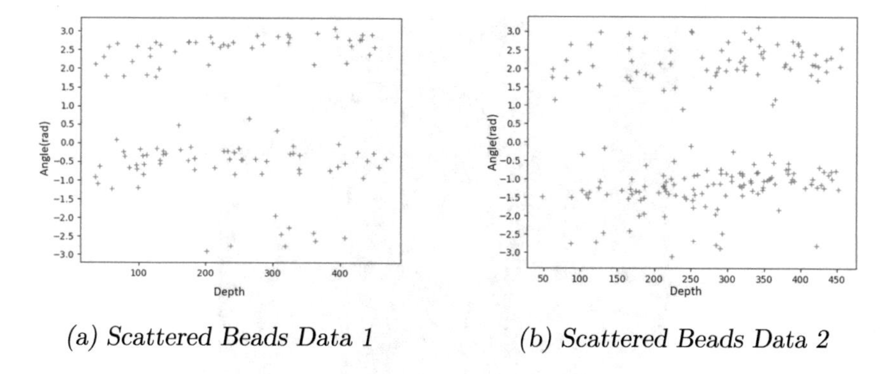

(a) Scattered Beads Data 1 *(b) Scattered Beads Data 2*

Fig. 3. Second principal component of beads images of two kinds datasets at each depth

We determined this assumption only from two beads datasets. This is because, when it is still rudimentary verification, it takes a time to take images by confocal microscope. Therefore there were not a large supply of beads data.

We divide cell pictures into 16 in the depth direction. This is because, convolution is used in RL method. It is useful to use parallel processing for RL method.

4 Experiment

We used two kinds datasets of beads images for predicting PSF. We did conventional method and proposal method. Conventional method takes only the depth direction into account for predicting PSF. Proposal method takes the depth and horizontal direction into account for predicting PSF. We manipulated RL method using PSF that is predicted by them. We used two different actin skeleton as degraded images. These size were $(x \times y \times z) = 512 \times 512 \times 111$ and $320 \times 452 \times 96$ (x, y, z are coordinates).

4.1 Choosing PSF

We used 702 beads to predict PSF. We removed too small and missing beads. Figure 4 is crowded beads picture where it is horizontally sliced at the 200 depth points.

4.2 Experimental Results

Conventional method varies PSF by taking only depth direction into account. Hence, the PSF is not estimated correctly at many points and the edge is too emphasized. From Fig. 8(a), the cell image that is restored by conventional

Fig. 4. Beads pictures that horizontally sliced at the 200 depth points

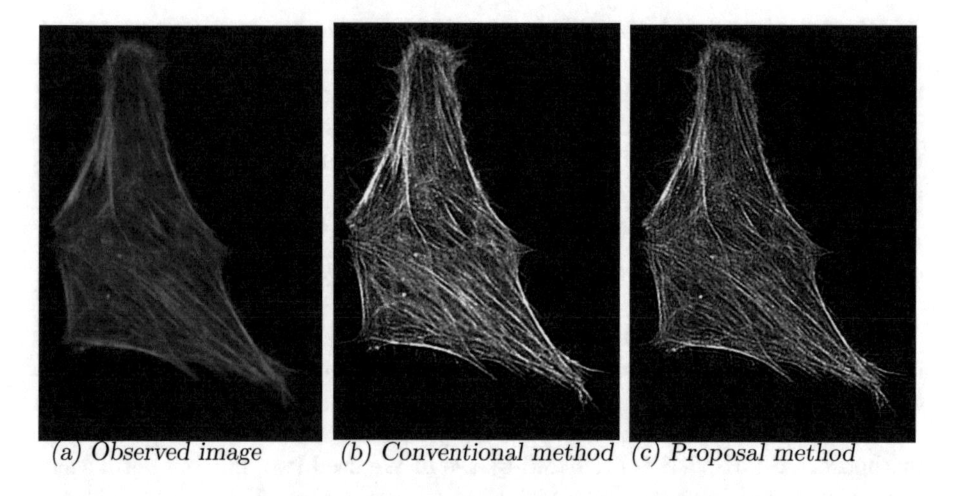

(a) Observed image *(b) Conventional method* *(c) Proposal method*

Fig. 5. Observed and restored images A

method has thick fibers. From Fig. 7(b), restoring high contrast images, the only parts that have high pixel values are emphasized and the parts that have low pixel values disappear. On the other hand, proposal method varies PSF by taking depth and horizontal direction into account. Therefore, the PSF is estimated more correctly and the edge emphasis becomes suppressed. It can be seen from Fig. 8(b). Even restoring high contrast images as shown in Fig. 7(a), the low pixel parts do not disappear. It can be seen form Fig. 7(c). From Fig. 8(b), in proposal method, it has a bad point that ringing is emphasized (Figs. 5 and 6).

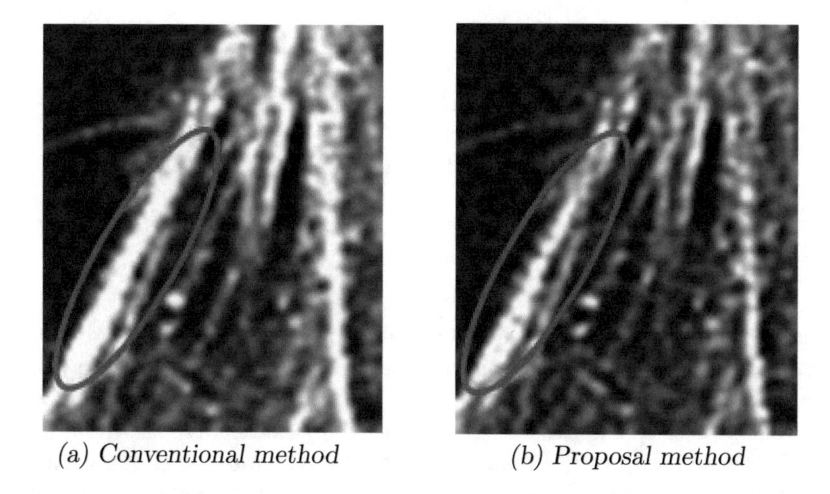

(a) Conventional method (b) Proposal method

Fig. 6. Details of restored images A results. The skelton of (b) that is surrounded by red circle is thinner than (b)

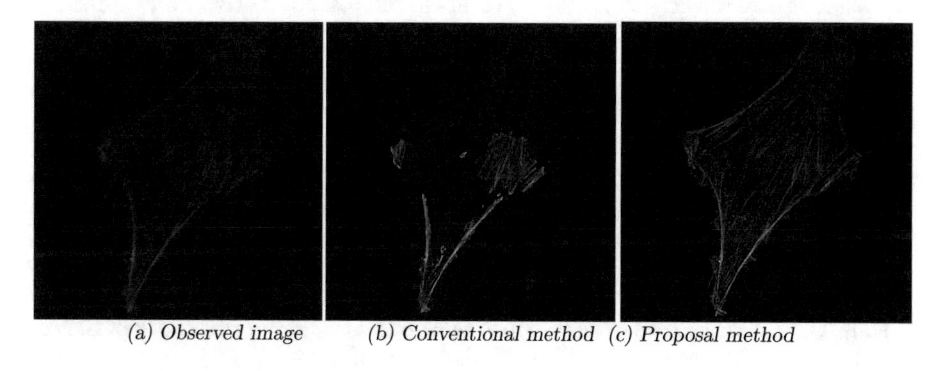

(a) Observed image (b) Conventional method (c) Proposal method

Fig. 7. Observed and restored images B

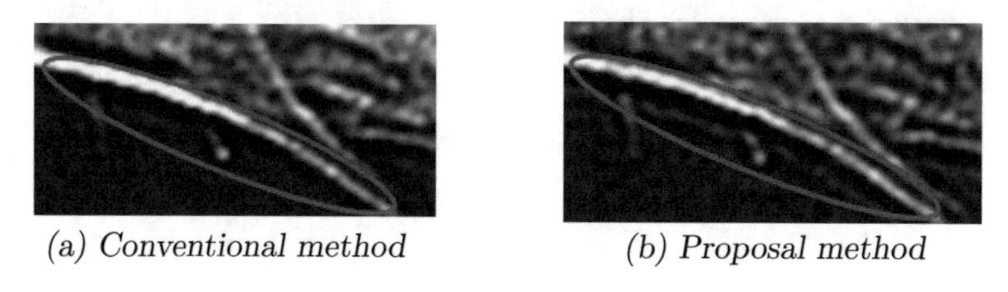

(a) Conventional method (b) Proposal method

Fig. 8. Details of restored images A results. The skelton of (b) that is surrounded by red circle is thinner than (b)

From Table 1, the parallel processing is faster 2 times than sequential processing, though it was not faster than we thought.

Table 1. Executing time

Processing method	Time
Sequential processing	8.7 h
Parallel processing	4.3 h

5 Conclusion

In this paper, we proposed an image restoring method using RL and PSF that is predicted taking the depth and horizontal effect of direction into account. When Considering not only the depth but also the horizontal direction, PSF is estimated smoothly, it is possible to predict PSF that is more adapted at each point. Consequently, even if restoring high contrast images, it does not make the part of the low pixel value disappear. We hope that this research will help the medical domain and that the various mysteries of cells will be elucidated. In future work, we will solve the ringing problem, and we will elucidate the reason why there is a regularity that PSF of confocal microscopy horizontally spread in a certain direction.

References

1. Pinaki, S., Nehorai, A.: Deconvolution methods for 3-D fluorescence microscopy images. Sig. Process. Magaz. IEEE **23**(3), 32–45 (2006)
2. Morioka, Y., et al.: Image restoration of confocal microscopy based on deconvolution algorithms depended on depth of focus. The Institute of Electrical Engineers of Japan IEEJ (2016)
3. Kino, G.S.: Intermediate opticsin Nipkowdisk microscopes. In: Handbook of Biological Confocal Microscopy, pp. 155–165. Springer, US (1995)
4. Richardson, W.H.: Bayesian based iterative method of image restoration. JOSA **62**(1), 55–59 (1972)
5. Lucy, L.B.: An iterative technique for the rectification of observed distributions. Astron. J. **79**, 745 (1974)

Static Dataflow Analysis for Soft Real-Time System Design

Alexander Kocian(✉) and Stefano Chessa

Department of Computer Science, University of Pisa, 56123 Pisa, Italy
kocian@di.unipi.it

Abstract. Synchronous (deterministic) dataflow (SDF) has been extensively used to model flow constraints of digital signal processing (DSP) applications executed on (hard) real-time (RT) operating system (OS). Modern internet-of-things are, however, are often equipped with (soft) RTOSs such as embedded Linux. To reduce design iterations for the latter, the paper proposes a stochastic approach to SDF graphs where the response time of each node is modeled as probability density function (pdf). With increasing number of iterations over the graph, the individual PDFs propagate through the network. The first and second central moments of the resulting joint pdf correspond to the expected system latency and jitter, respectively. The scheduler may execute the code sequentially or in parallel. The proposed analysis tool is helpful in identifying bottlenecks within the system.

Keywords: Probability · Soft real-time · Service time · Latency
Jitter · Synchronous density flow

1 Introduction

Distributed artificial intelligence (DAI) is an emerging discipline that deals with coordinating, distributing and reasoning the performance of tasks in a multiple agent environment. This research can be broadly classified into two areas: distributed problem solving and multi-agent systems. The emphasis on distributed problem solving is on dividing the *problem* among a number of cooperative agents which share knowledge, capability, information, and expertise given the agents would be able to agree. Multi-agent systems, in contrast, are autonomous *entities* that are able to agree intelligently with other entities given their sensory capabilities with ultimate goal to enhance the group intelligence beyond the sum of the individual agent's capability [2].

Focusing on distributed problem solving, Internet-of-Things (IoT) already interconnects billions of devices with projected non-polynomial growth [3]. To meet the vast demands of IoT in various application domains, such as smart home, wearable technology and connected cars, it is critical to connect mostly to the internet a large number of sensors and actuators, to integrate sensing, perform signal and data processing and control the device *in time*. Current IoT

S. Rodríguez et al. (Eds.): DCAI 2018, AISC 801, pp. 175–182, 2019.
https://doi.org/10.1007/978-3-319-99608-0_20

are equipped with (mostly open source) general purpose operating system (OS) or (mostly closed source) real-time operating system (RTOS).

Typical streaming applications on IoT are audio, video and control of robots. In particular, we want to realize an artificial audiologist that processes audio signals and communicates with the internet. In this application, time sequencing is critical. A loss of deadline results in distortion or loss of data which is unacceptable for the user. At the same time, the device shall exhibit plug-and-play mobility for the sensors and use COTS components possibly PC-compatible. To meet both constraints, we decided to implement the digital signal processing (DSP) application on a general purpose OS where the user code delivers deterministic execution but the real-time constraint is relaxed to be soft.

Data flow techniques have been extensively used to derive the end-to-end temporal behavior of tasks. We focus on the synchronous data flow (SDF) model by Lee and Messerschmitt [4] which is the most widely studied data model for DSP applications.

Current practice to design the temporal behavior of DSP tasks executed on *soft* RTOSs is a mixture of simulation and model based techniques ([1], Chap. 4). To close the gap, we propose a purely analytic method based on SDF. Specifically, we model each actor's response time as random variable with probability density function (pdf). With increasing iteration index, the PDFs propagate over the graph. In the end of a period the first and second central moments of the joint pdf represent the end-to-end latency and jitter, respectively. We do not consider interference from other tasks running on the same processor. The proposed method reduces the number of design iterations and hence, shortens the design time.

The paper is organized as follows. Section 2 reviews the SDF model. Section 3 outlines the proposed analysis tool for soft RTOS. Numerical examples in Sect. 4 illustrate the power of the method. Finally, Sect. 5 concludes the paper.

2 Background

In this section we define the SDF model, review its properties and how to analyze them mathematically. A typical DSP streaming task is composed of a number of sub-tasks that have to be executed in a certain order in a repetitive fashion.

Definition 1. *An SDF graph G is a tuple $G = (N, E, I, O, D)$ where*

- *N is the set of nodes in G;*
- *$E \subseteq N \times \mathbb{N}$ is the set of directed edges in G;*
- *$I : E \rightarrow \mathbb{N}$ describes the number of tokens consumed from edge $e \in E$;*
- *$O : E \rightarrow \mathbb{N}$ describes the number of tokens produced on edge $e \in E$;*
- *$D : E \rightarrow \mathbb{N}$ describes the number of initial tokens on edge $e \in E$.*

Properties - Each node in SDF is called an actor, representing a sub-task. The edge, in contrast, stands for a first-in first-out (FIFO) queue. When actor A_i

fires within a particular response time, it consumes a certain number of tokens from the FIFO queue, say $I_{i,j}$, executes a certain sub-task within a well-defined execution time, and produces a certain number of tokens, say $O_{i,j^{(\prime)}}$. The worst case response time (WCRT) T_i is the maximum time required by actor A_i, to process a single token. The edges in SDF are unidirectional communication links that passes tokens from one actor to another ([7], Chap. 2). Figure 1 illustrates the notation.

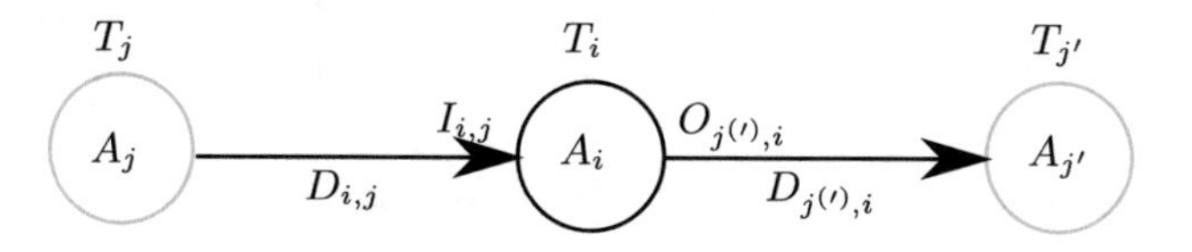

Fig. 1. Notation of SDF graph.

The number of nodes, edges, and tokens are known *a priori* and so are the individual WCRTs. To give intuition without the details, we only consider single-rate systems where a single token is produced and consumed when an actor fires. The analysis method can be easily extended to multi-rate systems.

The SDF graph can be characterized by the topology matrix $\boldsymbol{\Gamma}$. The (i, j)-th entry of Γ describes the amount of data produced on the edge between A_j and A_i. If the edge is not connected to the graph, its value is zero. Self nodes have no influence on the balance equation [4].

To compute the sequential schedule, we know that only *one* node can be invoked at iteration index ℓ. Scheduling does not take into account how long each node runs. However, the system latency does. The solution to the balance equation

$$\Gamma \boldsymbol{v} = \boldsymbol{0} \tag{1}$$

with $\boldsymbol{v}$ denoting the number of firing per actor, yields the optimum firing vector $\boldsymbol{v}^{\star}$. A blocked periodic admissible sequential schedule (PASS) exists with a unique integer solution to (1) if rank $(\boldsymbol{\Gamma}) < N$. This is the case when the graph is disconnected [4]. There are several PASS for a given $\boldsymbol{v}^{\star}$. The PASS is deadlock-free if there is on every cycle in the SDF graph at least one initial token. In this case, the latency $\tau[\ell+1]$ at iteration index $\ell+1$ reads

$$\tau[\ell+1] = \tau[\ell] + \boldsymbol{T}^{\mathsf{T}} \boldsymbol{v}[\ell] \tag{2}$$

where the column vector $\boldsymbol{T} = [T_1, \ldots, T_N]^{\mathsf{T}} \in \mathbb{R}^N$ contains the individual WCRTs with the initial latency $\tau[0] = 0$. Note that in the sequential case, the firing column vector $\boldsymbol{v}[\ell]$ comprises only one non-zero entry. Clearly, different PASS result in different evolution of $\tau[\ell]$. The periodic latency, however, remains unchanged.

The throughput can be increased substantially by distributing the load over more processors. For the sake of simplicity, we consider homogeneous parallel processors sharing memory without contention. A blocked periodic admissible parallel schedule is a set of lists $\{\Psi_i; i = 1, \ldots, M\}$ where M is the number of processors, and Ψ_i specifies a periodic schedule for processor i. When there exists a PASS, also a Periodic Admissible Parallel Schedule (PAPS) does exist [4]. For the latency $\tau[\ell + 1]$ of actor A_i at iteration index $\ell + 1$, it follows

$$\tau[\ell+1] = \max_i \left\{ \{\tau[\ell] + \boldsymbol{T}^\mathsf{T}\boldsymbol{v}[\ell]\}^{(1)}, \ldots, \{\tau[\ell] + \boldsymbol{T}^\mathsf{T}\boldsymbol{v}[\ell]\}^{(M)} \right\} \tag{3}$$

The sum-rule in (2) has turned into a max-rule because in the parallel implementation, actor A_i has possibly several M concurrent predecessors.

3 Synchronous Stochastic Data Flow

In soft RTOS, the scheduler meets deadlines only in a probabilistic way, implying that the WCRT of the actors is unbounded. The instantaneous response time varies over time but is finite. This fact motivates us to model the response time of actor A_i as random variable X_i with PDF $\mathrm{f}(x_i)$. The first moment of $\mathrm{f}(x_i)$ is the *expected response time*

$$\bar{T}_i \triangleq \mathrm{E}(X_i) = \int_{\mathbb{R}} x_i \mathrm{f}(x_i)\,\mathrm{d}x_i. \tag{4}$$

The second central moment is given by

$$\sigma_i^2 \triangleq \mathrm{Var}(X_i) = \int_{\mathbb{R}} (x_i - \bar{\tau}_i)^2 \mathrm{f}(x_i)\,\mathrm{d}x_i. \tag{5}$$

The square-root of (5) is defined as the *jitter* σ_i. Similarly, we model the latency at iteration index i as random variable $Y[i]$ with pdf $\mathrm{f}(y[i])$.

3.1 Sequential Implementation

To compute a PASS, the balance equation in (1) is independent of each actor's response time. For the latency, however, the sum in (2) has to be evaluated in the probabilistic sense. From the linearity of expectations, it follows for the expected latency at iteration index $\ell + 1$ that

$$\begin{aligned} \bar{\tau}[\ell+1] \triangleq \mathrm{E}(y[\ell+1]) &= \mathrm{E}(y[\ell] + \boldsymbol{x}^\mathsf{T}\boldsymbol{v}[\ell]) \\ &= \mathrm{E}(y[\ell]) + \mathrm{E}(\boldsymbol{x})^\mathsf{T}\,\boldsymbol{v}[\ell] \end{aligned} \tag{6}$$

regardless of whether the random variables are independent. In our case, they are and hence, it follows from (6) for the jitter at iteration index $\ell + 1$ that

$$\begin{aligned} \Sigma^2[\ell+1] \triangleq \mathrm{Var}(y[\ell+1]) &= \mathrm{Var}(y[\ell] + \boldsymbol{x}^\mathsf{T}\boldsymbol{v}[\ell]) \\ &= \mathrm{Var}(y[\ell]) + \sum_{i=1}^{N} v_i^2[\ell]\,\mathrm{Var}(x_i) \\ &= \mathrm{Var}(y[\ell]) + v_i^2[\ell]\,\mathrm{Var}(x_i) \end{aligned} \tag{7}$$

The latter equality results from the fact that only the i-th actor is active during iteration i. We have found expressions for the expected latency and jitter as a function of the iteration index.

Theorem 1. *Let X_1 and X_2 be two independent random variables with PDFs* $\mathrm{f}(x_1)$ *and* $\mathrm{f}(x_2)$. *Then the sum $Y = X_1 + X_2$ is a random variable with pdf* $\mathrm{f}(y)$ *being the convolution of* $\mathrm{f}(x_1)$ *and* $\mathrm{f}(x_2)$.

Proof. see ([6], Chap. 7).

Hence, the desired pdf at iteration index $\ell + 1$ can be written as

$$\mathrm{f}(y[\ell+1]) = \mathrm{f}(y[\ell]) * \frac{1}{|v_i|}\mathrm{f}(x_i)\,. \tag{8}$$

where $(\cdot) * (\cdot)$ denotes convolution between the former and the latter function.

3.2 Parallel Implementation

The PAPS executed on a soft RTOS is identical to that on a hard RTOS. The latency in (2), however, is unbounded. To obtain expressions for the expected latency and jitter, though, the maximization part in (2) can be solved as follows:

Theorem 2. *Let X_i, $i = 1, \ldots, M$ be M independent random variables with PDFs* $\mathrm{f}(x_i)$. *Then the maximum $Y = \max(X_i; i = 1, \ldots, M)$ is a random variable with pdf* $\mathrm{f}(y)$ *being the derivative of the product of the cumulative density functions (CDFs)* $\mathrm{F}(x_i)$.

Proof. We know that $Y < x$ is true only is every sample is less than x. As the random variables X_i are independent, the CDF $\mathrm{F}(y)$ yields

$$\mathrm{F}(y) = \mathrm{P}\left(X_1 \leq x, X_2 \leq x\right) = \prod_{i=1}^{M} \mathrm{P}\left(X_i < x\right) = \prod_{i=1}^{M} \mathrm{F}(x) \tag{9}$$

As the pdf is the first derivative of the cdf, we have

$$\mathrm{f}(y) = \frac{\mathrm{dF}(y)}{\mathrm{d}y} = \frac{\mathrm{d}\left\{\prod_{i=1}^{M} \mathrm{F}(x)\right\}}{\mathrm{d}x}\,. \tag{10}$$

This important result implies that the pdf of the latency at iteration index $\ell+1$ in (2) follows the following product-convolution rule:

$$\mathrm{f}(y[\ell+1]) = \frac{\mathrm{d}\left\{\prod_{i=1}^{M} \mathrm{F}(y[\ell+1])^{(i)}\right\}}{\mathrm{d}y[\ell+1]} \tag{11}$$

where $\mathrm{F}(y[\ell+1])^{(i)}$ is the cdf of (8) at processor i, and M is the number of parallel processes.

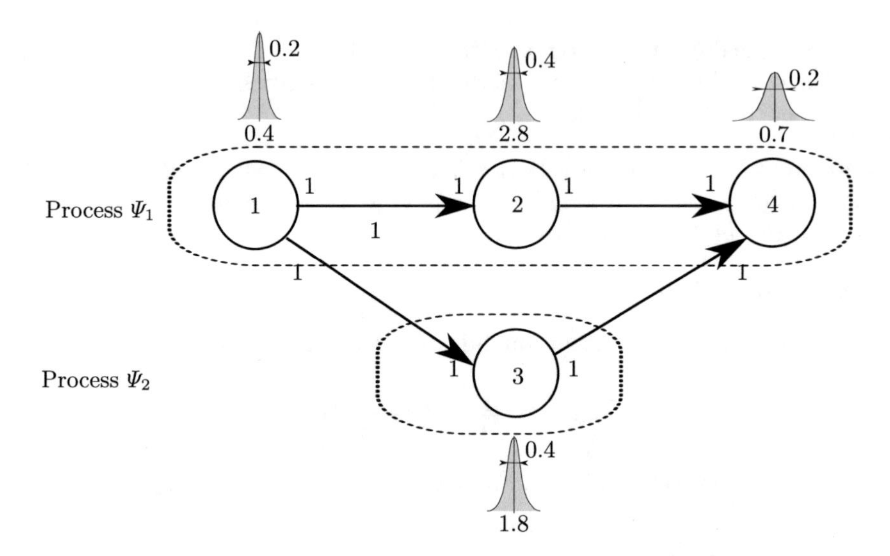

Fig. 2. SDF Example with $N = 4$ executed in hard and soft RTOSs.

4 Numerical Examples

To illustrate the proposed analysis tool, let us consider the SDF graph in Fig. 2 with $N = 4$ actors. The delay vector $\boldsymbol{D} \triangleq [D_{1,2}, D_{1,3}, D_{2,4}, D_{3,4}]^\intercal$ is given by $\boldsymbol{D} = [1, 0, 0, 0]^\intercal$ samples. For the hard RTOS implementation, let the vector containing the WCRT of the actors be given by $\boldsymbol{T} = [0.5, 3.0, 2.0, 0.8]$ ms. For the soft RTOS implementation, let the response time be Normal distributed with jitter $\boldsymbol{\sigma} = [0.1, 0.2, 0.2, 0.1]$ ms and expected value $\bar{\boldsymbol{T}} \triangleq \boldsymbol{T} - \boldsymbol{\sigma}$. In this case, the topology matrix Γ has the simple form

$$\boldsymbol{\Gamma} = \begin{bmatrix} 1 & -1 & 0 & 0 \\ 1 & 0 & -1 & 0 \\ 0 & 1 & 0 & -1 \\ 0 & 0 & 1 & -1 \end{bmatrix}. \tag{12}$$

Example 1 - Single processor implementation on hard and soft RTOSs. We know that rank $(\Gamma) = 3$. Hence, there exists an optimum firing vector $\boldsymbol{v}^\star$. Inserting (12) into (1) and solving with respect to $\boldsymbol{v}$ yields

$$\boldsymbol{v}^\star = \{1, 1, 1, 1\}.$$

Possible PASS sequences are $\Phi = \{1, 2, 3, 4\}$ and $\Phi' = \{1, 3, 2, 4\}$.

When the code is executed on *hard RTOS* and the schedule is purely sequential, the periodic latency $\tau[\phi]$ is bounded by the sum of the individual WCRTs. From (2), the periodic latency yields

$$\tau[4] = \sum_{i=1}^{4} T_i = 6.3 \text{ ms}. \tag{13}$$

When the code is executed on *soft RTOS*, it follows from (6) for the expected periodic latency and periodic jitter that

$$\bar{\tau}[4] = \sum_{i=1}^{4} X_i = 5.7 \text{ ms} = \tau[4], \quad \Sigma[4] = \sqrt{\sum_{i=1}^{4} \sigma_i^2} = 0.32 \text{ ms}.$$

Example 2 - Two-processor implementation on hard and soft RTOSs. Let us return to the SDF in Fig. 2. The topology matrix Γ in (12) has a null space. Therefore a PAPS does, too. Furthermore, it can be seen that this SDF has exactly one root vertex, namely Actor 1, and one final vertex, namely Actor 4. Hence, possible PAPS are

$$\Psi_1 = \{1, 2, 4\}; \Psi_2 = \{3\}. \tag{14}$$

Alternately,

$$\Psi_1' = \{1, 3, 4\}; \Psi_2' = \{2\}. \tag{15}$$

We choose the former schedule which is illustrated Fig. 2. In general, scheduling has impact on the periodic latency when executed in parallel. The optimum schedule has to be found by exhaustive search consuming NP complexity.

When the system is hard real-time, we readily obtain from (3)

$$\tau[3] = \max\{T_1 + T_2, T_1 + T_3\} + T_4 = 4.3 \text{ ms} \tag{16}$$

independent of the particular PAPS (14) and (15). Note that the period is only three. As the response time is Normal distributed, the recursion (8) yields

$$\begin{aligned} \mathrm{f}(y[1]) &\propto \mathcal{N}(x_1; \bar{\tau}[1], \Sigma[1]), \\ \mathrm{f}\left(y^{(\prime)}[2]\right) &\propto \mathcal{N}(x^{(\prime)}[2]; \bar{\tau}^{(\prime)}[2], \Sigma^{(\prime)}[2]), \\ \mathrm{f}\left(y^{(\prime\prime)}[2]\right) &\propto \mathcal{N}(x^{(\prime\prime)}[2]; \bar{\tau}^{(\prime\prime)}[2], \Sigma^{(\prime\prime)}[2]) \end{aligned}$$

with $\bar{\tau}[1] = 0.4$ ms, $\Sigma[1] = 0.1$ ms, $\bar{\tau}^{(\prime)}[2] = 3.2$ ms, $\Sigma^{(\prime)}[2] = \Sigma^{(\prime\prime)}[2] = 0.3$ ms and $\bar{\tau}^{(\prime\prime)}[2] = 2.2$ ms. From (11), the cdf of $Y[3]$ at iteration index 3 reads

$$\mathrm{F}(y[3]) = \Phi\left(\frac{y[3] - \bar{\tau}^{(\prime)}[3]}{\Sigma^{(\prime)}[3]}\right) \Phi\left(\frac{y[3] - \bar{\tau}^{(\prime\prime)}[2]}{\Sigma^{(\prime\prime)}[3]}\right)$$

with the cdf $\Phi(x) = 1/\sqrt{2\pi} \int_{-\infty}^{x} \exp(-t^2/2) \mathrm{d}t$. Moreover, $\bar{\tau}^{(\prime)}[3] = 3.9$ ms, $\bar{\tau}^{(\prime\prime)}[2] = 2.9$ ms and $\Sigma^{(\prime)}[3] = \Sigma^{(\prime\prime)}[3] = 0.4$ ms. For the first and second central moments of $Y[3] = \min(Y[2], X_i)$, there exists analytical expressions in [5]. We have

$$\begin{aligned} \bar{\tau}[3] = &\, \bar{\tau}^{(\prime)}[3]\, \Phi\left(\frac{\bar{\tau}^{(\prime)}[3] - \bar{\tau}^{(\prime\prime)}[3]}{\theta[3]}\right) + \bar{\tau}^{(\prime\prime)}[3]\, \Phi\left(\frac{\bar{\tau}^{(\prime\prime)}[3] - \bar{\tau}^{(\prime)}[3]}{\theta[3]}\right) \\ &+ \theta[3] \mathcal{N}(\bar{\tau}^{(\prime)}[3]; \bar{\tau}^{(\prime\prime)}[3], \theta[3]), \end{aligned}$$

$$\Sigma^2[3] = \left(\bar{\tau}^{(\prime)^2}[3] + \Sigma^{(\prime)^2}[3]\right)\Phi(\cdot) + \left(\bar{\tau}^{(\prime\prime)^2}[3] + \Sigma^{(\prime\prime)^2}[3]\right)\Phi(\cdot)$$
$$+ \left(\bar{\tau}^{(\prime)} + \bar{\tau}^{(\prime\prime)}\right)\mathcal{N}(\bar{\tau}^{(\prime)}[3]; \bar{\tau}^{(\prime\prime)}[3], \theta[3]) - \bar{\tau}^2[3]$$

with $\theta[3] \triangleq \sqrt{(\Sigma^{(\prime)^2}[3] + \Sigma^{(\prime\prime)^2}[3])}$. Ergo, the periodic latency and the periodic jitter are $\bar{\tau}[3] = 3.95$ ms and $\Sigma[3] = 0.74$ ms, respectively, in contrast to the instantaneous value $\tau[3] = 4.3$ ms in (16) for hard RTOS.

The example clearly demonstrates the power of our analysis tool. We are able to quantify all moments, including latency and jitter, of the target PDF at any iteration on the graph.

5 Conclusions

When the actors have bounded behavior, SDF graphs are ideally suited to estimate the performance of DSP applications executed on hard RTOS. On soft RTOS platforms, however, the worst case response time of actors is not bounded. To overcome the limitation, we have modeled the response time as (unbounded) probability density function (PDF). After one period of execution, the individual PDFs have propagate through the network. The first and second central moments of the joint PDF correspond to the expected system latency and jitter, respectively.

References

1. Bekooij, M., Hoes, R., Moreira, O., Poplavko, P., Pastrnak, M., Mesman, B., Mol, J.D., Stuijk, S., Gheorghita, V., van Meerbergen, J.: Dataflow Analysis for Real-Time Embedded Multiprocessor System Design, Philips Research, vol. 3. Springer, Dordrecht (2005)
2. Durfee, E., Rosenschein, J.: Distributed problem solving and multiagent systems: comparisons and examples. In: Proceedings of 13th International Workshop on Distributed Artificial Intelligence, pp. 94–104 (1994)
3. Hahm, O., Baccelli, E., Petersen, H., Tsiftes, N.: Operating systems for low-end devices in the internet of things: a survey. IEEE Internet Things J. **3**(5), 720–734 (2016)
4. Lee, E., Messerschmitt, D.G.: Synchronous data flow. Proc. IEEE **75**(9), 1235–1245 (1987)
5. Nadarajah, S., Kotz, S.: Exact distribution of the max/min of two Gaussian random variables. IEEE Trans. VLSI Syst. **16**(2), 210–212 (2007)
6. Papoulis, A.: Probability, Random Variables, and Stochastic Processes. McGraw Hill Inc. (1991)
7. Schaumont, P.R.: Data Flow Modeling and Transformation. Springer, Boston (2013)

Mobile Application for Smart City Management

Vivian F. López[1], María N. Moreno[1], Pablo Chamoso[1,2(✉)], Emilio S. Corchado[1], and Belén Pérez[1]

[1] Computer Sciences and Automation Department, Facultad de Ciencias, University of Salamanca, Plaza de Los Caídos s/n, 37008 Salamanca, Spain
{vivian,mmg,chamoso,escorchado,lancho}@usal.es

[2] BISITE Digital Innovation Hub, University of Salamanca, Edificio Multiusos I+D+I, 37007 Salamanca, Spain

Abstract. The wide acceptance by people when using mobile applications to communicate or interact, replacing traditional media, makes mobile applications a very useful tool today. Thanks to the interconnection that smartphones provide and the technological evolution that allows everyday objects to be connected with people, the concept of smart cities has gained strength. This paper presents an application that aims to connect the citizens of a city in a simple way with the administration and authorities of different sectors, in order to speed up communication and provide cities with a fast route (practically in real time) with which to obtain feedback from citizens.

Keywords: Smart cities · Mobile application · e-Government

1 Introduction

The development of mobile applications has become very popular in the last few years. This is because users are more keen to apply technology to different aspects of their daily life. Moreover, a large part of our society can afford to buy smartphones as their cost is much lower than in the past, they are easier to use but at the same time can perform increasingly complex tasks.

In today's society, people are almost always connected [1], that is, most of the time they move around with a data-connected device and use mobile connections to send and receive large amounts of information [2]. Smart cities are emerging as a result of the developments in technology and its growing capacity to improve all aspects of city life. Such cities generate large volumes of information that is generated by the users or the city elements (traffic lights, street lamps etc.) to be more efficient, cleaner, more productive and capable of finding sufficient conditions for faster development. One of the aims of smart cities is to use technology to adapt to the needs of the citizens that live in them, to do this they must find ways of connecting with citizens and of giving them the ability

S. Rodríguez et al. (Eds.): DCAI 2018, AISC 801, pp. 183–192, 2019.
https://doi.org/10.1007/978-3-319-99608-0_21

to work side by side with government entities in improving the quality of life in the city.

This is where the idea of creating a system for incident management came from. Moreover, a mobile application called Notif was designed for citizens living in urban areas; by installing this application on their mobile devices, they will be able to photograph and report the problems they encounter in the city. A management platform was created to help the administrations responsible for the city's services to resolve these incidents quickly. This application also helps to reduce costs as incidents that would have otherwise been unknown or unresolved for a long period of time will be reported by citizens from all parts of the city. Through the application they will be able to provide information with constant information and will ensure greater quality of life.

In addition, the competent administrations will have a greater volume of information on what is happening in their city, being able to benefit from the analysis and study of this data in order to establish strategies and projects for the improvement of city services and infrastructures.

The rest of the article is structured in the following way. The background analyses the concept of smart cities and its evolution over recent years. The proposed system section details the structure of the system and justifies the reasons for which it was followed. The results section presents the developed application. Finally, conclusions are drawn from the conducted work and future lines of work are discussed.

2 Background

The world is changing cities in a fast and long-term way. The cities themselves are constantly changing, evolving, without any particular destination. Today, we are at an important stage of this evolution. New technologies are emerging to change the way cities operate.

Social media is booming and thriving in this environment, revolutionizing the way in which leaders and businesses interact with citizens. As communities and states focus on increasingly national issues, cities must make use of the most advanced technologies to update the services they offer. New business models aim to create cutting-edge projects for greater efficiency in cities. Cognitive programming and its capacity to build citizen participation presents new opportunities to government organizations, to improve the lives of citizens and the business environment, as well as to offer personalized experiences and optimize the results of programs and services.

Smart cities are the result of the need to orient our lives towards sustainability in the urban environment. It is able to respond to the needs of institutions, businesses and citizens, both in economic terms and in social, environmental and coexistence aspects.

The concept of "smart city" appeared several years ago. In 1993 [3] the city of Singapore was already named an "intelligent city". Between 2000 and 2010, the concept of a "digital city" was prominent and was closely related to the smart city concept, although with different nuances. In [4], a digital city was defined as an open, complex and adaptable system, based on a computational network and urban information resources, that forms a virtual digital space for a city. While two different meanings for the concept of digital city are proposed in [5]: (i) a city that is being transformed or redirected through digital technology; (ii) a digital representation or reflection of some aspects of a real or imagined city. Digital cities were therefore a precedent to what is meant by smart city.

In the year 2007, a document [6] appeared which gave one of the first definitions for the term smart city as it is understood today, although it was highlighted that there was already a great ambiguity about the meaning of the concept at that time. The authors presented a smart city as a city that performs prospectively its activity in the industrial, educational, participative and technical infrastructure fields, combining them intelligently to serve its citizens.

However, it is not until 2010 that interest in smart cities increases exponentially and the number of definitions increases considerably.

The IBM company[1] published its own definition of the concept in the following section: [7]. Specifically, Harrison *et al.* defined a smart city as an "instrumented, interconnected and intelligent city", where by instrumented it refers to the capture and integration of real data in real time from the use of sensors, applications, personal devices and other sources. Interconnected refers to the integration of all such data into a computing platform that provides a set of services. Finally, the adjective intelligent refers to the component of complex analytical calculations, modeling, optimization and visualization of services in order to make better operational decisions.

Another definition can be found in [8]. The authors define a smart city as a city that operates in a sustainable and intelligent way, thanks to the integration of all its infrastructure and services as a cohesive whole and the use of intelligent devices for monitoring and control, guaranteeing sustainability and efficiency.

In order to understand the extent of the importance of smart cities, it is necessary to know their scope, since multiple aspects of citizens' daily life are addressed by this concept. In this article, the authors present the framework of smart cities in the following areas: transport, energy, education, health care, building, physical infrastructure, food, water and public safety.

Just as there are numerous definitions, there are also different approaches to the scope of smart cities. The first ever approach presented in this regard [9], indicates that the key dimensions of a smart city are: information technology (IT) in education, IT in infrastructure, IT in economics and quality of life. More recently, in Rudolf 2007smart, the authors propose economics, mobility, environment, people and government/administration as key areas. [10] states the following to be the key aspects of SCs: technology, economic development, more work opportunities and greater quality of life of citizens.

The wide range of areas that the smart city concept addresses and the technologies applies to them must be presented in a structured way; classified

[1] IBM - https://www.ibm.com.

according to their domain. In existing works such as the one published in [11], a classification according to domain is presented:

- Natural resources and energy [12–14]
- Transport and mobility [15,16]
- Smart building [17–19]
- Daily living [20–25]
- e-Government [26]
- Economy and society [27–30]

This article focuses on the category of e-Government, which in turn can be structured into three types of approaches:

- e-Government: digitisation of public administration through the management of documents and formalities using digital tools, in order to optimise work and provide new and faster services to citizens [31].
- Electronic democracy: use of information and communication systems for vote management [32].
- Transparency: allowing citizens to access official documents in a simple way. Decrease the likelihood of abuse of authorities that may use the system for their own interests or withhold relevant information [33].

3 Proposed System

Due to the importance and needs of smart cities to offer e-Government oriented solutions, all smart cities must have a system that offers a number of core services:

- Citizen participation: citizens must be provided with the ability to participate and decide on different aspects of city life. To this end, utilities or applications should enable direct access to governing entities, where they will be able to provide their opinions or report issues they encounter in the city.
- Information management: administrations or the people in charge of providing a service, must be able to access the information stored on the platform with a series of permits. Thus, there must be applications with restricted access that will offer, for example, a dashboard that intuitively controls and visualizes information without requiring specialized technical personnel.
- Real time notifications: whenever you work with information systems in real time, due to certain situations it may be necessary that a human take immediate action. Notification systems must be developed for such purpose.
- Information visualization: the information that is presented, either to the citizen or to administrative personnel, must be presented using different methodologies. The most common way for smart cities when the information is associated with a geographic component is to present it in a structured way according to its location.

In the first part of this first case study, the aim is to include all these basic elements in a system. It will integrate all the necessary functionalities to demonstrate the possibilities of citizen participation in the management of incidents from a dashboard. Enabling information to be displayed intuitively and issuing notifications to the managing authority.

When it comes to reaching citizens and allowing them to report incidents from any part of the city, it has been determined that the best way is to make a mobile application so that, from their own devices, they can easily report the incidents they encounter at any time.

The use of the mobile devices makes it easier to send incidents, since it is possible to automatically fill fields such as the location, read from the user's GPS position, or the use of the device's camera to justify or specify some aspect of the incident itself. One app for Android devices and one for iOS devices has been developed with the Xamarin[2] tool, which makes it possible to maintain an application for both operating systems with a single version of the code.

To create a dashboard for the monitoring and management of all the content, a web application has also been developed (by using HTML5, CSS, JavaScript, Bootstrap[3], Socket.io[4]). The web application allows to visualize and resolve incidents proposing the best solution to the human supervisor depending on the type of incidence. When a developer indicates that they want to get a copy of the project, the content of the web page is included within each node associated with the case study, so they can easily modify all the code to suit their design and functional needs.

4 Results

The result of this research is the designed mobile application, which allows citizens to report any problems that they encounter in the city. This mobile app is available for iOS and Android, the two most used mobile operative systems today. In Fig. 1 we can see two screenshots of the app. Users can create and administrate their own profile. When users report an incident, they help the corresponding administration to verify it. An incorporated metric measures the percentage of correctly reported incidences in the past, in order to establish the level of reliability of every profile.

Users can send a report with the help of just two simple clicks, if the user is reporting from the location of the incident, the mobile obtains the location automatically. In addition, users can attach graphic evidence to accompany the written text describing the incident, this can be done because the app provides access to the phone's camera and gallery.

These reports can be received as notifications in the browsers of the corresponding organizations, this depends on the category that the user has marked:

[2] Xamarin - https://www.xamarin.com/.
[3] Bootstrap - http://getbootstrap.com/.
[4] Socket.io - https://socket.io/.

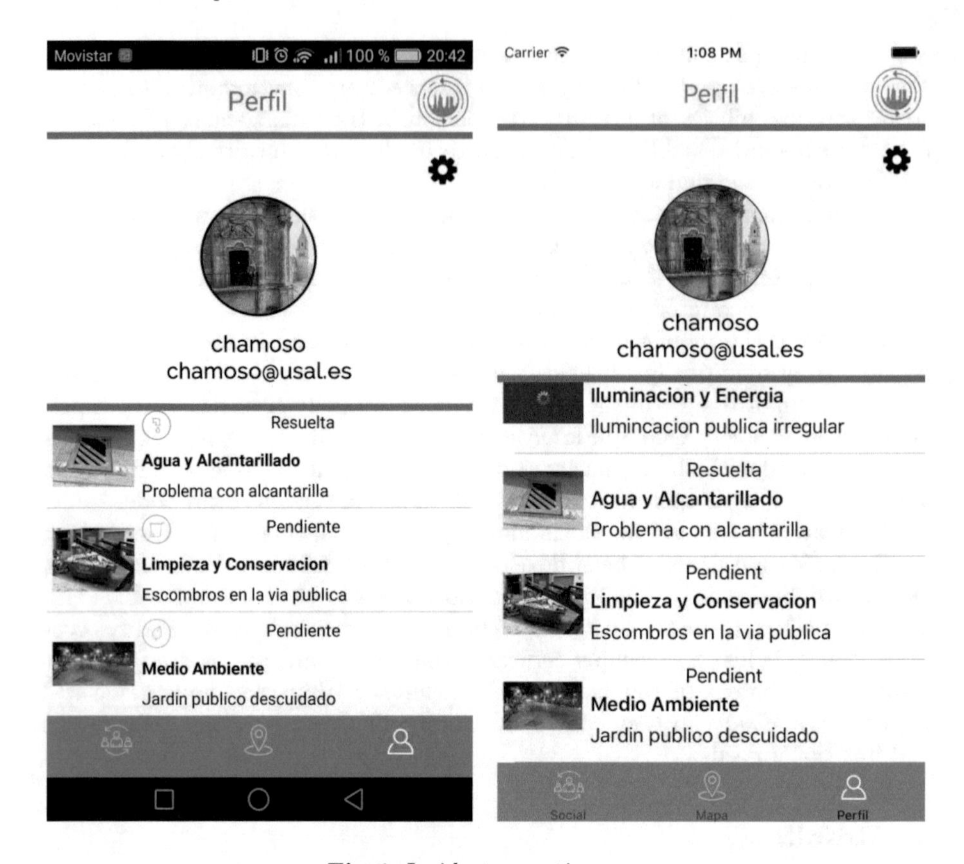

Fig. 1. Incident reporting app

water and sewerage, lighting and energy, cleaning and conservation, environment, pedestrians and cyclists, health, safety, traffic and roads, transport or urban planning. However, there may be a global organization profile that would include all incidents regardless of their categories.

The dashboard displays the reported incidents on the city map (Fig. 2) with an icon that shows its category, in addition there is a map that presents their status:

- Pending: when the incident has been received from the citizen, but has not yet been attended.
- Attended: when the corresponding notice has been sent to the maintenance entity responsible for solving the reported type of problem.
- Resolved: when the reported incident has been resolved.

The citizen who reported incidents can track their status from their profile on the app.

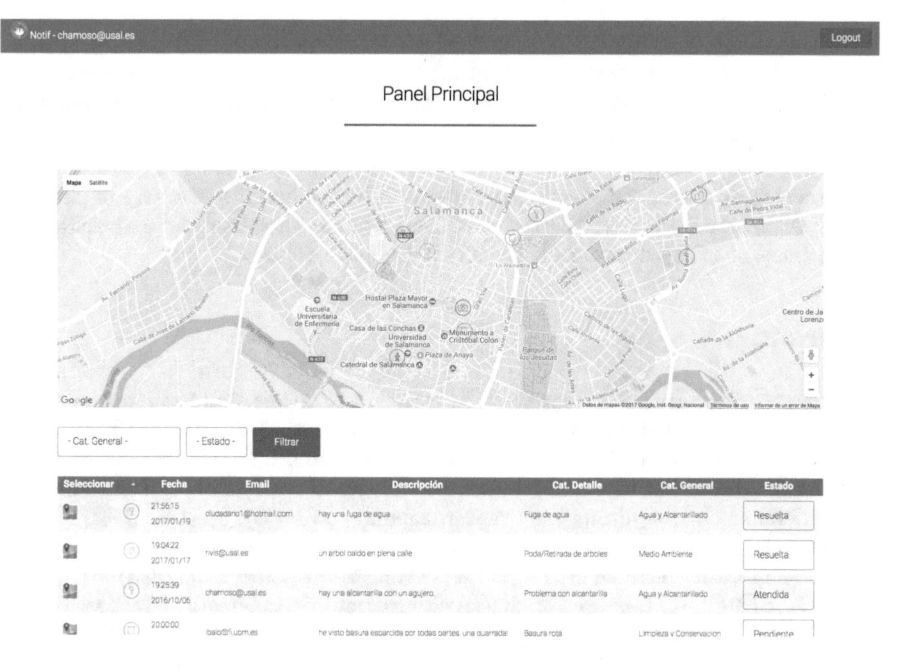

Fig. 2. Notifications dashboard

The system provides telephone or email contacts of the entities that are responsible for attending the different types of incidents, in this way the incident can be solved quickly.

The developed application acts as a tool that allows the city's citizens to report their needs, problems and inconveniances to the corresponding government entities, it also allows these entities to respond quicky and effectively to the problems encountered in the city.

Therefore, the app proposed in this work will give every citizen the same ability to contribute to improving different aspects of city life and to ensuring that no problems remain unresolved. Thus, the application assigns citizens the role of being stewarts of their own cities who, with the help of their smartphone can work side by side with governement entities.

5 Conclusion and Future Work

In conclusion, the proposed system allows citizens to report in real time the incidents they detect in the city. Depending on the type of problem, these reports are sent to the corresponding authority, with details that are not easy to provide when using traditional methods (phone, citizen service office, etc.), such as the exact location of the incident or even a photograph that accompanies the incident and corroborates that the report is true.

Future lines of work will be focused on the implementation of the application in different cities and on obtaining feedback that allows to expand on its functionalities. In addition, it will include a rating system that will encourage citizens to report problems.

Acknowledgements. This work was supported by the Spanish Ministry of Economy and FEDER funds. Project "SURF: Intelligent System for integrated and sustainable management of urban fleets" with ID: TIN2015-65515-C4-3-R.

References

1. Chamoso, P., Rivas, A., Rodríguez, S., Bajo, J.: Relationship recommender system in a business and employment-oriented social network. Inf. Sci. (2018)
2. Prieto, J., Chamoso, P., De la Prieta, F., Corchado, J.M.: A generalized framework for wireless localization in gerontechnology. In: 2017 IEEE 17th International Conference on Ubiquitous Wireless Broadband (ICUWB), pp. 1–5. IEEE (2017)
3. Heng, T.M., Low, L.: The intelligent city: Singapore achieving the next lap: practitoners forum. Technol. Anal. Strateg. Manage. **5**(2), 187–202 (1993)
4. Qi, L., Shaofu, L.: Research on digital city framework architecture. In: Proceedings 2001 International Conferences on Info-tech and Info-net ICII 2001, vol. 1, pp. 30–36. IEEE, Beijing (2001)
5. Ishida, T., Ishiguro, H., Nakanishi, H.: Connecting Digital and Physical Cities. Digital Cities II: Computational and Sociological Approaches, pp. 183–188 (2002)
6. Giffinger, R., Fertner, C., Kramar, H., Kalasek, R., Pichler-Milanović, N., Meijers, E.: Smart cities: ranking of European medium-sized cities: Centre of regional science (srf). Vienna University of Technology, Vienna, Austria, 15 Jan 2017. http://www.smart-cities.eu/download/smart_cities_final_report.pdf
7. Harrison, C., Eckman, B., Hamilton, R., Hartswick, P., Kalagnanam, J., Paraszczak, J., Williams, P.: Foundations for smarter cities. IBM J. Res. Develop. **54**(4), 1–16 (2010)
8. Hancke, G.P., Hancke Jr., G.P., et al.: The role of advanced sensing in smart cities. Sensors **13**(1), 393–425 (2012)
9. Mahizhnan, A.: Smart cities: the Singapore case. Cities **16**(1), 13–18 (1999)
10. Eger, J.M.: Smart growth, smart cities, and the crisis at the pump a worldwide phenomenon. I-WAYS J. E-Gov. Policy Reg. **32**(1), 47–53 (2009)
11. Neirotti, P., De Marco, A., Cagliano, A.C., Mangano, G., Scorrano, F.: Current trends in smart city initiatives: some stylised facts. Cities **38**, 25–36 (2014)
12. Faia, R., Pinto, T., Abrishambaf, O., Fernandes, F., Vale, Z., Corchado, J.M.: Case based reasoning with expert system and swarm intelligence to determine energy reduction in buildings energy management. Energy Build. **155**, 269–281 (2017)
13. Li, T., Sun, S., Bolić, M., Corchado, J.: Algorithm design for parallel implementation of the SMC-Phd filter. Sig. Process. **119**, 115–127 (2016). Cited by 14
14. Li, T., Sun, S., Corchado, J., Siyau, M.: A particle dyeing approach for track continuity for the SMC-Phd filter (2014). Cited by 10
15. Martín-Martín, P., González-Briones, A., Villarrubia, G., De Paz, J.F.: Intelligent transport system through the recognition of elements in the environment. In: International Conference on Practical Applications of Agents and Multi-Agent Systems, pp. 470–480. Springer (2017)

16. Corchado, J., Pavón, J., Corchado, E., Castillo, L.: Development of CBR-BDI agents: a tourist guide application. Lecture Notes in Computer Science (including subseries Lecture Notes in Artificial Intelligence and Lecture Notes in Bioinformatics), vol. 3155, pp. 547–559 (2004). Cited by 36
17. García, O., Chamoso, P., Prieto, J., Rodríguez, S., de la Prieta, F.: A serious game to reduce consumption in smart buildings. In: International Conference on Practical Applications of Agents and Multi-Agent Systems, pp. 481–493. Springer (2017)
18. Baruque, B., Corchado, E., Mata, A., Corchado, J.: A forecasting solution to the oil spill problem based on a hybrid intelligent system. Inf. Sci. **180**(10), 2029–2043 (2010). Cited by 31
19. Tapia, D., Fraile, J., Rodríguez, S., Alonso, R., Corchado, J.: Integrating hardware agents into an enhanced multi-agent architecture for ambient intelligence systems. Inf. Sci. **222**, 47–65 (2013). Cited by 23
20. Chamoso, P., De la Prieta, F., Pérez, J.B., Rodríguez, J.M.C.: Conflict resolution with agents in smart cities. In: Interdisciplinary Perspectives on Contemporary Conflict Resolution. IGI Global, pp. 244–262 (2016)
21. Corchado, J., Aiken, J.: Hybrid artificial intelligence methods in oceanographic forecast models. IEEE Trans. Syst. Man Cybern. Part C: Appl. Rev. **32**(4), 307–313 (2002). Cited by 31
22. Bajo, J., Corchado, J.: Evaluation and monitoring of the air-sea interaction using a CBR-agents approach, vol. 3620, pp. 50–62 (2005). Cited by 29
23. Bajo, J., Fraile, J., Pérez-Lancho, B., Corchado, J.: The thomas architecture in home care scenarios: a case study. Expert Syst. Appl. **37**(5), 3986–3999 (2010). Cited by 25
24. Corchado, J., Fyfe, C.: Unsupervised neural method for temperature forecasting. Artif. Intell. Eng. **13**(4), 351–357 (1999). Cited by 28
25. Tapia, D., Rodríguez, S., Bajo, J., Corchado, J.: Fusion* a SOA-based multi-agent architecture. Adv. Soft Comput. **50**, 99–107 (2009). Cited by 27
26. Angelopoulos, K., Diamantopoulou, V., Mouratidis, H., Pavlidis, M., Salnitri, M., Giorgini, P., Ruiz, J.F.: A holistic approach for privacy protection in e-government. In: Proceedings of the 12th International Conference on Availability, Reliability and Security, p. 17. ACM (2017)
27. Briones, A.G., Chamoso, P., Barriuso, A.L.: Review of the main security problems with multi-agent systems used in e-commerce applications. ADCAIJ: Adv. Distrib. Comput. Artif. Intell. J. **5**(3), 55–61 (2016)
28. Costa, A., Novais, P., Corchado, J., Neves, J.: Increased performance and better patient attendance in an hospital with the use of smart agendas. Logic J. IGPL **20**(4), 689–698 (2012). Cited by 11
29. Lima, A., De Castro, L., Corchado, J.: A polarity analysis framework for twitter messages. Appl. Math. Comput. **270**, 756–767 (2015). Cited by 12
30. Laza, R., Pavón, R., Corchado, J.: A reasoning model for CBR-BDI agents using an adaptable fuzzy inference system. Lecture Notes in Computer Science (including subseries Lecture Notes in Artificial Intelligence and Lecture Notes in Bioinformatics), vol. 3040, pp. 96–106 (2004). Cited by 14
31. Clohessy, T., Acton, T., Morgan, L.: Smart city as a service (scaas): a future roadmap for e-government smart city cloud computing initiatives. In: Proceedings of the 2014 IEEE/ACM 7th International Conference on Utility and Cloud Computing, pp. 836–841. IEEE Computer Society (2014)

32. Ali, S.M., Mehmood, C.A., Khawja, A., Nasim, R., Jawad, M., Usman, S., Khan, S., Salahuddin, S., Ihsan, M.A.: Micro-controller based smart electronic voting machine system. In: 2014 IEEE International Conference on Electro/Information Technology (EIT), pp. 438–442. IEEE (2014)
33. Janssen, M., van den Hoven, J.: Big and open linked data (bold) in government: a challenge to transparency and privacy? (2015)

Review of Technologies and Platforms for Smart Cities

Fernando de la Prieta[1,2](✉), Ana Belén Gil[1], María Moreno[1], and María Dolores Muñoz[1]

[1] Computer Sciences and Automation Department, Facultad de Ciencias, University of Salamanca, Plaza de Los Caídos s/n, 37008 Salamanca, Spain
{fer,abg,mmg,mariado}@usal.es
[2] BISITE Digital Innovation Hub, University of Salamanca, Edificio Multiusos I+D+I, 37007 Salamanca, Spain

Abstract. The importance that data have taken on in recent years, mainly due to technological evolutions (mainly in connectivity and processing capacity) and the reduction of associated costs, means that cities, generators of large volumes of data, invest and bet on infrastructures that analyze the data to obtain benefits. Such has been the importance that much of the efforts of computer scientists have focused on developing tools and platforms that allow cities to make the most of their information, becoming smart cities. This article presents a review of the definitions that the term smart city has received, as well as a review of the functionality of the most used platforms that give technological support to cities.

Keywords: Smart cities · Software platforms · Internet of Things

1 Introduction

Data has become a new raw material for any public or private organization, almost as important as material or labor. It is therefore clear that they can generate large profits, which have a positive and immediate impact on an increasingly socially aware society.

In recent years, the way in which information is generated, transmitted and analysed has changed enormously due to multiple advances in different areas. A great weight in these advances is directly linked to the evolution of technology. For example, there are smartphones with capacities very similar to desktop computers, high-speed Internet connection and geolocation systems. In addition to smartphones, today there are sensors of very different nature that allow you to optimize resources in industrial environments [1], monitor vehicles and roads, monitor vehicles, and monitor the Internet, monitor the condition of the elements of a building's elements [2] or integrate into hospital environments to provide new medical services (eHealth) [3–6]. Due to these advances, the term Internet of Things (IoT) appears, where sensors and actuators are integrated

S. Rodríguez et al. (Eds.): DCAI 2018, AISC 801, pp. 193–200, 2019.
https://doi.org/10.1007/978-3-319-99608-0_22

into the environment surrounding people on a day-to-day basis to use or provide information shared through technological platforms [7].

On the other hand, the emergence of new applications, such as social networks or open data sources, make a large amount of data of very diverse nature available not only to businesses or organizations, but also to the citizen. This fact has also favoured the above-mentioned change [8]. As a result, virtually all types of socio-economic, energy, public service and other data are readily available in real-time [9–14].

Finally, the concept of Smart City, as translated from English, appears to be a new paradigm of intelligent urban development and sustainable socioeconomic growth [15], although there is no shared and validated definition of exactly what a SC is, given the difficulty of identifying the set of common trends at global level, namely [16].

All of the above factors are closely related, but the latter concept, that of SC, is the central component that encompasses the rest.

This article presents a review of the state of the art on the smart city concept and the different platforms currently being used. The following section presents the background, then the existing platforms and finally a discussion and conclusions drawn.

2 Background

Over the last five years, the term SC has been widely used around the world and has had a major impact on urban strategies in cities of all sizes. The SC concept is easily associated with the application of technological advances by a city. In addition, it is common to misrelate this to energy efficiency alone, but while energy efficiency is a very important part of a SC, the whole idea of SC is not focused solely on energy or buildings. SC encompasses the entire human ecosystem: the benefits to society, the economic growth it brings and the opportunities it can create.

Therefore, it is a term that encompasses a large number of concepts or areas and, given its recent appearance, it is a trend on which there is still some confusion as to its true definition and extent. However, in the literature there are numerous works that propose definitions of a SC, the most relevant ones are presented below.

The concept of SC as an idea appeared several years ago. Already in 1993, in 1993, the city of Singapore was presented as a "intelligent city" in 1993. Between the years 2000 and 2010 appears the concept of "digital city" or digital city, closely related to the idea of SC, although with different nuances. In [17] a digital city is defined as an open, complex and adaptable system, based on a computational network and urban information resources, that forms a virtual digital space for a city. While two different meanings for the concept of digital city are proposed in [18]: (i) a city that is being transformed or redirected through digital technology; (ii) a digital representation or reflection of some aspects of a real or imagined city. Digital cities are therefore a precedent for what is meant by SC.

In the year 2007, in the document presented [19] appears one of the first definitions for the term SC as it is understood today is presented, although it is stressed that there is already a great ambiguity about the meaning of the concept at that time. The authors present a SC as a city that performs prospectively its activity in the industrial, educational, participative and technical infrastructure fields, combining them intelligently to serve its citizens.

However, it is not until 2010 that interest in SCs increases exponentially and the number of definitions increases considerably.

The company IBM[1], publishes its own definition of the concept in [20]. More specifically, Harrison *et al.* define a SC as an "instrumented, interconnected and intelligent city", where by instrumented it refers to the capture and integration of real data in real time from the use of sensors, applications, personal devices and other sources. Interconnected refers to the integration of all such data into a computing platform that provides a set of services. Finally, the adjective intelligent refers to the component of complex analytical calculations, modeling, optimization and visualization of services in order to make better operational decisions.

Another definition can be found in [21]. The authors define a SC as a city that operates in a sustainable and intelligent way, thanks to the integration of all its infrastructure and services to the citizen as a cohesive whole and the use of intelligent devices for monitoring and control, guaranteeing that sustainability and efficiency [22].

In order to understand how far the importance of SCs reaches, it is necessary to know their dimensions, since there are multiple areas of the daily life of citizens that SCs address. In this article, the authors present the framework of SCs in the following areas: transport, energy, education, health care, building, physical infrastructure, food, water and public safety.

As with the definition of the concept, there are different approaches to the dimensionality of SCs. The oldest of all, presented at [23], indicates that the key dimensions of a SC are: information technology (IT) in education, IT in infrastructure, IT in economics and quality of life. More recently, in [24], the authors propose economics, mobility, environment, people and government/administration as key dimensions. [25] defines that the key aspects of a SC are technology, economic development, job growth and increasing the quality of life of its citizens.

However, in all cases, the inclusion of intelligence in each sub-system of a city, individually, is not sufficient to create a SC or to make that city be considered intelligent, but a city should be considered as an organic whole [26], so that it fits perfectly with the definition presented in [21].

The final beneficiary in the vast majority of definitions is the citizen. In the SCs review recently presented in [27], it is concluded that the main objective of cities and regions is "people first and foremost".

In any case, it is clear that a SC requires an architecture with technological support, capable of storing and processing all the information in a connected and distributed way to provide a series of services in different areas that benefit

[1] IBM - https://www.ibm.com.

all its citizens. In terms of how these services are provided, there are several existing approaches that have been presented in recent years.

3 Existing Platforms

Throughout this section the different existing solutions created to support SC-oriented services are presented. The vast majority of them use technologies such as those previously presented.

When describing the different existing platforms, it is important to differentiate between platforms oriented to the use only on the part of a set determined by city managers (specific platforms) and generic platforms oriented to provide services in a general way, which can be used by any city that wishes to do so [10]. However, they can also be classified into public access platforms or private platforms, depending on whether any developer can use the platform to use its services or not, in which case the business model involves selling the private platform to different municipalities that decide to use its services when transforming their city into a SC.

- Sentilo[2]: Sentilo is an architectural part that isolates the applications developed to exploit the information generated by a city and the layer of sensors deployed throughout the city to collect and disseminate this information. The project began in 2012 with the Barcelona City Council and was used to put Barcelona at the forefront of SCs [28]. Nowadays, although it originated exclusively oriented towards the city of Barcelona, it has evolved and is being used by other cities such as Terrassa or Reus. It is also an open source software, which provides the source code through its own repository. It is therefore a generic and public system.
- SmartSantander[3]: the SmartSantander project proposes an urban-scale experimental research facility where different applications are supported and typical SC services are deployed. The project is not intended to be limited solely to the city of Santander, but is intended to extend to other cities such as Belgrade, Guildford or Lübeck. The project is private, however, it offers a free information access system for developers to use to make new applications. It is also adaptable to new cities, although privately, so it is considered specific to a closed group of certain cities.
- IBM Intelligent Operation Center[4]: is a private platform, owned by the IBM company, which is deployed in different cities around the world, such as Rio de Janeiro. It offers an environment that provides different default tools, but can be customized on demand. It is therefore a private and specific system, as it requires adaptation and maintenance by the owner company ýcitezhuhadar2017next.

[2] Sentilo - http://www.sentilo.io/.

[3] SmartSantander - http://www.smartsantander.eu/.

[4] IBM Intelligent Operation Center - http://www-03.ibm.com/software/products/es/intelligent-operations-center.

- CitySDK[5]: the CitySDK project aims to provide a programming structure to deploy systems for SCs, which has been tested in 8 cities in Europe: Amsterdam, Barcelona, Helsinki, Istanbul, Lamia, Lisbon, Manchester and Rome, with the participation of more than 5 private companies in collaboration with 5 universities. They allow the integration of new cities, but only from the use of the API they propose. It is therefore a private and general system.
- Open Cities[6]: is a platform that allows the data stored on it to be used (read and write access) in order to be used by developers to offer services in cities. It is a private system of free use of a generic nature (not oriented to specific cities).
- i-SCOPE[7]: is a platform that provides three types of services to SCs [29]: (i) improving the inclusion and mobility of citizens with routing systems and signalling barriers in the city; (ii) optimising energy consumption; (iii) environmental control. However, it is a private project, already completed and specific, of which the cities in which it is implemented are unknown.
- People[8]: is a platform that provides services, generally open source for the community to use and share those that develop, always oriented to SCs. For example, they have services in cities such as Bilbao, Bremen or Thermi. It is a public project, but can only be used in specific environments.
- IoT Open platforms[9]: is an initiative that provides a set of libraries, technical documentation, web services and protocols openly for use by the entire developer community. Among the tools they offer is VITAL-OS Smart City Platform, for example, which provides a set of visual tools to develop applications with reduced cost and effort. They use specific city-specific datasets.

But in addition to platforms directly oriented to cities, it is necessary to detail the different existing platforms that provide services to deploy infrastructures that can be used by cities, such as the following ones:

- FIWARE[10]: FIWARE is a platform that allows the development and deployment of Internet applications in multiple vertical sectors from a series of APIs. This allows many of these verticals to be related to the numerous services that can be offered in SCs, so one of the most interesting approaches that have been made with this platform is oriented towards this line.
- Carriots[11]: is a PaaS-type platform, designed to be used by IoT and M2M, so it can be used to connect the information-providing infrastructure to be used by a SC. However, the platform remains at this level, without offering user-oriented services, a layer that would be completely detached from the platform.

[5] CitySDK - https://www.citysdk.eu/.
[6] Open Cities - http://opencities.upf.edu/.
[7] i-SCOPE - http://www.iscopeproject.net/.
[8] People - http://www.people-project.eu/.
[9] IoT Open platforms - http://open-platforms.eu/.
[10] FIWARE - https://www.fiware.org/.
[11] Carriots - https://www.carriots.com/.

- Kaa[12]: It is an initiative that defines as an open and efficient cloud platform for providing IoT solutions. Frequent solutions include connecting all types of sensors that can be found or deployed in a SC.
- Sofia2[13]: Sofia2 is a middleware that allows the interoperability of multiple systems and devices, offering a semantic platform that makes real-world information available to intelligent applications, mainly oriented to the IoT.
- Webinos[14]: is a web application platform that allows developers to access native resources through APIs. This allows any device (IoT) to be easily connected.
- ICOS[15]: is an open repository of solutions for SCs, offering a set of existing applications and projects that can be reused for application creation.

Many of these tools use information from open data sources, including the following platforms:

- CKAN[16]: is a data management system that makes it accessible through the tools they provide to publish, share, search and use this data in streaming. This system is used by platforms that provide data such as Data.gov[17], which includes more than 190,000 data sets from multiple U. S. cities. They include topics such as local government, agriculture, climate, energy or education among others. It is also used by the Berlin Open Data[18] platform, which collects data from the city of Berlin.
- DKAN[19]: facilitates the publication of open-data with a system based on the famous Drupal[20].
- Socrata[21]: is a cloud-based solution that allows government organizations to publish their data online.
- OpenDataSoft[22]: includes 1,397 data sets from different countries on multiple topics.

4 Discussion

In conclusion, it can be seen that there are numerous service-oriented platforms that can be delivered in SCs, however, there are few solutions that have explicitly emerged directly oriented to support SCs. Among them, most of them are private although they allow free access or use (in exchange for giving the information)

---

[12] Kaa - https://www.kaaproject.org/.
[13] Sofia2 - http://sofia2.com/.
[14] Webinos - http://webinos.org/.
[15] ICOS - http://icos.urenio.org/.
[16] CKAN - https://ckan.org/.
[17] Data.gov - https://catalog.data.gov/dataset.
[18] Berlin Open Data - https://daten.berlin.de/.
[19] DKAN - http://getdkan.com/.
[20] Drupal - https://www.drupal.org/.
[21] Socrata - https://socrata.com/.
[22] OpenDataSoft - https://public.opendatasoft.com.

and only Sentilo allows you to download your code to be able to replicate in other cities freely, although it does not provide any high-level service by default.

There is, therefore, a lack of a generic and open platform that includes a set of high-level services that developers can use to take greater advantage of the available information and knowledge of the city that is encapsulated in the platform.

Acknowledgements. This work was supported by the Spanish Ministry of Economy and FEDER funds. Project "SURF: Intelligent System for integrated and sustainable management of urban fleets" with ID: TIN2015-65515-C4-3-R.

References

1. Shrouf, F., Ordieres, J., Miragliotta, G.: Smart factories in industry 4.0: a review of the concept and of energy management approached in production based on the internet of things paradigm. In: 2014 IEEE International Conference on Industrial Engineering and Engineering Management (IEEM), pp. 697–701. IEEE (2014)
2. Moreno, M., Úbeda, B., Skarmeta, A.F., Zamora, M.A.: How can we tackle energy efficiency in iot basedsmart buildings? Sensors **14**(6), 9582–9614 (2014)
3. Garcia-Ortiz, L., Perez-Ramos, H., Chamoso-Santos, P., Recio-Rodriguez, J., Garcia-Garcia, A., Maderuelo-Fernandez, J., Gomez-Sanchez, L., Martínez-Perez, P., Rodriguez-Martin, C., De Cabo-Laso, A.: [pp. 08.02] automatic image analyzer to assess retinal vessel caliber (altair) tool validation for the analysis of retinal vessels. J. Hypertension **34**, e160 (2016)
4. Bajo, J., Fraile, J., Pérez-Lancho, B., Corchado, J.: The thomas architecture in home care scenarios: a case study. Expert Syst. Appl. **37**(5), 3986–3999 (2010). Cited by 25
5. Tapia, D., Corchado, J.: An ambient intelligence based multi-agent system for alzheimer health care. Int. J. Ambient Comput. Intell. **1**(1), 15–26 (2009). Cited by 13
6. Costa, A., Novais, P., Corchado, J., Neves, J.: Increased performance and better patient attendance in an hospital with the use of smart agendas. Logic J. IGPL **20**(4), 689–698 (2012). Cited by 11
7. Gubbi, J., Buyya, R., Marusic, S., Palaniswami, M.: Internet of things (IOT): a vision, architectural elements, and future directions. Future Gener. Comput. Syst. **29**(7), 1645–1660 (2013)
8. Janssen, M., Charalabidis, Y., Zuiderwijk, A.: Benefits, adoption barriers and myths of open data and open government. Inf. Syst. Manage. **29**(4), 258–268 (2012)
9. Li, T., Sun, S., Bolić, M., Corchado, J.: Algorithm design for parallel implementation of the SMC-Phd filter. Sig. Process. **119**, 115–127 (2016). Cited by 14
10. Chamoso, P., De la Prieta, F., De Paz, F., Corchado, J.M.: Swarm agent-based architecture suitable for internet of things and smartcities. In: 12th International Conference Distributed Computing and Artificial Intelligence, pp. 21–29. Springer (2015)
11. Tapia, D., Fraile, J., Rodríguez, S., Alonso, R., Corchado, J.: Integrating hardware agents into an enhanced multi-agent architecture for ambient intelligence systems. Inf. Sci. **222**, 47–65 (2013). Cited by 23

12. Choon, Y., Mohamad, M., Deris, S., Illias, R., Chong, C., Chai, L., Omatu, S., Corchado, J.: Differential bees flux balance analysis with optknock for in silico microbial strains optimization. PLoS ONE **9**(7) (2014). Cited by 10
13. Lima, A., De Castro, L., Corchado, J.: A polarity analysis framework for twitter messages. Appl. Math. Comput. **270**, 756–767 (2015). Cited by 12
14. Mata, A., Corchado, J.: Forecasting the probability of finding oil slicks using a CBR system. Expert Syst. Appl. **36**(4), 8239–8246 (2009). Cited by 17
15. Hollands, R.G.: Will the real smart city please stand up? intelligent, progressive or entrepreneurial? City **12**(3), 303–320 (2008)
16. Neirotti, P., De Marco, A., Cagliano, A.C., Mangano, G., Scorrano, F.: Current trends in smart city initiatives: some stylised facts. Cities **38**, 25–36 (2014)
17. Qi, L., Shaofu, L.: Research on digital city framework architecture. In: Proceedings 2001 International Conferences on Info-tech and Info-net ICII 2001, vol. 1, pp. 30–36. IEEE, Beijing (2001)
18. Ishida, T., Ishiguro, H., Nakanishi, H.: Connecting digital and physical cities. Digital Cities II: Computational and Sociological Approaches, pp. 183–188 (2002)
19. Giffinger, R., Fertner, C., Kramar, H., Kalasek, R., Pichler-Milanović, N., Meijers, E.: Smart cities: ranking of European medium-sized cities: Centre of regional science (srf). vienna university of technology, Vienna, Austria, 15 Jan 2017. http://www.smart-cities.eu/download/smart_cities_final_report.pdf
20. Harrison, C., Eckman, B., Hamilton, R., Hartswick, P., Kalagnanam, J., Paraszczak, J., Williams, P.: Foundations for smarter cities. IBM J. Res. Develop. **54**(4), 1–16 (2010)
21. Hancke, G.P., Hancke Jr., G.P., et al.: The role of advanced sensing in smart cities. Sensors **13**(1), 393–425 (2012)
22. Chamoso, P., De la Prieta, F., Pérez, J.B., Rodríguez, J.M.C.: Conflict resolution with agents in smart cities. In: Interdisciplinary Perspectives on Contemporary Conflict Resolution. IGI Global, pp. 244–262 (2016)
23. Mahizhnan, A.: Smart cities: the singapore case. Cities **16**(1), 13–18 (1999)
24. Rudolf, G., Fertner, C., Kramar, H., Kalasek, R., Pichler-Milanovic, N., Meijers, E.: Smart cities-ranking of European medium-sized cities. Rapport technique, Vienna Centre of Regional Science (2007)
25. Eger, J.M.: Smart growth, smart cities, and the crisis at the pump a worldwide phenomenon. I-WAYS J. E-Gov. Policy Regul. **32**(1), 47–53 (2009)
26. Moss Kanter, R., Litow, S.S.: Informed and interconnected: a manifesto for smarter cities. Harvard Business School General Management Unit Working Paper (09-141) (2009)
27. Boulos, M.N.K., Tsouros, A.D., Holopainen, A.: Social, innovative and smart cities are happy and resilient: insights from the who euro 2014 international healthy cities conference. Int. J. Health Geogr. **14**(1), 3 (2015)
28. Bakıcı, T., Almirall, E., Wareham, J.: A smart city initiative: the case of barcelona. J. Knowl. Econ. **4**(2), 135–148 (2013)
29. De Amicis, R., Conti, G., Patti, D., Ford, M., Elisei, P.: I-Scope-Interoperable Smart City Services through an Open Platform for Urban Ecosystems. na (2012)

Virtual Organization for Fintech Management

Elena Hernández[1], Angélica González[2], Belén Pérez[2], Ana de Luis Reboredo[2], and Sara Rodríguez[1,2(✉)]

[1] BISITE Research Group, University of Salamanca, Edificio Multiusos I+D+I, Calle Espejo s/n, 37007 Salamanca, Spain
{elenahn, srg}@usal.es

[2] Computer Sciences and Automation Department, Facultad de Ciencias, University of Salamanca, Plaza de Los Caídos s/n, 37008 Salamanca, Spain
{angelica, lancho, adeluis, srg}@usal.es

Abstract. A review of the state of the art on Fintech and the most important innovations in the financial technology is presented in this article. It is proposed a social computing platform based on VOs which allow to improve user experience in all that is associated with the process of investment recommendation. Moreover, a case study is shown in which the VOs modules have been described graphically, the agent functionalities have been explain and the algorithms responsible for making recommendation have been proposed.

Keywords: Fintech · Digitalization · Virtual organization of agents Recommender system

1 Preliminary Concepts

Fintech is a term associated with The Financial Services Technology Consortium, a project initiated by Citigroup in 1990 to facilitate technological cooperation, however, by 2014 it aroused the interest of the financial industry, consumers, entities regulators, governments and academic researchers alike [2].

The emergence of Fintech is a result of the global economic-financial crisis that occurred in 2008. Companies known as Fintech distanced themselves from traditional banking in order to be able to offer the traditional services offered by banks, due to the cheapening of technology. In this way, small companies that grow in a technological environment have been able to use social networks to expand their market share [20]. Other aspects have also contributed to their expansion, such as the widespread use of smartphones, the bad reputation acquired by banks as well as the lack of transparency and the emergence of a new collaborative economy. With "new collaborative economy", we refer to services that allow individuals to share goods and services [32] instead of the usual business model in which goods and services are exchanged between the company and the individual. Collaborative Economy is defined as an economic model where ownership and access to goods and services are shared not only between companies but also between people who share ownership and give access to goods and services, thus promoting commercial growth [3].

S. Rodríguez et al. (Eds.): DCAI 2018, AISC 801, pp. 201–210, 2019.
https://doi.org/10.1007/978-3-319-99608-0_23

The research presented in this paper, focuses on an investment recommendation system for businesses in order to provide investment related suggestions. For this purpose, we identified different factors that could be extracted from the internet and from the information provided by the users. Perhaps, the biggest challenge is to gather relevant information to make through case-based reasoning (CBR), useful investment recommendations [4, 23].

The article is structured in the following way: in Sect. 2 we analyzed the concept of Financial Technology, and the data-oriented technology that this implies. We also describe Fintech's requirements, how it is being used to optimize business. The concept of Virtual Organizations is also described. VOs are our starting point in creating a recommendation system proposed in Sect. 3. Finally, we end the article with conclusions and with future lines of work that we have proposed in order to improve the recommender system in Sect. 4.

2 Financial Technology

Financial Technology (Fintech) can be considered as a consequence of the disruption of cloud computing, mobile devices, big data, cybersecurity and other Internet-related technologies, offering emerging business models that are more efficient, safer, innovative and more flexible than existing financial services [11].

In [16] the authors refer to how Fintech is sometimes used as a synonym of Digital Finance or e-finance when it comes to describing the processes of change in the financial sector, however, each of these terms suits different contexts. e-finance corresponds to the incorporation of information and communication technologies in financial sector companies, while the digital finance term is used to describe the digitization of the financial industry in general (electronic products, Chip credit cards, online banking, ATMs, electronic payment systems). IT Fintech companies present the following characteristics:

- Finance oriented
- Highly innovative companies
- New technologies are fundamental
- A challenging alternative to banking

However, beyond the generic characteristics that accompany Fintech companies, Molina (2016) in [31] shows that there are many types of companies (Fig. 1). Startups and small companies, unicorns and GAFAs, each has its own peculiarities but all have technology and it makes them more competitive than traditional financial system.

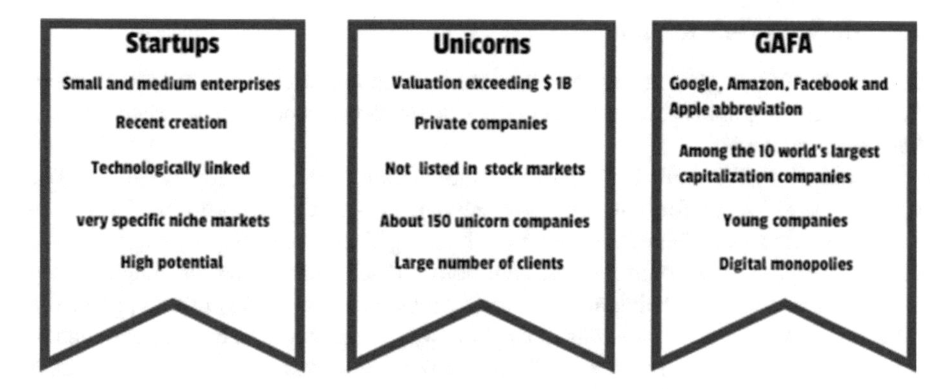

Fig. 1. Fintech companies' classification

2.1 Data-Oriented Financial Technology

Having defined the concept and motivations of Fintech, we should also mention the methodologies used to develop it. Taking into account that it is necessary to handle large amounts of data, the starting point of many authors is data processing and its security [20]. Data-oriented techniques begin with the mining of operational data in the context of Big Data. This is the main technique for obtaining valuable information, it allows to analyze large volumes of data. In the field of banking, Big Data techniques have been considered an essential tool when dealing with financial data [**¡Error! No se encuentra el origen de la referencia.**]. On the whole, the researchers developed on the handling of data in the field of Financial Technology are aimed at improving financial services or creating new ones. Obtaining datasets helps to distinguish processes, impacts and results and find solutions [16, 17]. In this regard, in order to improve performance and guarantee privacy, some authors have used machine learning for large size datasets [42].

2.2 Fintech Requirements

Due to the large amount of work, modern businesses use big data centers. For this reason, many businesses have been interested in the optimization of memory designs and energetic efficiency of the data centers [5, 15, 19, 32]. Researches intend to optimize computing performance with scalable and flexible systems. This tendency is a data mining challenge in a distributed environment, (Lu *et al.* 2008) [30]. Proposes a solution to the training problem in mining distributed data, a mechanism that could guarantee that different servers will process distributed data simultaneously, considering both the cost and efficiency. Yu *et al.* 2015 added another variable: availability. Furthermore, they stressed the importance of integrating the Data Base Management System with storage, security and performance requirements. The result they obtained was that the input/output operation was 27 time faster than the traditional method [43].

2.3 Business Optimization

In the field of management, the most commonly used approaches are optimization and machine learning at the time of making investment recommendations or when creating businesses strategies.

Li and Hoi [27] applied machine learning as online decision support system. The study consisted in performing an online survey on investment portfolios. They presented selection as a sequential problem, obtaining five group categorization of solutions to the problem of online investment portfolio selection. However, they affirmed that precision continued being an unresolved problem. Wang (2015) [40, 41] applied a different model to stock operations, it used fuzzy systems theory to transfer negotiation rules. Another approach to stock performance prediction was proposed by Hadavandi *et al.* en [22], they applied neural networks and integrated genetic fuzzy systems to predict performance on stock markets.

To sum up, many different data based techniques have been used in investment recommendation proposals: Machine learning, fuzzy logic algorithms, neural networks, etc. Nevertheless, it is necessary to include another approach to creating investment recommendations in the business sector. For this reason, the next section will overview the concept of Agent-based Virtual Organizations and the reasons for which they are a suitable recommendation model.

2.4 Virtual Organization (VO) of Agents

Agent technology is a branch of Distributed Artificial Intelligence (DAI). MAS (Multi-agent Systems) integrate different capabilities, of which the most notable are autonomy, reactivity, proactivity, learning, ubiquitous distributed communication and most importantly the intelligence of all their elements. These characteristics meet a large part of the requirements posed by Financial Technology, adapting to the needs of users in a ubiquitous, autonomous and dynamic manner [6, 18, 44].

In the field of computing, concretely in that of multi-agent systems, organizations are used to describe a group of agents who are coordinated through a series of behavioral patterns and roles aimed at achieving the system's objectives. A multi-agent system model has to be able to define organizations that can adapt dynamically to changes in the environment or to the specifics of the organization.

Dynamic adaptation includes adapting to changes in the structure and behavior of the MAS, as well as the addition, deletion and substitution of components during system execution, without affecting its correct functioning [8, 35].

Virtual Organizations (VOs) have a series of common characteristics:

- An organization of agents is made up of agents, roles and coordination and interaction rules.
- It pursues a common and global objective which is irrespective of the objectives of particular agents.
- Roles are assigned to the different agents. Thus, their task within the system is specialized in order to achieve the organization's global objectives.
- It divides the system in groups through departmentization, these groups are units of interaction between agents.

– It defines a series of limits for the agents belonging to the organization, their rules of interaction, its functionality and the services it offers.
– The entry and exit of agents in and out of the organization determines its dynamics, its roles can change depending on the objective of the organization.

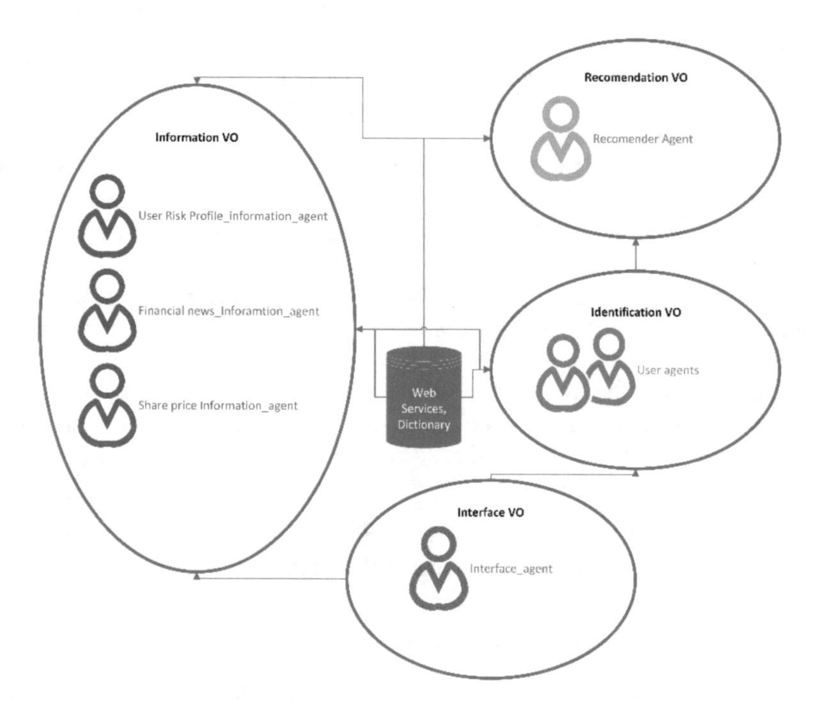

Fig. 2. Stock Investment recommender platform

3 Design Proposal

In this section is describe a design based on VOs with human-agent interaction modules which allow to improve user experience in all that is associated with the process of investment recommendation, taking into account the characteristics of Agent-based Virtual Organizations [9, 13, 18, 21, 24, 28].

Our proposal is designed as a heterogeneous system in languages, applications and characteristics. Figure 2 illustrates the different elements of the platform together with the modules that make it up. Below, each of these modules is described, each of them will individually compose a Virtual Organization, with distinctive characteristics, rules and structures.

– Information (V.O.): This group of agents is in charge of searching and processing information. In this case, we created different sub-organizations which are in charge of calculating the variables that are part of the recommendation system:

- User Risk Profile_Information_agent: this agent is responsible for collecting information on investment profiles, considering the level of risk that the investor is prepared to take (asset classes, profitability, interest rate, etc.).
- Share Price_Information_agent: it is in charge of obtaining the public process of shares.
- Financial news_ Information_agent: This agent is responsible for obtaining financial news published in the media which list the transactions of businesses, both internal and external. Thy will be included in the Recommendation System in order to be able to extrapolate patterns and provide users with accurate recommendations.

– Recommendation (V.O.): It is responsible for making the different investment recommendations. The Recommender_agent is in charge of calculating factor and weight, and of managing the CBR System where the suggestions of the users are stored.
– Identification (V.O.): User_agent is the interface that allows the user to Access recommendation functionalities. It is in charge of generating and updating a user profile.
– Interface (V.O.): the Interface_agent is responsible for showing the user the investment recommendation when the access the system, this recommendation will be based on the user profile previously created by the User_agent for personalized investment recommendations.

Once functionalities are defined, this distributed design is going to facilitate subsequent development and allow for future modifications and extensions.

3.1 Machine Learning (ML) Algorithms

In [37] the authors refer to Machine Learning (ML) algorithms like effective in fitting parameters automatically, avoiding over-fitting, and being capable of combining multiple inputs. Also mentioned that ranking investors' sentiment hence provides a natural way to select stocks based on the "portrayed performance" in news media.

Reviewing the literature, it is possible to find three kinds of machine learning algorithms for financial market prediction and trading strategies: price prediction, movement directions predictions and algorithms for rule-based optimization to determine optimal combinations. In our study, we will focus on those oriented to price prediction. Regression algorithms [10] and neural networks are able to perform approximations of the future performance of assets.

On the one hand, the concept of Support vector regression (SVR) is addressed in [29], the authors shown a typical regression problem to illustrate the SVR concept:

Consider a set of data $G = \{(x_i, q_i)\}_i^n$*, where* x_i *is a vector of the model inputs,* q_i *is actual value and represents the corresponding scalar output, and n is total number of data patterns. The objective of the regression analysis is to determine a function f(x), so as to predict accurately the desired (target) outputs (q). Thus, the typical regression function can be formulated as* $q_i = f(x_i) + \delta$*, where* δ *is the random error with distribution of* $N(0,\sigma^2)$*. The regression problem can be classified as linear and nonlinear regression problems. As the nonlinear regression problem is more difficult to deal with,*

SVR was mainly developed for tackling the nonlinear regression problem. On the other hand, regarding neural networks, the definition given is that comprehensive system that considers numeric inputs, performs computations on these inputs, and creates outputs for one or more numeric values [1]. Neural networks improve the traditional statistical methods such as linear regressions, using function approximations, discriminant analysis, and logistic regression [25, 39].

The proposal is developed through a hybrid system: HBP-PSO algorithm. BP networks with supervised learning rules is the most used in financial time series because can be capable of approximate any measurable function in a very precise manner [25]. However, the disadvantage of the BP neural network is its inability to search for the overall optimal value. That is why by using the PSO algorithm (intelligent optimization method to find the optimal value) together with the HBP an algorithm is obtained that maximizes the advantages of both [**¡Error! No se encuentra el origen de la referencia.**]. Therefore the algorithm will improve the accuracy of the Recommender_agent in charge of the CBR [7].

The data that will feed the CBR will be those collected by the Information V.O. and it corresponds to asset classes, profitability, interest rate, the public process of shares and financial news published in the media which list the transactions of businesses, both internal and external.

4 Conclusions

This article overviewed the different techniques that have been implemented to create Fintech services. After a close study of the state of the art, our proposal consisted in adding Agent-based Virtual Organizations for investment recommendation. VOs were implemented with the aim of creating a light, well-structured, scalable and user-adapted system.

In a future work, it is being considered to perform a sample analysis of the user's characteristics to identify the level of risk assumed. Once a significant sample has been obtained, the data collected from the IBEX35 history stock market could be used. In the case of the study, the hybrid algorithm HBP-PSO will be applied and the results obtained will be shown. Once the proposal has been tested in that market, and to give robustness to the work, the study the study could be replicated in other financial markets such as Dow Jones, NASDAQ, etc.

Acknowledgments. This work was supported by the Spanish Ministry of Economy and FEDER funds. Project "SURF: Intelligent System for integrated and sustainable management of urban fleets" with ID: TIN2015-65515-C4-3-R.

References

1. Abdou, H.A., Pointon, J., El-Masry, A., Olugbode, M., Lister, R.J.: A variable impact neural network analysis of dividend policies and share prices of transportation and related companies. J. Int. Financ. Markets Inst. Money **22**(4), 796–813 (2012)
2. Arner, D.W., Barberis, J., Buckley, R.P.: The evolution of Fintech: a new post-crisis paradigm. Geo. J. Int. **47**, 1271–1319 (2015)
3. Avital, M., Andersson, M., Nickerson, J., Sundararajan, A., Van Alstyne, M., Verhoeven, D.: The collaborative economy: a disruptive innovation or much ado about nothing? In: Proceedings of the 35th International Conference on Information Systems ICIS. Association for Information Systems. AIS Electronic Library (AISeL), pp. 1–7 (2014)
4. Bach, K.: Knowledge Engineering for distributed case-based reasoning systems. In: Synergies Between Knowledge Engineering and Software Engineering, pp. 129–147. Springer, Cham (2018)
5. Bajo, J., De la Prieta, F., Corchado, J.M., Rodríguez, S.: A low-level resource allocation in an agent-based cloud computing platform. Appl. Soft Comput. **48**, 716–728 (2016)
6. Chamoso, P., Rivas, A., Rodríguez, S., Bajo, J.: Relationship recommender system in a business and employment-oriented social network. Inf. Sci. 204–220 (2017)
7. Corchado, J.M., Lees, B.: Adaptation of cases for case based forecasting with neural network support. In: Soft Computing in Case Based Reasoning, pp. 293–319. Springer, London (2001)
8. Corchado, J.M., Bajo, J., de Paz, Y., Tapia, D.: Intelligent environment for monitoring alzheimer patients, agent technology for health care. Decis. Supp. Syst. **34**(2), 382–396 (2008). ISSN 0167-9236
9. Corchado, J.M., Laza, R.: Constructing deliberative agents with case-based reasoning technology. Int. J. Intell. Syst. **18**(12), 1227–1241 (2003)
10. De Paz, J.F., Bajo, J., González, A., Rodríguez, S., Corchado, J.M.: Combining case-based reasoning systems and support vector regression to evaluate the atmosphere–ocean interaction. Knowl. Inf. Syst. **30**(1), 155–177 (2012)
11. DeStefano, R.J., Tao, L., Gai, K.: Improving data governance in large organizations through ontology and linked data. In: 2016 IEEE 3rd International Conference on Cyber Security and Cloud Computing (CSCloud), pp. 279–284. IEEE, June 2016
12. Dombrowski, U., Wagner, T.: Mental strain as field of action in the 4th industrial revolution. Procedia CIRP **17**, 100–105 (2014)
13. Đurić, B.O.: Organisational metamodel for large-scale multi-agent systems: first steps towards modelling organisation dynamics. Adv. Distrib. Comput. Artif. Intell. J. **6**(3), 2017 (2017)
14. Elnagdy, S.A., Qiu, M., Gai, K.: Cyber incident classifications using ontology-based knowledge representation for cybersecurity insurance in financial industry. In: 2016 IEEE 3rd International Conference on Cyber Security and Cloud Computing (CSCloud), pp. 301–306. IEEE, June 2016
15. Gai, K., Du, Z., Qiu, M., Zhao, H.: Efficiency-aware workload optimizations of heterogeneous cloud computing for capacity planning in financial industry. In: 2015 IEEE 2nd International Conference on Cyber Security and Cloud Computing (CSCloud), pp. 1–6. IEEE, November 2015
16. Gai, K., Qiu, M., Sun, X.: A survey on FinTech. Comput. Appl. **103**, 262–273 (2017)
17. Gai, K., Qiu, M., Sun, X., Zhao, H.: Security and privacy issues: a survey on FinTech. In: International Conference on Smart Computing and Communication, pp. 236–247. Springer, Cham, December 2016

18. García, E., Rodríguez, S., Martín, B., Zato, C., Pérez, B.: MISIA: middleware infrastructure to simulate intelligent agents. In: International Symposium on Distributed Computing and Artificial Intelligence, pp. 107–116. Springer, Heidelberg (2011)
19. Georgakoudis, G., Gillan, C.J., Sayed, A., Spence, I., Faloon, R., Nikolopoulos, D.S.: Methods and metrics for fair server assessment under real-time financial workloads. Concurren. Comput. Pract. Exp. **28**(3), 916–928 (2016)
20. Gomber, P., Koch, J.A., Siering, M.: Digital finance and FinTech: current research and future research directions. Bus. Econ. **87**, 537–580 (2017)
21. González, C., Burguillo, J.C., Llamas, M., Rosalía, L.A.Z.A.: Designing intelligent tutoring systems: a personalization strategy using case-based reasoning and multi-agent systems. ADCAIJ: Adv. Distrib. Comput. Artif. Intell. J. **2**(1), 41–54 (2013)
22. Havandi, E., Shavandi, H., Ghanbari, A.: Integration of genetic fuzzy systems and artificial neural networks for stock price forecasting. Knowl. Based Syst. **23**(8), 800–808 (2010)
23. Hüllermeier, E., Minor, M. (Eds.): Case-based reasoning research and development. In: Proceedings of 22nd International Conference on ICCBR 2014, Cork, Ireland, vol. 8765. Springer, 29 September–1 October 2014
24. Isaza, G., Mejía, M.H., Castillo, L.F., Morales, A., Duque, N.: Network management using multi-agents system. ADCAIJ: Adv. Distrib. Comput. Artif. Intell. J. **1**(3), 49–54
25. Kaastra, I., Boyd, M.: Designing a neural network for forecasting financial and economic time series. Neurocomputing **10**(3), 215–236 (1996)
26. Lazarova, D.: Fintech trends: the internet of things, January 2018. https://www.finleap.com/insights/fintech-trends-the-internet-of-things/
27. Li, B., Hoi, S.C.: Online portfolio selection: a survey. ACM Comput. Surv. (CSUR) **46**(3), 35 (2014)
28. López Barriuso, A., Prieta Pintado, F.D.L., Lozano Murciego, Á., Hernández, D., Revuelta Herrero, J.: JOUR-MAS: a multi-agent system approach to help journalism management. Adv. Distrib. Comput. Artif. Intell. J. (2015)
29. Lu, C.J., Lee, T.S., Chiu, C.C.: Financial time series forecasting using independent component analysis and support vector regression. Decis. Supp. Syst. **47**(2), 115–125 (2009)
30. Lu, Y., Roychowdhury, V., Vandenberghe, L.: Distributed parallel support vector machines in strongly connected networks. IEEE Trans. Neural Netw. **19**(7), 1167–1178 (2008)
31. Molina, D.I.: Fintech: Lo que la tecnología hace por las finanzas. Profit Editorial, p. 150 (2016). ISBN: 9788416904020
32. Owyang, J., Tran, C., Silva, C.: The Collaborative Economy. Altimeter, New York (2013)
33. Qiu, M., Ming, Z., Li, J., Gai, K., Zong, Z.: Phase-change memory optimization for green cloud with genetic algorithm. IEEE Trans. Comput. **64**(12), 3528–3540 (2015)
34. Radziwon, A., Bilberg, A., Bogers, M., Madsen, E.S.: The smart factory: exploring adaptive and flexible manufacturing solutions. Procedia Eng. **69**, 1184–1190 (2014)
35. Rodriguez, S., Julián, V., Bajo, J., Carrascosa, C., Botti, V., Corchado, J.M.: Agent-based virtual organization architecture. Eng. Appl. Artif. Intell. **24**(5), 895–910 (2011)
36. Sagraves, A., Connors, G.: Capturing the value of data in banking. Appl. Market. Anal. **2**(4), 304–311 (2017)
37. Song, Q., Liu, A., Yang, S.Y.: Stock portfolio selection using learning-to-rank algorithms with news sentiment. Neurocomputing **264**, 20–28 (2017)
38. Syam, N., Sharma, A.: Waiting for a sales renaissance in the fourth industrial revolution: machine learning and artificial intelligence in sales research and practice. Industrial Marketing Management (2018)
39. Tkáč, M., Verner, R.: Artificial neural networks in business: two decades of research. Appl. Soft Comput. **38**, 788–804 (2016)

40. Wang, L.X.: Dynamical models of stock prices based on technical trading rules part I: the models. IEEE Trans. Fuzzy Syst. **23**(4), 787–801 (2015)
41. Wang, L.X.: Dynamical models of stock prices based on technical trading rules—part III: application to Hong Kong stocks. IEEE Trans. Fuzzy Syst. **23**(5), 1680–1697 (2015)
42. Xu, K., Yue, H., Guo, L., Guo, Y., Fang, Y.: Privacy-preserving machine learning algorithms for big data systems. In: 2015 IEEE 35th International Conference on Distributed Computing Systems (ICDCS), pp. 318–327. IEEE, June 2015
43. Yu, K., Gao, Y., Zhang, P., Qiu, M.: Design and architecture of dell acceleration appliances for database (DAAD): a practical approach with high availability guaranteed. In: 2015 IEEE 17th International Conference on High Performance Computing and Communications (HPCC), pp. 430–435. IEEE, August 2005. 2015 IEEE 7th International Symposium on Cyberspace Safety and Security (CSS), 2015 IEEE 12th International Conference on Embedded Software and Systems (ICESS)
44. Zato, C., Villarrubia, G., Sánchez, A., Bajo, J., Corchado, J.M.: PANGEA: a new platform for developing virtual organizations of agents. Int. J. Artif. Intell. **11**(A13), 93–102 (2013)

Special Session on Complexity in Natural and Formal Languages (CNFL)

Structured Methods of Representation of the Knowledge

Francisco João Pinto(✉)

Department of Computer Engineering, Faculty of Engineering, University Agostinho Neto, University Campus of the Camama, S/N, Luanda, Angola
fjoaopinto@yahoo.es

Abstract. This paper describes same of the structured methods of representation of the knowledge that we introduced as an alternative to the procedures of representation more formal. After a brief mention to the essential characteristics that we have to take into account to any structured method of representation of the knowledge, classified into declarative methods and procedural methods. Declarative methods give more importance to the facts and entities of the domain that to the mechanisms of manipulation of the same. By contrary, the procedural methods, although they operate on facts and entities of the domain of speech, gives greater attention to the mechanisms of relation between entities. The declarative methods studied in this paper are the semantic network in which the knowledge is represented like a collection of joined nodes among them by means of labeled arches, the frames that can be defined as complex semantic network that treat the problem of the representation from the optics of the reasoning for likeness. Moreover, we described the formalism of the production rules, putting special attention in their structure and their form of cooperating with some declarative of representation (for example frames).

Keywords: Structured methods · Representation of the knowledge
Artificial intelligence

1 Introduction

Knowledge representation and reasoning (KR) is the field of artificial intelligence (AI) dedicated to representing information about the world in a form that a computer system can utilize to solve complex tasks such as diagnosing a medical condition or having a dialog in a natural language. Knowledge representation incorporates findings from psychology about how humans solve problems and represent knowledge in order to design formalisms that will make complex systems easier to design and build. Knowledge representation and reasoning also incorporates findings from logic to automate various kinds of reasoning, such as the application of rules or the relations of sets and subsets. Examples of knowledge representation formalisms include semantic network, systems architecture, frames, rules, and ontology. Examples of automated reasoning engines include inference engines, theorem tester, and classifiers. The KR conference series was established to share ideas and progress on this challenging field [1].

S. Rodríguez et al. (Eds.): DCAI 2018, AISC 801, pp. 213–221, 2019.
https://doi.org/10.1007/978-3-319-99608-0_24

The earliest work in computerized knowledge representation was focused on general problem solvers such as the General Problem Solver (GPS) system developed by Allen Newell and Herbert A. Simon in 1959. These systems featured data structures for planning and decomposition. The system would begin with a goal. It would then decompose that goal into sub-goals and then set out to construct strategies that could accomplish each sub goal. It was the failure of these efforts that led to the cognitive revolution in psychology and to the phase of AI focused on knowledge representation that resulted in expert systems in the 1970s and 80s, production systems, frame languages, etc. Rather than general problem solvers, AI changed its focus to expert systems that could match human competence on a specific task, such as medical diagnosis. Expert systems gave us the terminology still in use today where AI systems are divided into a Knowledge Base with facts about the world and rules and an inference engine that applies the rules to the knowledge base in order to answer questions and solve problems. In these early systems the knowledge base tended to be a fairly flat structure, essentially assertions about the values of variables used by the rules [4]. In addition to expert systems, other researchers developed the concept of frame based languages in the mid 1980s. A frame is similar to an object class: It is an abstract description of a category describing things in the world, problems, and potential solutions. Frames were originally used on systems geared toward human interaction, e.g. understanding natural language and the social settings in which various default expectations such as ordering food in a restaurant narrow the search space and allow the system to choose appropriate responses to dynamic situations. It wasn't long before the frame communities and the rule-based researchers realized that there was synergy between their approaches. Frames were good for representing the real world, described as classes, subclasses, slots (data values) with various constraints on possible values. Rules were good for representing and utilizing complex logic such as the process to make a medical diagnosis. Integrated systems were developed that combined Frames and Rules.

Knowledge-representation is the field of artificial intelligence that focuses on designing computer representations that capture information about the world that can be used to solve complex problems. The justification for knowledge representation is that conventional procedural code is not the best formalism to use to solve complex problems. Knowledge representation makes complex software easier to define and maintain than procedural code and can be used in systems. For example, talking to experts in terms of business rules rather than code lessens the semantic gap between users and developers and makes development of complex systems more practical. Knowledge representation goes hand in hand with automated reasoning because one of the main purposes of explicitly representing knowledge is to be able to reason about that knowledge, to make inferences, assert new knowledge, etc. Virtually all knowledge representation languages have a reasoning or inference engine as part of the system [4].

2 Structured Methods of Representation of the Knowledge

The formal logic allows the utilization of procedures of resolution that make possible the reasoning facts. However, the objects of the real universe have properties and relate with other objects. Thus, it is useful to have representation structures that allow, on a part, group properties, and on the other hand, obtain unique descriptions of complex objects. In any case, the objects are not the only structured entities of the universe. Also it would be very useful to represent effectively stages and typical sequences of events. In order to give answer to such questions, in artificial intelligence we use diagrams no formal of representation of the knowledge. Such diagrams are fundamentally methods structured of representation, and to verify the following properties:

- Representational adequate: The diagram chosen has to be able to represent the distinct classes of knowledge of the domain.
- Inference adequate: The diagram chosen has to allow the manipulation of the knowledge to obtain new knowledge.
- Inference efficiency: The diagram chosen has to be versatile using that information that to allow optimize the process inferential.
- Effectiveness of acquisition: The diagram chosen has to supply roads that allow the representation of information and new knowledge.

Any structured diagram of representation of the knowledge can be classified in one of the following categories:

- Declarative methods
- Procedural methods

Un the declarative diagrams the knowledge represents like a static collection of facts for whose manipulation defines a generic group and restricted of procedures. The diagrams of this type present some advantages: The truths of the domain are stored only by one time. Besides, it is easy to increase and incorporate new knowledge without modifying neither alter the already existent. In the procedural diagrams the greater part of the knowledge represents as procedures, which confers the diagram of representation a dynamic character. Also the procedural diagrams present advantages [2]:

- When giving priority to the procedures do greater emphasis in the capacities of inferences of the system
- They allow to explore distinct models and technical of reasoning
- They allow to work with fault of information and with data of probabilistic character
- They incorporate of natural form knowledge of heuristic type.

Independently of the diagram of representation chosen, is very useful to consider a series of elements that allow us establish relations among distinct structures of knowledge. Such elements are:

- IS_A: That allows establishing relations among hierarchical taxonomies.

- PART_OF: That allows establishing relations between objects and components of an object.

An important property of both relations is the transitivity. See the following examples:

- If Milu is a Dog and a Dog is an Animal then Milu is an Animal.
- If the Nose is part of the Face and the Face is part of the Head then the Nose is part of the Head.

The transitivity is linked to the deductive reasoning. More important still is the fact that the transitivity of the IS_A relation allows us to establish a method for the obtaining of properties of the objects related. This configures a process of inheritance of properties by means of which, if an object belongs to a determinate class, through the IS_A relation, already said, the object inherits the properties of the class. From a formal perspective can define a correspondence between both relations and the logic of predicates. For example MAN (FRANCISCO), turns into IS_A (FRANCISCO, MAN).

The declarative methods that we will study in this work are the semantic networks. The semantic networks allow describing simultaneously events and objects.

3 Semantic Networks

The semantic networks are declarative structures of representation in which the knowledge is represented as a group of nodes connected among them by means of arches labeled. The arches represent linguistic relations between nodes (see Fig. 1).

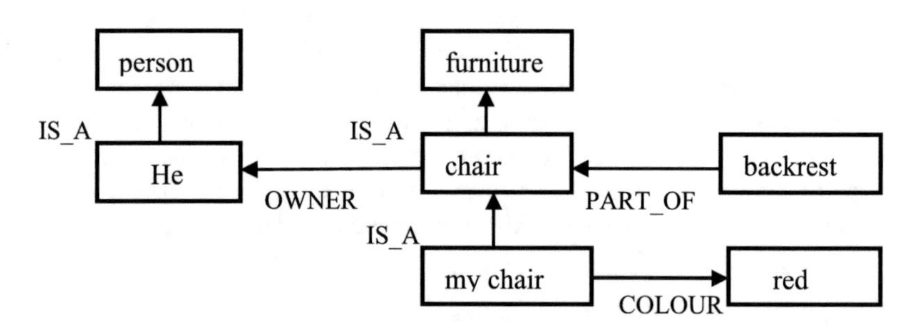

Fig. 1. A simple semantic network

Two of the binary relations more common in the semantic networks are the “IS_A” and “PART_OF” relations. In this context, the IS_A relation is applied in order to establish the fact that an element given is member of a class of elements that have in common a group of properties distinguishable. The concepts of class and of IS_A link can also be employed to represent situations, actions and events. The most frequent relations in semantic networks can be classified in one of the following categories:

- Occurrence: When it relates a member of a general category with the category to which belongs (is used to label “it belongs”).

- Generalization: When it relates an entity with another of character more general (IS_A).
- Aggregation: When they relate components of an object with the properly said object (PART_OF).
- Action: When they establish dynamic links among different objects.
- Properties: That they are relations between objects and characteristic of the objects.

4 Computational Perspective of a Semantic Network

From a computational perspective the implementation of a semantic network requires the construction of a table of tuples of the type: OBJECT-ATTRIBUTE-VALUE, so that:

- The father node is the OBJECT
- The arch is the ATTRIBUTE
- The destiny node is the VALUE

The Fig. 2 shows an example of construction of the table of tuples associated to a semantic network.

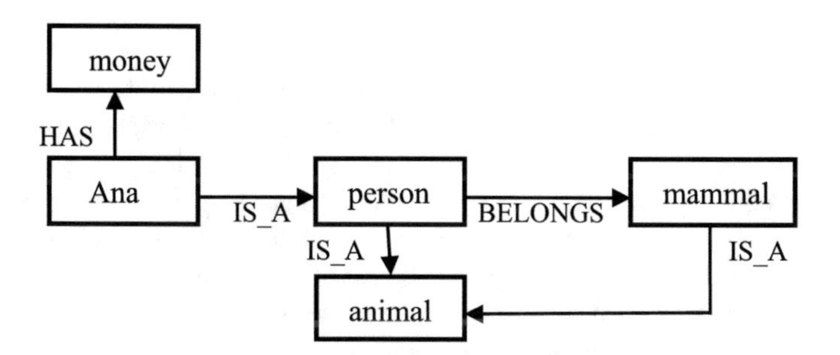

Fig. 2. An example of simple semantic network

There is a direct correspondence between the semantic networks and the formal logic. Like this, if we use predicate, the representation of the semantic network of the Fig. 6 would be: by means of the following predicates [2]:

- HAS (ANA, MONEY)
- IS_A (ANA, PERSON)
- IS_A (PERSON, ANIMAL)
- BELONGS (PERSON, MAMMAL)
- IS_A (MAMMAL, ANIMAL)

The Table 1 shows the Table of tuples associated to semantic network of the Fig. 2.

Table 1. Table of tuples

Object	Attribute	Value
Ana	has	Money
Ana	is_a	Person
Person	is_a	Animal
Person	belongs	Mammal
Mammal	is_a	Animal

Moreover, the internal representation of the knowledge of the semantic network of the Fig. 1 could be the following:

```
(ANA
        (HAS (MONEY))
        (IS_A (PERSON))
)
(PERSON
         (IS_A (ANIMAL))
                    (BELONGS (MAMMAL))
)
(MAMMAL (IS_A (ANIMAL)))
```

In semantic networks, the inheritance of properties says us that any property that consider some for a class of elements has to be true for any example of the class. This concept does that the semantic networks are particularly interesting to represent domains that can be structured like taxonomies. Regarding the form to reason with semantic networks, the model allows to obtain associations simply tracking the links of the system. Like this, in the following simple semantic network: 14 bigger than 9 and 9 bigger than 5, the track allows us conclude that 14 bigger than 5. Nevertheless, none rigorous semantic network guides the process. In a system of formal logic, the inferences are carried out on the base of handlings syntactic uniforms of symbols and, therefore, are always valid (although in occasions can be irrelevant).

5 Frames

The Frames can be described as complex semantic network that treat the problem of the representation from the optics of the reasoning for likeness. They describe classes of objects and they can be defined as structured representations of stereotyped knowledge. According to structure, a frame consists of a head that gives the name to the frame (and that it is a representative of the class of objects that are described), and of a slots group, each one of which represents a property or attribute of the generic element represented by the frame. Each slot can have different nested slots and without limitation of depth. Therefore right positions enable to locate the components of our previous experiences systematically in relation to the class of represented elements. Let us see an example:

AUTOMOBILE
TYPES
ALL LAND
SPORT
UTILITARIAN
COMPONENTS
CHASSIS
DOORS
MOTOR
. . .

Each one of the slots represents a level of knowledge and their content is a specialization of the previous level. A system in which the knowledge is represented by means of frames uses the inheritance concept. However they use the kind of slots "IS_A" that allow the entrance of information to a frame, in a level of certain knowledge starting from which the information of the corresponding class passes to the considerate object. In this way, a frame can be a simple of a frame of superior order. The frames usually incorporate also procedural information. In this case, certain slots take associated procedures that the most part in the time is inactive, but that when they are activated they unchain concrete actions [2].

6 Production Rules

The production rules (or simply rules), they are diagrams of representation of the knowledge that belongs to the procedural methods of representation. In the procedural methods, most part of the knowledge is represented as dynamic procedures. Structurally, the production rules are elements of representation of the dynamic knowledge constituted by a part "IF" (denominated condition or premise), a part "THEN" (denominated conclusion or action) and optionally, a part "ELSE" (or alternative conclusion-action that is executed when the premise is false) [5].

IF <CONDITION>
THEN <ACTION-CONCLUSION>
IF <CONDITION>
THEN <CONCLUSION>
ELSE <ALTERNATIVE>

Basically, the premise of a rule is constituted by a group of clauses elements that they can be nested through two relationship operators: AND, OR, NOT:

IF: A clause, AND B clause, OR C clause,…

A first problem that appears when we use production rules with nested clauses is that to get the premise represents, exactly and in a precise way, the heuristic one corresponding. This way, it is not the same thing:

- IF: [(AOR B) AND (C OR NOT D)]

- IF: [A OR (B AND C) OR NOT D]

Once we define the structure of the clause of the problem, the following step consists on finding an appropriate internal representation, so much for the clauses, like for the actions and the alternatives. Although this process is dependent of the tool that we use, or the language that we use, it is clear that the internal representation should be compatible with the elected representation for the static or declarative knowledge. Usually, this representation is carried out through ternary <parameter/relationship/value> , where the parameters are the characteristics that we want to investigate. The parameters should be compared with the corresponding values through the defined relationship operator, to the object of discovering if the clause is certain or not. On the other hand, the THEN part of the rule that can be multiple and therefore it can be constituted by several actions, it usually represents a working hypothesis (that it is verified when the premise is certain), or an action that can be executed [6]. Let us see the following example:

- IF: (1) The systolic arterial pressure is bigger than 160 mmHg,
- AND: (2) The diastolic arterial pressure is bigger than 95 mmHg;
- AND: (3) The average arterial pressure is bigger than 130 mmHg,
- THEN: (1) The patient presents arterial hypertension,
- AND: (2) To upgrade the database with the conclusion

The previous rule, described in natural language, it could be translated in the following way:

- IF: (1) (systolic arterial pressure) gt 160
- AND: (2) (diastolic arterial pressure) gt 95
- AND: (3) (average arterial pressure) gt 130
- THEN: (1) (Hemodynamic diagnostic arterial hypertension)
- AND: (2) (It UPGRADES, "hemodynamic diagnostic arterial hypertension", Database)

The previous example shows a rule with numeric parameters. It allows us to illustrate the rules can cooperate with other structures of declarative representation of the knowledge, for example frames. Effectively suppose that we have the following frame that represents the current situation of a certain case:

Arterial pressure

Systolic
177
Diastolic
99
Average
121

The representation of knowledge by means of production rules presents certain advantages. The conditions and the involved actions are explicit. Also, the knowledge is represented to modulate each rule of a given system constitutes a complete unit of knowledge. Lastly, the production rules allow to store and to use knowledge of a great specificity and the implied knowledge is generally of heuristic nature [3].

7 Conclusions

The formal logic allows the utilization of procedures of resolution that makes possible the reasoning facts. However, the objects of the real universe have properties and relate with other objects. It is useful to have structures of representation that allow, by a part, properties group, and on the other hand, obtain only descriptions of complex objects. In any case, the objects are not the only entities structured of the universe. Also it would be very useful can represent effectively stages and typical sequences of events. To treat to give answer to such questions, in artificial intelligence use diagrams no formal of representation of the knowledge. Such diagrams are fundamentally structured methods of representation of the knowledge. Any structured diagram of representation of the knowledge can be classified in one of the following categories: declarative methods and procedural methods. In the declarative diagrams of the knowledge is represented like a static collection of facts whose manipulation is defined as a generic group and restricted of procedures while in the procedural diagram the most of the knowledge is represented as procedures, which confers him the diagram of representation a dynamic character.

References

1. Baral, C., Delgrande, J., Wotter, F.: Principles of knowledge Representation and Reasoning. kr.org. KR Inc. Accessed 22 Nov 2017
2. Bonillo, V.M., Betanzos, A.A., Canosa, M.C., Berdiñas, B.G., Rey, E.M.: Fundamentos de Inteligencia Artificial, Capítulo 4, páginas 87–93, 97 e 98, 111–115, Universidad de la Coruña-España (2000)
3. Gonzales, D.: Verification and Validation. The engineering of Knowledge-Based Systems: Theory and Practice, Prentice-Hall International (1993)
4. Hayes-Roth, F., Waterman, D., Lenat, D.: Building Expert Systems. Addison-Wesley, Reading (1983). ISBN 0-201-10686-8
5. Rolston, D.W.: Principios de inteligencia Artificial y Sistemas expertos. McGraw-Hill (1990)
6. Rich, K.: Intelligence Artificial. McGraw-Hill, New York (1994)

An Approach to Measuring Complexity Within the Boundaries of a Natural Language Fuzzy Grammar

Adrià Torrens Urrutia(✉)

Universitat Rovira i Virgili, Tarragona, Spain
adria.torrens@estudiants.urv.cat

Abstract. This paper presents an approach to evaluate complexity by means of a Natural Language Fuzzy Grammar. Frequently, Linguistics has described a natural language grammar by means of discrete terms. However, a grammar can be explained in terms of degrees by following the concepts of linguistic gradience & fuzziness. Understanding a grammar as a fuzzy or gradient object allows us to establish degrees of grammaticality for every linguistic input. This shall be meaningful for linguistic complexity considering that the less grammatical an input is the more complex its processing will be. From this regard, an input's degree of complexity is always going to depend on its grammar. The bases of the natural language fuzzy grammar are shown here. Some of these are described by Fuzzy Type Theory. The linguistic inputs are characterized by constraints through a Property Grammar.

Keywords: Degrees of grammaticality · Degrees of complexity
Fuzzy grammar · Local complexity · Syntax

1 Introduction: What is Gradience and Fuzziness?

Fuzziness and gradience are pretty similar (if not the same). Gradience has appeared throughout the history of linguistics and can be defined as "a cover term to designate the spectrum of continuous phenomena in language, from categories at the level of the grammar to sounds at the level of phonetics" [1]. Some well-known studies approach gradience to lingusitic theory, such as Bolinger [2] or Keller [3]. However, it is in mathematics where we can find serious formal approaches to describe gradient relations, such as the gradient relation between *tall-short*, *big-small*. Nevertheless, the gradient phenomena in mathematics are called fuzzy phenomena and fuzzy logic is the right tool to formally describe these vague relations, which are also referred to as fuzziness. Zadeh's [4,5] mathematical description of gradient phenomena is well known. He describes the variable semantic values of words, or fuzzy phenomena, in terms of degrees.

S. Rodríguez et al. (Eds.): DCAI 2018, AISC 801, pp. 222–230, 2019.
https://doi.org/10.1007/978-3-319-99608-0_25

However, Zadeh did not develop a formal linguistic framework to describe fuzziness in natural language grammar. A brief methodological description distinguishing between both terms is shown:

- A *Fuzzy system (fuzzy grammar)* is a formal framework which defines any kind of linguistic information in any context (as humans do). This framework is set through a flexible constraint system which describes a natural language grammar. These constraints are known as properties. They work as logical operators that represent grammatical knowledge. They are flexible because they can be violated or satisfied to different degrees.
- *Processing gradience* refers to our capacity to judge fuzziness through a scale of degrees. The degree represents how hard or soft the violation is. In fuzzy logic, this might be referred as truth values, but since we are talking about language, we are going to talk about linguistic gradience as the truth value of an object.

2 Grammaticality as a Topic in Complexity

Nowadays the hypothesis of the "equi-complexity" is not as popular as in the 20th century. In fact, several authors such as Worther [6] or Dhal [7] have challenged this concept. Besides, usually, two different types of complexity are distinguished: absolute complexity and relative complexity. The absolute complexity is defined as a theoretically-oriented approach which evaluates the complexity of a language-system in a whole sense. On the other hand, the relative complexity takes into account the users of the language to identify the difficulty of processing, learning or acquisition. Other authors such as Blache [8] and Lindstrom [9] distinguish between Global complexity, Local complexity, and Difficulty. *Global complexity* is the absolute perspective of complexity. It aims to provide a number to rank a language as a whole system by means of a degree of complexity. This level is purely theoretical and it does not depend on any kind of linguistic realization. Blache in [8] claims that "in Chomskyan terms, this level concern competence", while the local complexity and difficulty belongs to the performance. In contrast, the degree of *local complexity* and *difficulty* are correlated to relative complexity, which is always provided once an input is given. However, local complexity is connected to the linguistic structure and its rules, whereas difficulty is an aspect to take into account for both psycholinguistic approaches and cognitive aspects, which have a role in the complexity evaluation. Within this classification, some authors place grammaticality in difficulty since it is considered a phenomenon of a cognitive aspect from the performance stage. The fact that grammaticality has an important role in the linguistic performance as well as in psycholinguistic approaches is not denied. Nevertheless, in this work, grammaticality is placed as an aspect of the local complexity for two reasons:

- (1) Local complexity is structure-sentence based, and difficulty is speaker-based. In this approach, grammaticality has a tight relation with the structures and the rules of a given input. Consequently, grammaticality belongs to local complexity. However, it has an impact on the difficulty since: the more complex a structure is in terms of grammaticality, the more difficult to process will be.
- (2) The theoretical bases of the Natural Language Fuzzy Grammar allow us to explain grammaticality by means of the grammar of a language itself, independently from the judgment of the speaker. In this instance, grammaticality is strictly based on the rules of the local complexity.

2.1 Grammaticality as an Element of Complexity

Linguistics has been highly influenced by the theoretical fragmentation of Competence - Performance from Chomsky's *Aspects* [10]. In general, grammaticality has been considered in two ways:

- *A categorical item:* since the competence is perfect, grammaticality can only be either satisfied or violated by means of the speaker or the receiver during the performance stage.
- *A matter of degrees:* grammaticality would be found as a part of an acceptability judgment. This regard considers that grammaticality is not equal to the whole value of an acceptability judgment, and yet it is an essential part which contributes to the total amount of the degree of acceptability from an input. As well as in the last case, here grammaticality belongs to the performance as well.

However, in the Natural Language Fuzzy Grammar (NFG) approach, the degree of grammaticality is something which is directly related to the grammar. Grammaticality in NFG does not necessarily come through the speaker, nor a performance. Once an input is given, the evaluation of the input is in contrast with the grammar of a language itself. The grammaticality value can be totally isolated from the acceptability judgment from either speaker or a receiver. Thus, in this regard, grammaticality is no longer only a psycholinguistic effect. It is also a direct consequence of a structure in relation to its grammar. In this sense, Grammaticality would play a role in the degree of relative complexity and local complexity. NFG might take into account the complexity of a linguistic structure and its features, such as: number of categories, number of words, number of rules in a structure and degree of grammaticality. In the following section, the base of the Fuzzy Grammar is going to be defined as well as described in a wider sense.

3 An Approach to a Fuzzy Grammar with Fuzzy Descriptions for Complexity

First of all, a brief introduction in which the formalisms used are going to be described. Secondly, what is understood as a fuzzy grammar will be explained.

3.1 Brief Introduction to Fuzzy Natural Logic

Fuzzy Natural Logic (FNL) [11] has been used here to describe what a grammar is. FNL is a fuzzy logic system which uses a Fuzzy Type Theory (FTT), a high-order mathematical fuzzy logic, with a Łukasiewicz algebra of truth values (Ł-TT).

Semantics of FNL. A few basic notions in relation to FNL are shown:

- FNL uses FTT. This is why this approach is based on describing elements (M) by types (ϵ). Every element has a type which represents the element M_ϵ. An element can be anything: people, objects, languages, etc.
- o: It is an essential type: *omicron*. It can be read as "degree of truth $\langle[0,1]\rangle$[1]". The degree of truth might represent different approaches such as degree of grammaticality, degree of complexity, and so on.
- In order to obtain a truth value, we need to contrast two arguments following the next rule: A: U $\rightarrow$ L. In a Universe (A) an element (U) has a truth value (L). The rule can be extended in a way that many connected elements which share properties can have the same truth value. A: U$\times$U $\rightarrow$ L.
- &: It represents the Łukaziewich conjunction and it can be read as an "and".
- $\equiv$: is an essential connective in this approach. It can be read as "equivalent to" and it is called fuzzy equivalence/equality.
- λ: Can be read as "each".

3.2 A Fuzzy Grammar Structure to Explain Degrees of Grammaticality and Complexity

A grammar is considered here as a set of rules. These rules allow the production of linguistic inputs. We show a simplified formula of a fuzzy grammar in a multi-modal sense:

$$GR(ph, m, x, s, l, pr, ps)$$

.

A Grammar (GR) is equivalent to the phonetic rules (ph), plus the morphological rules (m), plus the syntax rules (x), plus the semantic rules (s), plus the

[1] Note that the use of [ ] means any number/degree between 0 and 1. That could be 0.85512 and so on. Additionally, [0,1] can represent a piece of infinite, and not necessarily real numbers [30–10^{10}] = [0,1]. The discrete approaches represent $\langle 0,1\rangle$ as operators (without [ ]). In that way the use of $\langle 0,1\rangle$ means "either 0 or 1".

lexical rules (l), plus the pragmatic rules (pr), plus the prosodic rules (pr). In the formal approach, every element would have its type and would be contrasted with a universe, in this case, a dialect (d with type η). In this regard every dialect would be considered as a language. From the contrast of all the rules of each module in the universe, we would obtain the value of grammaticality.

$$Grammar \equiv \lambda d_\eta \lambda ph_\alpha \lambda m_\beta \lambda x_\gamma \lambda s_\delta \lambda l_\epsilon \lambda pr_\zeta \lambda ps_\kappa * (Ph_{(o\alpha)\eta} d_\eta) ph_\alpha \& (M_{(o\beta)\eta} d_\eta) m_\beta \& \\ (X_{(o\gamma)\eta} d_\eta) x_\gamma \& (S_{(o\delta)\eta} d_\eta) s_\delta \& (L_{(o\epsilon)\eta} d_\eta) l_\epsilon \& (pr_{(o\zeta)\eta} d_\eta) pr_\zeta \& (ps_{(o\kappa)\eta} d_\eta) ps_\kappa$$

The syntactic module is taken as an example to explain how this formula works. $(\mathrm{X}_{(o\gamma)\eta} d_\eta) x_\gamma$ means the syntactic module. Regarding this description, the syntax of a grammar has a type γ in a universe η. η is the type of universe. In this case, the universe is considered as a dialect. The syntax of this dialect is defined in terms of linguistic rules or constraints. The type o demonstrates the degree of truth according to the value of grammaticality provided by the fuzzy grammar. Thus, the syntactic module (x_γ) is in a dialect (d_η) the degree of truth according to the satisfied/violated syntactic constraints of the dialect $(\mathrm{X}_{(o\gamma)\eta})$. Consequently, the more constraints that are satisfied in a grammar by a given input, the more grammatical it will be. Therefore, a given input has a high value of grammaticality according to its grammar (and not by the speaker's perception). A given input which respects the structures and rules of its dialect will have a high grammaticality value. A given input which triggers a lot of violations will display more complex structures since those are not structures which belong to the grammar evaluating the input. Therefore, the higher the value of grammaticality in an input, the lower the value of its complexity in a determinate grammar.

4 Property Grammars: A Contraint-Based Theory for Dealing with Fuzziness and Gradience

Regarding fuzzy grammar, Blache's [12–14] Property Grammars have been chosen as the formal theoretical framework in defining natural language fuzziness and variability. This theory combines a full-constraint framework of independent and flexible constraints (or properties), with syntactic dependencies under the notion of construction from Construction Grammars. Constructions have been described in terms of their properties. Property Grammars display several constraints in order to describe the syntactic relations between local language phenomena. However, here we focus on the following ones:

- *Linearity* (>): Precedence order between two elements. A precedes B.
- *Requirement* ($\leftrightarrow$): Co-occurrence between two elements: A requires B.
- *Exclusion* (excl.): A and B never appear in co-occurrence in the specified construction.

5 An Example of Relative Complexity Within the Boundaries of a Fuzzy Grammar

Figure 1 is a sample of how gradient description of fuzziness and variability could work. We show the formal description of the PRON [pronoun]. Neutral Demonstrative, Relatives and Personal Pronouns are the canonical ones regarding our corpus (Universal Dependency Spanish Treebank Corpus 2.0). The most canonical structure is weighted as 1, a medium canonical is weighted as 0.5, a violation is weighted as −1 and recurrent variability has a 0.5 weight[2].

Pronoun in Subject Construction	
1	*CnW*1: [Neutral Demonstratives; Relatives Pronouns; Personal Pronouns]
Syntax Canonical Properties	
PRON *excl.* PREP∧ADJ∧ADV∧D∧PRON	
Syntax Variable Properties	
V1: PRON *excl.* ADJ	PRON SxPt 2 PRON↔ADJ: [solo] or [mismo]
V2: PRON *excl.* Det	PRON↔[yo] D↔[el] PRON↔fit NOUN SxPt
2	*CnW* 0.5: [lo]
Syntax Canonical Properties	
PRON>ADJ PRON↔ADJ PRON *excl.* PREP ∧ DET ∧ ADV	
VabW 0.5 [Non PRON in PRON fit]	
Syntax Variable Properties	
1	NPPF↔PREP [mod] NPPF > [de]
2	NPPF↔ADJ
Legend	
CnW: Canonical Weight *VabW*: Variability Weight V: Violation []: It marks the word-class or the lexical word SxPt: Syntactic Properties NPPF: Non PRON in PRON Fit	

Fig. 1. Pronoun's syntactic properties in subject construction.

[2] Note that these weights illustrate a basic idea of gradience. They are not related to the real weights of gradience in Spanish syntax. A precise value of gradience for each weight in each set or construction will be established in the future. We emphasize that this is currently in progress.

Our framework can describe inputs with grammatical violations and their variability. The fuzzy phenomenon is explained with a double analysis in the feature description:

- First Phase: *Syntax Canonical Properties*
- Second Phase: *Syntax Variable Properties.*

Firstly, a normal parsing is applied. This parser describes the Syntactic Properties considering only the canonicals. The result of this parsing describes both satisfied and violated canonical properties. The canonical deviations with its violations will be defined in terms of properties. The value of the addition between *CnW* and *VW* will be divided by the Total amount of Part of Speech (*TPS*). A value of complexity in terms of grammaticality is provided here (*VG1*: Value of Grammaticality 1):

$$VG_1 = \frac{{}^{+}CnW^{-}VW}{TPS} \tag{1}$$

Straightaway, the parser runs for a second time, taking into account the violations and defining the Syntax Variable Properties. In case some Syntax Property is violated, such as V1 or V2, SVP are triggered. Their weight of violability is going to be mitigated in case the violation respects these new properties. If the new properties are not satisfied, variability is not going to have any effect here and *VW* would remain as before. After this second analysis, a new value will be provided (*VG2*: Value of Grammaticality 2) following the formula in (2). This formula refers to our capacity to evaluate a linguistic input as a gradient-fuzzy object.

$$VG_2 = \frac{({}^{-}VW^{+}VabW)^{+}CnW}{TPS} \tag{2}$$

This system also works for explaining words which undergo a partial transition in terms of part of speech. These transitions concern fuzzy boundaries in parts of speech. The more transitions the more complex an input will be. Thus, we would assume that the word-class does not undergo a complete transition of membership, but more of context. This explains why other properties must be taken into account regarding variability. Several D [determiner] (especially articles and demonstratives) occur as PRON quite often, but never as often as they occur as a D (articles: 73.10%; demonstratives: 10,44% in more than 4000 occurrences). If those D ever appear as a PRON this framework detects a violation in the first parsing since, canonically, a D must precede N [Noun]. In the second parsing, the following SVP in the determiner will be triggered clarifying how it is possible to have a determiner without a N: Determiner: SVP: V1: D $>$ N: D $\leftrightarrow$ PRON *VabW* 0.5 NPPF 1 $\bigvee$ 2. Once this new fit is applied, and D is no longer considered a D but a PRON, the *VabW* impacts the *VW*, providing the above-mentioned a new grammaticality value (*VG2*).

Because the new fit in this case is a PRON, we describe their properties in the PRON. The same happens in V2 where PRON undergo a fit transition to the N syntactic properties and thus, their new properties are located in Noun Construction. In V1 occurs something similar but in a softer way, in which PRON undergo a transition to the properties of the canonical PRON case number 2 [lo].

6 Final Remarks

Local Complexity is dependent on an input's rules and structure. Our Natural Fuzzy Grammar (NFG) takes into account what happens when a sentence has rules which are satisfied or violated. A given input has a value of grammaticality according to its grammar (and not by the speaker's perception). The more constraints that are satisfied, the more grammatical it will be. An input which triggers a lot of violations is going to display more variable rules in the fuzzy grammar (as it was shown in the example of the pronoun). The process of a double parsing for variability rules would increase the complexity of the given sentences. In this sense, the lower the value of grammaticality, the higher the value of complexity for a determinate grammar. Besides, the input with violations would probably be more ambiguous, as shown in the example of the pronoun. Therefore, yet more complex.

Some theories in complexity establish that the more rules there are in a sentence, the more complex a sentence is. Actually, in this proposed approach, the complexity of a sentence might be mitigated or reduced in case the grammar rules are satisfied.

Acknowledgement. This research has been supported by the Ministerio de Economía y Competitividad and the Fondo Europeo de Desarrollo Regional under the project number FFI2015-69978-P (MINECO/FEDER, UE) of the Programa Estatal de Fomento de la Investigación Científica y Técnica de Excelencia, Subprograma Estatal de Generación de Conocimiento.

References

1. Aarts, B.: Conceptions of gradience in the history of linguistics. Lang. Sci. **26**(4), 343–389 (2004)
2. Bolinger, D.L.M.: Generality: Gradience and the All-or-None, 14th edn. Mouton & Company, Amsterdam (1961)
3. Keller, F.: Gradience in grammar: experimental and computational aspects of degrees of grammaticality. Ph.D. thesis. University of Edinburgh, Edinburgh (2000)
4. Zadeh, L.A.: Fuzzy sets. Inf. Control. **8**(3), 338–353 (1965)
5. Zadeh, L.A.: A fuzzy-set-theoretic interpretation of linguistic hedges. J. Cybern. **2**(3), 4–34 (1972)
6. Mc Worther, J.: The world's simpliest grammars are creole grammars. Linguist. Typology **5**(2), 125–166 (2001)
7. Dahl, Ö.: The Growth and Maintenance of Linguistic Complexity, vol. 71. John Benjamins Publishing, Amsterdam (2004)

8. Blache, P.: A computational model for linguistic complexity. Biol. Comput. Linguist. **288**, 155–167 (2011)
9. Lindstrom, E.: Language complexity and interlinguistic difficulty. Lang. Complex. Typology Contact Chang. **94**, 217 (2008)
10. Chomsky, N.: Aspects of the Theory of Syntax. MIT Press, Cambridge (1965)
11. Novák, V.: Fuzzy natural logic: towards mathematical logic of human reasoning. In: Towards the Future of Fuzzy Logic, pp. 137–165, Springer (2015)
12. Blache, P.: Property grammars and the problem of constraint satisfaction. In: Proceedings of ESSLLI 2000 workshop on Linguistic Theory and Grammar Implementation, pp. 47–56 (2000)
13. Blache, P.: Property grammars: a fully constraint-based theory. In: Constraint Solving and Language Processing, vol. 3438, pp. 1–16 (2005)
14. Blache, P.: Representing syntax by means of properties: a formal framework for descriptive approaches. J. Lang. Model. **4**(2), 183–224 (2016)

Syntax-Semantics Interfaces of Modifiers

Roussanka Loukanova(✉)

Stockholm University, Stockholm, Sweden
rloukanova@gmail.com

Abstract. The paper introduces a computational approach to syntax-semantics interface of human language. The goal is to develop an efficient, biologically inspired technique to computational processing of human language. The technique integrates two formal languages of syntax and semantics into generalised, type-theoretic grammar that efficiently incorporates lexicon, syntax, and semantics. Modifier expressions are among the most difficult classes of expressions in human language, that introduce great complexity both in their syntax and semantics. The focus of the paper is on providing efficient technique for syntax-semantics of modifier expressions.

Keywords: Syntax-semantics · Type theory · Recursion
Algorithms · Modifiers · Constraint-based grammar

1 Introduction

In this paper, we introduce a new technique to computational syntax-semantics interface in generalised Constraint-Based Lexicalized Grammar (CBLG). The technique integrates two formal languages, with corresponding theories, for computational syntax and semantics. Firstly, we use an approach to mathematical foundation of computational CBLG, introduced in details in Loukanova [3], which provides type-theoretic syntax of lexical and phrasal structures.

Different natural languages may have different surface syntax, i.e., appearances, precedences, and dominancies of syntactic categories. It is an open question in linguistics to what degree these different surface syntaxes can be based on fundamentally shared 'deep' syntax. There is general acknowledgement of innate grammar of human language that is a fundamental, biological facility for learning languages, on the fly in early age, and with conscious efforts later. The syntax-semantics rules in this paper represent some fundamentally shared features across syntactic categories via linguistic generalisations over the parts of

This research has been supported, by covering my participation in DCAI 2018, by the Ministerio de Economía y Competitividad and the Fondo Europeo de Desarrollo Regional under the project number FFI2015-69978-P (MINECO/FEDER, UE) of the Programa Estatal de Fomento de la Investigación Científica y Técnica de Excelencia, Subprograma Estatal de Generación de Conocimiento.

S. Rodríguez et al. (Eds.): DCAI 2018, AISC 801, pp. 231–239, 2019.
https://doi.org/10.1007/978-3-319-99608-0_26

speech of the head phrases. This facilitates computational efficiency and grammar learning. Our position is that semantics is inseparable from innate biological facilities.

We target computational approach to grammar of human language, that reflects on natural, inborn language capacity of humans. We consider that biological language capacities are fundamentally based on parametric blue-prints, which carry a major structural form in parameters. The blue-prints carry the common characteristics of structures that get instantiated dynamically, via syntax-semantics across lexicon and phrases, in an efficient way for modeling learning processes. We consider both its biological foundations, and integrating them into computational foundations of formal languages, especially, of programming. Similar style of syntax-semantics interfaces could be useful to represent core semantic representations associated with abstract, universal syntax that is shared among classes of human languages.

Individually, each of the techniques of syntax and semantics reduces computational complexity by using partial and parametric representations. Integrating both approaches results in a technique that reduces computational complexity of generalised, type-theoretic grammar that incorporates lexicon, syntax, and semantics, via syntax-semantics interfaces at lexical and phrasal levels.

In Sect. 2, we give an informal overview of generalised CBLG. Section 3 briefly provides the general technique of computational syntax-semantics interface from Loukanova [2].

In the rest of the paper, we introduce two major rules for syntax-semantics interface of modified phrases. The syntactic components of these rules are introduced in Loukanova [1]. By using these modifier rules, we provide a plausible syntactic medium for syntax-semantics interface of a range of modifiers, without implications that they represent any acclaimed syntactic analysis, or syntactic universality of modifiers.

2 Background on Syntax-Semantics in Constraint-Based Lexicalized Grammar

For a formal introduction to computational Constraint-Based Lexicalized Grammar (CBLG), see Loukanova [3]. In this section, we provide a brief overview of the approach of generalised CBLG.

The rules of CBLG are presented in feature-value descriptions. The notation in the grammar rules uses the symbol $\longrightarrow$, which reminds for the rules of a generative Context-Free Grammar (CFG). In its full formally, the relation between CBLG and CFG is not the subject of this paper. Here we shall mention that symbol $\longrightarrow$ in the rules expresses mother-daughter dominance of nodes in parse trees, and constraints on the well-formedness of the corresponding feature-structure descriptions.

The rules are constraints, which interact with other constraints, which are classified as grammar principles, and a lexicon, which has its own lexical rules.

The interrelations between the formal components are guided by the type hierarchy of the specifically chosen CBLG.

As in typical CBLG that employ features-value descriptions, we use features, such as: SYN, SEM, SPR, COMPS, etc. The feature SYN has values that are partial feature-structure descriptions of the syntax of the given structure. The feature VAL has values that are feature-structure descriptions representing the syntactic arguments of the enclosing structure. Major valence features are SPR, COMPS E.g., in verbal structures, these are correspondingly, SPR for the subject, and COMPS for the complements of the verb structure. The feature SEM has values that represent semantics of the structure in which it is enclosed. We note that the feature-value descriptions give partial information about the grammatical feature-structures under consideration. Typically, a feature-value description represents a class of fully specified feature-value structures of grammatical representations, e.g. of a node in a parse tree.

2.1 Phrasal Rules of CBLG

The rules for well-formed expressions are feature-value constraints. The syntactic portions of two of the major rules[1] of CBLG for formation of phrases are presented in (1), (2), by using the feature SYN with its values in feature-value descriptions. The rules saturate valance requirements, via the features VAL (for syntactic valences), SPR (for syntactic subject argument), COPMS (for syntactic complement arguments).

The syntactic components in the CBLG rules (1), (2) represent linguistically significant information, which is common across syntactic categories and filtered out from phrases of the major classic syntactic categories, i.e.: sentences – S, verb phrases – VP, noun phrases – NP, nominals – NOM, verb phrases – PP; adjectival phrases – AdjP, adverbial phrases – AdvP, determiner phrases – DP. Informally, each of the lexical categories for verbs V, nouns N, prepositions P, adjectives Adj, adverbials Adv, determiners Det, can have syntactic arguments, which are SPR and/or COMPS. Typically, the SPR of a verb is a NP; the SPR of a noun N is a determinar, etc.

2.2 Principles: HFP, VPr

Head Feature Principle (HFP): In any headed feature-structure, the HEAD value of the mother is identical to the HEAD value of the head daughter.

Valence Principle (VPr): Unless a rule dictates otherwise, the mother's values of the VAL features SPR, COMPS, are identical to those of the head daughter. Here, the VPr is simplified for expository purposes.

[1] Note that the syntax-semantics analyses of sentences with quantifier NPs in both specifier and complement positions require more complex rule, which is not in the subject of this paper. In addition, it would require space.

3 Syntax-Semantics Interface in CBLG by Type-Theory of Acyclic Recursion

In this section, we provide technical background of generalised CBLG with syntax-semantics interface. The semantic representations are provided by the formal language of acyclic recursion L_{ar}^{λ}. For more details on the technique, see Loukanova [2]. For a formal introduction to L_{ar}^{λ}, see, e.g., the original work of Moschovakis [4].

Here, for self-containment, we provide two of the major phrasal rules, for the syntax of phrasal structures. The major syntactic arguments of phrasal expressions are: (1) a specifier, SPR, e.g., the subject of a verbal phrase (VP), or the determiner of a noun phrase (NP), and (2) compliments COMPS. We add semantic representations via syntax-semantics interface provided to lexical items, which is then computationally propagated into the phrasal structures, formalised by using the formal language of feature-value structure descriptions. The compositional, semantic rendering of syntactic structures into semantic representations, is by L_{ar}^{λ}-terms.

The values of features T-HEAD and WHERE represent the head part and the set of where-assignments, respectively, of the L_{ar}^{λ}-term that is the rendering of the analysed syntactic expression, i.e., its logical form. The value of the feature L-TYPE is its L_{ar}^{λ}-type.

The values of T-HEAD and WHERE of the left hand side of the rules are the parts of L_{ar}^{λ}-terms, which are determined by: the types T_1 and T_2; the values of T-HEAD and WHERE in the daughters' feature structures on the right hand side; and the definition of the canonical form $\mathsf{cf}(A)$ of each term A (see the definition in 3.13, Moschovakis [4]). To avoid variable clashes, we also assume that, in each application of the grammar rules, the representations of the L_{ar}^{λ}-terms are such that all bound recursion variables are distinct and distinct from all the free recursion variables (and constants), by making appropriate renaming substitutions, where needed.

Head Specifier Rule (HSR): (1)

$$\begin{bmatrix} \textit{phrase} & \\ \text{SYN} & \begin{bmatrix} \text{VAL} & \begin{bmatrix} \text{SPR} & \langle\, \rangle \end{bmatrix} \end{bmatrix} \\ \text{SEM} & \begin{bmatrix} \text{L-TYPE} & T \\ \text{TERM} & \begin{bmatrix} \text{T-HEAD} & A_0 \\ \text{WHERE} & U \end{bmatrix} \end{bmatrix} \end{bmatrix} \longrightarrow \boxed{1} \begin{bmatrix} \text{SEM} & \begin{bmatrix} \text{L-TYPE} & T_1 \\ \text{TERM} & \begin{bmatrix} \text{T-HEAD} & A_{1,0} \\ \text{WHERE} & U_1 \end{bmatrix} \end{bmatrix} \end{bmatrix} \; \text{H} \begin{bmatrix} \text{SYN} & \begin{bmatrix} \text{VAL} & \begin{bmatrix} \text{SPR} & \langle \boxed{1} \rangle \\ \text{COMPS} & \langle\, \rangle \end{bmatrix} \end{bmatrix} \\ \text{SEM} & \begin{bmatrix} \text{L-TYPE} & T_2 \\ \text{TERM} & \begin{bmatrix} \text{T-HEAD} & A_{2,0} \\ \text{WHERE} & U_2 \end{bmatrix} \end{bmatrix} \end{bmatrix}$$

Head Complement Rule: Version 1 (HCR1): (2)

$$\begin{bmatrix} \textit{phrase} & \\ \text{SYN} & \begin{bmatrix} \text{VAL} & \begin{bmatrix} \text{COMPS} & \boxed{a} \end{bmatrix} \end{bmatrix} \\ \text{SEM} & \begin{bmatrix} \text{L-TYPE} & T \\ \text{TERM} & \begin{bmatrix} \text{T-HEAD} & A_0 \\ \text{WHERE} & U \end{bmatrix} \end{bmatrix} \end{bmatrix} \longrightarrow \text{H} \begin{bmatrix} \text{SYN} & \begin{bmatrix} \text{VAL} & \begin{bmatrix} \text{COMPS} & \begin{bmatrix} \text{FIRST} & \boxed{1} \\ \text{REST} & \boxed{a} \begin{bmatrix} \textit{list} \end{bmatrix} \end{bmatrix} \end{bmatrix} \end{bmatrix} \\ \text{SEM} & \begin{bmatrix} \text{L-TYPE} & T_2 \\ \text{TERM} & \begin{bmatrix} \text{T-HEAD} & A_{2,0} \\ \text{WHERE} & U_2 \end{bmatrix} \end{bmatrix} \end{bmatrix} \; \boxed{1} \begin{bmatrix} \text{SEM} & \begin{bmatrix} \text{L-TYPE} & T_1 \\ \text{TERM} & \begin{bmatrix} \text{T-HEAD} & A_{1,0} \\ \text{WHERE} & U_1 \end{bmatrix} \end{bmatrix} \end{bmatrix}$$

where in both rules (1)–(2):

$$T_i \equiv (\sigma \rightarrow \tau) \text{ and } T_j \equiv \sigma, \text{ for } i, j \in \{1, 2\} \text{ and } i \neq j,\ T \equiv \tau \text{ and} \tag{3}$$

There are two sub-cases for the term $[A_0 \ \mathsf{where}\ U]$ in[2] (1)–(2):

$$\begin{array}{l} \text{if } A_{j,0} \text{ is immediate, then} \\ \qquad A_0 \equiv A_{i,0}(A_{j,0}) \text{ and } U \equiv U_i \equiv U_1 \cup U_2 \end{array} \tag{4a}$$

$$\begin{array}{l} \text{otherwise (i.e., } A_{j,0} \text{ is proper),} \\ \qquad A_0 \equiv A_{i,0}(q_0) \text{ and } U \equiv \{q_0 := A_{j,0}\} \cup U_1 \cup U_2 \end{array} \tag{4b}$$

4 Modifier Rules

An expression that can serve as a modifier of other expressions has co-occurrence requirements of what expressions it can modify, and how this modifier can be syntactically attached to an expression to be modified. For example, an adjective phrase can modify an NP, or an adverb phrase can modify a sentence or a VP, and either can be pre- or post-positioned with respect to the modified expression. These co-occurrence requirements are grounded in the lexical entries by using the feature MOD of the lexical modifier. Then, the values of MOD are "passed up," by the lexical heads to the mother structures, in the co-occurrence rules. The appropriate rule, HMR in (5), for post-modification, or and MHR in (6), for pre-modification, can be applied depending on the value of the feature MOD of the lexical modifier. The lexical entries of modifiers include the feature POSTHM of the type *syn-cat*, inside the value of feature MOD. The possible values of POSTHM are + or - of type *bool*, for whether or not it can postmodify.

Head-Modifier Rule (HMR):

A phrase can consist of a (lexical or phrasal) head, labeled by some $\boxed{n}$ $(n \in \mathbb{N})$, followed by a compatible, co-labeled modifier: (5)

$$\begin{bmatrix} \textit{phrase} & \\ \text{SYN} & \begin{bmatrix}\text{POSTHM} & +\end{bmatrix} \\ \text{SEM} & \begin{bmatrix} \text{L-TYPE} & \tau \\ \text{TERM} & \begin{bmatrix} \text{T-HEAD} & A_0 \\ \text{WHERE} & U \end{bmatrix} \end{bmatrix} \end{bmatrix} \longrightarrow \text{H}\boxed{n} \begin{bmatrix} \text{SYN} & \begin{bmatrix}\text{VAL} & \begin{bmatrix}\text{COMPS} & \langle\,\rangle\end{bmatrix}\end{bmatrix} \\ \text{SEM} & \begin{bmatrix} \text{L-TYPE} & \tau \\ \text{TERM} & \begin{bmatrix} \text{T-HEAD} & A_{2,0} \\ \text{WHERE} & U_2 \end{bmatrix} \end{bmatrix} \end{bmatrix}$$

$$\begin{bmatrix} \text{SYN} & \begin{bmatrix} \text{VAL} & \begin{bmatrix} \text{COMPS} & \langle\,\rangle \\ \text{MOD} & \langle\, \boxed{n} \begin{bmatrix}\text{SYN} & \begin{bmatrix}\text{POSTHM} & +\end{bmatrix}\end{bmatrix} \rangle \end{bmatrix} \end{bmatrix} \\ \text{SEM} & \begin{bmatrix} \text{L-TYPE} & (\tau \rightarrow \tau) \\ \text{TERM} & \begin{bmatrix} \text{T-HEAD} & A_{1,0} \\ \text{WHERE} & U_1 \end{bmatrix} \end{bmatrix} \end{bmatrix}$$

[2] The additional sub-index 0 in $A_{i,0}$ and $A_{j,0}$ is unnecessary in these formulations, but allows easier re-formulations of the rules with constructs inside U_1 and U_2.

where, the parts of the term $[A_0 \ \ \mathsf{where} \ \ U]$ in (6), there are determined by (7a)–(7b).

Modifier-Head Rule (MHR):

A phrase can consist of a (lexical or phrasal) head, labeled by some $\boxed{n}$ ($n \in \mathbb{N}$), preceded by a compatible, co-labeled modifier: (6)

$$\begin{bmatrix} \textit{phrase} \\ \text{SYN } \textit{syn-cat} \\ \text{SEM} \begin{bmatrix} \text{L-TYPE} & \tau \\ \text{TERM} & \begin{bmatrix} \text{T-HEAD } A_0 \\ \text{WHERE } U \end{bmatrix} \end{bmatrix} \end{bmatrix} \longrightarrow \begin{bmatrix} \text{SYN} \begin{bmatrix} \text{VAL} \begin{bmatrix} \text{COMPS} & \langle\, \rangle \\ \text{MOD} & \langle \boxed{n} \begin{bmatrix} \text{SYN} \begin{bmatrix} \text{POSTHM} & - \end{bmatrix} \end{bmatrix} \rangle \end{bmatrix} \end{bmatrix} \\ \text{SEM} \begin{bmatrix} \text{L-TYPE} & (\tau \rightarrow \tau) \\ \text{TERM} & \begin{bmatrix} \text{T-HEAD } A_{1,0} \\ \text{WHERE } U_1 \end{bmatrix} \end{bmatrix} \end{bmatrix}$$

$$\text{H}\boxed{n} \begin{bmatrix} \text{SYN} \begin{bmatrix} \text{VAL} \begin{bmatrix} \text{COMPS} & \langle\, \rangle \end{bmatrix} \end{bmatrix} \\ \text{SEM} \begin{bmatrix} \text{L-TYPE} & \tau \\ \text{TERM} & \begin{bmatrix} \text{T-HEAD } A_{2,0} \\ \text{WHERE } U_2 \end{bmatrix} \end{bmatrix} \end{bmatrix}$$

where, the parts of the term $[A_0 \ \ \mathsf{where} \ \ U]$ in (6), are determined by (7a)–(7b).

If $A_{2,0}$ is immediate, then

$$A_0 \equiv A_{1,0}(A_{2,0}) \text{ and } U \equiv U_1 \cup U_2 \tag{7a}$$

otherwise (i.e., $A_{2,0}$ is proper),

$$A_0 \equiv A_{1,0}(a_{2,0}) \text{ and } U \equiv \{a_{2,0} := A_{2,0}\} \cup U_1 \cup U_2 \tag{7b}$$

5 Syntactic Analysis of Modifiers

Syntactically, according to the modifier rules, HMR in (5) and MHR in (6), the local, syntactic head of a phrase Φ acquires its pre-positioned modifiers at a lower level of the syntactic analysis of Φ than its post-positioned modifiers, as explained in Loukanova [1].

The HMR requires the value [POSTHM +] inside the mother's feature structure and serve as a marker that the phrase has been post-modified, or that it can be further post-head modified. This marker does not allow pre-head modification until some other rule, or constraint, "discharges" it. The major rules HSR and HCR can do that because they do not require any value of the POSTHM. If the HMR has not been applied, the POSTHM value inside the mother feature-structure can be underspecified or further specified by the upper levels of an enclosing syntactic analysis.

The constraint [MOD⟨$\boxed{n}$[SYN [POSTHM −]]⟩], inside of the modifier structure in the MHR, (6), via the shared structure $\boxed{n}$, gets enforced into the head daughter. This constraint requires that the head phrase H$\boxed{n}$, which is to be pre-modified by this rule, has not been (locally) post-modified, and thus, it is allowed to get a pre-positioned modifier. Note that the structure of H$\boxed{n}$ may have post-modifiers at a lower phrasal level. The value [POSTHM -] from the pre-modifier in the MHR, (6), is not propagated in the mother feature-structure and is left underspecified to [POSTHM *bool*], because POSTHM is not a HEAD feature. This phrase can be further pre- or post-modified.

6 Syntax-Semantics of Modifiers

Assume a generalised CBLG G with syntax-semantics interface using semantic representations by L_{ar}^{λ}, e.g., as in Loukanova [2], with a type-hierarchy, principles (see Sect. 2.2), a given lexicon with semantic representations, and phrasal rules, including HMR in (5) and MHR in (6).

Proposition 1. *The syntax-semantics rules of* G *render the* L_{ar}^{λ}*-term, which is the semantic representation, in a mother structure to its canonical form, assuming that the daughter terms are in canonical forms.*

Proof. It's a long proof, by induction on the syntax-semantics rules, using the definition of the canonical forms of the terms of L_{ar}^{λ} (see 3.13, Moschovakis [4]).

In Fig. 1, we present the syntax-semantics analysis of a quite complex nominal phrase of the syntactic category NOM, "young scholar responsible for the robot on Mars", which contains a complex modifier. For sake of space, we do not provide the syntax-semantics details of all the subexpressions. Typically, in a classic, higher-order λ-calculus, this expression would be represented by the term T_1 in (8a). T_1 is also a L_{ar}^{λ}-term, and, by using the reduction calculus of L_{ar}^{λ}, is reduced to its canonical form $\mathsf{cf}(T_1)$ in (8b)–(8d).

$$T_1 \equiv [\mathit{responsible\text{-}for}(\mathit{the}((\mathit{on}(\mathit{mars}))(\mathit{robot})))](\mathit{young}(\mathit{scholar})) \quad \text{(8a)}$$

$$\Rightarrow_{cf} \mathsf{cf}(T_1) \equiv [\mathit{responsible\text{-}for}(u)](y) \text{ where } \{y := \mathit{young}(s), \quad \text{(8b)}$$

$$s := \mathit{schiolar}, u := \mathit{the}(r_2), r_2 := [\mathit{on}(m)](r_1), \quad \text{(8c)}$$

$$m := \mathit{mars}, r_1 := \mathit{robot}\} \quad \text{(8d)}$$

The reduction to the term $\mathsf{cf}(T_1)$ in (8b)–(8d) involves a sequence of intermediate reductions steps, which we do not show here for sake of space. The grammar rules for syntax-semantics interface as we defined them in this paper, avoid such reductions. The syntactic parsing in Fig. 1 directly derives the canonical form $\mathsf{cf}(T_1)$ represented in the feature-structure description, by using the syntax-semantics grammar rules with appropriately formulated values of the feature SEM.

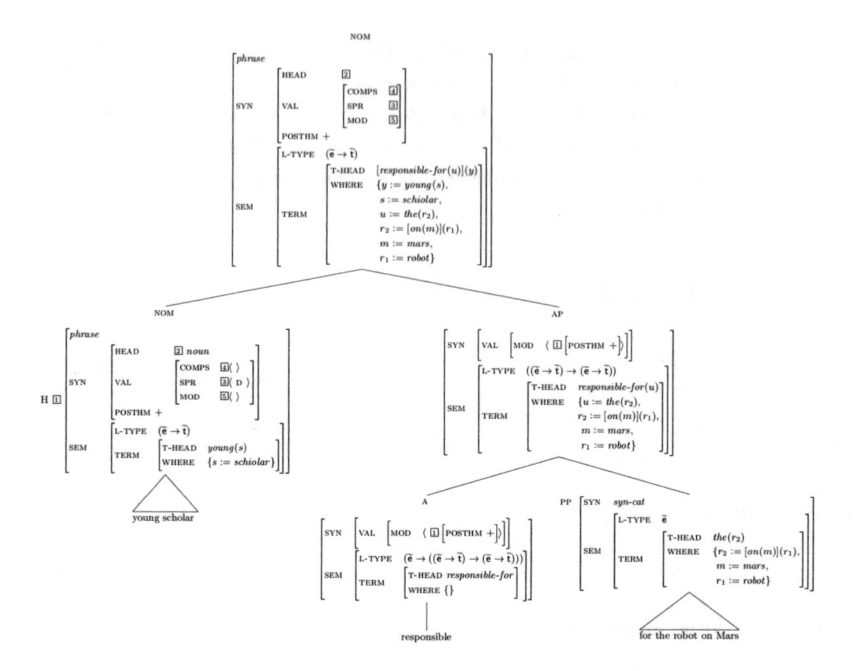

Fig. 1. Syntax-semantics interface of post-modified nominal expression (NOM)

7 Conclusions and Future Work

In this paper, we have provided analysis of modifier expressions of human language. We target adequate computational grammar that reflects biological fundamentals of syntax-semantic interface. In contrast to typical analysis, the introduced technique involves several layers of reducing the complexity of grammar, by using underspecified syntactic and semantic representations, and linguistically motivated generalisations.

We envisage investigation of the computational complexity of the provided syntax-semantics interface, including decision questions and complexity of algorithms provided by the formal languages that we use. That involves development of respective technique for complexity measurements.

References

1. Loukanova, R.: Constraint based syntax of modifiers. In: 2011 IEEE/WIC/ACM International Conferences on Web Intelligence and Intelligent Agent Technology, vol. 3, pp. 167–170 (2011). http://doi.ieeecomputersociety.org/10.1109/WI-IAT.2011.229
2. Loukanova, R.: Semantics with the language of acyclic recursion in constraint-based grammar. In: Bel-Enguix, G., Jiménez-López, M.D. (eds.) Bio-Inspired Models for Natural and Formal Languages, pp. 103–134. Cambridge Scholars Publishing (2011)

3. Loukanova, R.: An approach to functional formal models of Constraint-Based Lexicalized Grammar (CBLG). Fundamenta Informaticae **152**(4), 341–372 (2017). http://dx.doi.org/10.3233/FI-2017-1524
4. Moschovakis, Y.N.: A logical calculus of meaning and synonymy. Linguist. Philos. **29**(1), 27–89 (2006). http://dx.doi.org/10.1007/s10988-005-6920-7

Natural Language Complexity and Machine Learning

Leonor Becerra-Bonache[1] and M. Dolores Jiménez-López[2(✉)]

[1] CNRS, Laboratoire Hubert-Curien UMR 5516, Univ. Lyon, UJM-St-Etienne, Saint-Étienne, France
leonor.becerra@univ-st-etienne.fr

[2] Departament de Filologies Romániques, Research Group on Mathematical Linguistics, Universitat Rovira i Virgili, Tarragona, Spain
mariadolores.jimenez@urv.cat

Abstract. Eventhough complexity is a central notion in linguistics, until recently, it has not been widely researched in the area. During the 20th century, linguistic complexity was supposed to be invariant. In general, recent work on language complexity takes an absolute perspective of the concept while the relative complexity approach –although considered as conceptually coherent– has hardly begun to be developed. In this paper, we introduce machine learning tools that can be used to calculate natural language complexity from a relative point of view by considering the process of first language acquisition.

Keywords: Complexity · Natural language · Machine learning

1 Introduction

It is a fact that complexity has attracted a great deal of attention in linguistics since 2001. From the publication of the article by McWhorter [15] in the special volume of *Linguistic Typology*, there has been a big interest in the study of linguistic complexity. As pointed out by Mufwene [20], the number of books bearing complexity in their title is remarkable, suggesting that there may be a emergent research area whose focus is complexity in language. This interest contrasts with the indifference to complexity issues shown by the vast majority of linguists during the 20th century.

It has been defended for a long time that the linguistic complexity is invariant and that languages are not measurable in terms of complexity. Those ideas have been dubbed the *ALEC statement* ('All Language are Equally Complex') [12] or the *linguistic equi-complexity dogma* [14]. Many models have been proposed to confirm or refute the hypothesis of linguistic equi-complexity. Currently, there is not an unanimously accepted solution to quantify the linguistic complexity.

Considering the state-of-the-art of complexity in natural language and taking in account the current importance of research on complexity and complex systems in several scientific disciplines, we think it is necessary to provide an

S. Rodríguez et al. (Eds.): DCAI 2018, AISC 801, pp. 240–247, 2019.
https://doi.org/10.1007/978-3-319-99608-0_27

objective and meaningful method to calculate linguistic complexity. In order to reach this goal, an interdisciplinary solution -where computational models should be considered- is needed. Taking into account the necessity of interdisciplinarity, we propose a machine learning tool to calculate linguistic complexity.

The paper is organized as follows. Firstly, we present an overview of the meanings of complexity in Linguistics. Secondly, we introduce a machine learning tool to calculate linguistic complexity by showing two different approaches: one based on the Miniature Language Acquisition Task, and the other one on the Abstract Scenes Dataset. Thirdly, we discuss how these machine learning tools may deal with linguistic complexity. We conclude with some final remarks.

2 The Meanings of Complexity in Linguistics

As noted by Mufwene [20], it is surprising the scarcity of works that explain what complexity is when referring to language. The consequence of this situation is the lack of a unified definition of complexity. Instead, in the literature, we can find a variety of approaches that has led to a linguistic complexity taxonomy. In what follows, we will try to summarize some of the most common types of complexity found in the research on natural language.

One of the most used typology of complexity in linguistics is the one that distinguish between *absolute* complexity and *relative* complexity [16–18]. The *absolute –or objective– complexity* approach defines complexity as an objective property of the system and it is measured in terms of the number of parts of the system, the interrelations between the parts or the length of the description of the phenomenon. It is a usual complexity notion in cross-linguistic typology studies [11,15]. The *relative–or agent-related– complexity* approach takes into account the users of language and identifies complexity with difficulty/cost of processing, learning or acquisition. This type of approach is very common in the fields of sociolinguistics and psycholinguistics [14]. Some authors prefer to reserve the term *complexity* for absolute complexity and to use other terms such as *cost, difficulty* or *demandingness* to denote relative complexity [11].

Another common dichotomy in the literature is the one that distinguish *global* complexity from *local* complexity [17]. *Global complexity* is understood as complexity of a language as such. It is about the overall complexity of an entity. It calculates the total complexity of the linguistic system. *Local complexity* is understood as domain-specific complexity. This type of complexity is about some part of an entity. It analyzes the complexity of particular sub-domains of the language.

It is also very frequent in the linguistic complexity bibliography to distinguish between *system* complexity and *structural* complexity [11]. *System complexity* considers the properties of the language and calculates the content of the speaker's competence. It can be understood as a measure of the content that language learners have to master in order to be proficient in language. *Structural complexity* calculates the quantity of structure of a linguistic object, it analyzes the structure of the expressions. It is about the complexity of expressions at some level of description.

In a recent article, Pallotti [21] underlines the polysemy of the term complexity in the linguistic literature and summarizes the different notions of 'complexity' in this field by referring to three main meanings: (1) *Structural complexity*, a formal property of texts and linguistic systems having to do with the number of their elements and their relational patterns; (2) *Cognitive complexity*, having to do with the processing costs associated with linguistic structures; (3) *Developmental complexity*, the order in which linguistic structures emerge and are mastered in second (and, possibly, first) language acquisition. These three meanings cover the two conceptions that, according to Crystal [10], the concept has in linguistics, where 'complexity refers to both the internal structuring of linguistic units and psychological difficulty in using or learning them'.

In general, researchers agree that it is more feasible to approach complexity from an objective or theory-oriented viewpoint than from a subjective or user-related perspective. To approach complexity from the relative point of view constraints the researcher to face many problems. Firstly, what does complex mean?: more difficult, more costly, more problematic, more challenging? Secondly, it must be taken into account that different situations of language use may differ as to what is difficult and what is easy. And finally, a user-based approach would require focusing on one user-type over the others and this implies to decide which type of language use (and user) is primary. To address those problems is not easy, since there will always be some conflict between definitions of complexity based on different types of users, and no general user-type-neutral definition is possible [16]. All those matters concern the relative approach to complexity. Obviously, absolute definitions of complexity avoid these problems.

Generally, studies that have adopted a relative complexity approach have showed some preferences for L2 (second language) learners [14]. However, as pointed out by Miestamo [16], if we aim to reach a general definition of relative complexity, the primary relevance of L2 learners is not obvious. In fact, they could be considered the least important of the four possible groups that may be considered –speakers, hearers, L1 (first language) learners, L2 learners.

Besides the definition issues, when dealing with linguistic complexity we must tackle the metrics problem. There is no conventionally agreed metric for measuring the complexity of natural languages. The tools, criteria and measures to quantify the level of complexity of languages vary and depend on the specific research interests and on the definition of complexity adopted. In fact, in the field of linguistic complexity many *ad hoc* complexity measures have been proposed. The number of categories or rules, length of the description, ambiguity, redundancy, etc. are some examples [17]. Some researchers have attempted to apply the concept of complexity used in other disciplines in order to find useful tools to calculate linguistic complexity. Information theory [5,13], computational models [9], or the theory of complex systems [1] are examples of areas that provide measures to quantitatively evaluate linguistic complexity.

3 Machine Learning Tools

The field of machine learning [19] provides different tools that could be useful to study linguistic complexity. In this section, we present a computational model that has been developed with a different goal, but that could be a good tool to study this problem. The systems we introduce here are inspired by some previous work [2–4]. Our claim is that this computational system is linguistically well motivated and can be used to quantify the relative complexity of languages by focusing on the process of first language acquisition.

3.1 First Approach: Miniature Language Acquisition Task

In [6], Becerra-Bonache et al. presented a system based on inductive logic programming techniques that aims to learn a mapping between n-grams (i.e., sequences of words) and meanings, and also the structure of the language. The system was tested on a toy dataset based on the Miniature Language Acquisition task proposed in [22]; this task consists on learning a subset of a natural language from sentence-picture pairs that involve geometric shapes. This dataset contains simple noun phrases that refer to the color, shape, size and relative position of one or two geometric figures. For example, given a picture of a big red square to the left of a big green triangle (as the one depicted in Fig. 1), some of the possible noun phrases are: *a red square, a big triangle, a square to the left of a green triangle*, etc. The sentences were generated automatically using a very simple grammar.

Hence, the input of the learning algorithm is a dataset of pairs made up of phrases and the context in which these phrases have been produced. In contrast to previous approaches in grounded language learning, the meaning of the phrases are not provided in the training set. The context given to the learning algorithm is just a description of what the learner can perceive in the world. The learner has to really discover the meaning, making the learning problem far more complex.

In this approach, a phrase is represented as a sequence of words (n-grams) and a context as a set of ground atoms (first-order logic based representation). These atoms describe properties and relationships between objects. It is assumed that phrases cannot refer to something that is not in the context and they can give just a partial description of the context. Moreover, predicates and constants represent concepts that a child can recognize (predicates are used for properties, such as *color*, and constants for values of these properties such as *re*, which refers to "red"). Figure 1 shows an example of the input representation.

Becerra-Bonache et al. experimentally showed in [6] that their model can explain the gradual learning of simple concepts and language structure. Experiments with three different languages (English, Dutch and Spanish) demonstrated that the system learns a language model that can easily be used to understand, generate and translate utterances. For more details about the learning algorithm and results, see [6].

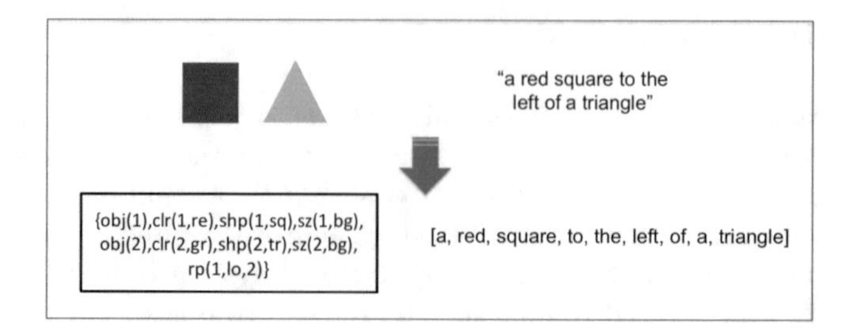

Fig. 1. Illustration of a pair (context, phrase) and its associated representation.

3.2 Second Approach: Abstract Scenes Dataset

In [7,8], Becerra-Bonache et al. presented an improvement of the model introduced in [6]. Instead of using a toy dataset that contains a few objects and simple noun phrases, they developed a system that deals with a more challenging dataset: the Abstract Scenes Dataset. This dataset was introduced in [23] and it contains clip-art pictures with sentences describing these pictures. The images contain children playing outdoors; they were created from 80 pieces of clip-art, representing 58 different objects, such as people, animals, toys, trees, etc. The sentences describing the images were generated by humans (3 sentences per image). The dataset was created by using Amazon's Mechanical Turk and contains a total of 10,020 scenes. Figure 2 shows an example of an image-sentence pair of this dataset.

Fig. 2. Illustration of an image-sentence pair extracted from the Abstract Scenes Dataset.

Like in the previous approach, the system learns from pairs consisting of a sentence and an image. Since the Abstract Scenes Dataset provides the exact list of objects present in an image, Becerra-Bonache et al. [6] could easily generate a logical representation of each image, similar to the one described in the previous section. It is worth noting that the use of abstract scenes allows to override the difficulties of using Computer Vision tools to automatically detect the objects that appear in the scene.

This dataset sets out different challenges for the system. For example, in the previous approach, it was assumed that a sentence only mentions things present in the picture. However, the Abstract Scenes Dataset contains *noise*, that is, a sentence may contain words whose corresponding meanings are not included in the description of the context. For instance, we can have an image of a tent and nobody around, and the sentence provided can be *"Mike is in the tent"*. Becerra-Bonache et al. showed in [7,8] that their system is able to learn in noisy environments. Moreover, the system was also improved to better learn the meaning of words and generate relevant sentences for a given scene (i.e., sentences that do not state things that are true for almost all the scenes, such as *"the sky is blue"*).

4 Measuring Linguistic Complexity with Machine Learning Tools

The machine learning tools we have proposed can calculate linguistic complexity from a relative point of view. Taking into account, the centrality of L1 learners, we defend that studies on developmental complexity may check differences among languages by considering child first language acquisition. Due to the problems that methods for studying language acquisition (observational and experimental) may set out to the study of linguistic complexity, we defend that machine learning models may be considered an important complementary tool that -by avoiding practical problems of analyzing authentic learner productions data– will make possible to consider children (or their simulation) as suitable candidates for evaluating the complexity of languages.

The two introduced approaches can calculate the number of interactions that are necessary to achieve a good level of performance in a given language by using a unique algorithm to learn any of the languages analyzed. Those approaches may show that not all the languages need the same number of linguistic interactions to reach the same level of performance. Therefore, these machine learning tools may be potentially adequate to measure linguistic complexity in *relative* terms. In fact, the unique algorithm used in the model could be equivalent to the innate capacity that allows humans to acquire a language. Moreover, the learner –this is, the machine– has no previously knowledge about the language. The machine represents, therefore, the child that has to acquire a language by just being exposed to this language.

To count the needed number of interactions for the machine to achieve a good level of performance in a specific domain of the language may be equivalent to calculate the child's cost/difficulty to acquire a language. Finally, to show that with the same algorithm not every language requires the same number of interactions may be interpreted (in terms of complexity) as an evidence to defend that the difficulty/cost to acquire different languages is not the same and, therefore, languages differ in relative complexity.

We claim, therefore, that those models may be used for calculating linguistic complexity. They are models that focused on the learning process. They do not require any prior language-specific knowledge and learn incrementally. Moreover, they use realistic data and psychologically plausible algorithms that include features like gradual learning, robustness to noise in the data, and learning incrementally.

5 Conclusions

Complexity is a controversial concept in linguistics. Until recently, natural language complexity has not been widely researched and still not clear how complexity has to be defined and measured. There have been attempts to apply the concept of complexity used in other disciplines in order to find useful tools to calculate linguistic complexity.

In this paper, we have presented a preliminary approach to the study of linguistic complexity with machine learning tools. Our proposal is an interdisciplinary solution that uses a machine learning model to quantify the cost/difficulty in the process of children first language acquisition. Those machine learning algorithms allow us to calculate the *cost* –in terms of the number of interactions– to reach a good level of performance in a given language and, therefore, offer the possibility to measure the *difficulty* of acquiring different natural languages. Machine learning models can be seen as an alternative to the methods that have been used so far in the area of linguistic complexity.

Linguistic complexity may be a key point in automatic natural language processing, since results in that field may condition the design of language technologies, for example. Therefore, Linguistics must tackle the problem of natural languages complexity. Researchers must propose tools for the analysis of linguistic complexity, since the results obtained from these studies may have important implications both from a theoretical and from a practical point of view.

Acknowledgments. This research has been supported by the Ministerio de Economía y Competitividad and the Fondo Europeo de Desarrollo Regional under the project number FFI2015-69978-P (MINECO/FEDER, UE) of the Programa Estatal de Fomento de la Investigación Científica y Técnica de Excelencia, Subprograma Estatal de Generación de Conocimiento.

The work of Leonor Becerra-Bonache has been performed during her teaching leave granted by the CNRS (French National Center for Scientific Research) in the Computer Science Department of Aix-Marseille University.

References

1. Andrason, A.: Language complexity: an insight from complex-system theory. Int. J. Lang. Linguist. **2**(2), 74–89 (2014)
2. Angluin, D., Becerra-Bonache, L.: A model of semantics and corrections in language learning. Technical report, Yale University (2010)
3. Angluin, D., Becerra-Bonache, L.: Effects of meaning-preserving corrections on language learning. In: Proceedings of the 15th International Conference on Computational Natural Language Learning, CoNLL 2011, Portland, pp. 97–105 (2011)
4. Angluin, D., Becerra-Bonache, L.: A model of language learning with semantics and meaning preserving corrections. Artif. Intell. **242**, 23–51 (2016)

5. Bane, M.: Quantifying and measuring morphological complexity. In: Chang, C., Haynie, H. (eds.) Proceedings of the 26th West Coast Conference on Formal Linguistics, pp. 69–76. Cascadilla Proceedings Project, Somerville (2008)
6. Becerra-Bonache, L., Blockeel, H., Galván, M., Jacquenet, F.: A first-order-logic based model for grounded language learning. In: Advances in Intelligent Data Analysis XIV - 14th International Symposium, IDA 2015, pp. 49–60 (2015)
7. Becerra-Bonache, L., Blockeel, H., Galván, M., Jacquenet, F.: Learning language models from images with regll. In: Machine Learning and Knowledge Discovery in Databases - European Conference, ECML PKDD 2016, pp. 55–58 (2016)
8. Becerra-Bonache, L., Blockeel, H., Galván, M., Jacquenet, F.: Relational grounded language learning. In: ECAI 2016 - 22nd European Conference on Artificial Intelligence, 29 August-2 September 2016, The Hague, The Netherlands - Including Prestigious Applications of Artificial Intelligence (PAIS 2016), pp. 1764–1765 (2016)
9. Blache, P.: A computational model for linguistic complexity. In: Bel-Enguix, G., Dahl, V., Jiménez-López, M. (eds.) Biology, Computation and Linguistics. New Interdisciplinary Paradigms, pp. 155–167. IOS Press, Amsterdam (2011)
10. Crystal, D.: The Cambridge Encyclopedia of Language. Cambridge University Press, Cambridge (1997)
11. Dahl, O.: The Growth and Maintenance of Linguistic Complexity. John Benjamins, Amsterdam (2004)
12. Deutscher, G.: Overall complexity: a wild goose chase? In: Sampson, G., Gil, D., Trudgill, P. (eds.) Language Complexity as an Evolving Variable, pp. 243–251. Oxford University Press, Oxford (2009)
13. Juola, P.: Assessing linguistic complexity. In: Miestamo, M., Sinnemäki, K., Karlsson, F. (eds.) Language Complexity: Typology, Contact, Change, pp. 89–108. John Benjamins, Amsterdam (2009)
14. Kusters, W.: Linguistic Complexity: The Influence of Social Change on Verbal Inflection. LOT, Utrecht (2003)
15. McWhorter, J.: The world's simplest grammars are creole grammars. Linguist. Typol. **6**, 125–166 (2001)
16. Miestamo, M.: On the feasibility of complexity metrics. In: Krista, K., Sepper, M. (eds.) Finest Linguistics. Proceedings of the Annual Finish and Estonian Conference of Linguistics, pp. 11–26. Tallinna Ülikooli Kirjastus, Tallinn (2006)
17. Miestamo, M.: Grammatical complexity in a cross-linguistic perspective. In: Miestamo, M., Sinnemäki, K., Karlsson, F. (eds.) Language Complexity: Typology, Contact, Change, pp. 23–42. John Benjamins, Amsterdam (2009)
18. Miestamo, M.: Implicational hierarchies and grammatical complexity. In: Sampson, G., Gil, D., Trudgill, P. (eds.) Language Complexity as an Evolving Variable, pp. 80–97. Oxford University Press, Oxford (2009)
19. Mitchell, T.M.: Machine Learning. McGraw Hill Series in Computer Science. McGraw-Hill, New York (1997)
20. Mufwene, S., Coupé, C., Pellegrino, F.: Complexity in Language. Cambridge University Press, New York (2017)
21. Pallotti, G.: A simple view of linguistic complexity. Second Lang. Res. **31**, 117–134 (2015)
22. Stolcke, A., Feldman, J., Lakoff, G., Weber, S.: Miniature language acquisition: a touchstone for cognitive science. Cogn. Sci. **8**, 686–693 (1994)
23. Zitnick, C., Parikh, D.: Bringing semantics into focus using visual abstraction. In: Proceedings of the International Conference on Computer Vision and Pattern Recognition, pp. 3009–3016. Portland (2013)

Special Session on Web and Social Media Mining (WASMM)

Twitter's Experts Recommendation System Based on User Content

Diego M. Jiménez-Bravo(✉), Juan F. De Paz, and Gabriel Villarrubia

BISITE Digital Innovation Hub, University of Salamanca, Edificio Multiusos I+D+I, 37007 Salamanca, Spain
{dmjimenez,fcofds,gvg}@usal.es
https://bisite.usal.es/en

Abstract. The Internet provides users with an overwhelming amount of information. For this reason they may not always be able to find the information they are looking for. Recommendation systems help users locate useful information and save time. Twitter is one of the social networks that implements this type of system in order to help its users in searching content. However, the traditional recommendation system implemented by Twitter only considers people from the user's surroundings or it suggests the followees/followers of the user's followees. Many use Twitter as a source of information, it is therefore necessary to create a recommendation system that would suggest experts profiles to other users. Experts must be capable of providing interesting information to users. The "expert" recommended to a users will be chosen on the basis of the content they publish and whether this content is of interest to the user. The proposed system offers accurate and suitable recommendations.

Keywords: Content-based recommendation · Information retrieval Recommendation system · Text mining · Twitter · Web mining

1 Introduction

Nowadays, computer systems contain a big amount of data. This makes it more difficult for platform users to view information that they find important and interesting. It is therefore necessary to manage all this information and decide for users what may be important for them. To this end, all platforms should use recommendation systems which consider user preferences to provide users with suitable information. Recommendations are based on user data and on the information that is related to them. Together with the application of machine learning techniques, the system can provide users with useful data.

Twitter is a micro-blogging social network, it has over 500 million users form all around the world. Consequently, it contains a huge amount of data on users and the interactions between them. For this reason, it was crucial for Twitter to include a recommendation system in its platform. Twitter called it

S. Rodríguez et al. (Eds.): DCAI 2018, AISC 801, pp. 251–258, 2019.
https://doi.org/10.1007/978-3-319-99608-0_28

"Who to follow". It is a feature that recommends followees (people/users that a user follows) who are similar to the user's existing followees, the system also recommends the followees of those followees. This Twitter tool also recommends users who are similar to the ones that they view/visit. However, the behavior behind this recommendation system is unknown [1]. Although this feature seems to work quite well, there are other types of alternatives that we will be described in Sect. 2. In this paper, we also propose an alternative recommendation system which makes recommendations on the basis of the content of tweets posted by users. It uses text mining techniques to process the words in the tweets and in this way it determines similarity between users. A case study was carried out with the proposed recommendation system, it provided recommendations to users interested in sports. The system achieved good results in recommending new users to the target user.

The following article is divided into the following sections. In the background Sect. 2, we talk about this line of research. The proposed system is outlined in the proposal Sect. 3. The case study Sect. 4 focuses on specific situations that have been considered during our research. In the results Sect. 5, the system's performance will be described and analyzed. Lastly, in the conclusions Sect. 6 we discuss the system and future lines of work related to the proposed study.

2 Background

We first discuss the techniques that are commonly used to make personalized recommendations to social network users and then describe recent works on followee recommendation for Twitter.

There are two main techniques in recommendation systems. These methodologies are collaborative filtering and content-based recommendation. First let us look at collaborative filtering (CF). There are two possible approaches in CF according to Kywe et al. [1], *user-to-user* or *item-to-item*. In the first approach, the recommender system tries to find users that are similar to the target person, based on their interactions and preferences. Then, the system predicts the target user's preferences based on the preferences of the users that are similar to them. On the other hand, *item-to-item CF* refers to items similarity. Two items are denoted as similar, if for example a large number of users buy or rate the two items. So, the target user can be recommended a an item, if it is similar to others that they have bought or rated. Su and Khoshgoftaar [2] and Jalili [3] explain that there are two different CF based techniques, memory-based CF and model-based CF. However, Su and Khoshgoftaar said that there is a third one, called hybrid recommender.

Secondly, content-based (CB) recommendation systems find similar items as well as CF. However, in this case the similarity of two items is found by comparing their features and characteristics. Consequently, recommended items are similar to other previously rated, bought or in Twitter's case, tweeted, retweeted, etc. [1]. An example of this is that, if a person rated one of Tom Cruise's action films 10 out of 10, the system will recommend him or her another action film or another Tom Cruise film. One important characteristic of this kind of recommendation system is that the user has to evaluate or rank items for them to be recommended.

Armentano et al. [4] make followee recommendation by using collaborative filtering and content-based recommendation. The CF approach assumes that the target user is similar to their followees' followers. They recommend the top ranked users according to the number of users its has in common with the target user. The CB study supposes that a target user will follow those who are similar to them. According to this assumption, the system will recommend those users whose followees' tweets are similar to the followees' tweets of the target user.

Also, Hannon et al.'s proposal [5] recommends users by using both CB and CF methods. In the CB technique users are represented by their tweets, their followees' tweets, their followers' tweets or a combination of them. If the system uses the target user's tweets, users with similar tweets will be recommended. In the CF approach users are represented by the IDs of their followees, followers or a combination of them. Then *tf-idf* weighting scheme is used to find users similar followers/followees and recommend them to the target user.

An article written by Garcia and Amatriain [6] proposes that the recommendation of followees to a target user is based on their followees' popularity and activity. If the target user follows popular users other popular users will be recommended. If the target user has popular and active followees, popular and active followees will be recommended. The approach is similar to the active followees approach.

Golder et al. [7] affirm that users are similar and share interests if they are following the same people. Also, that users are similar if they have the same followers. The system recommends those similar users. The study also proposes that a target user will be interested in the followees of their followees, so the platform recommends them as well.

More recent studies like Elmongui et al.'s [8] propose to create a user timeline based on their interest. Interests are based on social features, social interaction and the history of tweets. Interests change with time so the system takes that into account.

Our proposal makes use of a content-based recommendation system. We propose that relationship recommendations should be based on the content of the user tweets. Other users will be recommended to the target user on the basis of similarity of the top 20 words in the tweet, according to *tf-idf* weighting scheme. oto classify a user in a specific category. In 1 we can see a comparative between the different approaches.

3 Proposal

The proposal consists of a recommendation system for the social network, Twitter. The proposed system recommends new Twitter relationships to the target user, considering their preferences.

The system assumes that experts in a variety of fields exist within the social network. Those experts are usually followed by many people who wish to keep

Table 1. Comparison of the state of the art and the proposal

Features	[4]	[5]	[6]	[7]	[8]	Our proposal
Collaborative filtering	✓	✓	✓	✓	✓	
Content-based	✓	✓				✓
Assumes similarity's relationships						
Topic-based recommendations	✓					✓
Uses user's tweets		✓			✓	✓
Uses followers' tweets		✓				
Uses followees' tweets	✓	✓				
Uses *tf-idf*		✓				✓
Uses followees' behaviour			✓			
Uses followers or followees				✓		
Takes into account recent interests					✓	✓
Makes expert recommendation						✓

up to date on a certain issue. Thus, experts' tweets are a source of information for non-expert users. For this reason our proposal is aimed at recommending Twitter's expert users to standard users who may be interested in the information they provide.

The platform selects experts on Twitter. These experts are selected by an external person who knows about the fields and who determinates which user has to be stored in a database. Those users usually have followees that are also experts and they provide information to the original expert. This statement is used by the system in order to create a dataset of experts that could be recommended to users who want to get informed on a specific topic.

Once the system identifies expert users, it obtains the profiles of the expert's 200 most recent followees. These users are selected because they are representative of the expert user's interest. Once all the experts and their followees (experts as well) are compiled, the platform selects the 100 most recent elements on their time-lines. These elements also help the system determine the field in which the user is expert.

Then, tweets are analyzed individually. Words in the tweets are tokenized by the platform. Then, the words are analyzed using a *stopwords* analyzer, this allows to eliminate the most commonly used words in a given language, like articles, prepositions, etc. Once the words are separated and *stopwords* are eliminated, the system considers each word that every user uses in their tweets. A correlation factor is applied to all words in order to avoid very commonly used words in the user's field of knowledge being considered for similarity detection, this is what Shouzhong and Minlie did in [9]. So, for every word, we applied *term frequency-inverse document frequency*, better known as *tf-idf*. This statistical value reflects the importance of a word inside a corpus. Trstenjak et al. [10] and Aizawa [11] talk about this. *Tf-idf* increases the proportion of a word inside a document. However, this proportion is conditioned by the total number of times that the word appears in the documents. That helps to adjust the frequency of the most common words. There are different ways of calculating this statistical value. In our proposal we use the Eqs. 1 and 2.

$$tf = 1 + \log x \quad (1) \qquad idf = \log\left(1 + \frac{N}{n_t}\right) \quad (2)$$

Where x is the total number of times that a word appears in a document, N is the total number of documents and n_t is the number of documents that contain the word.

So, for every word in every document (tweet) *tf* and *idf* are calculated. This word receives the value of the multiplication of those two values, $tf \cdot idf$. Once this process has been applied to every word, the words are sorted by their *tf-idf* value. We select 20 words with the highest values. These words are elements of every user profile. The Eq. 3 as a whole defines the characteristics of a user. We should also point out that profiles with less than 20 words are eliminated.

$$user = \begin{cases} (id, w_1, v_{w_1}, label) \\ \vdots \\ (id, w_{20}, v_{w_{20}}, label) \end{cases} \quad (3)$$

Where *id* is the user name of every user, w_n refers to the words with highest *tf-idf* values, v_{w_n} is the *tf-idf* value of the word w_n and *label* is the expert's field of knowledge.

In this process the system composes the expert users' dataset which will be used for user recommendation. The platform will use machine learning algorithms on this dataset in order to create a recommendation system.

The technique that the platform uses is the k-nearest neighbors (k-NN). This algorithm is based on instance-based learning. This algorithm operates by processing each instance as if it was a vector in an n-dimensional space, where n is the number of attributes of the instance, and where each of instance has a class associated to it. When the system wants to classify a new instance, it calculates the distance from this instance to the rest of the instances in the dataset. The k-nearest instances determine the class of the new instance.

This methods allows the platform to recommend expert users on the basis of the content of the users' tweets.

4 Case Studies

This section describes the case study that has been conducted in order to evaluate the proposed system outlined in Sect. 3.

In this case study, the recommendation system suggested relationships with users who are experts in the field of Spanish sports. Spain has recently taken the lead in many sports, this has caused increased interest in this field on social networks. We selected several team sports that can be played as team sports. These sports are European football, basketball, athletics, handball, indoor European football, hockey, American football, rugby and rowing.

For every one of these sports we selected the Twitter accounts of the teams that play in the top league and their followees. Table 2 lists the selected leagues and the season.

Table 2. Sports leagues selected for the study

Sports	League	Season
European football	Santander's league	2017/2018
Basketball	Endesa's league	2017/2018
Athletics	First division and honor division	2017/2018
Handball	Asobal's league	2017/2018
European indoor football	LNFS' league	2017/2018
Hockey	OK's league	2017/2018
American football	A's serie	2017/2018
Rugby	Honor division	2017/2018
Rowing	San Miguel's league	2017/2018

These user's profiles were written in Spanish. Also, their followees were usually from Spain. Therefore, in this case study Spanish users had been recommended. The user profiles were processed using the methodology described in Sect. 3, we obtained a dataset of 270,500 instances which means that we had a total of 13,525 expert users that we were able to recommend to the target user.

5 Results

This section outlines the results and the validation of the recommender system presented in the article.

The system was evaluated with the measure of different statistical values. These values were analyzed in order to verify the performance of the presented platform.

Moreover, already mentioned in Sect. 3, the k-NN algorithm implemented by the software RapidMiner [12] was also used. Furthermore, in this section we are going to determine which k value is the optimal for the presented problem. We evaluated different k values. We evaluated the system with 5, 10, 20, 40 and 80 as k values. In Table 3 we show the results of the evaluation.

As can be seen, the different models implemented in the evaluation of the system present a large area under the curve (AUC) (see Huang and Ling [13] and Lobo et al. [14]), these are very good results. However, as we can see in 5, the MAP (Mean Average Precision) values decrease as the k values increase. For this reason, the best model that had been implemented was the one whose k is equal to 10.

Table 3. Model evaluation's results

Statistical measure	5-NN	10-NN	20-NN	40-NN	80-NN
AUC	0.969	0.970	0.975	0.977	0.978
prec@5	0.734	0.735	0.734	0.658	0.550
prec@10	0.408	0.409	0.409	0.378	0.322
prec@15	0.275	0.275	0.275	0.258	0.232
NDCG	0.978	0.979	0.978	0.911	0.823
MAP	0.976	0.978	0.976	0.879	0.757

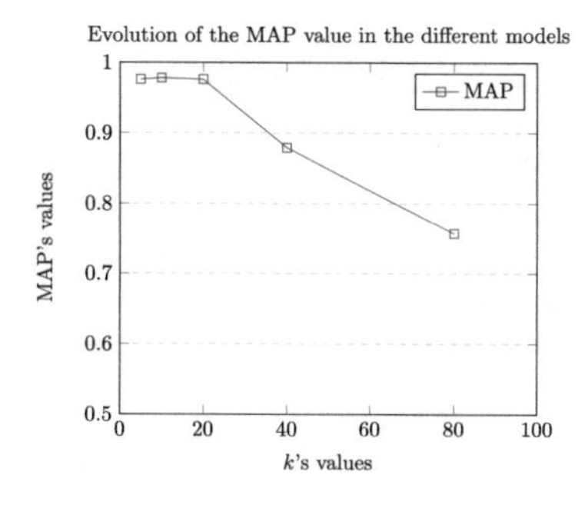

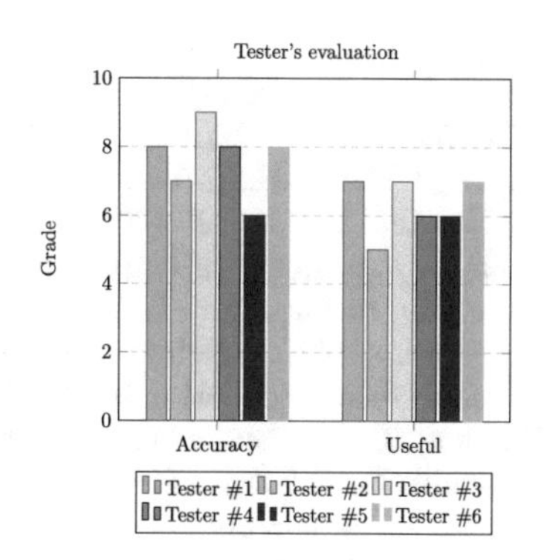

Table 3 shows that the recommendation system provides good results and works perfectly. With the selected model, the platform will be able to recommend expert users to a target user. We tested the system on different users. In 5 we can see the results of the evaluation.

6 Conclusions

Recommendation systems provide solutions to different user platforms and data related problems. They provide users with information or data that is suited to their interests. This paper has successfully implemented a recommendation system on Twitter. First, a methodology was defined in order to deal with the information that is stored on Twitter. Secondly, we designed a case study in order to evaluate the proposed methodology and the recommendation system.

The proposed methodology preprocesses the social network's data with several scripts located inside the presented system. Those scripts create the dataset used for the recommendation. The recommendation is done for the selected model. Despite the good performance of all the implemented models, we selected the one with the best results. This model has good results as shown in Sect. 5.

The presented recommendation system makes use of a technique called "content-based recommendation" which uses the content of the data to make recommendations. Our technique makes use of the most important words used by Twitter users in their tweets. These words show the content in which users are interested. Text mining techniques allow the platform to recommend expert users.

Future lines of research include the expansion and evaluation of the system to more expert areas and dealing with languages other than Spanish.

Acknowledgement. This work has been supported by project MOVIURBAN: Máquina social para la gestión sostenible de ciudades inteligentes: movilidad urbana, datos abiertos, sensores móviles. SA070U 16. Project co-financed with Junta Castilla y León, Consejería de Educación and FEDER funds.

References

1. Kywe, S.M., Lim, E.P., Zhu, F.: A survey of recommender systems in Twitter, vol. 7710, pp. 420–433 (2012)
2. Su, X., Khoshgoftaar, T.M.: A survey of collaborative filtering techniques. Adv. Artif. Intell. **2009**, 1–19 (2009)
3. Jalili, M.: A survey of collaborative filtering recommender algorithms and their evaluation metrics. Int. J. Syst. Model. Simul. **2**(2), 14 (2017)
4. Armentano, M.G., Godoy, D.L., Amandi, A.A.: A topology-based approach for followees recommendation in Twitter (2011)
5. Hannon, J., McCarthy, K., Smyth, B.: Finding useful users on Twitter: Twittomender the followee recommender, pp. 784–787. Springer, Heidelberg (2011)
6. Garcia, R., Amatriain, X.: Weighted content based methods for recommending connections in online social networks, pp. 68–71 (2010)
7. Golder, S.A., Yardi, S., Marwick, A., Boyd, D.: A Structural Approach to Contact Recommendations in Online Social Networks (2009)
8. Elmongui, H.G., Mansour, R., Morsy, H., Khater, S., El-Sharkasy, A., Ibrahim, R.: TRUPI: Twitter Recommendation Basedon Users' Personal Interests, pp. 272–284. Springer, Cham (2015)
9. Shouzhong, T., Minlie, H.: Mining microblog user interests based on textrank with TF-IDF factor. J. China Univ. Posts Telecommun. **23**(5), 40–46 (2016)
10. Trstenjak, B., Mikac, S., Donko, D.: KNN with TF-IDF based framework for text categorization. Procedia Eng. **69**, 1356–1364 (2014)
11. Aizawa, A.: An information-theoretic perspective of tf-idf measures. Inf. Process. Manag. **39**(1), 45–65 (2003)
12. Visual Workflow Designer for Data Scientists — RapidMiner Studio — RapidMiner (2008)
13. Huang, J., Ling, C.X.: Using AUC and accuracy in evaluating learning algorithms. IEEE Trans. Knowl. Data Eng. **17**(3), 299–310 (2005)
14. Lobo, J.M., Jiménez-valverde, A., Real, R.: AUC: a misleading measure of the performance of predictive distribution models. Glob. Ecol. Biogeogr. **17**(2), 145–151 (2008)

A Text Mining-Based Approach for Analyzing Information Retrieval in Spanish: Music Data Collection as a Case Study

Juan Ramos-González(✉) and Lucía Martín-Gómez

BISITE Digital Innovation Hub,
University of Salamanca, Edificio Multiusos I+D+i, 37007 Salamanca, Spain
juanrg@usal.es

Abstract. This paper presents a text mining-based search approach aimed at information retrieval in the Spanish language. For this purpose, a tool has been developed in order to facilitate and automate the analysis and retrieval, allowing the user to apply different analyzers when carrying out a query, to index and delete documents stored in the system and to evaluate the recovery process. To this extent, a dataset consisting in 27 songs has been used as a case study. Different queries have been made to investigate about the best fitting approaches to the Spanish language and their suitability depending on the query text.

Keywords: Text mining · Information retrieval · Stemming · Spanish

1 Introduction

Humans can easily interpret linguistic patterns, deal with changes of meaning depending on the context, with spelling variations and other "irregularities" in written language [4]. However, while we can face unstructured linguistic data, we lack the ability to process big amounts of textual information in an fast and efficient way. Information Retrieval (IR) and text mining techniques aim to solve this problem. However, algorithms for text analysis are language-dependent. In this regard, most works in the literature have focused on English as the study language, while other widespread Indo-European languages have received little attention, in spite of the great importance they have in the field of Linguistics.

Nowadays, most information is stored as text, which makes of text mining a field with high potential in several areas [2]. Text mining involve techniques that have been applied to Natural language processing, Information Extraction and Information Retrieval, as well as to solve Categorization problems [3,12].

In IR, a fundamental role is played by stemming algorithms, which are in charge of extracting the root of a word parting from inflected and derived forms. These algorithms are frequently used in this area to increase the recall rate

S. Rodríguez et al. (Eds.): DCAI 2018, AISC 801, pp. 259–266, 2019.
https://doi.org/10.1007/978-3-319-99608-0_29

and to obtain relevant results [10]. Though some efforts have been done for facilitating the implementation of stemming algorithms in other languages, but deeper comparative study of such techniques is needed in Spanish language.

In the present work, a text mining approach for information retrieval is presented. Different filters and analyzers for the Spanish language are applied and compared to evaluate their suitability. For this purpose, a web tool has been designed to efficiently perform different queries using the ElasticSearch engine [1]. To this extent, a data collection of Spanish music was built, involving 27 songs coming from 9 bands. Discussion about the performance of different filters and analyzers and their effectiveness for IR is provided.

2 A Web Tool for Data Retrieval

With the purpose of facilitating the study of retrieval process in Spanish an application has been developed. In this section, the main functionalities that this tool has for this purpose are explained. In first place, the system allows a user to index data. Then, a search can be made by choosing one of the different fields in which the documents are divided. After that, a set of operators, which involve different filters and analyzers also used in indexing stage are applied. The results are then shown to the user. Parallel, a set of evaluations queries are made, aiming to create a document of relevant results by expertise for each search. Figure 1 shows the structure of the retrieval process followed by the tool.

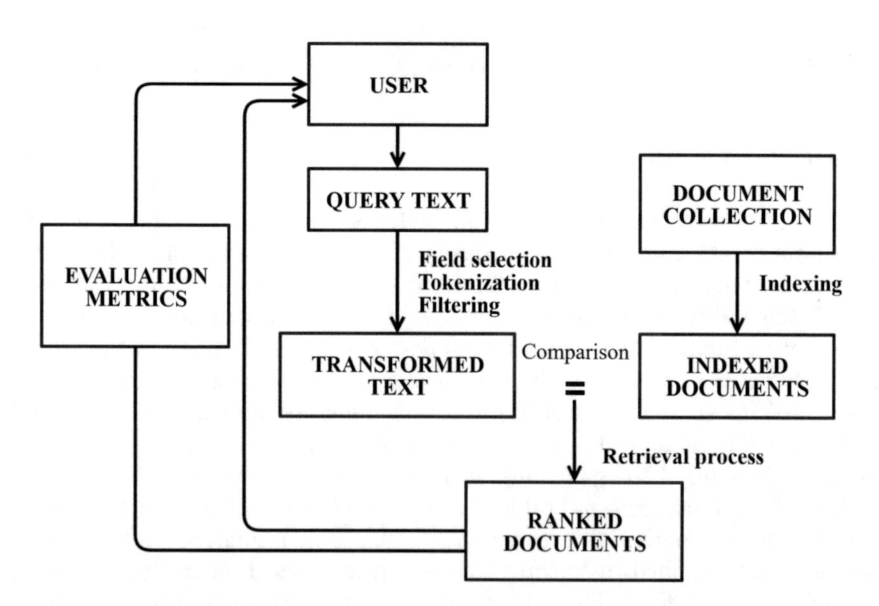

Fig. 1. Document retrieval system carried out by the developed web tool.

2.1 Indexing

When a document is created, the text it contains must pass through a tokenization process and a transformation to lower case. Then different operators that transform the text are applied over the selected fields of the data collection. An index consists in all the stored documents resulting after applying a certain transformation. There must be n different indexes where n is the number of filters or filter combinations we allow the user to select in a query.

2.2 Filters and Analyzers

As mentioned above, the set of operators can be applied to transform the text. Among these operators, two stemming algorithms have been included. Stemmers transform the text by reducing a word to its stem, so that the subsequent search can be sensible to tense, gender, number and case. We should clarify that, from the Computational Linguistics point of view it is important to establish a difference between stem and root. However, these concepts are hard to define. In this work we follow the ideas of Porter [6], which interprets stem as the resulting residue from stemming process while considers the root in an etymological way. In stemming, suffixes are sequentially removed according to a set of rules specifically designed for each language [9]. The following filters and analyzers have been included in the tool and used in our case study.

- **Stop words removal:** Common words with no meaning by themselves such as articles or prepositions can hinder information retrieval, as they can greatly influence the similarity scores. Consequently, these words are often removed. For this purpose, a list of stop words for each language is needed.
- **Standard analyzer:** The ElasticSearch default analyzer, does not performs transformation operations over the text in the query or the documents. It just involves a standard tokenization, thus its objective in this case is to search over the documents for coincidences with a given text string.
- **Light Spanish stemmer (LSS):** This algorithm removes certain inflectional suffixes such as singular and plural word forms or feminine and masculine forms. However, LSS does not account for person and tense variations used by verbs. [9]. Usually, light stemming is recommended when the search results are expected to be interpreted by humans, but other factors can affect this decision, like the data collection size or the next algorithm that will be applied [1].
- **Snowball Spanish stemmer (SS):** The Spanish stemmer, unlike LSS, is sensible to both inflectional an derivational constructions. It can face variations of tense and person in verbs. Therefore, this algorithm stems more aggressively than LSS. In many languages, when dealing with small data collections an aggressive stemming can be desirable, as the user may want to ensure that at least some documents are found. The SS is a snowball-generated stemmer algorithm. Snowball is a processing language for easily generating stemming algorithms [5].

2.3 Evaluation Process

As we can not evaluate the information retrieval process in every possible query, we performed various queries using different words and selected those documents that we considered relevant in each case. A file is then created containing this information. So that, some quality metrics can be calculated. In this case we provide the precision and recall of every evaluation query. Then, we can explore which filter combination is more appropriate for different queries. In addition, when carrying out a query (regardless of whether it is for evaluation purposes), a user normally looks for those documents of greater relevance. For this purpose, a scoring method allowing to show the resulting documents in order of importance is required. In the present case study, Term Frequency/ Inverse Document Frequency (TF-IDF) have been applied. TF-IDF is a model which determines the relative frequency of words in a document compared to the inverse proportion of such word over a document collection, defined as:

$$tfidf(t, d, D) = tf(t, d) \cdot idf(t, D) \tag{1}$$

where t is the term, d represents each document and D the collection of documents. TF-IDF provides information about the relevance of a concrete word in a document [7,8]. This method results from combining TF and IDF statistics and it is one of the most used term weighting approaches [13].

3 Case Study

In order to perform and evaluate the IR process in Spanish small data collections, a case study involving 27 songs is proposed. More concretely, 9 Spanish artists have been selected and from each of them, 3 songs. For each song, a document is created with the following fields: group, groupID, title, songID, style, year and lyrics. All songs used in the case study share the style "Spanish rock", but the "style" field was created anyway for future data. To store the information regarding fields in the documents, JSON format has been chosen as it is a widely used notation supported by several programming languages. For the present case study, the following indexes are created by applying the tool operators and combinations of them: Standard, LSS, SS, stop words, LSS + stop words and SS + stop words.

3.1 Results

As mentioned above, we can evaluate the retrieval process when searching for text strings that have been previously queried in order to store those results considered relevant in the evaluation document. In this regard, we performed various queries selecting different common words and combination of words present in the dataset including nouns, adjectives and verbs, as well as more complex test strings consisting of various words. In favor of comprehension, an English synonymous of each word is provided. However, it must be taken into account that usually there is no a perfect match translation. The results of any filtering containing the stop words removal filter can be ignored in those queries that do not contain stop words, as it does not transform the text in any way. Table 1 shows the results of the evaluation queries.

Table 1. Performance of selected operators. Precision and recall values for each search have been computed according to those results previously classified as relevant.

	Standard	LSS	SS	Stop words	LSS + Stop words	SS + Stop words
	Prec/rec	**Prec/rec**	**Prec/rec**	**Prec/rec**	**Prec/rec**	**Prec/rec**
"Mar"	1.000/0.800	1.000/1.000	1.000/1.000	1.000/0.800	1.000/1.000	1.000/1.000
"Calle"	0.800/1.000	0.667/1.000	0.667/1.000	0.800/1.000	0.667/1.000	0.667/1.000
"Morena"	1.000/1.000	1.000/1.000	1.000/1.000	1.000/1.000	1.000/1.000	1.000/1.000
"Loco"	1.000/0.333	1.000/0.333	1.000/0.667	1.000/0.300	1.000/0.300	1.000/0.667
"Comer"	1.000/0.667	1.000/0.667	0.300/1.000	1.000/0.667	1.000/0.667	0.300/1.000
"Pensar"	-	-	1.000/0.500	-	-	1.000/0.500
"Con los ojos cerrados"	0.600/0.750	0.600/0.750	0.600/0.750	1.000/0.625	1.000/0.750	1.000/0.750
"Salir de aquí"	0.700/0.700	0.700/0.700	0.600/0.600	1.000/0.800	1.000/0.800	0.900/0.900

- **Query#1 "Mar"** ("Sea"). When looking for this word, a maximum precision is obtained with all operators, reaching 100% recall value in most cases.
- **Query#2 "Calle"** ("Street"). In this case, all documents considered important are recovered, but precision present low values in those operator combinations involving stemmer analyzers.
- **Query#3 "Morena"** ("Brunette"). Results are optimum for every filtering and operator combination.
- **Query#4 "Loco"** ("Crazy"). We can observe that, in spite of the high precision, there are many relevant documents that have not been recovered, and recall is specially low in those filterings which do not include SS.
- **Query#5 "Comer"** ("To eat"). When SS is applied, all the relevant documents are recovered but we can see that precision decreases. Other analyzers and filter combinations than SS recovered only relevant documents.
- **Query#6 "Pensar"** ("To think"). With LSS, stop words removal and standard analyzer no document is recovered. The precision is maximum in the rest of operators, but non relevant documents are also recovered.
- **Query#7 "Con los ojos cerrados"**("with eyes closed"). In this case precision reaches 100% only in those cases where the stop words filter is applied.
- **Query#8 "Salir de aquí"**("getting out of here") As in the previous query, high precision values are achieved when stop words filter is selected. While LSS + stop words presents maximum precision, SS + stop words increases the recall rate.

3.2 Result Discussion

Considering the results given in Table 1, we analyze possible reasons beneath the behaviour of the proposed filters and analyzers. For queries consisting of a noun, we expected high precision. In this regard, we observed that for the noun “calle”, precision was low in those operators involving stemmers. Stemming analyzers reduce the word to its stem, in this case “call-”, which coincides with the root of verb “callar” (“to shut up”) so those documents containing verb forms of this verb were also recovered, resulting in a precision decrease. The word “mar” has no suffixes, but standard and stop words filters do not apply stemming so it is a textual search, omitting words derived from “mar” like “marinero” (“sailor”), explaining why the recall in these two filters presents a small reduction.

In the case of “loco”, in spite of the high precision values, little proportion of relevant documents was recovered, specially when applying operators that do not include SS. We expected that LSS would be able to recover more documents than the standard and stop words filter, as it can stem the word and find also the feminine form of “loco” (“loca”) like the SS algorithm does.

For the query “morena” recall and precision present optimum values. Since the masculine form of this word (“moreno”) does not appear in the data collection, standard analyzer also obtains maximum precision. If the masculine form appeared, neither of the two stemmers would see their capacity diminished as they are sensible to gender variations.

For the verb “comer”, LSS works in the same way as the standard operator, as it is not capable of stemming this word. Therefore, both operators reached 100% precision, but failed when recovering relevant documents. In contrast, SS recovered more documents but, the particle resulting from stemming process (“com-”) was also present in other words with different meaning, compromising the precision.

In the query “pensar”, no documents were recovered when using standard analyzer and LSS. We noted that different verb forms of “pensar” were present in the documents, so stemmers are supposed to work better in this situation. Nevertheless, recall is low in all cases. Only a little proportion of relevant documents is recovered, probably due to the fact that this verb is irregular. There are documents in which the verbs appear in forms like the present simple of first or third person (“pienso”, “piensa”) that have not been found by many operators. Only those containing SS recover documents, suggesting that a more aggressive stemming may be more effective when dealing with irregular verbs.

Two queries were done to test the retrieval process in text strings containing various words, including stop words. As expected, the stop words filter is fundamental for improving results in these cases. For the text string “con los ojos cerrados” such single filter obtained maximum precision values. As expected, when combined with a stemmer the recall value increased. The other complex string, “salir de aquí”, shows different results when applying LSS and SS combined with stop words filter: LSS increases precision while SS does the same with recall. As mentioned above, one of the main criteria for selecting an analyzer lies on the desired stemming degree and the data collection characteristics. LSS + stop words and SS

+ stop words operators have demonstrated to be the best options for the present data collection. In those cases where these approaches obtain different results, we observe that LSS + stop words reaches higher precision values while SS + stop words recovers a greater proportion of relevant documents.

According to this, we conclude that a small collection size does not imply that a light stemmer should be the analyzer of choice in Spanish. These results also lead us to suggest that, unlike in English, an aggressive stemmer may be a good option for human use, considering that Spanish counts with a richer morphology than English [11]. In other words, in Spanish, a greater amount of information lies on morphology, while English uses other methods to transmit such information like syntax or context formed by other words. For this reason, light stemming in English may be enough to successfully recover different words with the same meaning as "jump", while the same can not be expected in the case of Spanish. Semantically speaking, there are many cases in Spanish where a suffix removal have not the same effect. For example, English verbs have suffixes to express the first and third person (*-s* and *-es*), or the past and past perfect (*-ed* in regular verbs). However, the number is not expressed in the same word. In Spanish, one verbal tense involves more suffixes than the mentioned above for English ones. Take for instance the verbal form "saltaremos" (from verb "saltar", "to jump") in Spanish, which could be translated to "we will jump". A light stemmer which, for example, only removes plurals, will not produce any effect in the word "jump" while a plural removing stemmer for Spanish will transform "saltaremos" into "saltaré". However, this word is still expressing information about the person in that *-e* (first person). In addition, the stem "saltaré" does not match the etymological root, hence, when searching over a document collection, many forms of the verb "saltar" will not be found. In contrast, the English stem will match all words meaning the same. Many similar phenomena to these just described are occurring when stemming Spanish words. In addition, we have observed that Spanish morphology makes more difficult to hold a high precision and recall values at the same time. This is due to the fact that many particles resulting form stemming are present in other words with different meaning, consequently decreasing precision. Thus when applying a more aggressive stemmer, the recall will tend to increase, but at the cost of sacrificing precision. In contrast, a light stemmer will increase precision, but omitting many relevant documents. Therefore, further work is required for establishing some basic good practice recommendations for IR approaches in this language.

4 Conclusions

The goal of this paper is both to present a case study allowing the recovery process evaluation in the Spanish language as well as to develop a web tool for data retrieval. Discussion about the factors influencing the performance of selected operators and the query results is provided. In addition, relevant aspects affecting the retrieval process in the Spanish Language are explained.

The proposed system uses different filters and analyzers to index and transform the text string. It also allows the user to select the most appropriated analysis to recover relevant documents in each query. A concrete field can be selected to carry out a query, and also obtained results can be stored. As a future work, more filtering options will be developed, and a bigger dataset will be used to compare and test different analyzers.

Acknowledgments. This work has been supported by project MOVIURBAN Máquina social para la gestión sostenible de ciudades inteligentes: movilidad urbana, datos abiertos, sensores móviles (SA070U 16). Project cofinanced with Junta Castilla y Leon, Consejera de Educacion and FEDER funds. In addition, the research of Juan Ramos González has been co-financed by the European Social Fund and Junta de Castilla y León (Operational Programme 2014-2020 for Castilla y León, BOCYL EDU/602/2016).

References

1. Gormley, C., Tong, Z.: Elasticsearch: The Definitive Guide: A Distributed Real-Time Search and Analytics Engine. O'Reilly Media Inc., New York (2015)
2. Gupta, V., Lehal, G.S.: A survey of text mining techniques and applications. J. Emerg. Technol. Web Intell. **1**(1), 60–76 (2009)
3. Hotho, A., Nürnberger, A., Paaß, G.: A brief survey of text mining. Ldv Forum **20**, 19–62 (2005)
4. Patel, F.N., Soni, N.R.: Text mining: a brief survey. Int. J. Adv. Comput. Res. **2**(4), 243–248 (2012)
5. Porter, M.: Spanish stemming algorithm (2005). http://snowball.tartarus.org/algorithms/spanish/stemmer.html. Accessed 20 Jan 2018
6. Porter, M.F.: Snowball: a language for stemming algorithms (2001). http://snowball.tartarus.org/texts/introduction.html. Accessed 14 Jan 2018
7. Ramos, J., et al.: Using TF-IDF to determine word relevance in document queries. In: Proceedings of the First Instructional Conference on Machine Learning, vol. 242, pp. 133–142 (2003)
8. Salton, G., Buckley, C.: Term-weighting approaches in automatic text retrieval. Inf. Process. Manag. **24**(5), 513–523 (1988)
9. Savoy, J.: Report on CLEF-2001 experiments: effective combined query-translation approach. In: Workshop of the Cross-Language Evaluation Forum for European Languages, pp. 27–43. Springer (2001)
10. Sharma, D.: Stemming algorithms: a comparative study and their analysis. Int. J. Appl. Inf. Syst. **4**(3), 7–12 (2012)
11. Sproat, R.W.: Morphology and Computation. MIT press, Cambridge (1992)
12. Vijayarani, S., Ilamathi, M.J., Nithya, M.: Preprocessing techniques for text mining-an overview. Int. J. Comput. Sci. Commun. Netw. **5**(1), 7–16 (2015)
13. Wu, H.C., Luk, R.W.P., Wong, K.F., Kwok, K.L.: Interpreting TF-IDF term weights as making relevance decisions. ACM Trans. Inf. Syst. (TOIS) **26**(3), 13 (2008)

Social Influence-Based Similarity Measures for User-User Collaborative Filtering Applied to Music Recommendation

Diego Sánchez-Moreno, Javier Pérez-Marcos, Ana B. Gil González, Vivian López Batista, and María N. Moreno-García(✉)

Department of Computing and Automation, University of Salamanca, Plaza de los Caídos s/n, 37008 Salamanca, Spain
sanchezhh@gmail.com, {jpmarcos,abg,vivian,mmg}@usal.es

Abstract. Social characteristics present in current music streaming services allow to use methods for endowing these systems with more reliable recommendation functionalities. There are many proposals in the literature that take advantage of that information and use it in the context of recommender systems. However, in the specific application domain of music the studies are much more limited, and the methods developed for other domains cannot be often applied since they require social interaction data that are not available in the streaming systems. In this paper, we present a method to determine social influence of users uniquely from friendship relations. The degree of influence obtained is used to define new similarity metrics for collaborative filtering (CF) where more weight is given to more influential users.

Keywords: Music recommender systems · Social influence · Trust
Collaborative filtering · Streaming services

1 Introduction

One of the focus of recent research in recommender systems is the development of methods that take advantage of social information in order to provide more reliable recommendations. In the field of music, the increasing use of streaming services has facilitated this task since most of them are endowed with social utilities, such as music sharing, social tagging and friendship relations. However, social data obtained from these platforms are much more limited than those that can be obtained from other social networks as Facebook, Twitter, etc. As consequence, social features have not been sufficiently exploited in the area of music recommendation.

In some works, social influence has been derived from friendship relations and user interaction data in order to infer trust information, which can be used to improve recommendations. In most of them two assumptions are accepted, users are influenced by their friends in the network and users connected by friendship relations have similar preferences, hypothesis that are not always true.

S. Rodríguez et al. (Eds.): DCAI 2018, AISC 801, pp. 267–274, 2019.
https://doi.org/10.1007/978-3-319-99608-0_30

In this work, social influence is taken as a wider concept considering that some individuals have a greater power of influence than others. We use friendship relations to determine the degree of influence and introduce this information into the similarity measures, aiming at giving more weight to preferences of more influential users when applying user-user collaborative filtering recommender methods.

Our approach is tested in a music recommendation environment where the scarce social information available does not allow us to use well known graph-based methods for social influence analysis. In this application domain, we have also to deal with the drawback of the lack of explicit ratings, which are required to apply CF methods. Thus, it is necessary to resource to techniques for obtaining them in an implicit way. We have derived implicit ratings from the count of plays.

The rest of the paper is organized as follows: Sect. 2 presents a brief discussion about some representative related works. The following sections are devoted to the description of the proposal. Section 3 is focused in the method for obtaining similarity metrics, Sect. 4 in the procedure for deriving implicit ratings and Sect. 5 in the validation of the proposed approach by applying it to a music dataset. Finally, the conclusions are given in Sect. 6.

2 Related Work

There are several categories of methods that can be adopted in the development of recommender systems, although the most widely extended is collaborative filtering (CF). This kind of methods can be classified in two main groups, user-user (memory-based or user- based) CF techniques, where the predictions for the active user are based on the opinions of other users with similar preferences [2], and item-item (or item-based) CF [8], which provide recommendations by computing similarities between items. Both have advantages and drawbacks, which are usually addressed by means of hybrid approaches [10]. User-user CF methods are generally more reliable than the second ones, but they present scalability problems caused by the fact that the time for computing the similarity between users increases exponentially as the number of items and users in the system increases. Similar users are those with similar preferences since they have given similar scores (ratings) to the same items. Similarity between them can be obtained by means of measures such as cosine, Chebyshev, Jaccard, Pearson coefficient, etc.

A weakness common to all CF techniques is the requirement that users express explicitly their preferences by means of ratings assigned to items. However, in some domains it is very difficult to obtain this information from users, thus, the available ratings are not enough to make reliable predictions. This is known as the scalability problem and the most usual way to deal with it is to resource to procedures for obtaining implicit ratings. Another and most recent way is to complement user ratings with trust information obtained from social networks, which is used to give more relevance to the ratings of trusted users against others [6].

Trust has been used to improve the quality of recommender systems regardless of the presence of scalability problems. In the social context, it is a form of social influence that is often obtained from friendship relations, comments, messages, etc.

Social influence can be used locally, when only opinions of connected friends are taken into account, and globally, when reputed individuals in the entire network are considered [4]. On the other hand, there are approaches that use social influence without considering similarity between users, but in others it is used jointly with similarity values [1, 13] or even with additional factors such as different types of interactions in social networks [4]. Other aspects related with social influence that have been studied are both the time window in which influence is valid and the susceptibility of the users to the influence [12]. There are many works in the literature where diverse factors affecting social influence are addressed, but most of them are focused in social networks like Facebook or Twitter, from which a great variety of social information can be extracted.

To determine the level of social influence some techniques from graph theory and network analysis are commonly used. Well known measures such as centrality and algorithms similar to page-rank or HITS are among them. These procedures are based on directed graphs, so they cannot be used when only bidirectional friendship relations are available.

Although music was one of the first application field of CF methods, in the Ringo recommender system [9], the exploitation of social influence in this domain is practically inexistent. In this work, we take advantage of friendship relations, available in most music streaming platforms, to improve recommendations of that systems by incorporating social influence in CF.

3 Obtaining Social Influence-Based Similarity Metrics

User-based or user-user collaborative filtering methods base the recommendations on the similarity between users. Their objective is to predict some items that a user could like from the ratings that other users with similar preferences have given to those items. However, users' influence can be different although the similarity degree is the same. Therefore, opinions of more influential people should have more weight in the recommendations. This is the idea of our proposal, where similarity metrics used in CF are modified to take into account social influence, according to the following process.

Given a set of m users $U = \{u_1, u_2, \ldots, u_m\}$ and a set of n items $I = \{i_1, i_2, \ldots, i_n\}$, each user u_i have a list of k ratings that he has given to a set of items I_{ui}, where $I_{ui} \subseteq I$. In this context, a recommendation for the active user $u_a \in U$ involves a set of items $I_r \subset I$ that fulfill the condition $I_r \cap I_{ua} = \varnothing$, since only items not rated by him can be recommended. Ratings are stored in a $m \times n$ matrix called the rating matrix, where each element is the rating that a user u_i gives to an item i_j. Usually, this matrix has many empty elements since each user only rates a small percentage of available items.

The rating matrix is used to compute the similarity between users by means of different distance-based measures such as cosine, Chebyshev and Jaccard; or correlations coefficients such as Pearson, Kendall and Spearman.

Similarity between the active user u_a and another user u_i obtained by any of the above metrics is denoted as $sim(u_a, u_i)$. Once this measure is computed, it will be modified aiming at considering social influence.

Many social networks contain several resources from which social influence can be inferred. However, in most of music streaming services, social information is limited to friendship relations, which are always bidirectional. This social information is used in our proposal to obtain new similarity metrics. Given the compulsory bidirectionality of the relations, it is not possible to use popular social influence measures such as centrality, page-rank and other graph-based approaches.

In an analogous way to the determination of term importance in a document in text processing, given by its occurrence frequency, the social influence degree of a user can be obtained by friendship frequency. In this context, some frequency measures can be used. Absolute frequency is one of the most common, it is defined as follow.

Let's consider the set of friends $F = \{f_1, f_2, \ldots, f_l\} \mid F \subseteq U$ and the subset of friend of user u_i $F(u_i) \subseteq F$.

$$aFreq\,(u_i) = |F(u_i)| \tag{1}$$

Users who have greater absolute frequency represent more influential people. However, it cannot be assumed that the count of friends reflects the proportion of the influence. Thus, a common transformation of the absolute frequency, which can be more representative, is the following frequency function:

$$logFreq(u_i) = 1 + \log(aFreq(u_i)) \tag{2}$$

Both frequencies can be used to introduce social influence in the similarity computation. Then, we introduce the concept of similarity weighted with social influence ($SI - \mathrm{w} - sim$), which is a metric that gives more weight to more influencer users. We define this weighted similarity between active user u_a and another user u_i as:

$$SI - \mathrm{w} - sim(u_a, u_i) = sim(u_a, u_i) \times \frac{Freq(u_i)}{|U|} \tag{3}$$

Where $Freq(u_i)$ can be either $aFreq(u_i)$ or $logFreq(u_i)$.

After computing $SI - \mathrm{w} - sim(u_a, u_i) \forall i$, the prediction of the rating pr_{aj} that the active user would give to a certain item j is computed from the weighted sum of other users' ratings using the following equation:

$$pr_{aj} = \bar{r}_a + \frac{\sum_{i=1}^{m} SI - \mathrm{w} - sim(u_a, u_i)\left(r_{ij} - \bar{r}_i\right)}{\sum_{i=1}^{m} |SI - w - sim(u_a, u_i)|} \tag{4}$$

where

$$\bar{r}_i = \frac{1}{|I_i|} \sum_{j \in I_i} r_{ij} \tag{5}$$

4 Implicit Ratings

Most of the music datasets lack of explicit rating information, reason why it is usually estimated from the number of plays of the tracks, which is often the only available indication of user preferences. There have been proposed few methods for computing ratings from plays and most of them are based on simple frequency functions [5, 11]. However, these methods are not indicated in the context of artist recommendation where most of the artists have low number of plays and there are few highly played. As consequence, play frequencies have a clear power law distribution, also known as the "long tail" distribution. For that situation the method proposed by Pacula [7] is proved to be more suitable. Below we describe this approach, which is adopted in our work.

The play frequency for a given artist i and a user j is defined as follow:

$$pFreq_{i,j} = \frac{p_{i,j}}{\sum_{i'} p_{i',j}} \tag{6}$$

Where $p_{i,j}$ is the number of times that a user j plays an artist i.

On the other hand, $Freq_k(j)$ denote the k-th most listened artist for user j. Then, a rating for an artist with rank k is computed as a linear function of the frequency percentile:

$$r_{i,j} = 4\left(1 - \sum_{k'=1}^{k-1} \mathrm{p}Freq_{k'}(j)\right) \tag{7}$$

Once the ratings are calculated, collaborative filtering methods can be applied in the way it is done for dataset containing explicit user preferences.

5 Validation of the Proposed Method

5.1 Dataset

For the validation of the method in the context of music recommendation, we conducted a study where our CF approach was tested with a dataset obtained from Hetrec2011- lastfm [3]. This dataset contains social networking, and music artist listening information about 1000 users and 11680 artists of last.fm online music system. Although the original dataset has five files, we only used the following ones:

- artists.dat: Contains information about artists listened and tagged by the users.
- user_artists.dat: Contains the artists listened by each user. It also provides a listening count (number of plays) for each [user, artist] pair.
- user_friends.dat: Contains the friend relations between users in the database.

Those data are enough to apply the method proposed in this paper. It requires to compute ratings from number of plays according to the procedure described in Sect. 4, as well as, to compute both kinds of frequency and social influence weighted similarities defined in Sect. 3.

5.2 Results

The behavior of CF when it is used with the similarity metrics weighted with social influence was checked and compared with the one exhibited when using classical similarity metrics. $SI - \mathrm{w} - sim(u_a, u_i)$ was computed from both $aFreq(u_i)$ and $logFreq(u_i)$ friendship frequencies according to the procedure described in Sect. 3. Common metrics such as cosine similarity and other distance-based measures such as Chebyshev distance, Jaccard similarity, or Euclidean distance were tested. Euclidean and Chebyshev distances were normalized to have values between 0 and 1 and the corresponding similarity metrics were obtained subtracting this value from 1.

Table 1. NRMSE for user-user CF and user-user CF with weighted similarities

	CF	CF SI-w-Sim (aFreq)	CF SI-w-Sim (logFreq)
Cosine	0.2992	0.0599	0.0092
Jaccard	0.3013	0.0445	0.0055
Chebyshev	0.2527	0.0612	0.0122
Euclidean	0.3622	0.2745	0.2233

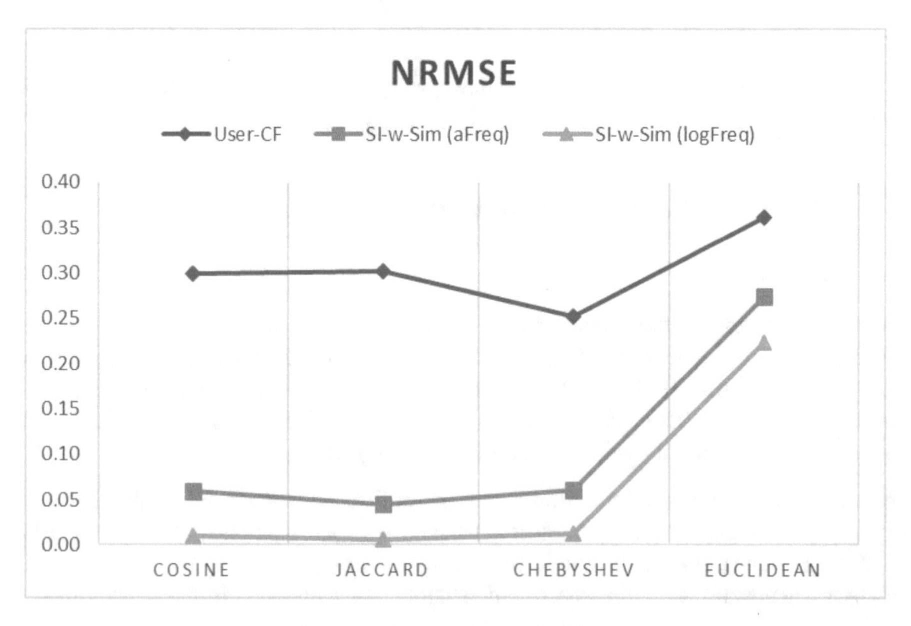

Fig. 1. Visualization of the difference between the NRMSE values of CF using similarity metrics in their original form (blue) and weighted with friendships frequencies (orange and green)

Table 1 shows the values of NRMSE (Normalized Root Mean Square Error) obtained when using CF with traditional similarity measures (first column), and when using CF with these similarity metrics weighted according to the social influence of users (second and third column). Values in the table prove that the introduction of friendship frequency functions leads to an important improvement of the results, especially for the first three measures (cosine, Jaccard and Chebysev), while the improvement for Euclidean distance is not so significant. On the other hand, it can be observed that the highest reduction of the error rate is achieved when using the logarithmic function of the frequency. In this case, a less remarkable improvement was also obtained with the Euclidean distance. These results can be better appreciated in Fig. 1.

6 Conclusions

In the las years, many recommendation methods based on data from social networks have been proposed. However, most of them cannot be applied in the domain of music recommender systems since they require social information that is not available in music streaming platforms.

In this work, we have proposed a method for determining social influence of users from their friendship relations. Two different frequency functions commonly used in the field of information recovery are applied for obtaining the influence level of each individual in the system. Then, similarity metrics used in collaborative filtering are weighted proportionally to that influence.

The results of the study carried out with a music dataset reveal an important reduction of the error rate when using the proposed metrics, especially in the case of the logarithmic function of the friendship relation frequency.

References

1. Akcora, C.G., Carminati, B., Ferrari, E.: User similarities on social networks. Soc. Netw. Anal. Min. **3**(3), 475–495 (2013)
2. Breese, J.S., Heckerman, D., Kadie C.: Empirical analysis of predictive algorithms for collaborative filtering. In: Proceedings of the Fourteenth Conference on Uncertainty in Artificial Intelligence, Madison, pp. 43–52 (1998)
3. Cantador, I., Brusilovsky, P., Kuflik, T.: 2nd workshop on information heterogeneity and fusion in recommender systems (HetRec 2011). In: Proceedings of the 5th ACM Conference on Recommender Systems, RecSys 2011, New York, NY, USA. ACM (2011)
4. Kalaï, A., Abdelghani, W., Zayani, C.A., Amous, I.: LoTrust: a social trust level model based on time-aware social interactions and interests similarity. In: 14th IEEE Fourteenth Annual Conference on Privacy, Security and Trust, Auckland, NewZeland, pp. 428–436 (2016)
5. Lee, K., Lee, K.: Escaping your comfort zone: a graph-based recommender system for finding novel recommendations among relevant items. Expert Syst. Appl. **42**(2015), 4851–4858 (2015)
6. Massa, P., Avesani, P.: Trust–aware recommender systems. In: ACM Conference on Recommender Systems, RecSys, Minneapolis, MN, USA, pp. 17–24 (2007)

7. Pacula, M.: A matrix factorization algorithm for music recommendation using implicit user feedback. http://www.mpacula.com/publications/lastfm.pdf
8. Sarwar, B., Karypis, G., Konstan, J., Riedl, J.: Item-based collaborative filtering recommendation algorithm. In: Proceedings of the Tenth International World Wide Web Conference, pp. 285–295 (2001)
9. Shardanand, U., Maes, P.: Social information filtering: algorithms for automating 'Word of Mouth'. In: Proceedings of the Conference on Human Factors in Computing Systems (CHI 1995), Denver, pp. 210–217 (1995)
10. Su, X., Khoshgoftaar, T.M.: A survey of collaborative filtering techniques. Adv. Artif. Intell. **2009**, 1–19 (2009)
11. Vargas, S., Castells, P.: Rank and relevance in novelty and diversity metrics for recommender systems. In: Proceedings of the Fifth ACM Conference on Recommender Systems RecSys 2011, New York, NY, USA, pp. 109–116. ACM (2011)
12. Yuan, T., Cheng, J., Zhang, X., Liu, Q., Lu, H.: How friends affect user behaviors? An exploration of social relation analysis for recommendation. Knowl.-Based Syst. **88**, 70–84 (2015)
13. Ziegler, C., Golbeck, J.: Investigating interactions of trust and interest similarity. Decis. Support Syst. **43**(2), 460–475 (2006)

Convolutional Neural Networks and Transfer Learning Applied to Automatic Composition of Descriptive Music

Lucía Martín-Gómez(✉), Javier Pérez-Marcos, María Navarro-Cáceres, and Sara Rodríguez-González

BISITE Digital Innovation Hub, University of Salamanca. Edificio Multiusos I+D+i, 37007 Salamanca, Spain
{luciamg,jpmarcos,maria90,srg}@usal.es

Abstract. Visual and musical arts has been strongly interconnected throughout history. The aim of this work is to compose music on the basis of the visual characteristics of a video. For this purpose, descriptive music is used as a link between image and sound and a video fragment of film Fantasia is deeply analyzed. Specially, convolutional neural networks in combination with transfer learning are applied in the process of extracting image descriptors. In order to establish a relationship between the visual and musical information, Naive Bayes, Support Vector Machine and Random Forest classifiers are applied. The obtained model is subsequently employed to compose descriptive music from a new video. The results of this proposal are compared with those of an antecedent work in order to evaluate the performance of the classifiers and the quality of the descriptive musical composition.

Keywords: Descriptive music · Automatic composition · Image Video · Transfer learning · Convolutional neural networks

1 Introduction

Human beings have always used art as a means of expressing their feelings, traditions and ideas, and often combine different types of art for this purpose [10]. This association has given rise to artistic creations such as poetry (blending of literature and music) or the combination of painting and photography.

The aim of this work is to fuse the visual and musical arts based on the concept of descriptive music [15]. The final goal is to compose a piece of music that describes a video. For this purpose, a video fragment from film Fantasia produced by Walt Disney [3] is analyzed. Fantasia is a concatenation of animated scenes designed for illustrating eight pieces of descriptive music. Character design and scene color are based on the emotions evoked by music. In the first stage of the system, the selected video is divided into a set of frames. Afterwards, a combination of convolutional neural networks [7] and transfer learning [14] is

S. Rodríguez et al. (Eds.): DCAI 2018, AISC 801, pp. 275–282, 2019.
https://doi.org/10.1007/978-3-319-99608-0_31

applied to extract some image descriptors from each video frame. Then, some classifiers are selected to establish a relationship between these image descriptors and the most important sound that is being played in each frame. Finally, this pattern is applied to the frames of a new video in order to translate each one of them into a sound. The musical result is a sequence of sounds that describes a video frame progression.

The most recurrent techniques used in this field are examined carefully. In addition, a past case study that have been conducted on the automatic composition of descriptive music is discussed. In order to validate the system, two different aspects are considered. On the one hand, the performance of the classifiers used in both cases is analyzed. On the other hand, the quality of the descriptive music is evaluated in a listening test taken by a set of users.

Section 2 reviews some methods that establish a relationship between image and sound in the composition of music. Section 3 describes a previous work that correlated image descriptors with sound. The techniques used to extract image descriptors in this proposal are outlined in Sect. 4. Section 5 presents a new approach to compose descriptive music from a video using convolutional neural networks. The performance of the classifier as well as the quality of the musical result are discussed in Sect. 6. Finally, Sect. 7 details the conclusions and outlines some future lines of research for this work.

2 Use of Visual Art for the Composition of Music

Throughout history many artists and researchers have tried to use visual arts to compose music. They have used several techniques which help link these two types of art; this includes synesthesia and descriptive music.

Synesthesia is a psychological phenomenon in which the stimulation of one sense triggers the activation of another one. It has been widely used in the study of the relationship between sound and image [10]. The most common synesthetic association in this field is the one that unites colour and sound, leaving out other image descriptors such as shape. Furthermore, this fusion technique poses a serious problem of subjectivity.

Another approach for the association of image and sound is descriptive music [15]. It is a musical genre where the sounds aim to narrate images, scenes or moods. Therefore, this concept requires a feature analysis of the preliminary image and its consequent translation to music. There are several techniques for the automated feature extraction of an image, such as Scale-Invariant Feature Transform [8] or color histogram [9]. However, the large number of image descriptors that must be taken into account hinders the process of musical composition.

To solve this problem it is necessary to employ a technique that extracts and analyzes immense amounts of image descriptors: Convolutional Neural Networks (CNN) [7]. The process of creating descriptive music using CNN and some additional classifiers is explained in detail in Sect. 5.

3 Earlier Proposal

The starting point of this work is an earlier work which studied the relationship between image and sound [12]. The workflow of this creative process was divided into two stages. The first step was the training stage, in which a preliminary video was divided into a set of frames. Then, two types of image descriptors were considered: shape and color features obtained by Scale-Invariant Feature Transform (SIFT) [8] in combination with Bag of Visual Words (BoVW) [17] and color histogram [9] respectively. Apart from these image descriptors, the most important note was obtained from the sound of each frame. Afterwards, three classifiers were applied to this data: Naive Bayes [6], Support Vector Machine [5] and Random Forest [1]. As a result, a model that established a relationship between image and sound was obtained at this stage. On the other hand, in the test stage, a new video was also divided into a set of frames. The same image descriptors as in the previous case were considered, and the model was applied to obtain a sound from each frame. Finally, the concatenation of the sounds led to the composition of a descriptive musical piece based on a video.

This proposal successfully built a system for the automatic composition of descriptive music from a preliminary video. However, different techniques can be applied in order to collect relevant data from each frame.

4 Techniques

Artificial vision tasks can be highly complex. Transfer learning is described in Sect. 4.1 as a technique to facilitate this task. Section 4.2 analyzes the suitability of Convolutional Neural Networks as tools for the extraction of image descriptors.

4.1 Transfer Learning

Transfer learning is a technique in which the knowledge acquired from solving a specific previous problem is used to solve another similar task [14]. This practice simulates the process of transferring knowledge from a teacher to a student. Thus, the knowledge acquired while learning to recognize dogs could be applied, for example, when trying to recognize specific dog breeds.

4.2 Convolutional Neural Networks

An Artificial Neural Network (ANN) is a computing system inspired by the functioning of biological neural networks [4]. Convolutional Neural Networks (CNNs) are a variation of ANNs [7]. They are very effective in artificial vision tasks. In combination with transfer learning, CNNs facilitate the adaptation of a previously trained model in order to solve more specific problems.

Inception is a CNN model trained for the ImageNet Large Visual Recognition Challenge using the data from 2012 [16]. Its architecture is based on Inception modules, which are smaller CNN architectures that process the input data in parallel. One of the CNN problems is the design of the network architecture. The Inception architecture uses different types of convolutional layers, then concatenates the results and let the model decide which convolutional layer fits the model. Each Inception module has 1×1, 3×3 and 5×5 convolutional layers in parallel, so Inception modules can extract local features from smaller layers and high-level features from larger layers. Also, a 1×1 convolutional layer precedes larger layers to perform a dimensional reduction of input data. With this reduction, the computational complexity of larger convolutional layers is lesser. These modules can be stacked to perform a deeper analysis of the input data, building a deep neural network. Finally, a softmax layer with an average pool and some fully connected layers were used to achieve the classification task. By removing these final layers and training some new ones, the model can be adjusted to solve new image classification problems.

5 Composition of Descriptive Music Using CNN

This work aims to compose a piece of descriptive music from a preliminary video. For this purpose, the workflow of the proposal was divided into two stages as can be seen in Fig. 1. In the training stage, a video was divided into a set of frames. The most important note that sounds when each frame is being played is extracted. This information was used as the input for the CNN, which obtains the image descriptors of each frame. Then, a classifier was applied to the visual and music data. Consequently, a model that relates image and sound was extracted. On the other hand, in the test stage, a new video was considered and once again a frame extraction process was carried out. Afterwards, the CNN was used to extract the image descriptors. Then, the model obtained in the previous stage was applied to obtain the sound that describes each frame. As a result, the concatenation of all the sounds gives rise to a musical composition that describes the video.

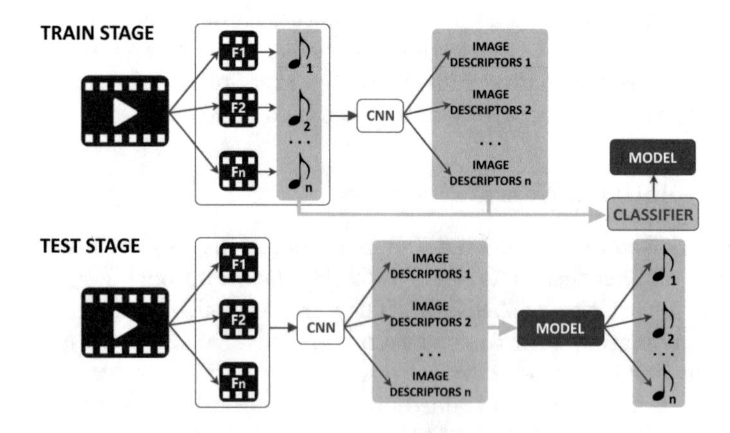

Fig. 1. Overview of the workflow applied in the composition process using CNN

According to this workflow, the specifications of the proposal are exposed. Section 5.1 describes all details related to the data: its extraction process, structure and preprocessing. In Sect. 5.2 all the information related to model induction is specified, and the composition process is explained in Sect. 5.3.

5.1 Data Description

Descriptive music aims to create some feelings or moods in the listener's mind [15]. Thus, its composition process is a creative task based on some scenes that the composer wants to detail. For this purpose, the fragment of "The Nutcracker Suite" played in the film Fantasia produced by Walt Disney is analyzed [2]. Three frames per second are extracted from this video. Only those frames that present relevant changes in color, shape or distribution of elements are considered in the next feature extraction process. This features can be divided into two groups:

- **Image descriptors**: The characteristics that define each frame are extracted by a CNN. Specifically, the previously trained model Inception-v3 is used in combination with transfer learning. Thereby, instead of using the CNN to perform a classification task, its final layers are removed and the vector of image characteristics obtained by its last pooling layer is considered. Thus, this characteristics are 2048 high-level image descriptors obtained by the CNN.
- **Sound**: The most important note that sounds when each frame is being played is extracted by a musician. This information is obtained from [13].

The data obtained by means of CNNs can be found in [11]. This work aims to relate this types of information to infer a pattern. The model induction process is explained in detail in the next section.

5.2 Model Induction

In order to keep coherence with the previous work, once again the classification algorithms applied in the model induction are those described in [12]: Naive Bayes (NB), Random Forest (RF) and Support Vector Machine (SVM). In this way, it was possible to compare the quality of the models obtained in both proposals. These three classifiers were applied to the data which consisted of the most important sounds and the image descriptors obtained from the CNN. As described in Sect. 6, in the current proposal SVM outperforms all other classifiers, but its results are lower than those obtained with RF in [12].

5.3 Music Composition Process

The process of composing descriptive music is based on applying the model to the frames of a new video in order to translate the concatenation of all frames into a sequence of sounds. For this purpose, the video fragment of Stravinski's "The Firebird" in film Fantasia 2000 [3] was considered at this stage. Due to its different range of colours and the movement of its characters, it was used in the test stage. First of all, the original music was removed from the video and three frames per second were extracted. Subsequently, the model obtained in the previous stage was applied to each frame to obtain a sound. Consequently, the concatenation of all the sounds led to the final musical piece. The musical prediction for"The Firebird" video fragment can be found in [11].

6 Discussion

The quality of the system is evaluated in this section. Since there are no similar and recent works in the literature which are related to the proposed method, this proposal is compared with the earlier work. For this purpose, two types of results are discussed: the performance of the algorithms applied in the classification task (Sect. 6.1) and the descriptive quality of the music (Sect. 6.2).

6.1 Performance of the Classifiers

The performance of NB, SM and RF in the earlier proposal and in the current one is shown in Table 1. The included metrics were described in [12].

Table 1. Comparison of the performance of NB, SVM and RF in the earlier (P1) and the current proposal (P2)

		Precision	Recall	F-Score	Kappa	RMSE	ROC
NB	P1	0, 474	0, 429	0, 421	0.389	0.324	0, 720
	P2	0, 552	0, 536	0, 533	0.464	0.289	0, 868
SVM	P1	0, 795	0, 793	0, 791	0.767	0.254	0, 950
	P2	0, 810	0, 807	0, 807	0.779	0.266	0, 948
RF	P1	0, 843	0, 832	0, 832	0.807	0.185	0, 983
	P2	0, 683	0, 615	0, 598	0.546	0.246	0, 930

As shown in Table 1, the results are quite different for both proposals. With regard to precision, recall and F-score metrics, NB is the worst classifier. SVM outperforms the other classifiers in the current proposal but in the previous one, RF was the best classifier. The Kappa metric shows that the best value for the current proposal is achieved by SVM (0.779); however, RF had a lower random agreement in the previous one (0.807). The best value for RMSE was obtained by RF in the previous proposal (0.185) demonstrating that the misclassified sounds are closer to the correct one than in the other cases. RF almost reached the maximum value for ROC metric in the previous case (0.983). As a result, in the current proposal SVM outperforms the other classifiers. However, in general terms, the performance of RF in the earlier work was even better.

6.2 Descriptive Quality of the Musical Composition

A listening test was designed to assess the association between image and sound. To simplify the assessment task, the video of "The Firebird" from Fantasia 2000 was divided into five short fragments. Both proposals, the earlier and the current one, were used to compose a piece of descriptive music related to each video fragment. As a result, 10 video fragments of descriptive music were obtained. The final goal of this test was to evaluate the descriptive quality of the composition based on the visual characteristics of the videos. 33 people were asked to take the test and to rate each fragment from 0 (very bad) to 10 (very good). The statistical mean of the valuations for each video fragment is shown in Fig. 2.

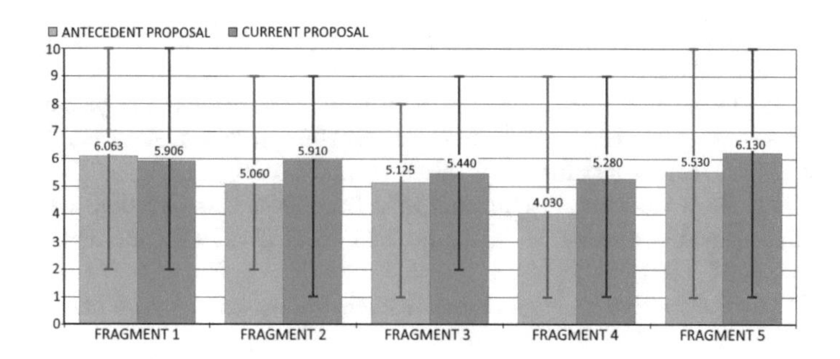

Fig. 2. Assessment of the descriptive quality of musical compositions.

After analyzing the results, two conclusions can be drawn. On the one hand, the mean of the ratings for all the video fragments was similar in all cases: the users think that the connection between image and sound is quite good. Despite this, the overall results are better in the current proposal than in the the previous one. On the other hand, the vertical lines that represent the rating range obtained for each fragment, were very wide in all cases. This means that the perceptive component of the descriptive music entails a subjectivity problem.

7 Conclusions and Future Work

This research work proposed a system that composes music on basis of the visual characteristics of a preliminary video. Thus, the concept of descriptive music is used as a link between visual and musical arts.

The complexity associated with artificial vision tasks has been reduced thanks to the use of transfer learning and CNN. In addition to pattern detection, the CNN architecture makes it possible to extract high-level image descriptors.

With regard to the classifiers, the earlier work had a better performance. However, the descriptive quality of the music is slightly better in the current proposal. Therefore, the two proposals solve the problem in a valid way.

Future lines of research include the use of other models of CNN such as Inception-v4 and harmonic composition in the creative process.

Acknowledgments. This work was supported by the Spanish Ministry, Ministerio de Economía y Competitividad and FEDER funds. Project. SURF: Intelligent System for integrated and sustainable management of urban fleets TIN2015-65515-C4-3-R.

References

1. Breiman, L.: Random forests. Mach. Learn. **45**(1), 5–32 (2001)
2. Clague, M.: Playing in 'Toon: Walt Disney's "Fantasia" (1940) and the imagineering of classical music. Am. Music **22**(1), 91–109 (2004)
3. Culhane, J.: Fantasia 2000: Visions of Hope. Disney Editions, Glendale (1999)
4. Haykin, S., Network, N.: A comprehensive foundation. Neural Netw. **2**(2004), 41 (2004)
5. Hsu, C.W., Chang, C.C., Lin, C.J., et al.: A practical guide to support vector classification (2003)
6. John, G.H., Langley, P.: Estimating continuous distributions in Bayesian classifiers. In: Proceedings of the Eleventh Conference on Uncertainty in Artificial Intelligence, pp. 338–345. Morgan Kaufmann Publishers Inc. (1995)
7. Krizhevsky, A., Sutskever, I., Hinton, G.E.: ImageNet classification with deep convolutional neural networks. In: Advances in Neural Information Processing Systems, pp. 1097–1105 (2012)
8. Lowe, D.G.: Distinctive image features from scale-invariant keypoints. Int. J. Comput. Vis. **60**(2), 91–110 (2004)
9. Lu, G., Phillips, J.: Using perceptually weighted histograms for colour-based image retrieval. In: 1998 Fourth International Conference on Signal Processing Proceedings, 1998. ICSP 1998, vol. 2, pp. 1150–1153. IEEE (1998)
10. Marks, L.E.: On colored-hearing synesthesia: cross-modal translations of sensory dimensions. Psychol. Bull. **82**(3), 303 (1975)
11. Martín-Gómez, L., Pérez-Marcos, J.: Image and sound data from film Fantasia produced by Walt Disney (2018). https://figshare.com/articles/FantasiaDisney_ImageSound/5999207
12. Martín-Gómez, L., Pérez-Marcos, J., Navarro-Cáceres, M.: Automatic composition of descriptive music: a case study of the relationship between image and sound. In: Proceedings of the Workshop Computational Creativity, Concept Invention, and General Intelligence (C3GI) 2017 (2017)
13. Martín-Gmez, L., Pérez-Marcos, J.: Data repository of fantasia case study (2017). https://github.com/lumg/FantasiaDisney_data
14. Pan, S.J., Yang, Q.: A survey on transfer learning. IEEE Trans. Knowl. Data Eng. **22**(10), 1345–1359 (2010)
15. Seeger, C.: Prescriptive and descriptive music-writing. Music. Q. **44**(2), 184–195 (1958)
16. Szegedy, C., Vanhoucke, V., Ioffe, S., Shlens, J., Wojna, Z.: Rethinking the inception architecture for computer vision. In: Proceedings of the IEEE Conference on Computer Vision and Pattern Recognition, pp. 2818–2826 (2016)
17. Yang, J., Jiang, Y.G., Hauptmann, A.G., Ngo, C.W.: Evaluating bag-of-visual-words representations in scene classification. In: Proceedings of the International Workshop on Workshop on Multimedia Information Retrieval, pp. 197–206. ACM (2007)

Classifying Emotions in Twitter Messages Using a Deep Neural Network

Isabela R. R. da Silva(✉), Ana C. E. S. Lima, Rodrigo Pasti, and Leandro N. de Castro

Universidade Presbiteriana Mackenzie, São Paulo, SP, Brazil
isabelaruizroque@gmail.com, aceslima@gmail.com,
rodrigo.pasti@gmail.com, lnunes@mackenzie.br

Abstract. Many people use social media nowadays to express their emotions or opinions about something. This paper proposes the use of a deep learning network architecture for emotion classification in Twitter messages, using the six emotions model of Ekman: happiness, sadness, anger, fear, disgust and surprise. We collected the tweets from a labeled dataset that contains about 2.5 million tweets and used the Word2Vec predictive model to learn the relations of each word and transform them into numbers that the deep network receives as input. Our approach achieved a 63% accuracy with all the classes and 77% accuracy on a binary classification scheme.

Keywords: Deep learning · Emotion classification · Sentiment analysis

1 Introduction

There are many studies about *emotion* in psychology. It is a confusing topic because some words, like love and fear, may have different meanings from person to person. Many researchers defined emotion as a chemical response of the brain when it breaks its normal state [11–13]. Ekman et al. [1] proposed a model of six basic emotions after analyzing facial expressions observed in distinct cultures: *happiness*, *sadness*, *anger*, *fear*, *disgust* and *surprise* [1].

The *sentiment analysis* area studies the opinion, sentiment or emotions from humans and it is one of the most studied fields in social media mining. It consists of inferring a sentiment from a set of documents, e.g. *tweets* [4]. This classification can be based on polarity, for example, determining if a message is positive, negative or neutral [4]. However, there are other types of analyses, such as SentiStrength, which detects the strength of the sentiment to its positive or negative character [7]. More recently, [8] classified tweets based on Ekman's model of emotions [1, 9] classified messages using another model of emotions proposed by Plutchik [10].

This paper proposes the use of Deep Learning techniques to perform the classification of emotions in Twitter messages, using the basic Ekman's emotions model [1]. The Word2Vec predictive model, which has two learning algorithms, CBOW and Skip-gram, was used to perform the word representation to create a matrix of numbers that is the input of the deep neural network. The dataset used in this work was originally created by [3] and it was sampled to 28,598 tweets. The deep neural network implemented has two convolutional layers, one dropout layer and one dense layer.

S. Rodríguez et al. (Eds.): DCAI 2018, AISC 801, pp. 283–290, 2019.
https://doi.org/10.1007/978-3-319-99608-0_32

The paper is organized as follows. Section 2 describes how the emotion analysis was performed using a deep neural network and Sect. 3 presents the results obtained. The paper is concluded in Sect. 4 with some comments about the results obtained and future investigations.

2 Emotion Analysis Using a Deep Neural Network

The sentiment analysis area studies people's opinion, sentiment or emotion, usually through social media texts, regarding several subjects, like brands, products and politicians [4]. This area is in constant growth, since most people express their feelings or emotions through social media texts, like Facebook and Twitter, and many novel classification methods are proposed and assessed continuously. Within sentiment analysis, this paper focuses on the classification of Twitter messages into one of the five Ekman's emotions: anger, sadness, fear, disgust and happiness.

There are different definitions of what is an emotion. Damásio [5] argues that an emotion is a set of bodily reactions that happens in a very short period of time based on certain events, but that humans have difficulties in controlling. Most often, they can affect how people act and cannot be easily hidden. The sentiment, however, is the mental experience of how the brain interprets the emotions [5]. This paper uses the six basic emotions model of Ekman [1] to analyze emotions in Twitter texts.

This section describes the methodology used in our experiments. First, we describe the dataset used, then how Word2Vec was employed to generate the feature vectors and then the deep neural network architecture chosen.

2.1 The Emotions Dataset

The dataset used was created by [3] and contains 2,488,982 tweets labeled with seven emotions: joy, sadness, fear, anger, thankfulness, surprise and love, collected between November 10th and December 22nd, 2011, while monitoring hashtags like "#excited", "#surprised" and many others. The authors believed that hashtags are used to express emotions. The full dataset was sampled and the subset used in the experiments contains about 28,598 tweets.

Since we are using the Ekman's model of emotions, the emotion thankfulness was removed from our analysis, the emotions "joy" and "love" were coupled into a single emotion and, as the dataset does not have the emotion "disgust", we used the hashtags "#disgusted" and "#disgusting" to capture this emotion. Table 1 brings a summary of the total number of tweets per emotion on the sampled dataset.

In our experiments, we trained and tested the whole dataset and then removed disgust and surprise, since these two emotions do not have many tweets and represent less than 5% of the whole data set. We split the dataset into train and test, as described in Table 2.

Table 1. Number of tweets per emotion of the sampled dataset.

Emotion	Number of tweets	%
Happiness	12,460	43,57
Sadness	7,440	26,02
Anger	6,228	21,78
Fear	1,642	5,74
Disgust	539	1,88
Surprise	289	1,01
Total	28,598	100%

Table 2. Number of tweets in the training and test sets.

	Number of tweets	%
Train	20,018	70.00
Test	8,580	30.00
TOTAL	28,598	100%

2.2 Word Embeddings

Deep learning neural networks (and most machine-learning algorithms) cannot receive texts as inputs, only numbers. *Word embeddings* are a way of representing words as numbers [6]. Words that have similar contexts have similar representations, which are vector representations of the semantic and syntactic information of words. With such representation, most machine-learning algorithms and neural networks can receive them as input.

Since word embeddings are important to represent words in a way that neural networks can process as input, we used the Word2Vec to generate a numerical (vector) representation of words. Word2Vec is a predictive model that receives words as inputs and process them by means of two learning algorithms [6]: *skip-gram* and *continuous bag-of-words* (CBOW). The main difference between the two algorithms is that skip-gram identify words that are more semantically similar and CBOW detects the missing word on a sentence and can extract the meaning of bi-grams [6].

Since Word2Vec will build the vocabulary based on the words of the tweets on the training set, we cleaned the dataset removing hashtags, symbols like '@' and links to websites. Also, we lowercased all words to keep a pattern, then we tokenized all the tweets to split the tweets into words, which is the way Word2Vec reads the input.

2.3 Neural Network Architecture

In the approach proposed here, we set the window size to 10, the dimensionality of the feature vectors to 800, and we used the skip-gram algorithm to train the word vectors using Word2Vec. After training, we obtained 2,006 words for the vocabulary and the output is a weight matrix with 800 dimensions and 2,006 lines, for each word of the vocabulary based on their similarities calculated inside the Word2Vec model.

The similarity between the words are calculated using the cosine similarity as we can see in Eq. 1, where A and B are the vectors that we want to calculate the similarity and i represents the component of each vector.

$$cosine(\theta) = \frac{\sum_{i=1}^{n} A_i \times B_i}{\sqrt{\sum_{i=1}^{n} (A_i)^2} \times \sqrt{\sum_{i=1}^{n} (B_i)^2}} \quad (1)$$

The network was implemented using the Keras API on Python language and has two convolutional layers and one dense layer. The input of the convolutional layer has the shape of the output of the Word2Vec. To prevent overfitting, we added one Dropout layer between the convolutional layers and a L2 kernal regularizer on the second convolutional layer; by doing this, our network performed better. We trained the neural network using 100 epochs and set the batch size to 32.

The proposal of using deep learning on this study is that convolutional neural networks (CNNs) have been used on natural language processing (NLP) problems and showed their efficacy on sentiment analysis problems [2, 14].

We run our experiments on an Intel i5 processor and used the Intel Python 3.5.3, which has an easy to use optimized performance for data science libraries for packages like SciPy, Numpy, scikit-learn (Intel Data Analytics Acceleration Library) and neural networks enhancements [15].

3 Results Obtained and Discussion

3.1 Classifying Emotions

To assess the performance of the proposed deep network, three experiments were run:

1. The deep network trained with all the six emotions (happiness, sadness, anger, fear, disgust and surprise);
2. The deep network trained without disgust and surprise;
3. The network trained with only the bipolar emotions happiness and sadness, resulting in a binary classification problem.

Using all six emotions (Experiment 1) we achieved an average emotion classification of 63%, which performed well, compared to other networks that use convolutional layers [2]. Figure 1 shows the confusion matrix after training the neural network with 100 epochs.

Table 3 below presents the precision, recall and F1-score for each emotion after training the neural network. Although the global emotion classification achieved 63%, we can see on the metrics that the emotions that have less tweets are the ones that did not perform well.

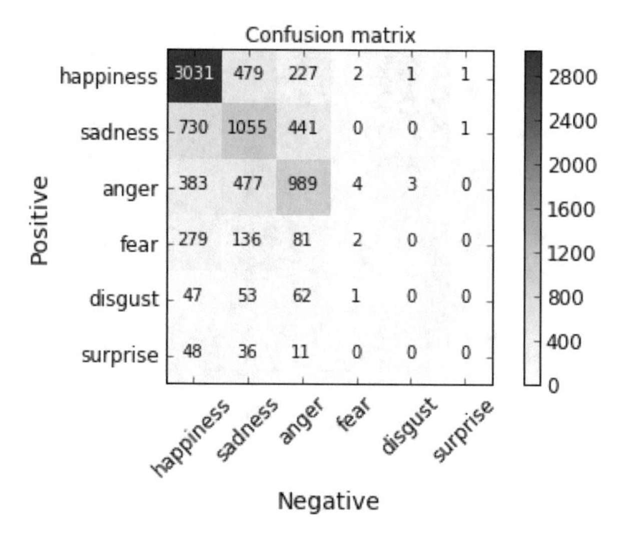

Fig. 1. Six emotions confusion matrix after training 100 epochs.

Table 3. Precision, Recall and F-score on six emotions.

Emotion	Precision	Recall	F-Score
Happiness	67%	81%	73%
Sadness	47%	47%	47%
Anger	55%	53%	54%
Fear	22%	0%	1%
Disgust	0%	0%	0%
Surprise	0%	0%	0%

Without the emotions disgust and surprise in the dataset, we achieved an average emotion classification rate of 64%. This experiment was made to check if the network could achieve a better result removing the emotions that represent less than 2% of the whole dataset. Figure 2 shows the confusion matrix after training the neural network with 100 epochs.

As can be observed in Table 4, even after removing disgust and surprise from the analysis, there was no significant change in the overall network performance. Those classes with a small number of samples are little representative in the network.

Since the emotions happiness and sadness are the ones that have the larger number of tweets, we decided to train the network only with them in Experiment 3. In this scenario, the network achieved an average emotion classification rate of 77%, as depicted in the confusion matrix of Fig. 3.

Table 5 shows the Precision, Recall and F1-score of the emotions happiness and sadness, showing an improvement over the other two experiments.

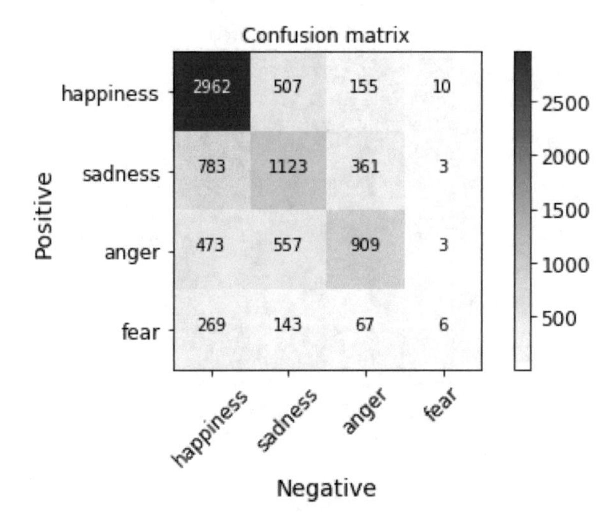

Fig. 2. Confusion Matrix after training 100 epochs without disgust and surprise.

Table 4. Precision, Recall and F-Score after 100 epochs with happiness, sadness, anger and fear.

Emotion	Precision	Recall	F1-Score
Happiness	66%	82%	73%
Sadness	48%	49%	49%
Anger	61%	47%	53%
Fear	27%	1%	2%

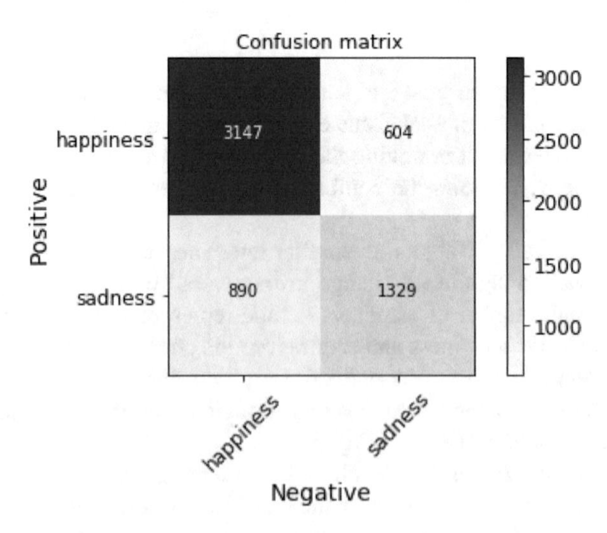

Fig. 3. Confusion Matrix after training 100 epochs only with happiness and sadness.

Table 5. Precision, Recall and F-Score after 100 epochs on a binary classification.

Emotion	Precision	Recall	F1-Score
Happiness	78%	84%	81%
Sadness	69%	60%	64%

After running all the experiments and analyzing the results, we can see very clearly that the accuracy we obtained can be improved, since the dataset we used on this study is imbalanced, which means that it has classes that have much more tweets than other classes.

4 Conclusions and Future Trends

In this paper we proposed the use of a deep learning network to classify emotions from microblog texts and analyze the output of Word2Vec to make clusters to check if words have some relation between emotions. To transform the texts into a data matrix to train the network we used Word2Vec, which is a sort of network that creates word embeddings and is trained to reconstruct the linguistic contexts of words. To assess our proposal we used an emotion classification dataset from the literature and proposed three experiments: all six Ekman's model emotions; four out of the six emotions; and a bipolar classification emotion set. The results showed that by suppressing the less representative classes we gained in classifier performance.

As future research, we will use other emotion classification datasets from the literature; we will balance the dataset to verify how it affects the classifier performance, we will try, improve the network setup and architecture and will make a further investigation on relations between words and emotions.

Acknowledgements. The authors thank CAPES, CNPq, Fapesp, and MackPesquisa for the financial support. The authors also acknowledge the support of Intel for the Natural Computing and Machine Learning Laboratory as an Intel Center of Excellence in Machine Learning.

References

1. Ekman, P., Friesen, W.V., Ellsworth, P.: Emotion in the Human Face, 1st edn. Pergamon, New York (1972)
2. Santos, C.N.D., Gatti, M.: Deep convolutional neural networks for sentiment analysis of short texts. In: 25th International Conference on Computational Linguistics: Technical Papers, Dublin, Ireland, pp. 69–78 (2014)
3. Wang, W., Chen, L., Thirunarayan, K., Sheth, A.P.: Harnessing twitter "Big Data" for automatic emotion identification. In: Proceedings of the 2012 AE/IEEE International Conference on Social Computing, Washington, DC, USA, pp. 587–592 (2012)
4. Liu, B.: Sentiment Analysis and Opinion Mining. Morgan & Claypool Publishers, San Rafael (2012)

5. Damasio, A.R.: Descartes Error: Emotion, Reason and the Human Brain. G.P. Putnam's Sons, New York (1994)
6. Mikolov, T., Chen, K., Corrado, G.S., Jeffrey, D.: Efficient estimation of word representations in vector space. In: Proceedings of Workshop at ICLR
7. Thelwall, M., et al.: Sentiment strength detection for the social web. J. Am. Soc. Inf. Sci. Technol., 2544–2558 (2010)
8. Balabantaray, R.C., Mohammad, M., Sharma, N.: Multi-class twitter emotion classification: a new approach. Int. J. Appl. Inf. Syst. **4**(1), 48–53 (2012)
9. Suttles, J., Ide, N.: Distant supervision for emotion classification with discrete binary values. In: Computational Linguistics and Intelligent Text Processing, pp. 121–136 (2013)
10. Plutchik, R.: Emotion: Theory, Research and Experience, pp. 370–372. Academic Press, New York (1980)
11. Damasio, A.: Looking for Spinoza: Joy, Sorrow, and the Feeling Brain. Harvest, San Diego (2003)
12. Scherer, K.R.: What are emotions? And how can they be measured? Soc. Sci. Inf. **44**, 695–729 (2005)
13. Stets, J.: Emotions and sentiments. In: Handbook of Social Psychology, pp. 309–335. Springer, US (2006)
14. Tang, D., Wei, F., Qin, B., Liu, T., Zhou, M.: Coooolll: a deep learning system for twitter sentiment classification, pp. 208–212 (2014)
15. Intel Distribution for Python. https://software.intel.com/en-us/distribution-for-python. Accessed 26 Mar 2018

Tree-Structured Hierarchical Dirichlet Process

Md. Hijbul Alam[1], Jaakko Peltonen[1,2(✉)], Jyrki Nummenmaa[1], and Kalervo Järvelin[1]

[1] University of Tampere, Tampere, Finland
{hijbul.alam,jaakko.peltonen,jyrki.nummenmaa, kalervo.jarvelin}@uta.fi
[2] Aalto University, Espoo, Finland

Abstract. In many domains, document sets are hierarchically organized such as message forums having multiple levels of sections. Analysis of latent topics within such content is crucial for tasks like trend and user interest analysis. Nonparametric topic models are a powerful approach, but traditional Hierarchical Dirichlet Processes (HDPs) are unable to fully take into account topic sharing across deep hierarchical structure. We propose the Tree-structured Hierarchical Dirichlet Process, allowing Dirichlet process based topic modeling over a given tree structure of arbitrary size and height, where documents can arise at all tree nodes. Experiments on a hierarchical social message forum and a product reviews forum demonstrate better generalization performance than traditional HDPs in terms of ability to model new data and classify documents to sections.

Keywords: Hierarchical Dirichlet Processes · Topic modeling
Message forum

1 Introduction

Modeling online discussions is important for studies of discussion behavior, for tracking trends of ideas and consumer interests, for recommendation of discussion content or targeted advertising, and for intelligent interfaces to browse discussions. Online discussion often occurs in venues having a prominent hierarchical organization such as hierarchical forums (message boards). General-interest forums cover a broad range of interests such as politics, health, product reviews, and so on. As a case study we use a popular Finnish forum Suomi24 (www.suomi24.fi) spanning 16 years and 6.5 million threads. Forums are organized into hierarchical sections created by administrators for prototypical interests. Hierarchical organization also occurs in online reviews at, e.g., websites such as Amazon.com, where reviews follow the hierarchy of the products.

Md. H. Alam and J. Peltonen had equal contributions. The work was supported by Academy of Finland decisions 295694 and 313748.

S. Rodríguez et al. (Eds.): DCAI 2018, AISC 801, pp. 291–299, 2019.
https://doi.org/10.1007/978-3-319-99608-0_33

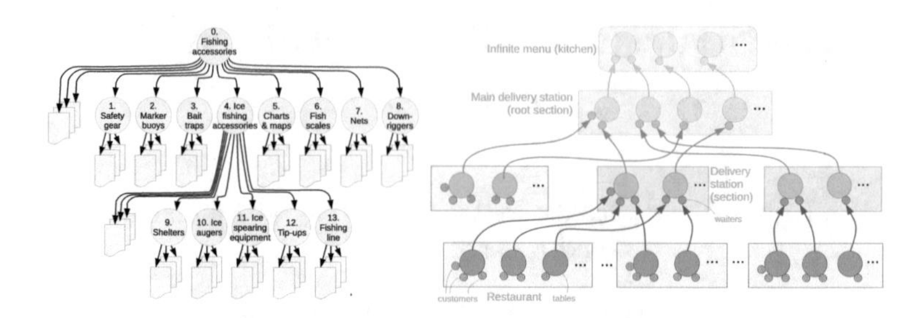

Fig. 1. (Left) Hierarchical document organization in part of the Amazon product hierarchy. Yellow icons denote documents which here are reviews (threads); they can appear under any section (blue circles) at any hierarchy level. (Right) An illustration of an Imperial Chinese Banquet.

Administrator-created sections are simplified divisions that do not suffice to describe the variety of semantic content in discussions; an important task in data analytics of online forums is to extract latent topics of discussion. Modeling text data is often done by generative topic models such as Latent Dirichlet Allocation [1] and Dirichlet Processes [2], which represent unstructured text as a bag of words arising out of a mixture of latent topics. In this paper we give a solution for the challenge of effectively taking hierarchical structure of data collections into account in such modeling. User interests need not match the administrator-created structure. Issues touching on multiple interests (say food and health) may have no dedicated section, and users may discuss them in multiple sections. Users may digress from the section theme; threads with many users follow a mixture of their interests. Thus, forum sections need not correspond to section themes.

Recent work on text mining has attempted hierarchical text analysis: most works [3–5] build an unsupervised hierarchy of topics from a document set; they ignore predefined organization of documents in a section hierarchy and cannot extract the topic distributions of a section in the hierarchy. HDP and its variations aim to take data division into account [2] but cannot readily be extended where sets of documents can arise at any node in the tree, as in Fig. 1 (Left).

We introduce the Tree-structured Hierarchical Dirichlet Process (THDP), a new model which identifies latent topics of each section in a hierarchy. THDP is a generative model for the documents in any position of a hierarchy and can be applied to data in hierarchies of arbitrary size and height. Our contributions: **1.** We develop a new nonparametric hierarchical topic model to model forum texts which can come from any place of the section hierarchy. The key is a new nonparametric generative process, the Imperial Chinese Banquet, representing a top-down percolation of topics to documents at different hierarchy levels. **2.** We develop a Gibbs sampling algorithm that extracts topics and their usage across threads and hierarchical sections. **3.** In experiments, evaluated with various metrics and use cases, our model outperforms the state-of-art models.

2 Related Work

A topic model [1] is a parametric Bayesian model for count data such as bag-of-words representations of text documents. Teh et al. [2] propose HDP (Fig. 2, Left), a non parametric model where the number of topics does not need to be pre-specified. The crucial difference to our work is that HDP by Teh et al. is not designed for deep hierarchies; as presented in Teh et al., their model was mainly used for a "flat" division of documents into groups: Dirichlet processes (DPs) of each document were only connected by one DP for each group, under an overall DP. In such a flat model, documents always occur at the groups and the parent level is unobserved; in our model, documents can occur under any node in the deep hierarchy. Alternative models exist e.g. placing additional sparsity priors for topic sharing [6] but again not for deep hierarchies. Some variants involve a hierarchy: in the nested Chinese restaurant process [3] and knowledge-based hierarchical topic model [7], a document is modeled as a distribution over a path from the root to the leaf node; in the recursive Chinese restaurant process [8], a document has a distribution over all of the nodes of the hierarchy; in the tree-structured stick-breaking process [5], a document is modeled by a node of the tree. In these models, a tree structure is learned to represents topics; whereas in THDP we do not need to learn the structure as our model is based on a known hierarchy; we focus on modeling using the given deep hierarchy as the model structure.

3 Tree-Structured Hierarchical Dirichlet Process

We describe a generative process, THDP, given a tree-structured hierarchy of sections, where documents can arise at any section. A global distribution G^0_{root} over topics is first drawn from a Dirichlet process (DP) with base distribution H and concentration parameter α^0 for the root node of a given tree, denoted $G^0_{root} \sim DP(\alpha^0, H)$. The root node corresponds to the root section. We index nodes as v. For each child section v of the root, a discrete distribution G^1_v is drawn from a DP with base distribution G^0_{root} and concentration parameter α^1, denoted $G^1_v \sim DP(\alpha^1, G^0_{root})$. This is repeated recursively for every child node to generate its grandchild sections: a node v at level l in the hierarchy (l steps down from the root) has a discrete distribution G^l_v generated from a DP with base distribution $G^{l-1}_{p(v)}$ and concentration parameter α^l, where $p(v)$ is the parent node of v, denoted $G^l_v \sim DP(\alpha^l, G^{l-1}_{p(v)})$. Lastly, G_j for a document j under a node v at level l is drawn from a DP with base distribution G^l_v and concentration parameter α^{l+1}, denoted $G_j \sim DP(\alpha^{l+1}, G^l_v)$. Document content is then generated: for each word i in a document j, draw the topic $\theta_{ji} \sim G_j$ and draw the observed word from the topic's word distribution as $x_{ji} \sim F(\theta_{ji})$. Figure 2 (Middle) shows the plate representation graphical model of THDP, with an instantiation for an example hierarchy in Fig. 2 (Right).

We describe a metaphor for THDP which we call the *Imperial Chinese Banquet* (ICB); we will use it for inference. A banquet is arranged in a multilevel

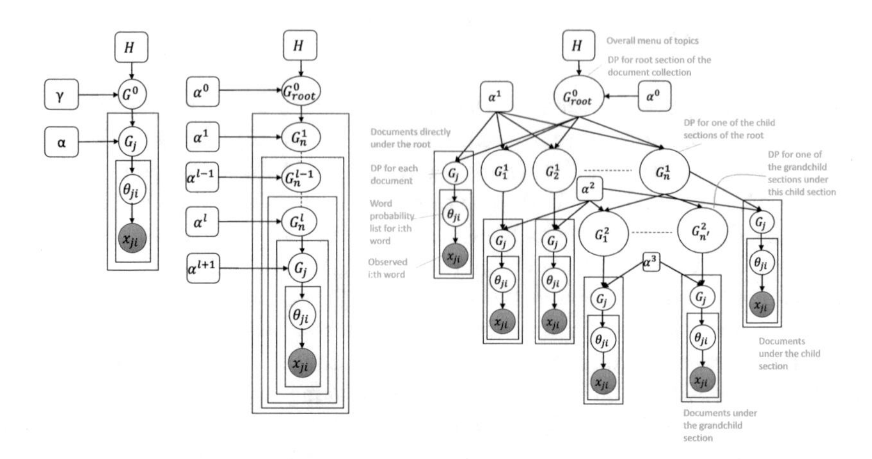

Fig. 2. (Left) Hierarchical Dirichlet Process. (Middle) Tree-structured Hierarchical Dirichlet Process (THDP). (Right) Detailed plate model instantiation example of the THDP, for an example hierarchy having three levels and documents arising at each level.

palace: each level has several *food-delivery stations*, each serving several *restaurants* (dining rooms) at that level. Attendees (i.e., customers) visit dining rooms (i.e., restaurants) to eat popular dishes: each restaurant has tables for customers, and there is a responsible *waiter* at every table who brings a dish to the table, fetching it from a table in a food-delivery station. At food-delivery stations, the tables also have responsible waiters who bring the dishes from an upper-level delivery station, recursively. Each time a customer/waiter chooses a table, they prefer popular tables that other customers/waiters have also picked.

4 Inference

We introduce a Gibbs sampling scheme for THDP, based on the ICB representation. We sample tables, pointers to ancestor tables, and dishes for tables. Let $f_k^{-x_{ji}}(x_{ji})$ denote the conditional density or likelihood of x_{ji} given all data items except x_{ji}, where k is the dish at the table of x_{ji}, and j is a document index. We have $f_k^{-x_{ji}}(x_{ji}) \propto n_{kw}^{-ji}/n_{k\cdot}^{-ji}$ for an existing dish and $f_{k_{new}}^{-x_{ji}}(x_{ji}) \propto 1/V$ for a new dish [2], where w is the word index of x_{ji}, V is the vocabulary size, n_{kw}^{-ji} is the number of occurrences of w from dish k (other than x_{ji}), and $n_{k\cdot}^{-ji}$ is the sum over different word indices. We denote $f_k^{-\boldsymbol{x}_{jt}}(\boldsymbol{x}_{jt}) = (\prod_w(\beta + n_{kw} - 1)...(\beta + n_{kw}^{-jt}))/((V\beta + n_{kw} - 1)...(V\beta + n_{k\cdot}^{-jt}))$ as the conditional density of data $\boldsymbol{x}_{jt}$ at table t given all data items associated with mixture component k leaving out $\boldsymbol{x}_{jt}$, where β is a hyperparameter.

Part 1. Sampling Table t for a Customer x_{ji} at a Restaurant: For an individual customer the likelihood for a new table $t_{ji} = t^{new}$ can be calculated by integrating out the possible values of the new dish $k_{jt^{new}}$: $p(x_{ji}|\boldsymbol{t}_{-ji}, t_{ji} =$

$t^{new}; \boldsymbol{k}) = \sum_{k=1}^{K} \frac{m_{.k}}{m_{..}+\alpha^0} f_{k_{jt}}^{-x_{ji}}(x_{ji}) + \frac{\alpha^0}{m_{..}+\alpha^0} f_{k_{jt}^{new}}^{-x_{ji}}(x_{ji})$. Here the m values are total counts of tables from restaurants at all leaf nodes (observed documents), and α is a hyperparameter. We make a computationally efficient approximation in the right-hand term (corresponding to a new table at the parent node) by evaluating its word probabilities directly from the root instead of recursively traveling up. Therefore, at a restaurant the conditional distribution of t_{ji} is: $p(t_{ji} = t) \propto (n_{jt.}^{-ji}/(n_{.j.} + \alpha^{l+1})) f_{k_{jt}}^{-x_{ji}}(x_{ji})$, and $p(t_{ji} = t^{new}) \propto (\alpha^{l+1}/(n_{j..} + \alpha^{l+1}))p(x_{ji}|\boldsymbol{t}_{-ji}, t_{ji} = t^{new}; \boldsymbol{k})$, where $n_{jt.}$ is the number of customers in restaurant j at table t.

Part 2. Sampling a Table t from Delivery-Station v for a New Waiter with First Customer x_{ji}: When a customer x_{ji} sits at a new restaurant table, it has no dish yet: the waiter at that table must fetch a dish for this first customer from the delivery station for the restaurant, and must then choose some table t_{jt} from delivery-station v. The delivery-station table can be either a table that other waiters have also picked, or a new delivery-station table; in the latter case a new dish must then be brought from the upper-level delivery station. The likelihood for $t_{jt} = t^{new}$ can be calculated as follows: $p(t_{jt}|\boldsymbol{t}_{-jt}, t_{jt} = t^{new}; \boldsymbol{k}) = \sum_{k=1}^{K}(c_{vt.}/(c_{v..} + \alpha^l)) f_{k_{jt}}^{-x_{ji}}(x_{ji}) + (\alpha^l/(c_{v..} + \alpha^l)) f_{k_{jt}^{new}}^{-x_{ji}}(x_{ji})$, where $c_{vt.}$ is the number of tables point to table t in node v and $c_{v..}$ is the number of tables point to tables in node v. Therefore, the conditional distribution of t_{jt} (with a customer at a restaurant) is $p(t_{jt} = t) \propto (c_{vt.}^{-jt}/(c_{v..} + \alpha_j)) f_{k_{jt}}^{-x_{ji}}(x_{ji})$ and $p(t_{jt} = t^{new}) \propto (\alpha_j/(c_{v..} + \alpha_j))p(t_{jt}|\boldsymbol{t}_{-jt}, t_{jt} = t^{new}; \boldsymbol{k})$.

Part 3. Sampling a Delivery-Station Table t for a Waiter with Several Existing Customers: The likelihood for $t_j = t^{new}$ for many customers in a table can be calculated as: $p(t_{jt}|\boldsymbol{t}_{-jt}, t_{jt} = t^{new}; \boldsymbol{k}) = \sum_{k=1}^{K}(c_{vt.}/(c_{v..} + \alpha^l)) f_k^{-\boldsymbol{x}_{jt}}(\boldsymbol{x}_{jt}) + (\alpha^l/(c_{v..} + \alpha^l)) f_{k_{new}}^{-\boldsymbol{x}_{jt}}(\boldsymbol{x}_{jt})$. Therefore, the conditional distribution of t_j, given all customers in the table, is $p(t_{jt} = t) \propto (c_{vt.}^{-jt}/(c_{v..} + \alpha^l)) f_k^{-\boldsymbol{x}_{jt}}(\boldsymbol{x}_{jt})$ and $p(t_{jt} = t^{new}) \propto (\alpha^l/(c_{v..} + \alpha^l))p(t_j|\boldsymbol{t}_{-jt}, t_{jt} = t^{new}; \boldsymbol{k})$. If the sampled value of t_{ji} is t^{new}, we create a new table at the upper level, and recursively sample its dish. If the upper level is the root level, a topic is sampled k_{jroot_t} with respect to k_{jt} and propagated to all its descendants.

Part 4. Sampling k: The conditional probability of a dish at the root level k_{jroot_t} i.e., k_{jt} is: $p(k_{jt} = k) \propto (m_{.k}^{-jt}/(m_{..}+\alpha^0)) f_k^{-\boldsymbol{x}_{jt}}(\boldsymbol{x}_{jt})$ and $p(k_{jt} = k_{new}) \propto (\alpha^0/(m_{..} + \alpha^0)) f_{k_{new}}^{-\boldsymbol{x}_{jt}}(\boldsymbol{x}_{jt})$.

Part 5. Sampling k for a New Table: If a customer is given a new table ($t_{ji} = t^{new}$) we sample $k_{jt^{new}}$ as follows: $p(k_{jt^{new}} = k) \propto (m_{.k}/(m_{..} + \alpha^0)) f_k^{-x_{ji}}(x_{ji})$ and $p(k_{jt^{new}} = k_{new}) \propto (\alpha^0/(m_{..} + \alpha^0)) f_{k_{new}}^{-x_{ji}}(x_{ji})$.

We summarize the Gibbs sampling algorithm for THDP inference: sample a table assignment for each word in a document with a recursive procedure as follows. For a word, sample a table as in Part 1; if it's a new table, move to the parent node to sample a table from the parent node as in Part 2; repeat until the root node is reached; then select a topic for the table in the root as in Part

Table 1. Data set properties. Section counts at level 2–4 below the root given in parentheses.

	#sections	#Train docs	#Test docs	#terms	Avg. doc len
Suomi24 politics	49 (16 + 16 + 17)	980	245	50217	323.5
Suomi24 health	15 (5 + 9 + 1)	300	75	14700	209.9
Suomi24 relationship	18 (14 + 4 + 0)	360	90	17804	264.1
Amazon fishing acc.	13 (0 + 8 + 5)	260	65	3206	256.6

5, and update the topic of all tables in the descendant's nodes of the table in the root. Similarly, for each table (i.e., a group of words associated with a table) in a document, we sample a parent table i.e. a table from the parent using as in Part 3. We repeat the process until the root is reached and eventually sample a topic for the root table using as in Part 4.

5 Experimental Results

We first describe the data sets, summarized in Table 1. We begin by qualitative comparisons, and then present quantitative comparisons.

We evaluate the THDP model against the baseline HDP [2] on difficult modeling tasks where relatively little observation data is available, and a well-chosen model structure can thus help. We used two different data sources, *Suomi24* and Amazon. Suomi24 has in total 2434 sections in its hierarchy. The data set (https://www.kielipankki.fi/corpora/) is publicly available in original and lemmatized forms. From this source, we created several data sets for our experiments. The second data source is *reviews on Amazon.com*, a major shopping site with numerous shopping sections, for example, 1933 sections under Sports and Outdoors department [9]. We select the Fishing Accessories data set which is under Sports and Outdoors → Sports → Hunting and Fishing → Fishing category. The data set contains products at different levels of hierarchy. For each section containing products, at whatever level of hierarchy, we select 20 threads for training and 5 threads for testing. Therefore, the Amazon Fishing Accessories data set contains in total 260 reviews for training and 65 reviews for testing. We lemmatized the words in all reviews. Table 1 shows the numbers of sections and total training and test set sizes.

Qualitative Analysis. We verify that extracted THDP topics in a section are related to the topic of the section. The analysis could be carried out for different alpha values; we present results for an example alpha value 1. The top words of THDP topics for the Fishing Accessories data set are shown in Table 2. For many sections, we observe that extracted latent topics correspond to the section themes. For example, top words of topics 6 and 10 are names of Ice Spearing Equipment and verbs for using them (e.g., spear, gun, load, shoot etc.) and details of Fishing Scales equipment (e.g., scale, battery, weight etc.). Similarly, Topics 3, 5, 12, and 13 are about Shelters, Marker Buoys, Nets, and Fishing Line,

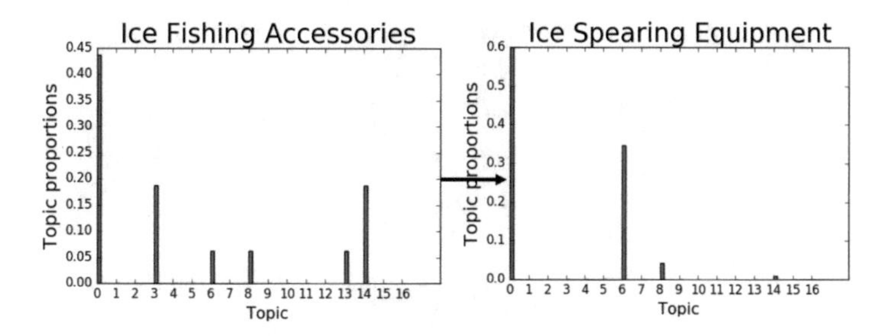

Fig. 3. THDP topic proportions in two example sections of the fishing accessories data set.

Table 2. THDP topics for Amazon data set, sections where they are active, and top words

Topic	Sections	Stemmed top words of the topic
0	All	Fish work great good ice easi product bought make time made line order
1	0, 1, 5	Oar clam gun collar book chart lock detail oarlock razor map guid
2	0, 5	Fli cast video joan dvd learn watch wulff fish great teach instruct
3	0, 4, 9, 12	Sled shelter shack chair wind warm set frame plenti heater pak front
4	0, 3, 8	Planer board rapala belt fight releas descript pro tension troll brown
5	0, 2	Buoy clamp float holder rod anchor outrigg kayak crab sand umbrella
6	0, 4, 11	Gun spear band speargun load shoot dive shaft shot jbl cressi spearfish
7	0, 1	Glass cabl wear sunglass cablz neck snow pair retain face read goggl
8	0, 4, 11	Glove batteri heat provid hand pair warm finger cell pack wear chemic
9	0	Spear frog sharpen tine gig hook sharp barb point head bend weld file
10	0, 6	Scale weigh accur measur batteri scaler lip gripper digit pound tape
11	0, 3	Trap crab bait pot door wire tank caught tie danielson fold blue
12	0, 7	Net minnow cast throw return sink hook sein foot tradit styrofoam bow
13	0, 4	Tini yard firelin leader bead invis knot sufix reel strong cast crystal
14	0, 4, 8, 10–13	Grabber tag grab equip flop chip slipperi slimi soft slip northern
15	0, 1, 6	Helmet complet stingray pad roman buyer paint gaiter foam hurt insid
16	0, 1	Alarm loud sound night brother wrap backward speaker china lit led

respectively. THDP topics are also shared across different sections. For example, the THDP Topic 1 discusses both Safety Gear (e.g., oar, oarlock, lock etc.) and Charts & Maps (e.g., chart, maps, book etc.). Topic 4 is another example where the discussion is about both Downriggers and Bait Traps section with fish catching related equipment keywords such as rapala, belt and troll. However, Bait Traps is also specifically discussed in Topic 11 (e.g., trap, bait, wire, tank, crab etc.).

We also analyze topic proportions at sections in the hierarchy. Figure 3 shows THDP topic proportions for the Ice Fishing Accessories section and one of its child sections Ice Spearing Equipment. In both charts Topic 0 is about fishing or

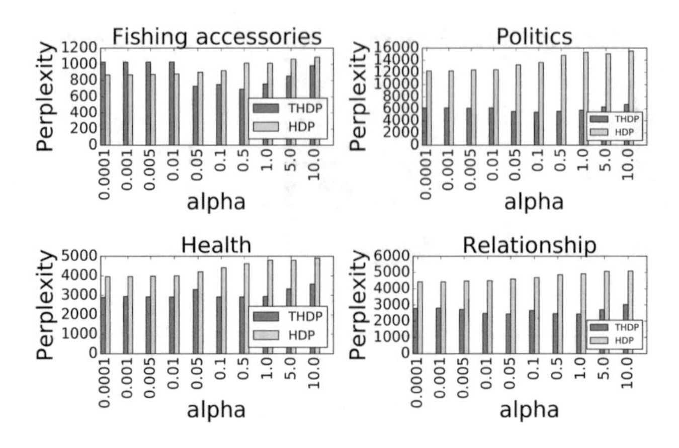

Fig. 4. Perplexity on different test data sets with different alpha values

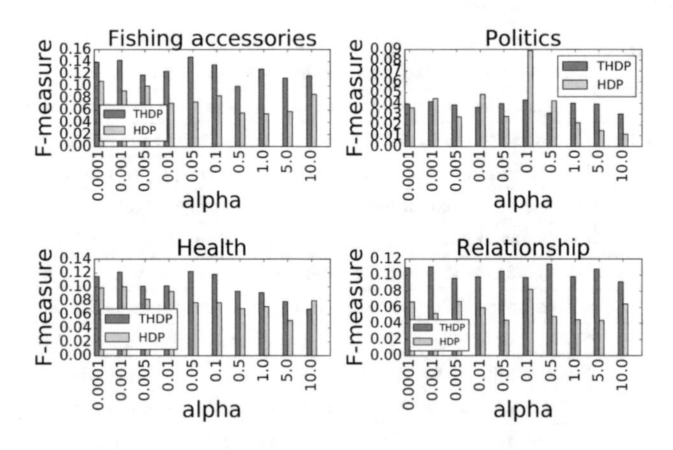

Fig. 5. Section prediction performance in terms of the F-measure of retrieving the correct section in different data sets with different alpha values.

price in general as shown in Table 2; the topic has a large portion in all sections. Ice Spearing Equipment section activates Topics 0, 6, and 8, which are also present in the parent section.

Quantitative Analysis. First, we evaluate the ability of THDP to represent new incoming documents with perplexity of held-out test documents [1]. Figure 4 shows the results, lower perplexity is better. THDP outperforms HDP in perplexity for most of the data sets for most of the tried alpha values (except for some alpha values in Fishing accessories). Next, for section prediction, we train a HDP model for each section in the dataset. For THDP, we train a single model for each dataset. To predict the section for each test document, we compute perplexity for the test document under the model for each section, and assign the document to the section or sections that yield the lowest perplexity. Figure 5 shows the resulting F-measures for different α values for different datasets, averaged over 5 runs. THDP outperforms HDP with higher F-measure for most data sets for most alpha values (except for some alpha values in Politics) since it incorporates section hierarchy information in the model.

6 Conclusions

We introduced the Tree-structured Hierarchical Dirichlet Process (THDP), a generative model for documents in deep tree-structured hierarchies such as online discussion forums. THDP extracts latent topics (discussion themes) shared across discussion sections, and outperforms the state-of-the-art model HDP in modeling new documents (measured by perplexity) and section prediction (measured by F-measure). Unlike previous work, THDP can incorporate a truly multilevel hierarchy. It can be adapted to many topic modeling applications to take into account their hierarchical data structure.

References

1. Blei, D., Ng, A., Jordan, M.: Latent Dirichlet allocation. JMLR **3**, 993–1022 (2003)
2. Teh, Y., Jordan, M., Beal, M., Blei, D.: Hierarchical Dirichlet processes. J. Am. Stat. Assoc. **101**, 1566–1581 (2006)
3. Blei, D., Griffiths, T., Jordan, M.: The nested Chinese restaurant process and Bayesian nonparametric inference of topic hierarchies. J. ACM **57**, 7:1–7:30 (2010)
4. Li, W., McCallum, A.: Pachinko allocation: DAG-structured mixture models of topic correlations. In: Proceedings of ICML, pp. 577–584. ACM (2006)
5. Adams, R., Ghahramani, Z., Jordan, M.: Tree-structured stick breaking for hierarchical data. In: Proceedings of NIPS, pp. 19–27. Curran Associates Inc. (2010)
6. Faisal, A., Gillberg, J., Leen, G., Peltonen, J.: Transfer learning using a nonparametric sparse topic model. Neurocomputing **112**, 124–137 (2013)
7. Xu, Y., Yin, J., Huang, J., Yin, Y.: Hierarchical topic modeling with automatic knowledge mining. Expert Syst. Appl. **103**, 106–117 (2018)
8. Kim, J., Kim, D., Kim, S., Oh, A.: Modeling topic hierarchies with the recursive Chinese restaurant process. In: Proceedings of CIKM, pp. 783–792. ACM (2012)
9. He, R., McAuley, J.: Ups and downs: modeling the visual evolution of fashion trends with one-class collaborative filtering. In: Proceedings of WWW, pp. 507–517 (2016)

Estimated Rating Based on Hours Played for Video Game Recommendation

Javier Pérez-Marcos[1(✉)], Diego Sánchez-Moreno[2], Vivian López Batista[2], and María Dolores Muñoz[2]

[1] BISITE Digital Innovation Hub, University of Salamanca, Edificio Multiusos I+D+i, 37007 Salamanca, Spain
jpmarcos@usal.es

[2] Departamento de Informática y Automática, University of Salamanca, Salamanca, Spain
sanchezhh@gmail.com, {vivian,mariado}@usal.es

Abstract. This work presents a method to estimate ratings for video games based on the user's playing hours. Based on these ratings, through collaborative filtering techniques, it is possible to make recommendations for video games without taking into account their popularity, solving the problem of long tail. The item-based k-NN algorithms and SVD++ are the ones that obtains the best results with the proposed estimation method, improving the original one and obtaining similar results in the rest of cases.

Keywords: Rating estimation · Rating prediction · Video games Collaborative filtering · Recommender systems

1 Introduction

The videogame sector is one of the few markets that did not suffer from the past world crisis, moreover, it was one of the few that grew during it. The number of games published each year is increasing steadily. The emergence of indie studies and greater accessibility for the development of indies games makes it easier for platforms like Steam to increase the number of games published each month, as reflected on the steamspy website[1]. However, recommendation systems that platforms like Steam[2] contain favor popular games, to the detriment of lesser-known games from smaller studies. This phenomenon is known as Long Tail.

In this work we propose a method for estimating valuations based on the hours a user has played a video game. This method is intended to deal with the problem of Long Tail while providing quality predictions for video game recommendation. Therefore, this rating estimation method will be tested in different prediction algorithms to evaluate its predictive quality.

---

[1] http://steamspy.com/year/.

[2] http://store.steampowered.com/.

S. Rodríguez et al. (Eds.): DCAI 2018, AISC 801, pp. 300–307, 2019.
https://doi.org/10.1007/978-3-319-99608-0_34

2 Related Work

There are no previous works related to the recommendation of video games, nor on the rating estimation from the number of hours played to a video game by a user. However, there are two works of which this paper is based. The first is [1] where the authors propose two methods for rating estimation for television programs, one linear and the other logarithmic. However, these methods do not take into account the viewing hours character of a program (that is, according to the duration of the program). The second work [5] is the one that this article takes as a reference. In this paper, the author proposes a method based on quartile distribution for the estimation of song ratings.

3 Proposal

Collaborative filtering (CF) is a recommendation methodology based on user ratings. From these ratings, it is possible to predict future items in which users may be interested. However, these ratings may not be explicitly available to the system. Therefore, it is necessary that from some record of the user interaction with the system (e.g. hours played to a video game) it is possible to obtain these ratings implicitly. Our proposal for estimating valuations based on hours played is described in Subsect. 3.1. The CF algorithms used to test this proposal are detailed in Subsect. 3.2.

3.1 Estimating Ratings from Payed Hours

In this work we propose a variation of the estimation of valuations based on the number of times a user listens to a song, proposed in [5]. Unlike [5], and although the number of hours played per video game follows a power law distribution, the nature of these differs from that of the number of plays of the music. First in domain: the number of times a song is played belongs to the $[0, \infty) \in \mathbb{N}$ interval, while the hours played belong to the $[0, \infty) \in \mathbb{R}$ interval. Second in character: a song does not have an estimated number of plays per user, while a video game does have an estimated number of hours of playtime. This means that the duration in hours of the games varies among them, some last 10 h, others last 30 h, and even the case of multiplayer games whose estimation in hours is infinite.

Pacula's proposal is based on the definition of the frequencies of hours played per user, according to the following formula:

$$f_{i,u} = \frac{h_{i,u}}{\sum_{j \in I} h_{j,u}}. \tag{1}$$

Where $u \in U$ is the user and $i \in I$ is the game of which the frequency is estimated, $h_{i,u}$ is the hours played to the game i by the user u. Based on these frequencies, the rating is estimated for an item i of a user u as follows:

$$r_{i,u} = 4\left(1 - \sum_{k=1}^{k-1} f_{k,u}\right). \tag{2}$$

Where $k \in K$ is the k-thest user item u, sorted by frequency, and $f_{k,u}$ is the frequency for the item k of the user u. In this way you get valuations between $(0, 4] \in \mathbb{R}$ for each user item.

However, this approach is wrong, especially in the case of video games. Estimating frequency by means of user items can make video games with hours played above their estimated hours of duration less valuable than games with hours played below their estimated hours of duration. For example, a game with a duration of 10 h that has been played for 20 h will be worth less than a game played for 30 h with an estimated duration of 50 h.

In this work we propose a variation of the method proposed by Pacula, in such a way that it takes into account the estimation in hours of play of a video game. Therefore, the frequency of a game i for a user u is defined as:

$$f_{i,u} = \frac{h_{i,u}}{\sum_{v \in U} h_{i,v}}. \tag{3}$$

Where $h_{i,u}$ is the hours played to the game i for the user u. Unlike Pacula, the frequency is calculated on a game and not on a user. In addition, the rating estimate is amended as follows:

$$r_{i,u} = 4\left(1 - \sum_{k=1}^{k-1} f_{i,k}\right) + 1. \tag{4}$$

Where k is the k-th user of item i, ordered by frequency, and $f_{i,k}$ is the frequency for item i of user k. Therefore, the rating for the video game will depend on the hours played by the other users. In addition, unlike Pacula, the rating range is $(1, 5] \in \mathbb{R}$, with the lowest rating being 1. This method of rating estimation is intended to improve the behaviour of CF algorithms, especially those that seek relationships between items.

3.2 Prediction Algorithms

The following is a description of the CF algorithms used in this work, the results of which are discussed in Sect. 4. These algorithms will take as input the ratings estimated in the previous subsection.

k-Nearest Neighbors. CF algorithms based on k-Nearest Neighbors (k-NN), estimate their predictions by searching for the k nearest neighbors according to some measure of similarity. In this work a variation of the k-NN has been used which takes into account a base-line rating [2]. The mean square error (MSE), cosine distance and Pearson's similarity coefficient have been used as a measure of similarity. In addition, the algorithm has been applied to both users and items.

Singular-Value Decomposition. Singular-Value Decomposition (SVD) is a matrix dimensional reduction method. Its application to CF results in the search for latent factors in items to make predictions to users. In this work we have used the SVD++ version proposed in [3], which combines the search for latent factors taking into account implicit user valuations.

Slope One. Slope One is a CF algorithm proposed in [4] that is based on the pre-calculation of the mean difference between the ratings of one item and another for users who rated both, by defining predictors in the form $f(x) = x + b$.

4 Evaluation

The proposed method has been evaluated and compared with the one proposed by Pacula based on the Steam Video Games data set[3] of the Kaggle platform[4]. This dataset collects real information about the playing habits of Steam users. Each record contains the user's id, the name of the video game and the hours played. In total the dataset contains 70489 records, 11350 users and 3600 video games.

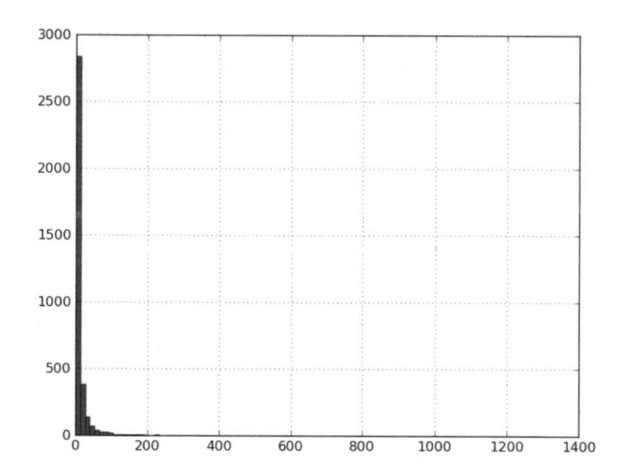

Fig. 1. Frequency of average played hours per video game

As can be seen in the Fig. 1, the frequency of the average hours spent on a video game follows a power law distribution. This indicates that there are very few video games with lots of hours of gameplay while most video games have been

[3] https://www.kaggle.com/tamber/steam-video-games.
[4] https://www.kaggle.com.

played much less. This may result in skewed recommendations where popular games are favored. The proposed method solves this problem by estimating the rating of a video game based on the hours played by users.

4.1 Model Selection

In order to evaluate which model best predicts for each rating estimation, a 10 fold cross validation has been performed for each algorithm. In this way the results are independent of training and testing sets. The Mean Absolute Error (MAE) has been calculated from each iteration of the cross validation according to Eq. 5 and the Root Mean Square Error (RMSE) has been calculated according to the Eq. 6. Both values are derived from the error in the predictions, either the first of absolute error and the second of quadratic error.

$$MAE = \frac{1}{|B_i|} \sum_{b_k \in B_i} |r_i(b_k) - p_i(b_k)|. \tag{5}$$

$$RMSE = \sqrt{\frac{1}{|B_i|} \sum_{b_k \in B_i} (r_i(b_k) - p_i(b_k))^2}. \tag{6}$$

In order to be able to compare both rating estimation methods and decide which method obtains better results, the Normalized Mean Absolute Error (NMAE), defined in the Eq. 7, and the Normalized Root Mean Square Error (NRMSE), defined in Eq. 8 are calculated. By normalising the MAE and RMSE coefficients it is possible to compare methods whose ranges of values differ.

$$NMAE = \frac{MAE}{r_{max} - r_{min}}. \tag{7}$$

$$NRMSE = \frac{RMSE}{r_{max} - r_{min}}. \tag{8}$$

Finally, for each prediction algorithm, average NMAE and NRMSE are calculated from the 10 folds. The following subsection shows and discusses the results obtained.

4.2 Results

This subsection shows and discusses the results obtained in the experiments carried out. In the Table 1 the NMAE and NRMSE coefficients obtained for each algorithm and estimation method are shown. The Fig. 2 shows the NMAE comparison of each estimation method for each algorithm, while the Fig. 3 shows the NRMSE comparison between both methods.

As shown in the Table 1, item-based k-NN algorithms with the proposed method improve the results compared to those obtained with the Pacula's method. In other words, the proposed method has a better behavior in algorithms that look for relationships between items. The SVD++ algorithm with

Table 1. NMAE and NRMSE for both estimating ratings method

	NMAE		NRMSE	
Algorithm	Proposal	Pacula	Proposal	Pacula
KNN User MSD	0.1844	0.1844	0.2677	**0.2678**
KNN User Cosine	0.1846	0.1859	0.2667	0.2709
KNN User Pearson	0.1865	**0.1829**	0.2720	0.2690
KNN Item MSD	0.1788	0.2019	0.2582	0.3054
KNN Item Cosine	**0.1785**	0.2022	0.2577	0.3056
KNN Item Pearson	0.1820	0.2070	0.2664	0.3011
SDV++	0.1803	0.2147	**0.2543**	0.3123
Slope One	0.2050	0.2370	0.2843	0.3587

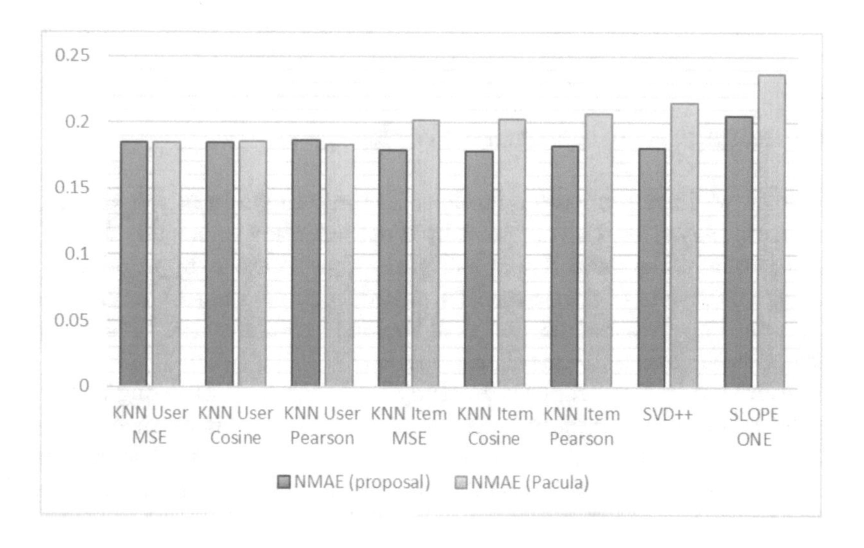

Fig. 2. NMAE for every predicting algorithm

the proposed estimation method is the best performing algorithm for NRMSE, i.e. it has fewer errors of greater range in the prediction. The k-NN algorithm based on items with the cosine distance and the proposed method is the one that shows the best results for NMAE, so the error frequency in the prediction is lower than the rest.

On the other hand, in user-based k-NN algorithms both methods obtain similar results, with the best results obtained by the method proposed by Pacula. This can best be seen in the Figs. 2 and 3, where both methods obtain similar results for user-based k-NN, and yet the method proposed by Pacula worsens significantly in the rest, while our proposal improves slightly.

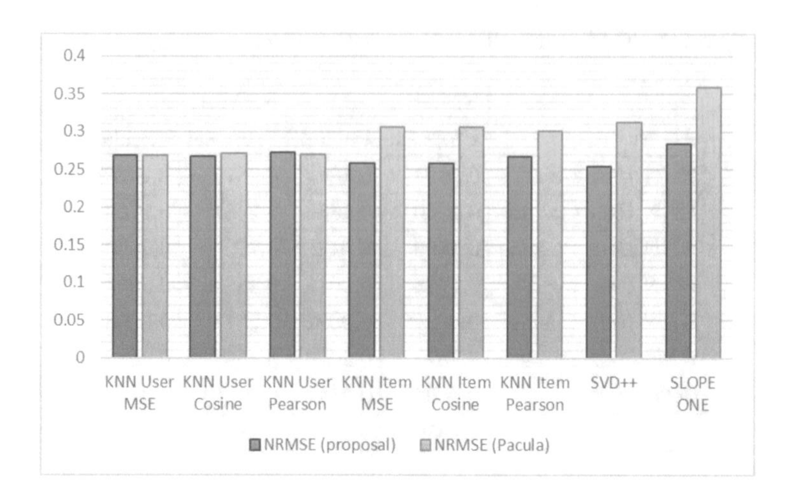

Fig. 3. NRMSE for every predicting algorithm

5 Conclusions and Future Work

In this work we have proposed a method for the estimation of valuations based on the hours played to a video game by a user. As has been said, this method takes into account the casuistica of each game to estimate an appropriate rating, improving the results of item-based CF algorithms. As the results have shown, our proposal improves Pacula's original method, obtaining the best results for item-based k-NN and SVD++ algorithms.

As future work, it is proposed to study other methods of rating estimation, especially those carried out for services based on time spent such as TV. In addition, taking into account attributes of video games, such as gender; or taking into account social characteristics of users such as friendship; would improve the prediction results.

Acknowledgements. This work was supported by the Spanish Ministry, Ministerio de Economía y Competitividad and FEDER funds. Project. SURF: Intelligent System for integrated and sustainable management of urban fleets TIN2015-65515-C4-3-R.

References

1. Hu, Y., Koren, Y., Volinsky, C.: Collaborative filtering for implicit feedback datasets. In: 2008 Eighth IEEE International Conference on Data Mining, pp. 263–272. IEEE, December 2008. http://ieeexplore.ieee.org/document/4781121/
2. Koren, Y., Research, Y.: Factor in the neighbors: scalable and accurate collaborative filtering. ACM Trans. Knowl. Discov. Data **4**(1) (2010). http://courses.ischool.berkeley.edu/i290-dm/s11/SECURE/a1-koren.pdf

3. Koren, Y.: Factorization meets the neighborhood. In: Proceeding of the 14th ACM SIGKDD International Conference on Knowledge Discovery and Data Mining - KDD 2008, p. 426. ACM Press, New York (2008). http://dl.acm.org/citation.cfm?doid=1401890.1401944
4. Lemire, D., Maclachlan, A.: Slope One Predictors for Online Rating-Based Collaborative Filtering, February 2007. http://arxiv.org/abs/cs/0702144
5. Pacula, M.: A matrix factorization algorithm for music recommendation using implicit user feedback (2009)

Doctoral Consortium and Short Papers

Towards an Adaptive and Personalized Assessment Model Based on Ontologies, Context and Collaborative Filtering

Oscar M. Salazar[1(✉)], Demetrio A. Ovalle[1(✉)], and Fernando de la Prieta[2(✉)]

[1] Departamento de Ciencias de la Computación y la Decisión, Universidad Nacional de Colombia – Sede Medellín, Medellín, Colombia
{omsalazaro,dovalle}@unal.edu.co

[2] Departamento de Ciencias de la Computación, Universidad de Salamanca, Salamanca, Spain
fer@usal.es

Abstract. The assessment phase plays a very important role during the students' teaching-learning processes since from this phase the knowledge acquired by them are validated and the shortcomings and/or strengths from students are detected. However, to do so, the questions selection made by the teacher or the learning platform does not always respond to the needs, limitations and/or cognitive characteristics of the students. In this context, it become necessary the incorporation of mechanisms that allows to obtain the main student features in a better way in order to use them during the process of question selection. In fact, this brings several benefits such as a better acquired knowledge measurement, an increase in the students' interests, a better fail detection for new educational resource recommendation, among others. In order to make a better question selection that fulfil the student's needs, this paper aim at proposing a characterization of the most relevant techniques and models for question selection. Likewise, an ontological model of personalized adaptive assessment is proposed, supported by Artificial Intelligence techniques that incorporate relevant cognitive and contextual information of the student to carry out a better selection and classification of questions during the e-assessment process.

Keywords: e-Assessment · Adaptive e-Assessment
Automatic question selection and classification · Context
Collaborative filtering · User profiles

Problem Statement. The assessment process plays a crucial role in the student's learning process since it allows to diagnose the previous state of knowledge, to monitor the progress and to validate the level of knowledge obtained at the end of the learning activity [1]. Some efforts have explored different ways of integrating assessment within virtual environments such as generation of automatic grades, the diversification of question types, the generation of tracking statistics, continuous feedback, among others. However, most of current learning environments focus on generic assessment models that do not consider the student features, in other words, all students are evaluated with the same scheme without contemplating likes, limitations, learning styles, cognitive levels, social contexts, spatial contexts, etc. [2].

S. Rodríguez et al. (Eds.): DCAI 2018, AISC 801, pp. 311–314, 2019.
https://doi.org/10.1007/978-3-319-99608-0_35

The majority of current assessment schemes focus on building large question banks that are then randomly selected to students, regardless of their difficulty level. Wauters et al. [3] argue that students learn better and feel more motivated when the exercises are selected according to the difficulty level. In fact, really easy or extremely complicated exercises generate disinterest and frustration in the students. Unfortunately, measuring the difficulty level of the questions to be selected within an assessment activity is not an easy task, since it is a subjective and highly personal exercise. Some research works argue that teachers tend to overestimate the difficulty of easy questions and underestimate the difficulty of those difficult questions. In contrast, other works argue that teachers generally classify the difficulty of hard questions better. Something similar happens with the students, some of them consider certain questions easy while others find them difficult and complexed [4].

Related Work. A proposal for an adaptive assessment system is presented by Baneres et al. [5], wherein activities can be adapted based on evidences collected during the student's learning process. The concept of evidence concerns to previous scores, amount of performed exercises, plagiarism detection, and forum interactions.

Pérez et al. [6] analyze the ability of teachers to determine the difficulty level of the questions associated with an e-Assessment, in contrast to the perception of difficulty expressed by the students. The proposed system has the ability to capture the question difficulty level when the assessment instructional design is performed. On the other hand, an expert system is responsible for capturing the difficulty perceived by the students during the test depending on several items such as the spent time, the obtained grade, and the number of accesses or readings that the student made before sending the answer. The work concludes that students perceive the questions with a higher difficulty level in contrast to the level initially defined by the teacher. Singhal et al. [4] propose a difficulty model for the question generation through formal domains according to the level of difficulty granted by the user. The model is based on lexicographical ordering to compare the difficulty of questions based on a factor order defined by the user and an associative algorithm to manage these factors. The developed system allows generating a large number of questions with different difficulty degrees considering only the structure nature of the questions.

The previously described research works evidence efforts for developing mechanisms that allow a better questions selection to build the assessments. However, most of them do not consider student features to select the difficulty levels of the assessments, or they do not establish a formal structure to describe the entities associated with the assessment process, neither do they include student contextual characteristics (Fig. 2).

Model Proposed. This section presents the development of an adaptive and personalized assessment model based on the measurement of the difficulty level of questions in accordance with the student's profile, to be used for question selection within the e-assessment process. First of all, a formal mechanism based on the concept of ontologies is defined; this allows the representation of the entities related to the domain of e-assessment along with their relationships.

During the first phase of the ontological model definition, the entities linked to the domain that we wanted to represent were described. The first element to describe refers to the concept of the question, to which we were adding traditional attributes such as: title, description, options, creator, etc. One of the great difficulties found in current models is that they try to define the question difficulty level based on the attributes of it, our approach focuses on describing the difficulty level according to the student characteristics and their perceptions. Based on this, the next concept to be described was the student profile; for doing so, we consider three dimensions: (1) the basic data such as name, document, mail, gender, languages, learning style, among others. (2) the reputation, which is described based on the performed assessments, the obtained grades, the program average, the performance achieved in each of defined knowledge areas, etc. (3) the student context, which in turn considers three sub-dimensions, a spatial context (location where the assessment interaction takes place), a social context (life quality, schooling level, etc.), and a learning context (defines the kind of assessment: formative, summative, training, etc.). From the definition of the structure of the student's profile, new attributes were incorporated into the question concept, which help to improve the question selection process. Some of these attributes were: difficulty level perception (teacher and students), target audience, spatial context, country, average time for solution, target schooling level, among others. In the same way, the ontology captures the question perception in terms of perceived difficulty, readings made, and individual solution time. Figure 1 presents the taxonomy of the developed ontology, the attributes of the concepts and their relationships.

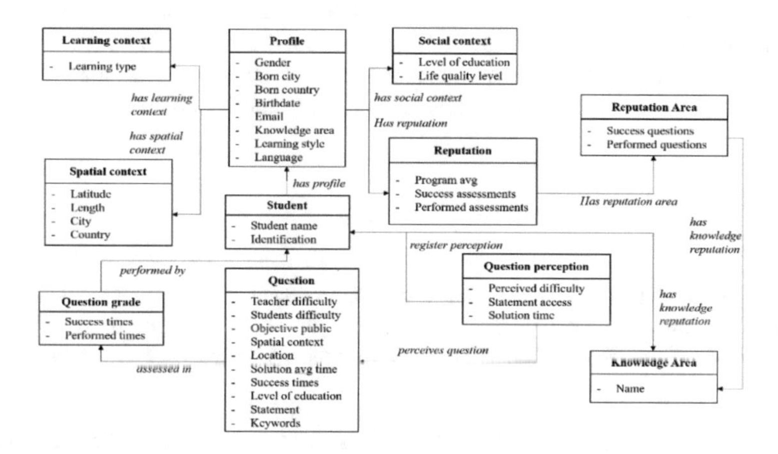

Fig. 1. Ontology taxonomy

The question categorization process using difficulty level is performed by means of a Fuzzy Inference System, which is responsible for collecting some data recorded in the ontology in order to process them. Based on this, the system builds an initial question set having different difficulty levels. In the same way, the ontology records the individual perception of the student regarding each of the questions that the student has previously answered. Based on this information, the system tries to find similarities between student profiles. This collaborative filtering module allows selecting questions that similar users have previously qualified, thus feeding the initial described question bag. Subsequently, a module intended for the difficulty categorization process according to the user profile is implemented. This module is responsible for defining the difficulty using student's characteristics that establish an appropriate relationship

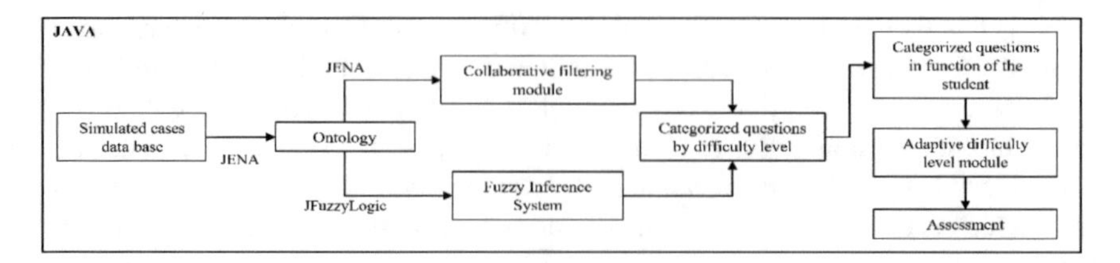

Fig. 2. Adaptive e-assessment system flow.

between each of the questions and the student's profile. Finally, a difficulty adaptive module is responsible for receiving the student's answers and, based on the obtained result is able to select the following question. This means that the system adjusts the level based on the user's interaction with the assessment.

Reflections. The e-assessment adaptation in accordance with the student's characteristics is a fundamental process in the assessment activities within Virtual Courses since it allows to better identify the students' fails and strengths. In addition, varying and customizing the question difficulty degree during their development process helps to maintain the student's interest and to make their learning process more dynamic and personalized. The adaptive e-assessment model proposed allows to control the question difficulty level according to the student's features and the context in which the learning activity takes place, capturing this kind of information allows a better adaptation of the questions and improves the appropriate selection of them in future cases.

References

1. Monteiro, D., Alturas, B.: The adoption of e-Recruitment: The Portuguese case: study of limitations and possibilities by the point of view from candidates and from recruiters. In: Proceedings of the 7th Iberian Conference on Information Systems and Technologies (CISTI 2012) (2012)
2. Hajjej, F., Hlaoui, Y.B., Ayed, L.J.: Ben: personalized and generic e-assessment process based on cloud computing. In: 2015 IEEE 39th Annual Computer Software and Applications Conference, pp. 387–392. IEEE (2015)
3. Wauters, K., Desmet, P., Van den Noortgate, W.: Adaptive item-based learning environments based on the item response theory: possibilities and challenges. J. Comput. Assist. Learn. **26**, 549–562 (2010)
4. Singhal, R., Goyal, S., Henz, M.: User-defined difficulty levels for automated question generation. In: 2016 IEEE 28th International Conference on Tools with Artificial Intelligence (ICTAI), pp. 828–835. IEEE (2016)
5. Baneres, D., Rodríguez, M.E., Guerrero-Roldán, A.-E., Baró, X.: Towards an adaptive e-assessment system based on trustworthiness. In: Formative Assessment, Learning Data Analytics and Gamification, pp. 25–47. Elsevier (2016)
6. Perez, E.V., Santos, L.M.R., Perez, M.J.V., de Castro Fernandez, J.P., Martin, R.G.: Automatic classification of question difficulty level: teachers' estimation vs. students' perception. In: 2012 Frontiers in Education Conference Proceedings, pp. 1–5. IEEE (2012)

Visualizing History: A *Virtual Timeline* for Teaching and Learning Historical Sciences

Elena Llamas-Pombo(✉)

Departamento de Filología Francesa and IEMYRhd, University of Salamanca, Salamanca, Spain
pombo@usal.es

Abstract. This article discusses the features of traditional infographics used for representing data in the course of a chronological axis and the development of innovative learning in virtual environments for Historical Sciences. We are namely interested in implementing a prototype of an online learning platform that should represent chronology in a dynamic timeline. An adequate solution for the implementation of this proposal is a multi-agent system. This is because it distributes knowledge, employs problem-solving techniques and makes it easier to access databases, internet information and any other type of resources required by the system. The use of this technology in the prototype will allow it to evolve and adapt to changes, as well as interact with users smoothly and effortlessly.

Keywords: Timeline · Chronology · Counter · Historical Sciences
Learning objects

1 Representation of Time Factor in Sciences

1.1 Traditional Chronology Visualization

Numerous natural and human sciences represent the knowledge along a *chronological axis* that is namely essential for disciplines as geological stratigraphy, human history, art history and historical linguistics.

The data that scientists represent around a time factor are divulgated through symbolical and schematic visual patterns that are not neutral. They predetermine our way of learning and teaching sciences, from childhood to the highest level of University education. They also condition the very configuration of knowledge itself. Let's consider some examples.

1. The purpose of the *Chronostratigraphic Chart* is to fix "the basis for the units (periods, epochs and age) of the International Geologic Time Scale; thus setting global standards for the fundamental scale for expressing the history of the Earth", as presented in Fig. 1 (International Chronostratigraphic Chart 2017). This digital image allows us to visualize both chronology and terminology in Geology, but it is a bi-dimensional digital image of a document with static and closed diagrams, like in a traditional page.

S. Rodríguez et al. (Eds.): DCAI 2018, AISC 801, pp. 315–321, 2019.
https://doi.org/10.1007/978-3-319-99608-0_36

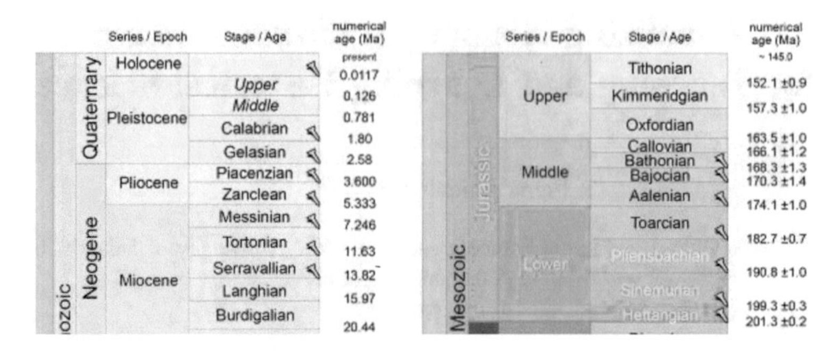

Fig. 1. Detail of international chronostratigraphic chart, 2017/02.

2. The time factor is also the main component of historical linguistics, as a science that studies languages evolution and linguistic change mechanisms. Different patterns have allowed linguists to represent the chronology symbolically[1]. For instance, since the nineteenth century, and following the evolutionary model of natural sciences, historical grammar has represented the language diversification by a tree diagram, using terms such as *The genealogical tree of Indo-European Languages*, as shown in Fig. 2. Similar infographic images are representing the historical depth of languages in a more abstract but always static way, as exemplified in Fig. 3 (Gray and Atkinson 2003).
3. Since the nineteenth century, in both historical phonetics and historical semantics, language changes for each word are usually represented in isolation, either in a linear structure or in a sequential diagram. In Fig. 4, we can read the typical expression of the phonetic evolution of a word.
4. The *stemma codicum* or genealogical relation between different copies of the same text has been schematized with similar type of infographics, where hierarchy is presented as an inverted tree diagram. See the *stemma* shown in Fig. 5.
5. The progress of history is usually visualized itself within lineal schematizations and always designed either in a bi-dimensional page format, as presented in Fig. 6, or in a multiple-page format, as the famous *Time Chart History of the World* (Adam 1999 [1871]). This chart was created in 1871 following the model of Jacques Barbeu-Du Bourg's *Machine chronographique* (1753) and has been continuously reprinted until present day.

1.2 New Timelines and Teaching Tools

In several human sciences, data visualization has not been updated for the last two centuries, and educational models reproduce on screens the very same illustrations used in books during the nineteenth century. In the age of digital globalization, the teaching of human sciences in schools shouldn't be left behind regarding the use of

[1] Junker (1992), Junker (1994), Landheer (2002), Llamas-Pombo (2011), Coulson and Pagán Cánovas (2014).

Fig. 2. Detail of family tree of the Indo-European Languages, © K. Scarfe Beckett 1972. Image from Deutscher (2005, p. 57).

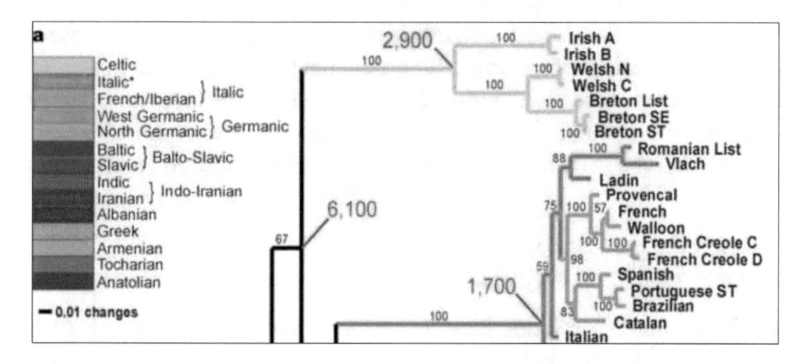

Fig. 3. Detail of language-tree divergence times (Gray and Atkinson 2003, p. 437).

GAUDIA > *gaŭyya* > ... *ĵọ̀yya* > *ĵọ́yya* > *ĵọ́ya* > ... *ĵọ́ię*;

Fig. 4. Phonetic evolution from Latin GAUDIA fo French *joie*.

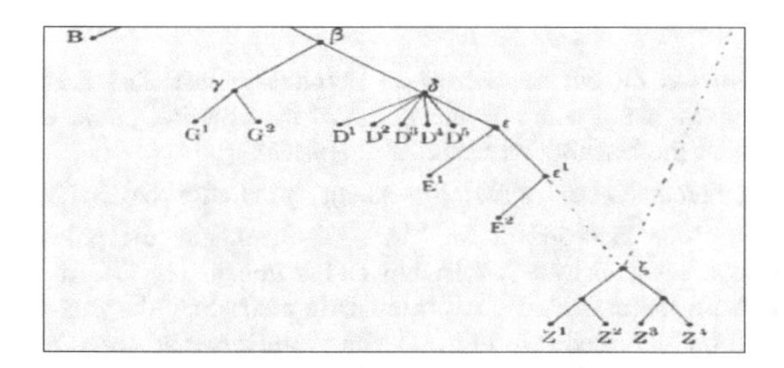

Fig. 5. Detail of a *Stemma codicum*

Fig. 6. Detail of Adam's *Time Chart History of the World.*

technological tools when compared with other disciplines (Genevois 2013). It should be a requirement in the entire educational system, including higher education.

Specifically, historical linguistics, as being taught today at universities, should advance at the same pace as natural sciences in the way they have already been digitized and visualized[2]. We should have at our disposal new supports for knowledge, without renouncing the previous ones, drawing upon the educational creativity offered by:

1. Instructional design theories (Spector 2016).
2. Digital tools created by specialists in Technology Enhanced Learning (TEL).

Multiple research laboratories have taken up the design of teaching tools, which improve and diversify the representation of data in a chronology. For instance:

1. Several contributions[3] offer a complete history of the *timelines* in the visualization of History and have demonstrated the great usefulness of the virtual environments for enhancing the learning of historical chronology.
2. The MIT's *Visualizing Cultures* repository purpose is to help in the reconstruction of the past and training students as active historians[4].

[2] Our observation is based on our experience as lecturers in historical French linguistics at the University of Salamanca, along with our knowledge of the materials produced by English, French and Spanish editors for the teaching on historical linguistics.

[3] Boyd-Davis (2013), Foreman et al. (2008), Rosenberg and Grafton (2010), Staley (2015).

[4] "Visualizing Cultures was launched at MIT in 2002 to explore the potential of the Web for developing innovative image-driven scholarship and learning. The VC mission is to use new technology and hitherto inaccessible visual materials to reconstruct the past as people of the time visualized the world (or imagined it to be) […] *The Visualizing Cultures Curriculum* offers a full complement of standards-compliant lessons, providing a pathway for teachers and students to become active historians and knowledgeable readers of images" (MIT 2015). See also *Clio Visualizing History* which objective is to create "innovative online history exhibits designed to attract students and educators" (Clio 2018).

3. The time factor often appears as a mere numerical indication by an automatic counter in excellent *animated maps,* as the ones elaborated by *Paleomap project* for the divulgation of planet Earth geological evolution[5], as well as in representations of human history[6] or the evolution of Indo-European languages[7].
4. In other schemes, the time is shown on a numbered and static horizontal line[8].
5. Many *new philologists* are nowadays implementing computer-assisted analysis of textual traditions for the detection of contamination and polygenesis, two major issues for genealogical textual analysis[9].

2 A Virtual Timeline for Mapping the Chronology

The current time visualization tools that are available for data analysis in teaching have the following features:

1. They are static. The time factor appears either numbered or as a horizontal line.
2. The user is not allowed to interact with the tools or complete the data.
3. The course of time is reduced to a brief linear scheme.

In our project *Virtual timeline*, we propose the creation of a digital tool for the representation of chronology, which would allow the incorporation and visualization of cultural contents. This new tool would have the following features:

1. Continuous digital format page, with a chronological axis.
2. Different from fixed infographics, as it would be dynamic and shapeable. The representation of the chronological course of time in a "time tunnel" would enable a larger mobile vision, both retrospective and prospective.
3. The timeline would include tools to build a repository with links to websites, documents or images saved in chronological order and not only in "folders" or in "boxes".
4. The user would be allowed to gather contents from different subjects (Human History, Art History, History of language, etc.) in a single visual object.

The learning objectives of this tool would be:

1. To support the student knowledge acquired in a dispersed or disconnected way in different subjects or stages of the learning process.

[5] Tools as the spectacularly *animated maps* made for *Paleomap Project* should be an indispensable part of the school (Scotese 2003), (Scotese 2015). A recent *Spanish Conference of Deans of Geology*, held at the University of Salamanca, has stated so in a manifest: Geology has little space in the education system and an effective elementary science education on the subject is imperative (CEDG 2018).

[6] See *The History of the World: Every Year* (Bye 2016).

[7] See *Indo-European Migrations & Language Geographical Evolution* (Pietrobon 2016).

[8] Like in an animated map showing *How Language Spread across Europe & Asia* (Kuzoian and Atkinson 2015) or like in some timelines of human evolution (Martínez 2013).

[9] Camps and Cafiero (2018), following Froger (1968) and Poole (1979).

2. To allow active learning or learning by making. The student enriches the information of his/her platform.
3. To improve learning of facts in their own chronology, hence to improve historical knowledge.

3 Future Work

In order to innovate the teaching of the historical linguistics, a first prototype could be progressively expanded and improved as a three-dimensional visual object. The final goal is to create a tool which would enable both simultaneous and parallel temporal axes. These possibilities of representation could allow a chronological registration of different levels in the evolution of language (phonetics, semantics, syntax, spelling, etc.) with axes overlapping different languages at various levels.

An adequate solution for the implementation of this proposal is a multi-agent system. This is because it distributes knowledge, employs problem-solving techniques and makes it easier to access databases, internet information and any other type of resources required by the system. The use of this technology in the prototype will allow it to evolve and adapt to changes, as well as interact with users smoothly and effortlessly. For data analysis and transformation, it is necessary to design machine learning and hybrid distributed systems. Specific algorithms will be required to perform correlation and feature extraction, which will facilitate the classification of data[10].

References

Adam, S.: The Timechart History of the World. The third Millennium Trust, Chippenham (1999 [1871])

Bajo, J., De la Prieta, F., Corchado, J.M., Rodríguez, S.: A low-level resource allocation in an agent-based Cloud Computing platform. Appl. Soft Comput. **48**, 716–728 (2017)

Barriuso, A., De la Prieta, F., Rodríguez-González, S., Bajo, J., Corchado, J.M.: Social simulations through an agent-based platform, location data and 3D models. In: Alonso-Betanzos, A., et al. (ed.) Agent-Based Modeling of Sustainable Behaviors, pp. 99–120 (2017)

Boyd-Davis, S.: Inventing the timeline: a history of visual history. A presentation to the Information Design Association, London, 29 January 2013 (2013). http://www.infodesign.org.uk/Events/Recent-Events/20130129stephen-boyd-davis

Bye, O.: The History of the World: Every Year, History Channel (2016). https://www.youtube.com/watch?v=-6Wu0Q7x5D0. Last Accessed 30 Jan 2018

Camps, J.-B., Cafiero, F.: Stemmatology: an R package for the computer-assisted analysis of textual traditions. In: Proceedings of the Second Workshop on Corpus-Based Research in the Humanities CRH-2, 25–26 January 2018, Vienna, pp. 65–74 (2018). https://www.oeaw.ac.at/fileadmin/subsites/academia ecorpora/PDF/CRH2.pdf

CEDG. Conferencia española de decanos de Geología, La falta de cultura geológica: un problema social (2018). http://saladeprensa.usal.es/node/111750. Last Accessed 07 Feb 2018

[10] Barriuso et al. (2017), Bajo et al. (2017), González-Briones et al. (2018).

Clio. Clio Visualizing History (2018). https://www.cliohistory.org/. Last Accessed 30 Jan 2018
Coulson, S., Pagán Cánovas, C.: Understanding timelines: conceptual metaphor and conceptual integration. Cogn. Semiot. **5**(1–2), 198–219 (2014)
Deutscher, G.: The Unfolding of Language: An Evolutionary Tour of Mankind's Greatest Invention. Henry Holt, New York (2005)
Foreman, N., Boyd-Davis, M., Korallo, L., Chappell, W.: Can virtual environments enhance the learning of historical chronology? Instr. Sci. **36**(2), 155–173 (2008)
Froger, D.J.: La critique des textes et son automatisation. Deinot, Paris (1968)
Genevois, S.: Culture numérique et citoyenneté mondiale: quels enjeux pour l'École. Tréma, 1–14 (2013). http://journals.openedition.org/trema/3036
González-Briones, A., Villarrubia, G., De Paz, J.F., Corchado, J.M.: A multi-agent system for the classification of gender and age from images. Comput. Vis. Image Underst. (2018, in Press)
Gray, R., Atkinson, Q.: Language-tree divergence times support the Anatolian Theory of Indo-European origin. Nature **426**, 435–438 (2003)
International Chronostratigraphic Chart. International Commission on Stratigraphy 2017/02 (2017). http://www.stratigraphy.org/index.php/ics-chart-timescale
Junker, M.-O.: Metaphors we live by: the terminology of linguistic theory. Nat. Lang. Linguist. Theory **10**, 141–145 (1992)
Junker, M.-O.: Les métaphores de la théorie linguistique. Présence franco-phone **45**, 75–84 (1994)
Kuzoian, A., Atkinson, Q.: How Language Spread across Europe & Asia, Insider Science Channel (2015). https://www.youtube.com/watch?v=KdQwalCPNAs. Last Accessed 07 Feb 2018
Landheer, R.: Le rôle de la métaphorisation dans le métalangage linguistique. Verbum **24**, 283–294 (2002). http://www.info-metaphore.com/index.html. (N. A. Flaux, ed.)
Llamas-Pombo, E.: La machine du temps. Méditation sur la métaphore dans le discours de l'Histoire de la langue. In: Losada Goya, J.M. (ed.) Temps: texte et image. Universidad Complutense, Madrid, pp. 169–178 (2011)
Martínez, J.: Atlas histórico (2013). http://epilatlashis.blogspot.com.es/. Last Accessed 07 Feb 2018
MIT. Massachusetts Institute of Technology, Visualizing Cultures (2015). https://ocw.mit.edu/ans7870/21f/21f.027/home/index.html
Pietrobon, M.: Indo-European Migrations & Language Geographical Evolution (2016). https://www.youtube.com/watch?v=6YGZ-AAYYZg. Last Accessed 07 Feb 2018
Poole, E.: L'analyse stemmatique des textes documentaires. In: Colloque International CNRS, Paris, 29–31 mars 1978. Éditions du CNRS, Paris, pp. 151–161 (1979)
Rosenberg, D., Grafton, A.T.: Cartographies of Time. A History of the Timeline. Princeton Architectural Press, New York (2010)
Scotese, C.R.: Paleomap Project (2003). http://www.scotese.com/. Last Accessed 30 Jan 2018
Scotese, C.R.: Plate Tectonics and Paleography (2015). https://www.youtube.com/watch?v=uLahVJNnoZ4. Last Accessed 30 Jan 2018
Spector, J.: Foundations of Educational Technology: Integrative Approaches and Interdisciplinary Perspectives. 2nd edn. Rouledge, New York, London (2016)
Staley, D.H.: Computers, Visualization and History. How New Technology Will Transform Our Understanding of the Past. Rouledge, London, New York (2015)

Analysis of Quality of Live in the Deaf Community from a Blog

María Cruz Sánchez-Gómez(✉), Ariana Castillo Alzuguren, and María Victoria Martín-Cilleros

University of Salamanca, Paseo Canalejas 169, 37008 Salamanca, Spain
mcsago@usal.es

Abstract. According to the report presented by the National Institute of Statistics for 2008, there are 1,064,100 people with hearing impairment in Spain. This group is daily challenged by communication difficulties and isolation, which severely affects their participation in society leading them to perceive their quality of life as unsatisfactory. This community has traditionally gathered in associations as spaces where they are able to interact and find social support, thus creating a deaf culture. However, the advance of new technologies offers new and better prospects, since it bridges the difficulties posed by traditional space-time communication. The purpose of this study is to analyse the concerns, expressed in a deaf blog, of people with different degrees of hearing impairment from the perspective of quality of life. The sign language discourses contained in 44 video messages expressed in a Spanish public blog created by deaf people are analysed using qualitative methods. The results show general dissatisfaction with their quality of life, social exclusion, emotional discomfort and self-determination being the central topics around which the analysed discourses revolve.

Keywords: Deaf community · Quality of life · Blog · Qualitative research

1 Introduction

Hearing impairment or complete deafness has an impact on virtually all areas of life of those who suffer such conditions. Previous studies reveal that the quality of life of this community is significantly lower than that of the average population, both in terms of physical and psychological standards, and also because they suffer from higher degrees of mental distress [1]. Communication barriers can lead to isolation and derive in severe psychological problems, characterized by higher scores in different psychopathological symptoms, when compared to hearing people [1, 2], including depression, anxiety, insomnia, stress, paranoid ideation and interpersonal sensitivity [3]. Those who feel socially isolated have a lower capacity to cushion the impact of stressors, which increases the risk for harmful effects on their health [4].

The literature offers a number of definitions of quality of life, but they all share the common concept of multidimensionality due to it being influenced by contextual and personal factors and their interaction [5]. The World Health Organization [6] defines it as a concept that reflects an individual's perception of his/her position in the cultural

S. Rodríguez et al. (Eds.): DCAI 2018, AISC 801, pp. 322–329, 2019.
https://doi.org/10.1007/978-3-319-99608-0_37

context and that of the system of values he/she lives in, and puts it in relation to personal goals, expectations, standards and concerns. Like other definitions, such as the one contributed by Schalock [7], it refers to perceived life satisfaction. These author defines it as a desired condition of personal wellbeing built on eight dimensions, which are emotional wellbeing, interpersonal relationships, material wellbeing, personal development, physical wellbeing, self-determination, social inclusion and rights, all of which can be measured using qualitative or quantitative approaches, based on certain indicators, such as satisfaction, self-concept and absence of stress or negative feelings for emotional well-being, social networks, having clearly identified friends, family relationships for interpersonal relationships, etc.

Associations of people with diverse degrees of hearing loss have traditionally provided an ideal space and context for the development of the deaf community and its identity [8]. Nevertheless, the unrelenting advance of new technologies provides new and broader avenues for this community, overcoming the space-time limitations inherent to signed communication [9].

The deaf community association movement is “moving” to virtual spaces where information, dissemination and mobilization activities are farther reaching and occur at breakneck speed [1]. So far, research, mostly qualitative, reveals that these web surfers, whether deaf or hearing, have been using new technologies for quite a long time [10, 11]. It is widely known that the development of mobile telephony and instant messaging has noticeably promoted communication in this community; however, the Internet is the clearest breaker of these communicative and geographical barriers, and, consequently, young people visit face-to-face associations lesser and lesser, leaving them to those in older age groups [8]. However, even if the Internet is threatening the survival of onsite associations of deaf people, it is not a menace to their community, since it works as an extension of it, broadening it and providing greater cohesion, as well as being a significant source of information [8, 12, 13]. It is somehow as if they were freed from their hearing impairment to reassert their condition of deaf people, resulting in a phenomenon of “internationalization of the deaf world” [8, 11, 12, 14]. However, those who have not developed adequate language skills, due to poor quality education, will still find it difficult to manage in these online spaces, which, to a greater or lesser extent, require reading and writing skills [11, 12].

Social networks, message boards and blogs are the most common media for communication used by deaf people [10–12, 15]. Vlogs are a version of traditional blogs, characterized by their use of multimedia material such as audio and video [16]. Over the past few years, the deaf community has created its own individual or group “deaf blogs” using sign language as the communication system. These deaf blogs broach a wide variety of topics, such as culture, education, history, dissemination, leisure, travel, news, deaf community, etc. [13, 16]. The procedure (Fig. 1) consists of the user, known as blogger, uploading a video in sign language to the blog so that the rest of members can watch it and make asynchronous comments (in traditional text or as a sign language video), meaning that an immediate response is not required and they can, therefore, take as long as they need to make their contributions [15].

While the literature available focuses on examining the written contents of different blogs and message boards [12, 13], there is still a major stumbling block to be overcome by these studies: the general challenges deaf people face when it comes to

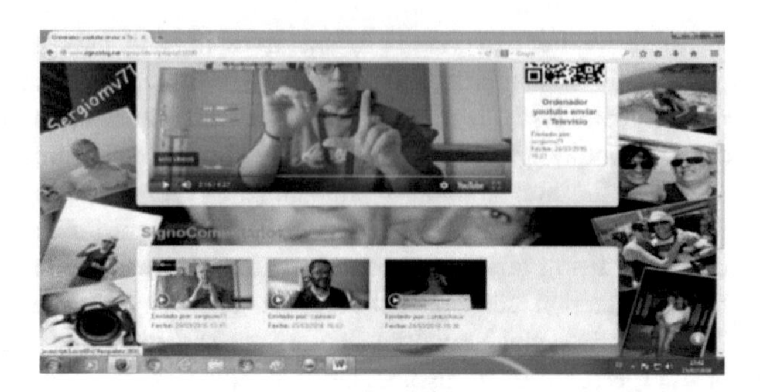

Fig. 1. Screenshot of the working of a deaf blog

expressing themselves through writing [11, 12]. The literature devoted to virtual deaf communities is only beginning to be produced; some studies have already been conducted in different countries, but there is no research on this phenomenon in Spain, so that, using a qualitative design, the purpose of this study is to approach such field drawing on the richness and diversity of virtual environments, where the deaf themselves are in charge of the information flow [17]. In an attempt to go one step further, and so that the data could be as natural as possible, only those transmitted in sign language were gathered, this being the language preferred by the deaf community to communicate, since they find it easier to make richer, longer and more elaborate discourses than through writing [9, 11, 14, 17].

Accordingly, the purpose is to use the information delivered in a deaf blog to analyse the deaf community's concerns from a quality-of-life perspective.

2 Design

The study relies on a qualitative approach, drawing upon the free availability of multimedia documents on the Internet and taking advantage of their corresponding spontaneity, ease and relevance [9].

Without aiming to generalise the experience to the rest of the population, as would a more quantitative method, the purpose of using a research strategy based on a qualitative approach is to obtain reliable knowledge through a precise and meticulous description of the phenomenon in its context, thus ensuring the objectiveness of the findings [17, 18].

The technique used consisted of peripheral and asynchronous observation: peripheral because of the researcher's passive role in the data collection process, to minimize possible interference with participants; and asynchronous because of the characteristics of the deaf blog itself and because this technique makes it possible to collect videos already loaded in the past to carry out an ex post facto analysis [9].

2.1 Participants and Data Collection

The realm is made up of the videoposts, in this case sign language posts, that are embedded on a Spanish blog intended for the deaf community: www.signoblog.net (Fig. 2). The platform, which has been working uninterruptedly for over five years, is community-based and open to all Internet users. It is the only one in Spain with these characteristics. Hearing impairment is not a requirement to be able to access it but, although the content is fully public, only registered users can post sign language messages and answer with comments and sign language comments. While communication on deaf blogs is asynchronous, there have been tentative attempts in recent years to hold live discussions using sign language posts and commentaries via streaming.

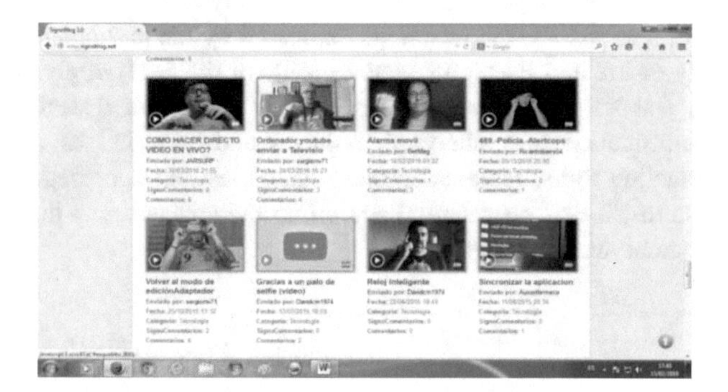

Fig. 2. Screenshot of the deaf blog's interface

To gather information about the platform, it was necessary to get in touch with the creator and coordinator of the website, a deaf person, through a sign language video produced by one of the researchers and sent to the former's email address. There are 284 deaf bloggers, 48 of whom are either not Spanish nationals, or are living abroad. Since this study is aimed at gaining insight on the situation in a given context, namely Spain, the latter were removed, leaving a total of 236 deaf bloggers.

The platform's administrator reports that more than 10,235 sign language posts and 5,316 sign language comments have been uploaded from its creation in 2011 until the beginning of 2016, and that there are 34 different thematic categories into which sign language posts are classified according to their content.

2.2 Data Transformation

Once selected, the data were interpreted into Spanish. After this stage, the data transformation and reduction process was carried out; the word was established as the unit of analysis, that is, the minimum meaningful fraction, and the sentence was determined as the coding unit [17]. A categories system is drawn from the discourse of the sign language posts for its correct coding. To ensure methodological rigour and overcome any weaknesses that may arise in the study, two types of interjudge assessment tests were carried out. The first one, related to interpreting the sign language videos, was carried out by a currently employed sign language interpreter, with a professional career of over 15 years, and with the deaf administrator of the website. After both of them had stated their agreement to collaborate in the study, they were emailed the relevant information. Nevertheless, the researcher received no response, so that, in agreement with the director, the decision was made to delete seven videoposts whose interpretation posed difficulties, due to their being in Catalan sign language (CSL) and the low-quality recording of some of them. Finally, after a rigorous and systematic process, 44 participants (sign language posts) remained. The second type of interjudge test required the participation of expert teachers in disability matters of the Disability Research master's degree and the Language and Communication Disorders master's degree of the University of Salamanca.

The deaf posts were analysed through content analysis using NVIVO11 software. Following the studied theory, the meanings expressed by the participants were used to build the quality-of-life and E-empowerment profile of the deaf people participating in the deaf blog, elaborated taking into consideration previous research by different authors. The information was analysed, reduced and arranged as a concept map including information units, meta-categories, categories and subcategories. After the experts' suggestions, it was restructured again into the concept map that serves as the basis for the content analysis (see Fig. 3).

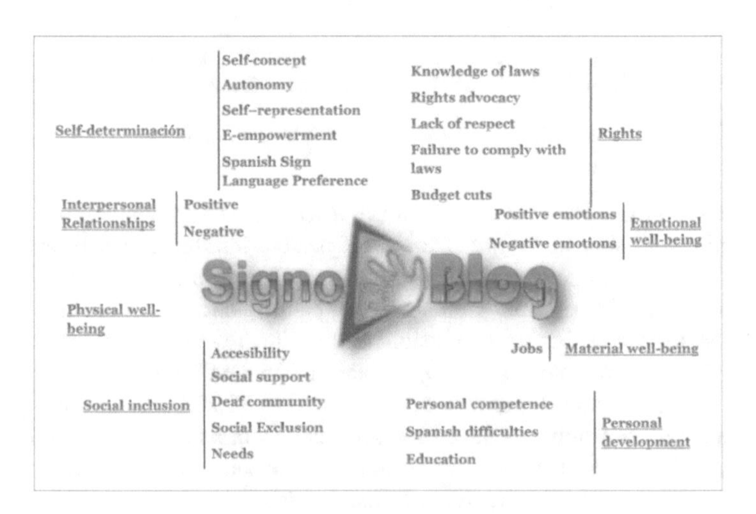

Fig. 3. Concept map

3 Results

First of all, a brief description of the words that appear most often in the videoposts will be provided, followed by the presentation and description of the content related to quality of life that is shown in the concept map included above.

Most of the feelings and emotions expressed on the videos have to do with social inclusion. In terms of emotions related to social inclusion, 75.7% of the cases express negative feelings, amongst which the most frequent are, in this order: indignation, offense and disappointment. Amongst the positive emotions expressed, the most noteworthy are thankfulness and curiosity.

The category of self-determination is among the three most referred to in the videoposts. It appears as related to offence and suffering. If the analysis is taken further to what is between the "self-determination" and "rights" dimensions (Fig. 4), a number of relationships can be detected: "illiterate" is perceived as a disrespectful designation, and there is evidence of a struggle for adopting the communication system they think best, in this case, Spanish sign language over oral language (OL).

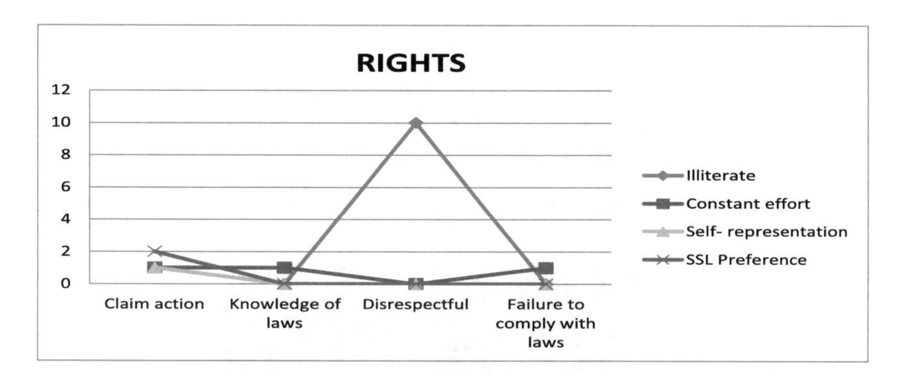

Fig. 4. Relationship between pronouncements on rights and self-determination

Figure 5 shows how the analysed comments defend that accessibility should be a protected right, be it with adaptations or through the use of sign language. The associative movement and community association are perceived as important tools to claim their rights, the aim being for the government and other institutions to gain awareness of the communication barriers that deaf people face.

Social inclusion accounts for almost 50% of the analysed content. People with different degrees of hearing impairment speak negatively about it, complaining about social exclusion and alleging discrimination and the existence of communication barriers (Fig. 6). What they most demand in terms of needs are accessibility through the figure of the interpreter, the use of Spanish sign language and the raising of awareness among both citizens and institutions.

4 Discussion and Conclusions

Schalock's quality of life model not only provides extremely valuable information to learn the social reality of the community studied, but it is also affords guidance for creating professional programmes that can better meet deaf people's needs, offering support wherever it is required in order to improve their wellbeing [7].

Social inclusion is their main concern, especially communication barriers, since lack of information and isolation lead to psychological and emotional distress [1, 3]: they are forced to put up with feelings of indignation, offence and suffering through all stages of their lives. Therefore, breaking down such barriers, without exception, should

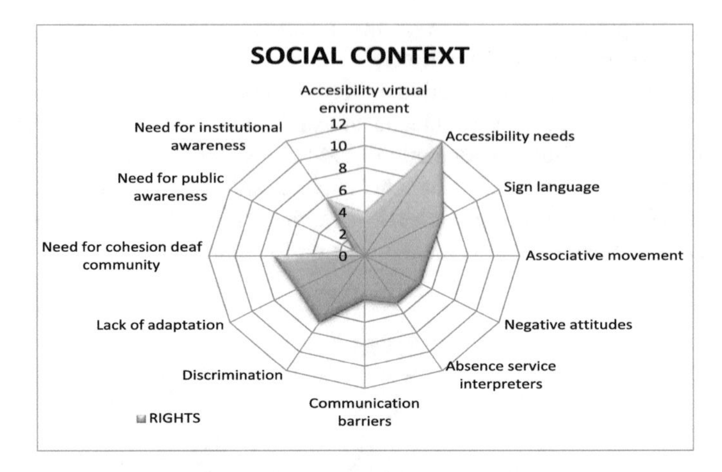

Fig. 5. Relationship between pronouncements about rights and social inclusion

Fig. 6. Sign language bloggers expressing the "discrimination" sign

be the backbone of any programme launched for the inclusion of people with hearing impairments: sign language must be given its full place in society, and communication, either through spoken or sign language, should be provided with the different support means that may be required, since these people, like any other citizen, are entitled to full communication. In this regard, education is a particularly sensitive area since, as explained by Echeita [19], educational exclusion is usually a prelude to social exclusion [19,5]. Decades ago, Oliver Sacks contemplates in his work that "It is not just language, but thought, that must be introduced. Otherwise, the child will remain helplessly in a concrete and perceptual world" [20]. Indeed, statements have been gathered explaining how the learning and use of sign language has favoured reasoning and the development of reading and writing. This could encourage education professionals towards bilingual (sign language and Spanish) curricular proposals to ensure successful literacy for deaf children.

Inaccessibility to information is a violation of their rights, and because of this, they express strong feelings of discomfort on the web. New technologies work as a new way for group empowerment, offering possibilities of mutual support and strengthening bonds, while encouraging the community's members to defend their rights in an organized and joint way [12, 13]. Several studies claim that the Internet promotes group identity processes by increasing people's network of contacts [8, 10, 11].

References

1. Fellinger, J., Holzinger, D., Dobner, U., Gerich, J., Lehner, R., Lenz, G., et al.: Mental distress and quality of life in a deaf population. Soc. Psychiatry Psychiatr. Epidemiol. **9**(40), 737–742 (2005)
2. Tambs, K.: Moderate effects of hearing loss on mental health and subjective well-being: Results from the Nord-Trondelag hearing loss study. Psychomatic Med. **66**, 776–782 (2004)
3. Fellinger, J., Holzinger, D., Pollard, R.: Mental Health of Deaf People. Lancet London, England (2012)
4. Gerich, J., Fellinger, J.: Effects of social networks on the quality of life in an elder and middle-aged deaf community sample. J. Deaf. Stud. Deaf. Educ. **17**(1), 102–115 (2012)
5. Martín-Cilleros, M.V., Sánchez-Gómez, M.C.: Qualitative analysis of topics related to the quality of life of people with disabilities. Cienc. Saude Coletiva **21**(8), 2365–2374 (2016)
6. World Health Organization Quality of Life Assessment. La gente y la salud Qué calidad de vida? Foro Mundial de la Salud 17, pp. 385–387 (1996)
7. Schalock, R.L.: The concept of quality of life: what we know and do not know. J. Intellect. Disabil. Res. **48**(3), 203–216 (2004)
8. Valentine, G., Skelton, T.: Changing spaces: the role of the internet in shaping deaf geographies. Soc. Cult. Geogr. **9**(5), 469–485 (2008)
9. López, D.M., Sánchez-Gómez, M.C.: Técnicas de recolección de datos en entornos virtuales más usadas en la investigación cualitativa. Revista de Investigación Educativa **24**(1), 205–222 (2006)
10. Blom, H., Marschark, M., Vervloed, M.P., Knoors, H.: Finding friends online: online activities by deaf students and their well-being. PLoS ONE **9**(2), e88351–e88351 (2014)
11. Power, D., Power, M.R.: Communication and culture: signing deaf people online in Europe. Technol. Disabil. **21**(4), 127 (2009)
12. Hamill, A.C., Stein, C.H.: Culture and empowerment in the Deaf community: an analysis of Internet weblogs. J. Community Appl. Soc. Psychol. **21**(5), 388–406 (2011)
13. Shoham, S., Heber, M.: Characteristics of a virtual community for individuals who are d/Deaf and hard of hearing. Am. Ann. Deaf **157**(3), 251–263 (2012)
14. Carr, D.: Constructing disability in online worlds: conceptualising disability in online research. Lond. Rev. Educ. **8**(1), 51–61 (2010)
15. Barak, A., Sadovsky, Y.: Internet use and personal empowerment of hearing-impaired adolescents. Comput. Hum. Behav. **24**(5), 1802–1815 (2008)
16. Álvarez Sánchez, P.: Reconsideraciones sobre el diseño inicial de un corpus digital de lengua de signos española. Language Windowing through Corpora. Visualización del lenguaje a través de corpus, pp. 49–60 (2010)
17. Anguera, T.: Evaluación de programas desde la metodología cualitativa. Acción Psicológica **5**(2), 87–101 (2008)
18. Sánchez-Gómez, M.C.: La calidad en la investigación cualitativa. In: García-Valcárcel, A. (ed.) Investigación y tecnologías de la información y comunicación al servicio de la innovación educativa, pp. 241–266, Aquilafuente, Salamanca (2013)
19. Echeita, G.: Inclusión y exclusión educativa. De nuevo "voz y quebranto". Revista Iberoamericana sobre Calidad, Eficacia y Cambio en Educación **11**(2), 99–118 (2013)
20. Sacks, O.: Seeing voices: A Journey into the World of the Deaf. Harper Perennial, New York (1990)

An On-Going Framework for Easily Experimenting with Deep Learning Models for Bioimaging Analysis

Manuel García(✉), César Domínguez, Jónathan Heras, Eloy Mata, and Vico Pascual

Dpto. de Matemáticas y Computación, Universidad de La Rioja, Edificio CCT, Madre de Dios, 53, 26006 Logroño, La Rioja, Spain
{manuel.garcia,cesar.dominguez,jonathan.heras, eloy.mata,vico.pascual}@unirioja.es

Abstract. Due to the broad use of deep learning methods in Bioimaging, it seems convenient to create a framework that facilitates the task of analysing different models and selecting the best one to solve each particular problem. In this work-in-progress, we are developing a Python framework to deal with such a task in the case of bioimage classification. Namely, the purpose of the framework is to automate and facilitate the process of choosing the best combination of feature extractors (obtained from transfer learning and other techniques), and classification models. The features and models to test are fixed by a simple configuration file to facilitate the use of the framework by non-expert users. The best model is automatically selected through a statistical study, and then it can be employed to predict the category of new images.

Keywords: Deep learning · Machine learning · Parallelization Bioimaging · Image processing

1 Problem Statement

Nowadays, there exists an increment in the use of deep learning methods in a wide variety of computer vision applications. These methods can be used together with more traditional techniques in image processing. There is not a silver bullet solution to solve the problem. In addition, these methods can be successfully applied in areas such as Bioimaging [1–3] where the specialists could not have the knowledge to use them. Therefore, these techniques are beyond the possibilities of some of the researches. The aim of this work consists in trying to facilitate the use of these methods to non-expert users in object bioimage classification problems.

This work was partially supported by Ministerio de Economía, Industria y Competitividad [MTM2014-54151-P, MTM2017-88804-P], and Agencia de Desarrollo Económico de La Rioja [ADER-2017-I-IDD-00018].

S. Rodríguez et al. (Eds.): DCAI 2018, AISC 801, pp. 330–333, 2019.
https://doi.org/10.1007/978-3-319-99608-0_39

2 Related Work

This project can be framed in the context of democratising artificial intelligence and more particularly deep learning techniques. In this line, we can highlight two projects. The first one [4] compares five different model classifiers with your dataset; and the model with the best accuracy is used to carry out the next comparisons. The user provides the dataset as an input and get automatically created a high-performance ready-to-use pattern recognition system. The second one [5] helps users to create their own classifiers from a two-dimensional visual interface. This project guides to non-experts users to create their own models without any help of a learning algorithm. This kind of users can get good classifiers using that simple user interface. Expert users can also use the tool to produce the models. They can build models with better results than those generated by a learning algorithm and thus these models be more intelligible. Both of them compare the clasiffiers with the dataset that you want to study and give us the classifier with the best accuracy.

Other related framework has just launched by Google. This platform is called Google Cloud AutoML [6]. It helps not-expert users to generate automatic learning models in a simple way. At the moment it is in a testing phase and you need permission to have access.

All these projects have some drawbacks. For instance, they focus only on deep learning models. They are also not extensible in the sense that they do not allow user to add new models.

3 Proposal

We propose a framework that tries to solve some of the previous drawbacks. Figure 1 includes the workflow of the intended framework proposal. It tries to facilitate the task of defining models for bioimage classification using classical image processing and deep learning techniques (using transfer learning with neural networks in Keras such as VGG16, VGG19, Resnet, Inception, etc.). The workflow can be divided in four different phases: dataset and configuration file, feature extraction, statistical analysis and training the best model and predicting. We are going to detail such as phases in the following subsections.

3.1 Dataset and Configuration File

The input of the framework is a dataset of images and a configuration file. The dataset consists of folders of images where each folder represents a category. In addition, a configuration file containing the parameters of the program is necessary. We have to specify the path of the dataset, a list of feature extractors and a list of classification models.

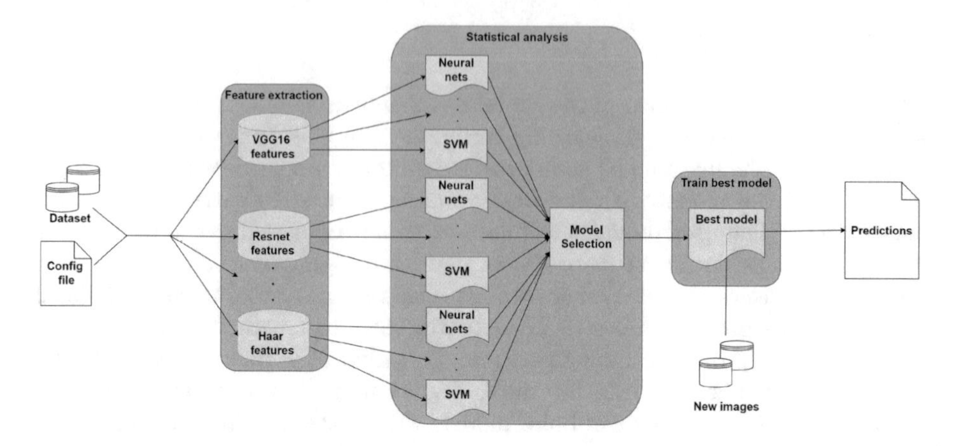

Fig. 1. Workflow of the framework proposal

3.2 Feature Extraction

In this step, the framework collects the features of each image of the dataset using both transfer learning and classical computer vision techniques. For each feature extractor provided in the configuration file, the framework goes through each image and extracts the features. Afterwards, this information is saved in an h5py file for later use.

3.3 Statistical Analysis

Based on the features gathered in the previous step, the framework carries out a statistical study [7], to compare several classification models and obtain the best one. This study measures the accuracy of each combination of a feature extractor and a classification model using cross-validation. Namely, for each extractor, the best classification model is obtained. Subsequently, those models are compared to identify the best combination of feature extractor and classification model.

3.4 Training the Best Model and Predicting

In the last step, the best featureExtractor-modelClassifier combination is trained. The framework trains the model with all the dataset images.

Once the model is trained, it is exported to a file that can be used later to predict the classes of the images that are passed to the model. The whole process is automatically performed by the framework.

4 Preliminary Results

At this moment the framework is under development using Python. It is used OpenCV [8] to extract classification features using classical image techniques and Keras [9] to extract classification features using deep learning. Besides, a combination of these features extracted with several classification models is performed; and the best one is selected using a statistical analysis based on the accuracy of the models. In the near future, a full test on a real dataset will be carried out to study the suitability of the proposal. Since the techniques including in these processes are very time consuming, we could apply parallelization techniques.

5 Reflections

As we can see, this project tries to bring artificial intelligence closer to those users without notions about what a deep learning model is or how to configure the models correctly to obtain good predictions. The aim consists in trying to bring this powerful technology to most people so they can easily make use of its benefits.

It is expected that the best model will be incorporated automatically into a web application that will guide those non-expert users in their bioimage classification problems.

Apart from the above, it is planned to look for new deep learning libraries to increase the number of models. In addition, a new option will be added to the configuration file to choose different measures (beyond accuracy) to calculate the performance of the models so that we can improve the user's options and select the model that best suits the problem in each case.

References

1. Ronneberger, O., Fischer, P., Brox, T.: U-Net: Convolutional Networks for Biomedical Image Segmentation (2015). http://arxiv.org/abs/1505.04597
2. Akkus, Z., Galimzianova, A., Hoogi, A., Rubin, D.L., Erickson, B.J.: Deep learning for brain MRI segmentation: state of the art and future directions. J. Digit. Imaging **30**(4), 449–459 (2017)
3. Nauman, M., Ur Rehman, H., Politano, G., Benso, A.: Beyond Homology Transfer: Deep Learning for Automated Annotation of Proteins. bioRxiv (2017). https://www.biorxiv.org/content/early/2017/07/25/168120
4. Reif, M., Shafait, F., Goldstein, M., Breuel, T., Dengel, A.: Automatic classifier selection for non-experts. Pattern Anal. Appl. **17**(1), 83–96 (2012)
5. Ware, M., Frank, E., Holmes, G., Hall, M., Witten, I.H.: Interactive machine learning: letting users build classifiers. Int. J. Hum. Comput. Stud. **55**(3), 281–292 (2001)
6. Google: Google cloud automl (2018). https://cloud.google.com/automl/
7. Sheskin, D.: Handbook of Parametric and Nonparametric Statistical Procedures. CRC Press, Boca Raton (2011)
8. Intel: Opencv (2016). https://opencv.org/
9. Chollet, F., et al.: Keras (2017). https://keras.io/

Towards Integrating ImageJ with Deep Biomedical Models

Adrián Inés[(✉)], César Domínguez, Jónathan Heras, Eloy Mata, and Vico Pascual

Dpto. Matemáticas y Computación, Universidad de La Rioja, Edificio CCT, Madre de Dios, 53, 26006 Logroño, La Rioja, Spain
{adrian.ines,cesar.dominguez,jonathan.heras,eloy.mata, vico.pascual}@unirioja.es

Abstract. Nowadays, deep learning techniques are playing an important role in different areas due to the fast increase in both computer processing capacity and availability of large amount of data. Their applications are diverse in the field of bioimage analysis, e.g. for classifying and segmenting microscopy images, for automating the localization of proteins or for automating brain MRI segmentation. Our goal in this project consists in including these deep learning techniques in ImageJ – one of the most used image processing programs in this research community. To do this, we want to develop an ImageJ plugin from which to use the models and functionalities of the main deep learning frameworks (such as Caffe, Keras or Tensorflow). It would be feasible to test the suitability of different models to the problem that is being studied at each moment, avoiding the problems of interoperability among different frameworks. As a first step, we will define an API that allows the invocation of deep models for object classification from several frameworks; and, subsequently, we will develop an ImageJ plugin to make the use of such an API easier.

Keywords: Bioimage · Deep learning · Image processing · ImageJ Interoperability · Object classification

1 Problem Statement

Currently, computer vision and artificial intelligence have many different applications in diverse fields: security, biology or medicine for example. Also, deep learning is playing an important role due to both the fast increase in computer processing capacity and the availability of large amount of data. Specifically, in the field of bioimage analysis, several applications of these deep learning techniques are being applied for classifying and segmenting microscopy images [1], for

This work was partially supported by Ministerio de Economía, Industria y Competitividad [MTM2014-54151-P, MTM2017-88804-P], and Agencia de Desarrollo Económico de La Rioja [ADER-2017-I-IDD-00018].

S. Rodríguez et al. (Eds.): DCAI 2018, AISC 801, pp. 334–338, 2019.
https://doi.org/10.1007/978-3-319-99608-0_40

automating the localization of proteins [2] or for automating the segmentation of brain MRI [3] for example.

Applying these deep learning techniques is not easy for a non-expert user. It being required to have basic knowledge of programming and to know different concepts of deep learning. Because of this, our idea is to make the use of these deep learning techniques easier integrating them in a processing program, in this case ImageJ [4] - one of the most used image processing programs in the biological research community.

In addition, there are a lot of different deep learning frameworks (such as Caffe [5], Keras [6], TensorFlow [7] or DeepLearning4J [8]) with different aims and functionalities. So, we consider to improve the interoperability among these frameworks by means of integrating all of them in ImageJ.

2 Related Work

As we have indicated, the goal of this project is to improve the interoperability between different deep learning frameworks and image processing programs - like ImageJ for example. There are some other projects that try to connect ImageJ with deep learning or machine learning frameworks. One of these projects, *ImageJ - TensorFlow* [9], connects ImageJ with TensorFlow. Specifically, this project translates between ImageJ images and TensorFlow tensors. Our approach to integrate deep learning techniques with ImageJ is different. We want to include an intermediate layer between the ImageJ program and the deep learning frameworks. In this layer we want to abstract the characteristics of the different frameworks to classify an image. In addition, we do not focus only on one deep learning framework, our aim is to integrate the main deep learning frameworks and to define a way to include any other framework in an easy way.

Due to the fact that there are a lot of deep learning frameworks with different characteristics, there exist several active projects with the aim to transform a model from one deep learning framework into other. The project *Deep learning model converter* [10] deals with this problem, allowing to use models in different frameworks easily. Other project in the same line is the *Open Neural Network Exchange (ONNX)* [11]. There exists a lot of different frameworks with their own specification and proprietary representation of the graph of the models. In this project, a common representation of the computation graph is provided and different converters have been created. Our approach tries to be different. We want to provide the users several models and frameworks. In this way, the users can select the most suitable option for their problems.

3 Proposal

Our goal in this project consists in providing ImageJ with different deep learning techniques. To do this, we want to develop an ImageJ plugin that allows the users to have the models and functionalities of the main deep learning frameworks (such as Caffe, Keras or Tensorflow) available. In this way, it will be possible to test the suitability of different models to the problem that is studied at each moment, avoiding the problems of interoperability between different frameworks. As a first step, we will define an API that allows the invocation of deep models for object classification from several frameworks, as we can see in Fig. 1. This API, will gather the main frameworks of deep learning and will offer an abstraction layer that allows us to use these frameworks through a set of methods. Our idea is to integrate in this API the main deep learning frameworks for object classification as well as including other functionalities such as object detection, object localization or image segmentation. Also, we want to allow the users to integrate other frameworks in the API or integrate their own models in a simple way.

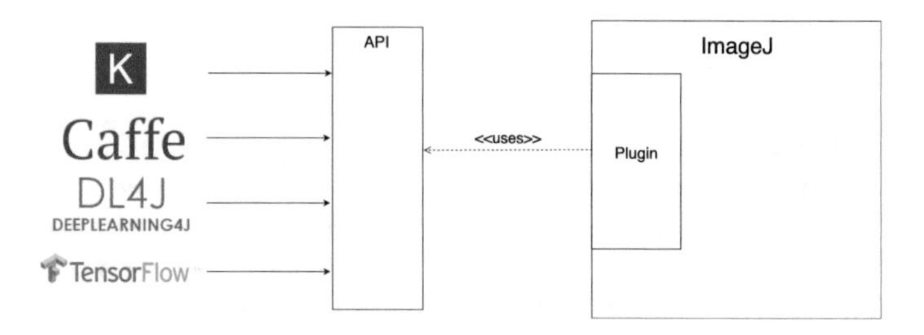

Fig. 1. Structure of the project.

In addition, due to ImageJ is one of the most used image processing programs in research community, specifically in biology, we want to develop an ImageJ plugin to make the use of such an API easier. The idea is that this plugin allows the user to select both the framework and the model used to classify an image and then communicates with the API to obtain the correct class.

4 Preliminary Results

As a first step of our project, we have defined an API that integrates the main deep learning frameworks like Keras, Caffe and DeepLearning4J. This API, has one method called *predict* that allows the user to classify an image with one model included in one of these frameworks. Also, we have developed an ImageJ plugin that allows us to use this API in a simple way. An example of using this plugin is included in Fig. 2, for classification of bioimages. In this case we have classified melanomas images.

With this plugin, the users only have to select the framework they want to use. Then the plugin shows the models included in the API for this framework. Once, the users choose the model, the plugin communicates with the API and shows the class that the image belongs.

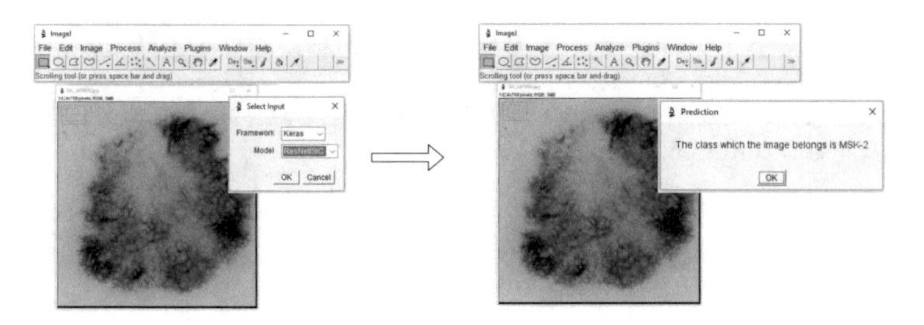

Fig. 2. Interface of the ImageJ plugin for select the input and show the output.

5 Reflections

Today, deep learning techniques are very useful and powerful to solve different image processing problems like object classification, object detection, object localization or image segmentation. In addition, the use of these techniques is becoming commonplace. Because of our experience we have detected that the use of these techniques is not easy for a common user, this project has been considered to make the use of these techniques easier.

With this project we try to provide deep learning techniques to the image processing users, abstracting deep learning techniques details. The users will only have to select the framework and the model that best suits their problem.

We think that this work will allow users to improve the results of some researches by applying the latest deep learning techniques in a simple way.

References

1. Ronneberger, O., Fischer, P., Brox, T.: U-Net: convolutional networks for biomedical image segmentation (2015). http://arxiv.org/abs/1505.04597
2. Nauman, M., Ur Rehman, H., Politano, G., Benso, A.: Beyond homology transfer: deep learning for automated annotation of proteins. bioRxiv (2017). https://www.biorxiv.org/content/early/2017/07/25/168120
3. Akkus, Z., Galimzianova, A., Hoogi, A., Rubin, D.L., Erickson, B.J.: Deep learning for brain MRI segmentation: state of the art and future directions. J. Digit. Imaging **30**(4), 449–459 (2017)
4. Rueden, C.T., Eliceiri, K.W.: The ImageJ ecosystem: an open and extensible platform for biomedical image analysis. Microsc. Microanal. **23**(S1), 226–227 (2017)
5. Jia, Y., Shelhamer, E., Donahue, J., Karayev, S., Long, J., Girshick, R., Guadarrama, S., Darrell, T.: Caffe: convolutional architecture for fast feature embedding. arXiv preprint arXiv:1408.5093 (2014)
6. Chollet, F., et al.: Keras (2015). https://github.com/keras-team/keras
7. Abadi, M., et al.: TensorFlow: large-scale machine learning on heterogeneous distributed systems (2016). http://arxiv.org/abs/1603.04467
8. Eclipse Deeplearning4J Development Team: Deeplearning4j: Open-source, Distributed Deep Learning for the JVM (2018). https://deeplearning4j.org/
9. Google Brain team: ImageJ/TensorFlow integration library plugin (2018). https://imagej.net/TensorFlow

10. Yuan, S.: Deep learning model convertors (2018). https://github.com/ysh329/deep-learning-model-convertor
11. Microsoft, Facebook open source & AWS: ONNX: Open Neural Network Exchange (2018). http://onnx.ai/

Integrating Biological Context into the Analysis of Gene Expression Data

Cindy Perscheid(✉) and Matthias Uflacker

Hasso Plattner Institute, University of Potsdam, Potsdam, Germany
cindy.perscheid@hpi.de

Abstract. High-throughput RNA sequencing produces large gene expression datasets whose analysis leads to a better understanding of diseases like cancer. The nature of RNA-Seq data poses challenges to its analysis in terms of its high dimensionality, noise, and complexity of the underlying biological processes. Researchers apply traditional machine learning approaches, e.g. hierarchical clustering, to analyze this data. Until it comes to validation of the results, the analysis is based on the provided data only and completely misses the biological context.

However, gene expression data follows particular patterns – the underlying biological processes. In our research, we aim to integrate the available biological knowledge earlier in the analysis process. We want to adapt state-of-the-art data mining algorithms to consider the biological context in their computations and deliver meaningful results for researchers.

Keywords: Gene expression · Machine learning · Feature selection
Association rule mining · Biclustering · Knowledge bases

1 Introduction and Research Statement

Gene expression is the process by which information from specific sections of the DNA, i.e. genes, is synthesized to functional products like proteins, which are catalyzing the metabolic processes in our cells. Data on gene expression thus provides valuable insights for researchers on the molecular processes of a cell, e.g. gene functions and interactions, which lead to a better understanding of diseases like cancer. RNA-Sequencing (RNA-Seq) delivers a complete snapshot of gene expression in a cell, with a single experiment containing expression levels of tens of thousands of genes from multiple hundred samples [15].

Currently, researchers apply traditional machine learning approaches in the analysis to identify patterns in RNA-Seq data, e.g. k-means or hierarchical clustering to derive disease-specific expression profiles [2]. However, gene expression data follows particular predefined patterns namely, the underlying biological processes. Curated and most recent knowledge on such processes is provided by a growing number of online repositories, e.g. about gene-gene/gene-disease interactions, gene functions and co-expressions, signaling pathways, or recent scientific

S. Rodríguez et al. (Eds.): DCAI 2018, AISC 801, pp. 339–343, 2019.
https://doi.org/10.1007/978-3-319-99608-0_41

publications [4–7,13,18,25]. Even meta knowledge bases integrating information from various well recognized resources are emerging [9,12,20,26,27]. In the current analysis of gene expression data, however, information from external resources is only used at the end of the process to validate the biological significance of the results. Literature also suggests to include the biological context earlier in the analysis pipeline to achieve better results [3,19]. With this said, we formulate our *Research Hypotheses* as follows:

> We can improve the analysis of gene expression data by integrating external knowledge on biological processes early into the process. This will have the following two positive effects for the overall analysis:
> (a) Reduction of computational complexity of the applied algorithms and thus the overall process, and
> (b) Delivery of interpretable and biologically meaningful results.

We consider (a) and (b) to come into effect for the following reasons: First, biological processes are highly complex. Identifying them properly in the data is a major issue due to overlapping, weak or irrelevant signals and requires the application of complex learning models. Currently, researchers need to find a balance between the computational complexity, i.e. runtime duration, of their analysis and result set accuracy. By integrating existing knowledge on biological processes from external resources, we can support algorithms in identifying these processes in the data without applying highly complex models or computations.

Second, gene expression data covers the complete gene activity of a cell. However, only a small part of the signals covered is relevant for the posed research question. In addition, some signals can correlate by accident in the data but do not participate in the same biological processes at all. Algorithms that are completely data-driven are not able to differentiate those signals and can thus deliver results that have no biological meaning or rather a result set where the biologically relevant results are hidden in the overall result set. Separating these biologically relevant results from the rest is a major issue in result set validation.

2 Research Plan

We concentrate on integrating the automatic retrieval of biological context information from public knowledge bases into the complete analysis process of gene expression data. Our work is structured according to the analysis steps of dimensionality reduction, pattern identification, and validation.

2.1 Dimensionality Reduction

The curse of dimensionality of gene expression data increases computational complexity of analysis algorithms and thus requires measures for dimensionality reduction [16,23]. In addition, gene expression data contains a lot of noise because not all genes are relevant for the specific research question and thus could interfere with pattern identification. Our research concentrates on feature selection approaches. These are classified into filter, wrapper, embedded, hybrid, and integrative approaches [3,10,17]. While more simple filter approaches are often applied for feasibility reasons, the computationally more complex wrapper, embedded, and hybrid approaches deliver more accurate results [14]. Integrative feature selection approaches have emerged in recent years in bioinformatics [1,8,21,22]. They incorporate domain knowledge from external knowledge bases and thus enable a selection of biologically relevant features [24]. In our research, we combine selected external data sources with traditional statistical feature selection approaches, hence reducing computational complexity and runtime, e.g. by preselecting feature sets from external data sources.

2.2 Pattern Identification

For analyzing gene expression data, researchers apply traditional clustering and classification techniques, e.g. kMeans, hierarchical clustering, self-organizing maps (SOMs), or statistical approaches [10,11]. However, biological processes are highly complex and genes can be involved in multiple cell processes. Hence, the data should be viewed from different perspectives and alternative pattern identification approaches should be applied. In our research, we aim at employing two approaches for pattern identification and enhancing them with biological domain knowledge: Biclustering and Association Rule Mining (ARM) [24]. Both of them bear challenges with regards to their application to gene expression data, which we concentrate on in our research. On the one hand, the computational complexity of such algorithms is higher than for traditional single clusterings, and scalability to large data sets is an important issue. On the other hand, both biclustering and ARM output large numbers of biclusters and association rules, respectively. These must be filtered according to their biological relevance in order to be beneficial for the user.

2.3 Validation

Assessing the biological correctness and validity of analysis results is a crucial task, as researchers draw the actual conclusions for the clinical interpretation at this point. Nowadays, this involves judgment by domain experts with long-term experience and/or manual screening of the many existing knowledge bases [4–7,13,18,25]. While we are not able to supersede domain experts, we can improve the manual screening process. In our research, we concentrate on the automatic assessment of analysis results by combining external knowledge bases by means of conventional information retrieval techniques. We aim at defining a "reliability score" that combines the information from the different knowledge bases in a meaningful way, e.g. by giving stronger weights to information from manually curated knowledge bases. This task poses challenges in dealing with the strong data heterogeneity both in terms of data formats, i.e. structured, semi-structured, and unstructured, but also in terms of data identifiers, e.g. a gene has separate identifiers across knowledge bases.

3 Preliminary Results

Figure 1 shows a first study on existing state-of-the-art data-driven feature selection approaches (IG - Information Gain, VB - Row Variance, ReliefF, SVM-RFE) and compares them with our own hybrid approach using DisGenet and KEGG. We analyzed eight TCGA data sets (GBM, HNSC, KIRC, LAML, LUAD, SARC, THCA, UCEC) consisting of 3,189 samples and 55,572 genes in total. Our study reveals that all approaches deliver good clustering accuracy above 90% at 10 genes, which even increases to more than 95% at 16 genes. Embedded approaches applying more complex algorithms, e.g. SVM-RFE, perform best. However, applying an integrative approach by combining a simple statistical measure, e.g. Information Gain, with an external knowledge base, e.g. DisGeNET, delivers equally good results as machine learning approaches - at one third of runtime.

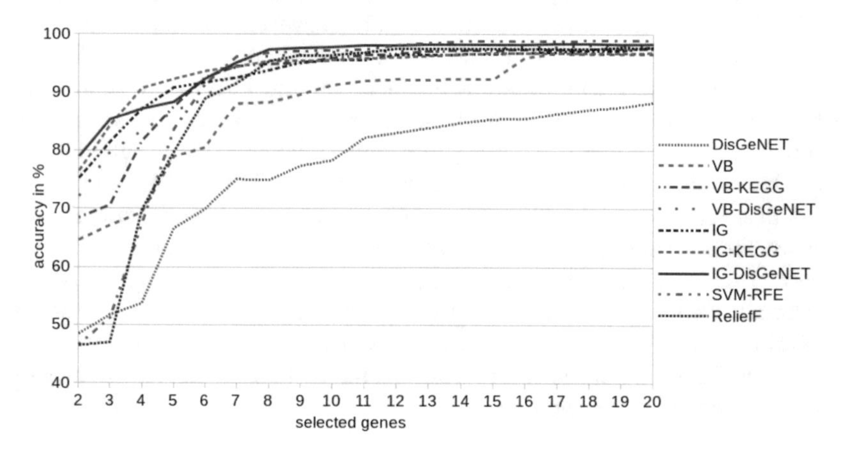

Fig. 1. Accuracy comparison of data-driven, knowledge-driven, and combined feature selection approaches. Our combined approach performs equally to computationally complex ones.

References

1. Acharya, S., Saha, S., Nikhil, N.: Unsupervised gene selection using biological knowledge: application in sample clustering. BMC Bioinform. **18**(1), 513 (2017)
2. Babu, M.M.: Introduction to microarray data analysis. Comput. Genomics Theory Appl. **17**(6), 225–249 (2004)
3. Bellazzi, R., Zupan, B.: Towards knowledge-based gene expression data mining. J. Biomed. Inform. **40**(6), 787–802 (2007)
4. Gene Ontology Consortium: expansion of the gene ontology knowledgebase and resources. Nucleic Acids Res. **45**(D1), D331–D338 (2016)
5. UniProt Consortium: UniProt: the universal protein knowledgebase. Nucleic Acids Res. **45**(D1), D158–D169 (2016)

6. NCBI Resource Coordinators: database resources of the national center for biotechnology information. Nucleic Acids Res. **44**(Database issue), D7 (2016)
7. van Dam, S., Craig, T., de Magalhaes, J.P.: GeneFriends: a human RNA-seq-based gene and transcript co-expression database. Nucleic Acids Res. **43**(D1), D1124–D1132 (2014)
8. Fang, O.H., et al.: An integrative gene selection with association analysis for microarray data classification. Intell. Data Anal. **18**(4), 739–758 (2014)
9. Farkas, I.J., Szántó-Várnagy, Á., Korcsmáros, T.: Linking proteins to signaling pathways for experiment design and evaluation. PloS ONE **7**(4), e36202 (2012)
10. Inza, I., et al.: Filter versus wrapper gene selection approaches in DNA microarray domains. Artif. Intell. Med. **31**(2), 91–103 (2004)
11. Jiang, D., Tang, C., Zhang, A.: Cluster analysis for gene expression data: a survey. IEEE Trans. Knowl. Data Eng. (TKDE) **16**(11), 1370–1386 (2004)
12. Kamburov, A., et al.: ConsensusPathDB: toward a more complete picture of cell biology. Nucleic Acids Res. **39**(suppl_1), D712–D717 (2010)
13. Kanehisa, M., Goto, S.: KEGG: kyoto encyclopedia of genes and genomes. Nucleic Acids Res. **28**(1), 27–30 (2000)
14. Kohavi, R., John, G.H.: Wrappers for feature subset selection. Artif. Intell. **97**(1–2), 273–324 (1997)
15. Kukurba, K.R., Montgomery, S.B.: RNA sequencing and analysis. Cold Spring Harbor Protocols **2015**(11) (2015). pdb–top084970
16. Lazar, C., et al.: A survey on filter techniques for feature selection in gene expression microarray analysis. IEEE/ACM Trans. Comput. Biol. Bioinform. (TCBB) **9**(4), 1106–1119 (2012)
17. Mahajan, S., Singh, S., et al.: Review on feature selection approaches using gene expression data. Imperial J. Interdisc. Res. **2**(3) (2016)
18. Okamura, Y., et al.: COXPRESdb in 2015: coexpression database for animal species by dna-microarray and rnaseq-based expression data with multiple quality assessment systems. Nucleic Acids Res. **43**(D1), D82–D86 (2014)
19. Pasquier, N., et al.: Mining gene expression data using domain knowledge. Int. J. Softw. Inform. (IJSI) **2**(2), 215–231 (2008)
20. Piñero, J., et al.: DisGeNET: a discovery platform for the dynamical exploration of human diseases and their genes. Database **2015** (2015)
21. Qi, J., Tang, J.: Integrating gene ontology into discriminative powers of genes for feature selection in microarray data. In: SAC, pp. 430–434. ACM (2007)
22. Raghu, V.K., et al.: Integrated theory-and data-driven feature selection in gene expression data analysis. In: ICDE, pp. 1525–1532. IEEE (2017)
23. Saeys, Y., Inza, I., Larrañaga, P.: A review of feature selection techniques in bioinformatics. Bioinformatics **23**(19), 2507–2517 (2007)
24. Shao, B., Conrad, T.: Epithelial-mesenchymal transition regulatory network-based feature selection in lung cancer prognosis prediction. In: IWBBIO, pp. 135–146. Springer (2016)
25. Stark, C., et al.: BioGRID: a general repository for interaction datasets. Nucleic Acids Res. **34**(suppl_1), D535–D539 (2006)
26. Szklarczyk, D., et al.: STRING v10: protein-protein interaction networks, integrated over the tree of life. Nucleic Acids Res. **43**(D1), D447–D452 (2014)
27. Uhlén, M., et al.: Tissue-based map of the human proteome. Science **347**(6220), 1260419 (2015)

Applying Scale-Invariant Dynamics to Improve Consensus Achievement of Agents in Motion

Ilja Rausch(✉), Yara Khaluf, and Pieter Simoens

Ghent University - imec, Technologiepark-Zwijnaarde 15, 90520 Ghent, Belgium
ilja.rausch@ugent.be

Abstract. In order to efficiently execute tasks, autonomous collective systems are required to rapidly reach accurate consensus, no matter how the group is distributed over the environment. Finding consensus in a group of agents that are in motion is a particularly great challenge, especially at larger scales and extensive environments. Nevertheless, numerous collective systems in nature reach consensus independently of scale, i.e. they are scale-free or scale-invariant. Inspired by these natural phenomena, the aim of our work is to improve consensus achievement in artificial systems by finding fundamental links between individual decision-making and scale-free collective behavior. For model validation we use physics-based simulations as well as a swarm robotic testbed.

Keywords: Consensus achievement · Scale invariance
Swarm robotics

1 Introduction

Many modern technologies and scientific efforts in the field of artificial intelligence demand problem-solving approaches in which distributed collective systems (CS) play the key role. In a CS, a large group of agents autonomously takes decisions based only on information from nearby agents or the environment. With proper agent decision models, a collective response to external stimuli may appear spontaneously, with no centralized state or control. Reaching consensus is a type of collective response that is important for autonomous systems to be able to perform tasks. Often, consensus on the best option must be found as fast as possible, no matter how the group is distributed over the environment. The latter condition imposes special attention when agents are in motion, however its realization is vital because motion enables the agents to explore, interact and shape their environment. The challenge is particularly apparent at wider scales, when information must be propagated over many agents or when a large group of agents is distributed over a large environment so there are fewer interactions.

Our ambition is to improve consensus achievement (CA) by finding a fundamental link between individual computational models for decision-making (DM)

S. Rodríguez et al. (Eds.): DCAI 2018, AISC 801, pp. 344–348, 2019.
https://doi.org/10.1007/978-3-319-99608-0_42

and scale-free CB. The underlying hypothesis is that it is possible to overcome the problems of scale in dynamic multi-agent systems by realizing scale-free CB. The latter is invariant to the rescaling of system parameters, such as the size of the space, the number of individuals or the frequency of interactions. This idea is inspired for instance by flocking starlings, where starlings from either edge of the flock may influence each other no matter the size of the flock [3]. Hence, their collective response is not characterized by any scale of the group, i.e. it is scale-free. Optimization of CS response to external stimuli—that require modifications of the global system behavior—is one of the most promising advantages that may result from our research. In particular, the speed of the collective response can be increased by optimizing information propagation. Two promising approaches to the latter challenge are (i) the use of scale-free network topologies for agent communication which, as mentioned above, leads to a denser node connectivity [4], and (ii) defining the agent behavior such that the correlation length between any two agents is longer than the group diameter and is scale-invariant with respect to the group size, as it has been observed in bird flocks [3].

2 Related Work

There are various unresolved challenges to CA that can be identified in literature. One particular challenge is that the individual models in the studies on the emergence of collective behavior (CB) mostly rely on two assumptions: (i) the number of individuals per unit of surface is high enough to ensure frequent interactions [7], (ii) the system is well-mixed [9]. Although often useful for mathematical modeling, these assumptions can be rude simplifications because in real CSs the spatial distribution of agents is non-uniform and the system may fail (i.e., no collective consensus is achieved) if the swarm density is below a critical value [8]. Hence, when agents are moving in large environments, the system may split into disjoint subgroups or the CB may dissociate, as it was observed, among others, in locust swarms [2]. Therefore, to keep the traits of CSs, such as scalability, robustness and flexibility [1], it is important that a scale-independent (or scale-free) behavior emerges spontaneously from the individual DM. Several models for realizing scale-free dynamics via self-organization can be found in the domains of natural systems (e.g. starling flocks, speciation, ant search patterns) or statistical physics (e.g. directed percolation, forest fire model, preferential attachment) [5]. Two main mechanisms that were identified in literature to generate system dynamics with scale-free or scale-invariant properties are (i) dynamics of scale-free networks and (ii) dynamics that emerge from self-organized criticality [7]. Scale-free networks constitute complex topologies, with a few nodes (called hubs) having an exceptionally high connectivity degree. This high connectivity allows the nodes to share their information over only a few hops, on average [4]. Self-organized criticality denotes the notion of the system to stabilize near a critical point: the point at which the system is on the border of two phases, such as ordered vs. disordered electron spin ensembles, chaos vs. organization or free vs. congested flow [6]. A consequence of criticality is that some features (e.g. correlation length) become scale-free [7].

3 Proposal

Our research hypothesis is that integrated scale-free dynamics can help maintain a coherent CB and response, even under highly dynamic and unpredictable conditions. However, for scale-invariance (SI) to emerge, the appropriate individual DM processes need to be well understood. Therefore, our approach consists of the following three objectives:

1. Investigate the conditions and DM processes that promote spontaneous emergence of SI in artificial systems. A fundamental question in any study on collective behavior is the link between the microscopic behavioral model of the individuals and the macroscopic dynamics. Often SI is observed under very specific conditions at the macroscopic level and it is not well investigated how these observations link to the microscopic behavior. For instance, while it is known that scale-invariant features emerge in statistical physics when the system is near a phase transition, how these phases should be defined in an artificial CS in motion and what individual DM processes may push that system towards a phase transition, are two open research questions. Therefore, determining the appropriate mechanism that governs the system and is responsible for its SI is an important first step.

2. Integrate scale-free dynamics into collective systems facing CA problems in static environments. The second objective is to exploit scale-invariant propagation of information in order to improve, both, the speed and accuracy of multi-agent CA—without losing on flexibility, scalability and robustness. One possible mechanism is by transferring the agent interaction network into a scale-free network, whose key property is the occurrence of hubs, i.e. nodes with an exceedingly above-average node degree. However, the common generating mechanisms (such as preferential attachment) cannot be readily adopted with agents in motion whose node degree is constantly changing. Instead, we will need to conceive hubs in a novel way: as best-informed and most influential units. This can be achieved by a problem-specific feedback mechanisms such as self-promotion or election of an agent.

3. Apply SI to collective systems facing CA problems in dynamic environments. The third objective is to validate if scale-invariant dynamics may improve the flexibility, adaptivity and responsiveness of collective behavior in environments with multiple stimuli which vary in quality/intensity. To address this objective, we need to reformulate the microscopic behavioral models of the agents such that the spontaneous and autonomous emergence of scale-invariance becomes adaptable to environmental changes. This stage advances the insights gained from the last two objectives towards more sophisticated case studies.

4 Preliminary Results

Our research hypothesis is grounded on the insights from an exploratory study that we have already conducted[1]. The results were submitted to the Journal of

[1] https://drive.google.com/open?id=1RdsZt2j7Qogcmm31oJOBjb1XiNpgpmRO.

Swarm Intelligence and are currently under revision. The goal of this work was to verify, by means of physics-based simulations, the spontaneous occurrence of scale-invariant features emerging from existing individual behaviors that were not designed with SI in mind. As a case study, we implemented the task of collective foraging in swarm robotics. In this task, a robot swarm is required to maximize the number of food items retrieved from the foraging area into the nest, with respect to the number of exploration attempts. Hence, at the individual level, every robot needs to decide frequently between resting and foraging. For certain parameter configurations of the agent's behavioral model, we found scale-invariant features, such as the number of interactions or the time agents spent in resting state. We also revealed a positive correlation between the occurrence of SI and the swarm performance. Additionally, in the same study, a scale-free interaction network has been imposed onto the robot communication to examine its influence on collective response to external stimuli. In comparison to agents having only local interactions with a random number of neighbors, the scale-free communication led to faster but more fluctuating swarm response.

5 Reflections

SI has been observed in many natural CS where it is known to have beneficial impact. Inspired by these studies, our aim is to explore and exploit the emergence of SI in artificial CS. Our models will be validated by simulations and on a swarm robotic testbed to close the reality gap. On the long run, our work can help understand the role of SI in complex CS as well as improve the development of large robotic systems for rescue missions or (extra)terrestrial exploration.

References

1. Brambilla, M., Ferrante, E., Birattari, M., Dorigo, M.: Swarm robotics: a review from the swarm engineering perspective. Swarm Intell. **7**(1), 1–41 (2013)
2. Buhl, J., Sumpter, D.J.T., Couzin, I.D., Hale, J.J., Despland, E., Miller, E.R., Simpson, S.J.: From disorder to order in marching locusts. Science **312**(5778), 1402–1406 (2006)
3. Cavagna, A., Cimarelli, A., Giardina, I., Parisi, G., Santagati, R., Stefanini, F., Viale, M.: Scale-free correlations in starling flocks. Proc. Nat. Acad. Sci. **107**(26), 11865–11870 (2010)
4. Cohen, R., Havlin, S.: Scale-free networks are ultrasmall. Phys. Rev. Lett. **90**, 058701 (2003)
5. Herrero, C.P.: Ising model in clustered scale-free networks. Phys. Rev. E **91**, 052812 (2015)
6. Hu, M.B., Wang, W.X., Jiang, R., Wu, Q.S., Wu, Y.H.: Phase transition and hysteresis in scale-free network traffic. Phys. Rev. E **75**, 036102 (2007)
7. Khaluf, Y., Ferrante, E., Simoens, P., Huepe, C.: Scale invariance in natural and artificial collective systems: a review. J. Royal Soc. Interface **14**(136), 20170662 (2017)

8. Khaluf, Y., Pinciroli, C., Valentini, G., Hamann, H.: The impact of agent density on scalability in collective systems: noise-induced versus majority-based bistability. Swarm Intell. **11**(2), 155–179 (2017)
9. Reina, A., Valentini, G., Fernndez-Oto, C., Dorigo, M., Trianni, V.: A design pattern for decentralised decision making. PLOS ONE **10**(10), e0140950 (2015)

Towards to Secure an IoT Adaptive Environment System

Pedro Oliveira[1,2], Tiago Pedrosa[2], Paulo Novais[1(✉)], and Paulo Matos[2]

[1] Department of Informatics, Algoritmi Centre/University of Minho, Braga, Portugal
pjon@di.uminho.pt

[2] Department of Informatics and Communications, Institute Polytechnic of Bragança, Bragança, Portugal

Abstract. This paper, deals with the actual problem of secure an IoT adaptive system, namely using secure techniques to secure a Smart Environment System, and the privacy of their users. On a new era of interaction between persons and physical spaces, users want that those spaces smartly adapt to their preferences in a transparent way. This work wants to promote a balanced solution between the need of personal information and the user's privacy expectations. We propose a solution based on requiring the minimal information possible, together with techniques to anonymize and disassociate the preferences from the users.

Keywords: Adaptive-system · AmI · Security · Privacy · IoT

1 Introduction

Systems that deal with personal data always bring privacy and security issues. And also the balance of these issues, with the need that persons have in interact with spaces in a transparent way and that those spaces smartly adapt to their preferences.

That said, in this project, is proposed a solution to overcome these issues, and don't compromise the balance between security and personal comfort.

In addition to the physiological conditions mentioned above, there are two critical/essential dimensions, these are the space (user location) and time. In the case of the space can be as an example, the differences between the preferences of a personal, professional, recreational or other environment. Contextualize the user location is essential to optimize the conditions of comfort and contribute to the performance and effectiveness of the solution.

Figure 1, shows the scenario of an environment where it intends to develop this work. Explaining this figure, it can be seen the user who through its different devices (smartphone, wearable, and other compatible) communicates with the system, and for that can be used different technologies, like Near Field Communication (NFC) [1], Bluetooth Low Energy (BLE) [2] and Wi-Fi Direct [3]. Next, the system performs communication with the Cloud, to validate the information. And then the system will perform the management of the different components in the environment (climatization systems, security systems, other smart systems).

S. Rodríguez et al. (Eds.): DCAI 2018, AISC 801, pp. 349–352, 2019.
https://doi.org/10.1007/978-3-319-99608-0_43

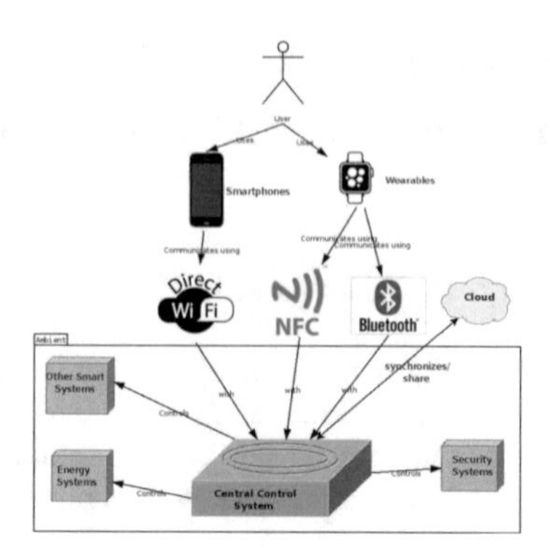

Fig. 1. Problem statement

2 Materials and Methods

2.1 Security and Privacy

The technological revolution that is felt, particularly in behavioral analysis fields, IoT or big data, brings significant new challenges, including those related to the type of user information that can be collected, and the knowledge that can be obtained derived from the compilation of this information. Although not necessarily existing the user's authority to make this kind of information collection.

This IoT revolution has yet clearly identified problems. In particular, the privacy and security of user data. Foreseeing the dissemination of intelligent spaces, of which the user can, and want to take advantage of the interaction between systems, and the consequent sharing of personal data, this is a theme that needs resolution in a short-term [4].

Obviously at this point there will be the requirements for concessions and commitments on the part of the user.

Because it is understood that the solution will include the user's authority in relation to the autonomous information sharing with the system and what is the information that it believes that it should be shared, block, and in concrete with which systems.

The IoT increases the risk of personal privacy, and the confidentiality and integrity of data in organizations. Some IoT applications for consumers, particularly those related to health and wellness, which store sensitive personal information, and that consumers may not want to share.

Normally, it is known that on latter case, there must be commitments by the user, so it has access to all system capabilities. But it is up to each one, set his own commitment threshold, between privacy and comfort that it intends to have by using the system, in the environments that it uses.

3 Results

All attack vectors identified, are minimized using the techniques identified in this section. Consequently increasing significantly the degree of complexity so that an attacker can gain access to useful information, or can link this information to take advantage, or even affect the system users.

As mentioned in Sect. 2.1, one of the priorities of this project will be ensure privacy and data confidentiality. To achieve this goal, several mechanisms are designed in order to minimize the possible attack vectors.

- **Use of universally unique identifier (UUID)**, to identify the user. The user identification process, it is necessary in this context to relate him to his preferences card, and is performed by generating a UUID in the first use of the system application. This unique UUID is randomly created by the application and is then validated their nonexistence on the server, if the validation is positive, the UUID is associated with the user's preference card. If the validation is not positive is generated a new random UUID and the validation process will be held again.
  The application will allow the user to export the UUID created for his personal email or store it locally in another way, so that if it wants to use more than one device in the system or switch the device, this can be done. Note that only the randomly generated UUID and the preferences card are transmitted to the server, so there was no possibility of identifying the user [5].
- **Servers and component isolation**, two physical servers will be used. In order to separate the logic and data layer (database). Therefore possible individual attacks, which enable access to the servers do not compromise the entire system.
- **Data encryption**, all data transmitted between the servers are encrypted using SHA-256 hash mechanisms, which introduces an extra security layer in protection of the data stored in the system [6].
- **Server hardening**, both servers only allow access through key mechanisms. Communication processes will be based on HTTPS and TLS [6]. Other most common mechanisms for server hardening will be applied [7].
- **Communication with the local system**, as explained above, the communication between the user's smartphone and the local system, can be performed using BLE, NFC or Wi-Fi Direct. These technologies have their own security mechanisms implemented at the stack level, which will be properly configured in the local systems to maximize security. However, the UUID is also ciphered with the server A public key before is sent to the Local System. With that we can guarantee that the UUID can't be captured in clear, and is only known by the smartphone application and the Server A.
- **Mask of GPS coordinates**, even though the user anonymization process is covered, for greater safety and because issues related to the user's location storage are critical. It is planned to convert the GPS coordinates of the local systems. This process is achieved by associating the coordinates to a randomly and periodically change of the Local System ID. Therefore the user's location

information from a system, will be stored using the UUID of the user and the system ID, which due to its periodic change will not relate any information that can allow to achieve the user tracking.

The implementation of these mechanisms allows to significantly reduce the attack vectors identified. At the user data privacy level, this work allowed to don't store any user information. So even if the data is compromised, will not be possible identify the user, or make any relationship with that information.

4 Discussion and Conclusions

Currently IoT systems are in a big security risk. Especially because the developers, are not worried enough about the safety of such systems. However, with the growing trend of such systems and is integration in our everyday lives, this concern will have to increase as they start to appear isolated cases which have harmed the users, both financially and in their safety and welfare. The proposed security architecture, to one of these IoT systems, wants to avoid any of the presented risks, to the users of this system.

Acknowledgements. This work has been supported by COMPETE: POCI-01-0145-FEDER-007043 and FCT Fundação para a Ciência e Tecnologia within the Project Scope: UID/CEC/00319/2013.

References

1. Want, R.: Near field communication. IEEE Pervasive Comput. **3**, 4–7 (2011)
2. Bluetooth, S.: Bluetooth core specification version 4.0. In: Specification of the Bluetooth System (2010)
3. Camps-Mur, D., Garcia-Saavedra, A., Serrano, P.: Device-to-device communications with wi-fi direct: overview and experimentation. IEEE Wirel. Commun. **20**(3), 96–104 (2013)
4. Khan, R., Khan, S.U., Zaheer, R., Khan, S.: Future internet: the internet of things architecture, possible applications and key challenges. In: 2012 10th International Conference on Frontiers of Information Technology (FIT), pp. 257–260. IEEE (2012)
5. Leach, P.J., Mealling, M., Salz, R.: A universally unique identifier (uuid) urn namespace (2005)
6. Dierks, T.: The transport layer security (tls) protocol version 1.2 (2008)
7. White, D., Rea, A.: Server hardening tactics for increased security. Working Paper, Technical Report (2003)

Matlab GUI Application for Moving Object Detection and Tracking

Beibei Cui(✉) and Jean-Charles Créput

Le2i FRE2005, CNRS, Arts et Mtiers, Univ. Bourgogne Franche-Comté, 90010 Belfort Cedex, France
beibei.cui@utbm.fr

Abstract. In this paper a novel tool for moving object detection and tracking is presented. The main contribution of the proposed application is the achievement of a simple and intuitive graphic interface during the extraction the silhouette of targets by means of a new algorithm. This proposed algorithm which combined frame difference method, background subtraction method, Laplace filter and Canny edge detector together can realize a way to achieve sparse detection fast. Some modular architecture in this Graphical User Interface has been developed in order to enhance the user's experience. The experiment was tested by using sequence images from the MULTIVITION dataset, and experimental results showed that our proposed method has more validity and flexibility to get the desired result than conventional algorithm.

Keywords: Graphical user interface · Frame difference
Background subtraction · Laplace filter · Canny detector

1 Introduction

Moving object detection is a course of image processing used to extract moving objects with relatively obvious background motion in the image sequences, usually according to the features of images such as the edge, color and texture. For intelligent monitoring, human behaviors tracking, autonomous vehicle tracking and many other applications, it is undoubtedly an indispensable research filed. Nowadays there are many researchers in-depth study under different research directions methods. Through reading and summarizing the literature, most researchers are more likely to focus on three categories: Rectangle box representation method [1]; Contour representation method [2] and Silhouette representation method [3]. This paper will focus on Silhouette representation method. Figure 1 shows the different results obtained by these three different representation methods in the same scenario.

GUI which means Graphical User Interfaces [4,5] is a convenient user toolkit among many of MATLAB products. It is indirectly related to computer programs, implicitly related equipments and instruments through the use of characterization symbols, interfaces and windows to achieve direct communication

S. Rodríguez et al. (Eds.): DCAI 2018, AISC 801, pp. 353–356, 2019.
https://doi.org/10.1007/978-3-319-99608-0_44

Fig. 1. (a) Rectangle box representation method; (b) Contour representation method; (c) Silhouette representation method.

between people and computers. The realization of the process is quick and easy. Thanks to its advantages, it can be manipulated easily by clicking on a specific command or task without typing in a command, so that, the use of GUI is getting more and more popular among many users nowadays.

The organization of this paper can be summarized as follows: We first present some related work about moving object detection and GUI, then the implementation of our proposed object tracking approach are presented. Thirdly, GUI operating interface design is demonstrated. Lastly the conclusion is presented.

2 Implementation

Currently, ones of the core algorithms used for tracking include frame difference method (FD), background subtraction method (BS), and optical flow method. Here, we are looking at the first two approaches since very adequate for very fast real-time treatments whereas optical flow has higher computation cost since related to a dense estimation.

Three-frame difference method is based on two-frame difference method by calculating difference between every two consecutive images. Among them, a common frame k_{th} taken from the sequence images is defined as the current frame. The current frame color image is m_k. After gray scale conversion it convert into gray scale image n_k, then through Laplace filter it is translated into f_k. The basic difference operation formula is defined as below:

$$D_k(x, y) = |f_k(x, y) - f_{k-1}(x, y)| \tag{1}$$

where the current frame image is f_k, the adjacent frame image is f_{k-1}, and D_k is image after difference operation between f_k and f_{k-1}. We define R_k as the binary translation of the difference image. If D_k(x, y)$> T$, R_k(x, y) belongs to object and is set to 1, on the contrary, it belongs to the background and is set to 0, where T is the threshold with a fixed value. Background subtraction method is similar to frame difference method. Whereas background subtraction is to carry out the differential operation between the current frame and background frame.

Combination of FD and BS with filters and edge detectors is a way to achieve sparse detection fast. This paper presents a tracking algorithm based on a new combination of FD and BS, using Canny edge detector and Laplace filter. Laplace filter occupies a leading role to sharpen the outlines and details. Canny edge

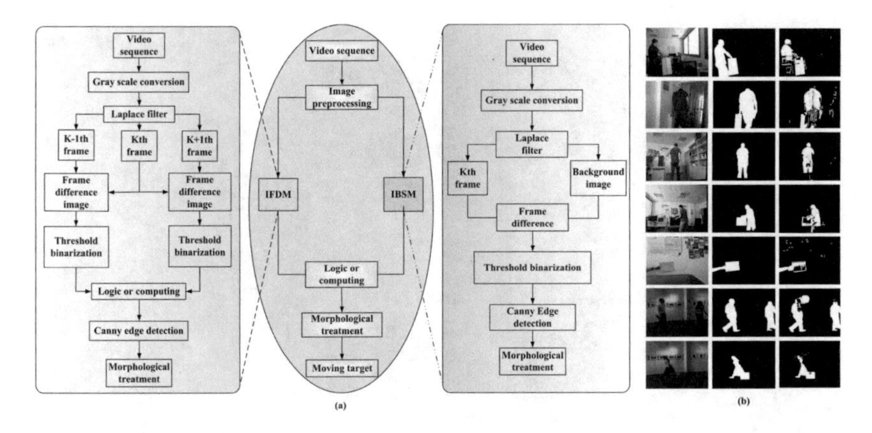

Fig. 2. (a) The flow chart of our improved algorithm. (b) Experiment results.

detector identifies and extracts edge information. Morphology processing is used to eliminate interfering items finally. The flow chart of our proposed approach is shown in Fig. 2(a).

This proposed method is tested by using seven sequence images from the MULTIVITION dataset [6]: ChairBox, Hallway, LabDoor, LCDScreen, Wall, Crossing and Suitcase. Results of object detection are demonstrated in Fig. 2(b). We display each sequence with three pictures, the first picture is the current frame image, the second one is the ground truth image and the third one is the extracted result which had been processed by our method.

3 Operating Interface Design

Matlab GUI interface contains several major components, they are: push button, slider, pop-up menu, listbox, table, axes, edit text and so on. As we getting started with GUI interface design, we need to set a panel with appropriate size at first, and then add the function button in this panel. When the entire layout page is well-designed, we need to set properties for GUI components and write the MATLAB language callbacks. To sum up, the entire process of GUI design can be summarized as the following steps: Open a new GUI in the layout editor; Set the GUI panel size; Add the components; Align the components; Compilation callbacks code. This application enables the user to generate tracking target straightforward drag-and-drop and context-menu operations. As you can see from Fig. 3, this GUI operating interface mainly contains basic image processing and experimental evaluation modules. A comparison of application between our proposed approach and conventional tool (frame different and background subtraction) were performed, which showed that our proposed method can achieve a higher degree of recognition. In the evaluation module, Accuracy, Precision, Recall and F-measure are adopted as evaluation criteria.

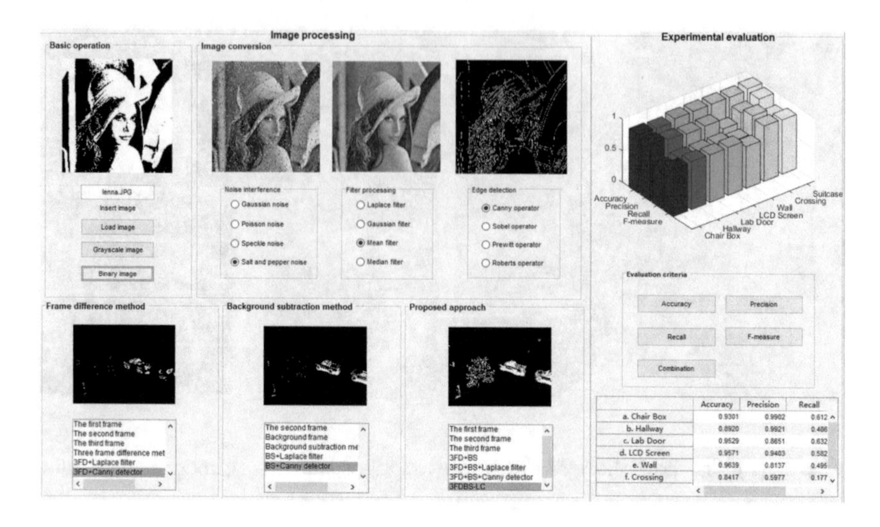

Fig. 3. The GUI display for object detection system.

4 Conclusion

In this paper, an improved object tracking algorithm is proposed. Several techniques for mainstream detection algorithm (frame difference method and background subtraction method) and basic operation (Laplace filter, Canny edge detector, morphological dilation and erosion operators) have been integrated, while still allowing all of their programmatic be assembled representative through Graphical User Interface to make the whole complex process simple and straightforward. For the future development, we would like to supplement more applications about image matching and develop multiple moving objects tracking by using rectangle box representation method.

References

1. Milan, A., Rezatofighi, S.H., Dick, A.R., Reid, I.D., Schindler, K.: Online multi-target tracking using recurrent neural networks. In: AAAI, pp. 4225–4232 (2017)
2. Shotton, J., Blake, A., Cipolla, R.: Contour-based learning for object detection. In: Tenth IEEE International Conference in Computer Vision, vol. 1, pp. 503–510 (2005)
3. Zhong, Z., Zhang, B., Lu, G., Zhao, Y., Xu, Y.: An adaptive background modeling method for foreground segmentation. IEEE Trans. Intell. Transp. Syst. **18**(5), 1109–1121 (2017)
4. Sulaiman, S., Hussain, A., Tahir, N.M., Samad, S.A.: Graphical user interface (GUI) development for object tracking system in video sequences. World Appl. Sci. J. **4**(2), 244–249 (2008)
5. Kumar, G.N.: Moving Object Detection and Tracking Using MATLAB GUI with ARDUINO
6. Fernandez-Sanchez, E.J., Rubio, L., Diaz, J., Ros, E.: Background subtraction model based on color and depth cues. Mach. Vis. Appl. **25**(5), 1211–1225 (2014)

Software Defined Networks and Data Distribution Service as Key Features for the 5G Control Plane

Alejandro Llorens-Carrodeguas(✉), Cristina Cervelló-Pastor, and Irian Leyva-Pupo

Deparment of Network Engineering, Universitat Politècnica de Catalunya (UPC), Esteve Terradas, 7, 08860 Castelldefels, Spain
{alejandro.llorens,cristina,irian.leyva}@entel.upc.edu

Abstract. The latency and flexible requirements of the 5G network are challenging telecommunication operators to have a more flexible, scalable, faster and programmable architecture. To solve this problem, this paper proposes a hybrid hierarchical set of Software Defined Networks (SDN) controllers as the control plane for 5G networks. The architecture is based on a federation of hierarchically-superior controllers which use Data Distribution Service (DDS) to communicate among each other and coordinate multiple sub-network controllers.

Keywords: SDN · 5G · Area controller · Global controller · DDS

1 Introduction

The necessity of offering services that satisfy 5G requirements such as latency, scalability and resilience entails numerous challenges. Software Defined Networks (SDN) offer the opportunity to develop, deploy and operate networks with new design principles. Using distributed SDN controllers instead of using a single controller ensures performance and scalability in large-scale networks. Although, the **problem** of sharing network information among controllers fulfilling the 5G latency requirement should be solved.

Several **related works** have been done to distribute SDN controllers. In [1], the authors propose a distributed event-based control plane called HyperFlow which provides scalability while keeping network control logically centralized. Similarly, Koponen et al. in [2] propose a distributed system which runs on a cluster of physical servers with multiple Onix instances. These approaches impose a consistent network-wide view in all the controllers and generate large control traffic. In addition, in [3,4] the authors propose two level of controllers formed by the domain and the area controllers. Both kinds of controllers use a scalable NoSQL database to store global host and switch information. In [5,6], the authors propose DDS to establish a communication between switches and SDN controllers. This work uses Data Distribution Service (DDS) to communicate the top controllers of our SDN hierarchical controllers of the 5G Control Plane.

S. Rodríguez et al. (Eds.): DCAI 2018, AISC 801, pp. 357–360, 2019.
https://doi.org/10.1007/978-3-319-99608-0_45

The **hypothesis** of our research is that the 5G latency requirement could be achieved using an SDN controllers hierarchy where the upper layer of controllers uses DDS to exchange network information and manages the bottom layer of controllers, while the latter is responsible for the User Plane Function (UPF) and flow management.

2 Proposal of the 5G Control Plane Architecture

Our proposal is based on the 5G 3GPP standardized architecture [7]. The design of this architecture consider two major requirements: scalability and latency. In order to satisfy both conditions, we propose an SDN controllers hierarchy formed by Area Controllers (ACs) in the bottom level and Global Controllers (GCs) in the top level. This sort of architecture guarantees scalability because we can define specific functions to each kind of controllers in order to determine their role within the overall network. Figure 1 illustrates the SDN controllers hierarchy and the relation among the different modules that integrate the GC and the AC.

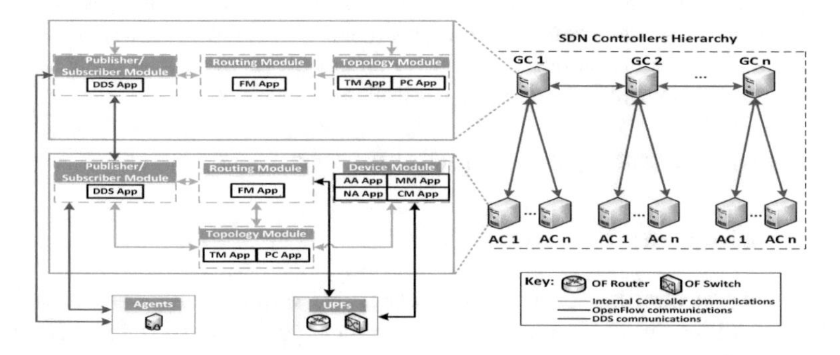

Fig. 1. Hierarchical architecture of federated SDN controllers.

The GCs communicate among each other using DDS to keep a consistent network state, establish inter-domain flow routes, etc. Meanwhile, the ACs update their GCs by means of this application when occurs a change in their topology or in the state of their assigned nodes. Similarly, the GCs inform their ACs when there is a change in the global topology that affects the communication among nodes under the control of different ACs. Not only do GCs and ACs use DDS to communicate among each other, but the Agent also uses it to communicate with GCs and ACs. In this way, both GCs and ACs are notified of a service request and they can reply it in a leisure time. The concept of Agent is introduced taking into account the Edge Computing paradigm, as it seeks to place processing and storage resources closer to end users. Hence, these Agents will be the nearest as possible to the users in order to communicate with them and collect network information. Through the use of DDS, the latency requirement could be satisfied since both kind of controllers would be able to execute commands in a short time frame.

2.1 Preliminary Results

The Publisher/Subscriber Module is the most important one in our proposal as it allows to send or receive network information to or from others GCs and ACs. For this reason, we have focused on developing this module to guarantee its performance in our proposal. To evaluate its behavior, we make some simulations with real topologies from the Internet Topology Zoo [8]. We run the Abilene network with two Opendaylight controllers that have installed this module. A different group of switches was assigned to each controller. Figure 2 shows the node tables of each controller before and after being federated.

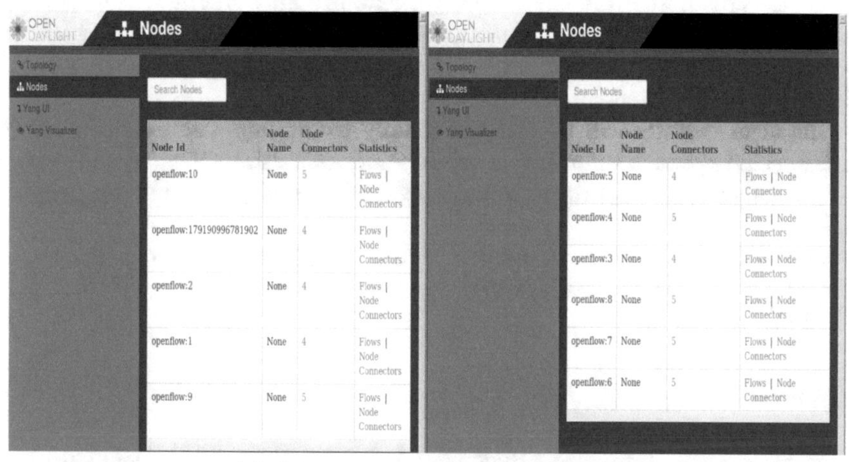

(a) Controllers before being federated

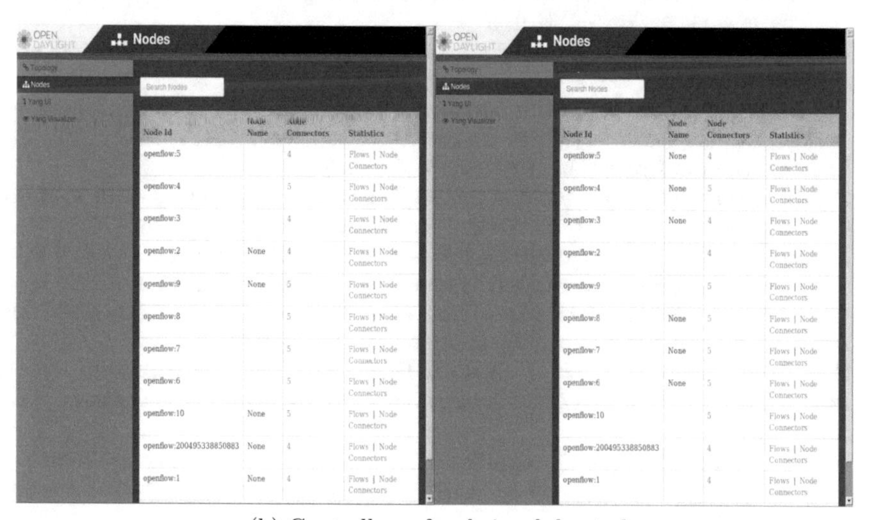

(b) Controllers after being federated

Fig. 2. Node tables.

Using DDS in this scenario has shown that it is possible to exchange network information between both controllers with microsecond performance. This result is a first step to accomplish the 5G latency requirement, while the reliability and resilience of the network are improved.

3 Reflections

A hybrid hierarchy of SDN controllers was proposed in this paper as a new architecture for the 5G Control Plane. In this architecture, a set of distributed and federated SDN controllers manage a group of SDN controllers in the bottom level. This design offers scalability, flexibility and programmability to the network, some of the 5G requirements. Moreover, the DDS was utilized to share network information between two SDN controllers as the first step to solve the 5G latency requirement.

Our future efforts will be directed mainly to develop a software agent that sends specific network information to GCs and ACs using DDS. In addition, we want to elaborate an algorithm to optimize some functionalities in our proposal.

Acknowledgment. This work has been supported by the Ministerio de Economia y Competitividad of the Spanish Government under the projects TEC2016-76795-C6-1-R and AEI/FEDER, UE.

References

1. Tootoonchian, A., Ganjali, Y.: Hyperflow: A distributed control plane for OpenFlow. In: Proceedings of the 2010 internet network management conference on Research on enterprise networking. (2010) 3
2. Koponen, T., et al.: Onix: A distributed control platform for large-scale production networks. OSDI. **10**, 1–6 (2010)
3. Fu, Y., et al.: Orion: A hybrid hierarchical control plane of software-defined networking for large-scale networks. In: Network Protocols (ICNP), 2014 IEEE 22nd International Conference on, IEEE (2014) 569–576
4. Fu, Y., et al.: A hybrid hierarchical control plane for flow-based large-scale software-defined networks. IEEE Trans. Netw. Serv. Manag. **12**(2), 117–131 (2015)
5. Bertaux, L., Hakiri, A., Medjiah, S., Berthou, P., Abdellatif, S.: A DDS/SDN based communication system for efficient support of dynamic distributed real-time applications. In: Distributed Simulation and Real Time Applications (DS-RT), 2014 IEEE/ACM 18th International Symposium on, IEEE (2014) 77–84
6. Choi, H.Y., King, A.L., Lee, I.: Making DDS really real-time with openflow. In: Proceedings of the 13th International Conference on Embedded Software, ACM (2016) 4
7. 3GPP: TS 23.501- System Architecture for the 5G System; Stage 2. http://www.3gpp.org/ftp/Specs/archive/23_series/23.501/23501-f00.zip [Online; accessed 01-January-2017]
8. Knight, S., Nguyen, H.X., Falkner, N., Bowden, R., Roughan, M.: The internet topology zoo. IEEE J. Sel. Areas Commun. **29**(9), 1765–1775 (2011)

Towards Computer-Aided Differential Diagnosis: A Holonic Multi-agent Based Medical Diagnosis System

Zohreh Akbari[1(✉)] and Rainer Unland[1,2]

[1] Institute for Computer Science and Business Information Systems (ICB), University of Duisburg-Essen, Essen, Germany
{zohreh.akbari, rainer.unland}@icb.uni-due.de
[2] Department of Information Systems, Poznan University of Economics, Poznan, Poland

Abstract. The state of the art of the Medical Diagnosis Systems (MDSs) has demonstrated an exciting advancement in recent years. Clearly, the success of these systems is very much dependent on the quality of their input, however, so far no computer-aided decision support system has been introduced to address this issue. Such a system should be capable of performing the Differential Diagnosis (DDx) process, in which upon receiving the chief compliant, some potential diagnoses are considered, according to which enough evidence and supporting information will be gathered in order to shrink the probability of the other candidates. This paper shows that DDx domain is a holonic domain, and hence, this process can be implemented using a holonic multi-agent based MDS.

Keywords: Holonic Multi-agent system (HMAS) · Machine Learning (ML) Medical Diagnosis System (MDS)

1 Introduction

Medical Diagnosis Systems (MDSs) are Decision Support Systems (DSSs) capable of providing an ordered list of potential diagnoses for given signs and symptoms. The state of the art of the MDSs, including the IBM Watson Health [1] and the Isabel [2], mainly focus on finding the perfect link between the patient's medical history and their health knowledge. However, no matter how powerful they are, it is always possible that their final strong deduction is worthless due to an incomplete input. In order to provide the all-encompassing input, a physician should literally perform the Differential Diagnosis (DDx) [3], in which (S)he would carefully listen to the symptoms explained by the patient, considers some potential diagnoses and then tries to gather enough evidence and supporting information to shrink the probability of the other candidates. In a patient encounter, this method is used in a process called the History and Physical examination (H&P) [4]. Only physicians and in some institutions, in order to compensate the shortage of the physicians, some specially trained nurses are qualified to perform this process. A system capable of guiding a focus H&P, however, will allow less experienced nurses to perform this process, and furthermore, can provide second

S. Rodríguez et al. (Eds.): DCAI 2018, AISC 801, pp. 361–365, 2019.
https://doi.org/10.1007/978-3-319-99608-0_46

opinions in critical cases. As the DDx domain is in fact a holonic domain (see Sect. 2), the development of a Holonic Multi-Agent System (HMAS), which is capable of performing DDx, is the practical contribution of this work (see Sect. 3). On the other hand, the introduction of the Machine Learning (ML) techniques that support the functionality of this HMAS, is the theoretical contribution of this research (see Sect. 4).

2 The Differential Diagnosis Domain and the Holonic Domain

One of the well-known Multi-Agent System (MAS) architectures is the HMAS architecture. The term holon was first introduced in [7] in order to name recursive and self-similar structures. A holonic agent of a well-defined software architecture may join several other holonic agents to form a super-holon, which is then seen as a single holonic agent with the same software architecture [8]. The organizational structure of a holonic society, i.e. the holarchy, offers advantages such as robustness, efficiency, and adaptability [8]. Obviously, all these advantages do not mean that this architecture is more effective than the others, however, specifically for the domains, it is meant for. As stated in [8] such domains should (1) involve actions of different granularities, (2) induce abstraction levels, (3) demand decomposable problem settings, (4) require efficient intra-holonic communications, (5) include cooperative elements, and (6) may urge real-time actions. The DDx problem can be decomposed recursively into subproblems of weighting the probability of the presence of the possible diseases. These subproblems may induce different abstraction levels and can be of different granularities. According to the nature of DDx, the problem solvers are collaborative and those dealing with similar diseases need to have more communications, which are to be conducted in timely manner. Hence, this domain meets the characteristics of holonic domains [6].

3 The Holonic Medical Diagnosis System (HMDS)

The HMDS consists of two types of agents: comparatively simple Disease Representative Agents (DRA) as the system fundamentals on the lowest level and more sophisticated Disease Specialist Agents (DSA) as decision makers on the higher levels of the system [5]. DRAs are atomic agents, and form the leaves of the holarchy. Each DRA is an expert on a specific disease and maintains the Disease Description Pattern (DDP) – an array of possible signs, symptoms, and test results, i.e. the holon identifier. In order to join the diagnosis process, these agents only need to perform some kind of pattern matching, i.e. calculating their Euclidean distance to the diagnosis request. DSAs are holons consisting of numbers of DRAs and/or DSAs with similar symptoms; i.e., representing similar diseases. This encapsulation, in fact, enables the implementation of the DDx. The DSAs are designed as moderated groups [8], where agents give up part of their autonomy to their super-holon. To this end, for each DSA, a head is defined, which provides an interface and represents the super-holon to its environment. This head is created for the lifetime of the super-holon, based on agent cloning. The holon identifier of a DSA will be the average of the holon identifiers of its members.

The holarchy has one root, i.e. a DSA, which will play the role of the exclusive interface to the outside world. Initially, this DSA takes all the DRAs as its members, and lets the DSAs form automatically. To this end, it clusters them based on the affinity, i.e., the similarity, and defines a head for each of the clusters. This is repeated recursively until no further clustering is necessary (see Sect. 4). Later on, the system can still refine its holarchy using its self-organization technique (see Sect. 4).

The HMDS works as follows: The head of the holarchy receives the diagnosis request and places it on its blackboard. Any member of this DSA, can then read this message. A DRA's reaction would be to send back its similarity to the request. However, a DSA will join the diagnosis process only if the request is not an outlier considering its members. This means that it will place the diagnosis request on its own blackboard. Then the same process repeats recursively until the request reaches the final level of the holarchy. Results obtained by participating agents now flow from the bottom to the top of the holarchy and are sorted according to their similarity and frequency. Literally, each agent will send its top diagnoses together with all of the signs, symptoms or test results that are relevant from its point of view, but their presence or absence has not been specified. This implies that originally not provided relevant information might be requested from the user in a second step [5]. For a more comprehensive description of the system and the demonstration of its functionality please refer to [5, 6].

4 Machine Learning in the HMDS

As mentioned in Sect. 3, the initial holarchy of the system can be created using clustering, as an unsupervised learning technique. Thereafter, it is still essential to support the system in learning based on the new observations and the feedback. In HMDS, holon identifiers are updated applying the exponential smoothing method, as a supervised learning method, and the self-organization of the holarchy is supported by Q-learning, as a reinforcement learning technique. The rest of this section covers these techniques, however, the demonstration of the learning in HMDS and the discussion on the quality of the learning feedback for the last two methods can be followed in [6].

1. **Clustering:** The Density-Based Spatial Clustering of Applications with Noise (DBSCAN) [9] is one of the best algorithms for clustering in HMDS. In [10] a simple and effective method for automatic detection of the input parameter of DBSCAN is presented, which helps best to deal with the complexity of the problem at hand.
2. **Exponential Smoothing:** This technique is a scheme for producing smoothed time series [11]. Using this technique, the past observations are assigned exponentially decreasing weights and recent ones are given relatively higher weights:

$$s_t = \alpha . x_t + (1 - \alpha) . s_{t-1} \tag{1}$$

where α is the smoothing factor ($0 < \alpha < 1$), s_t is the smoothed statistic, x_t represents the current observation, and s_{t-1} stands for the previous smoothed statistic.

3. **Holonic-Q-Learning (HQL):** The HQL is a Q-learning technique introduced for self-organization in the HMDS [5]. In HQL, the Q-value is in fact measuring how good it is for a holon to be a member of another holon. In this case, the states are the existing holons $\{h_i\}$ and action h_i indicates becoming a sub-holon of holon i:

$$\begin{aligned} &Q_t(sub(h), h) \leftarrow \\ &(1-\alpha_t)Q_{t-1}(sub(h), h) + \alpha_t(R_t(sub(h), h) + \\ &\gamma \operatorname{argmax}_{Q_{t-1}(h, \sup(h))}(Q_{t-1}(h, \sup(h)).Aff(sub(h), sup(h)))) \end{aligned} \tag{2}$$

where, $\alpha_t = \frac{1}{1 + visits_t(sub(h), h)}$, $\gamma \epsilon [0, 1)$ is the discount factor, and $Aff(sub(h), h) = 1 - \frac{d(sub(h), h)}{\max d(sub(h), h)}$. For more information regarding the reward engineering and the proof of convergence please refer to [5, 6].

5 Conclusion

This paper explained that one of the main limitations of the available MDSs is the lack of support for the H&P. The H&P applies DDx, which represents a holonic domain. This implies that a MDS with holonic architecture could be able to support DDx. This research also suggested appropriate ML techniques that may maintain and improve the functionality of the HMDS. The simulation of the system is now in progress and future work includes the complete simulation and validation of the system.

References

1. IBM Watson Health: Available https://www.ibm.com/watson/health/
2. Fisher, H., Tomlinson, A., Ramnarayan, P., Britto, J.: ISABEL: support with clinical decision making. Pediatr. Nurs. **15**(7), 34–35 (2003)
3. Merriam-Webster: Differential Diagnosis. https://www.merriam-webster.com/dictionary/differential%20diagnosis
4. Segen, J.: Concise Dictionary of Modern Medicine. McGraw-Hill, New York (2006)
5. Akbari, Z., Unland, R.: A holonic multi-agent system approach to differential diagnosis. In: Berndt, J., Unland, R. (eds.) Multiagent System Technologies. MATES 2017. LNCS, vol. 10413, pp. 272–290 (2017)
6. Akbari, Z., Unland, R.: A holonic multi-agent based diagnostic decision support system for computer-aided history and physical examination. In: Submitted to the 16th International Conference on Practical Applications of Agents and Multi-Agent Systems (2018)
7. Koestler, A.: The Gost in the Machine. Hutchinson, London (1967)
8. Gerber, C., Siekmann, J., Vierke, G.: Holonic Multi-Agent Systems, Technical report DFKI-RR-99-03. German Research Centre for Artificial Intelligence (1999)

9. Ester, M., Kriegel, H.-P., Sander, J., Xu, X.: A density-based algorithm for discovering clusters in large spatial databases with noise. In: Simoudis, E., Han, J., Fayyad, U. (eds.) The 2nd International Conference on Knowledge Discovery and Data Mining, pp. 226–231 (1996)
10. Akbari, Z., Unland, R.: Automated determination of the input parameter of the dbscan based on outlier detection. In: Artificial Intelligence Applications and Innovations. IFIP Advances in Information and Communication Technology, vol. 475, pp. 280–291 (2016)
11. NIST/SEMATECH e-Handbook of Statistical Methods. http://www.itl.nist.gov/div898/handbook/. Accessed Mar 2018

Automatic Screening of Glaucomatous Optic Papilla Based on Smartphone Images and Patient's Anamnesis Data

Jose Carlos Raposo da Camara(✉)

Universidade Aberta, R. Escola Politécnica 141-147, 1269-001 Lisbon, Portugal
1701367@estudante.uab.pt

Abstract. Glaucoma can cause irreversible damage to the optic nerve and lead to blindness. Current treatments can prevent vision loss if lesion are detected in early stages. Our research proposal has as main goal the development of a system for automatic screening of glaucomatous excavation in the optic nerve based on images obtained by cellular associated to patient's anamnesis data. The system uses a multi-agent approach in order to combine the results of different detection algorithms and other data related to glaucomatous disease risk factors. It is expected that the system is able to assist professionals with less experience and in distant places or places with little technical availability to screen and monitor the progression of the loss of fibers of the optic disc and thus expedite the diagnosis and treatment.

Keywords: Glaucoma screening · Smartphone · Multi-agents

1 Introduction

Glaucoma is one of the main groups of diseases that can cause irreversible damage to the optic nerve and lead to blindness and may not present symptoms in the early stages. Risk factors for the disease include age range over 40, race, family history, eye pressure, myopia, which, together with visual field diagram, may be included in the multi-agent system for comparison, supplementation and diagnostic assistance in suspected cases where the optic disc has not yet demonstrated the development of the disease, favoring the initial diagnosis mainly for professionals with less experience and in more distant places or with little technical availability.

The automatic segmentation methods can provide reliable, reproducible, fast and economical results [1]. Photo capture is one of the most important steps in the construction of a methodology for image segmentation. To verify the quality of the segmentation a base is needed that provides a correct marking of the region to be segmented and is an important prerequisite for the quantitative evaluation of the excavation caused by the loss of nerve fibers (ganglion cells) can be established through the relation between the diameter of the papilla and the center of the optical disk [2] extraction of color and entropy resources in the segmented area [3].

S. Rodríguez et al. (Eds.): DCAI 2018, AISC 801, pp. 366–369, 2019.
https://doi.org/10.1007/978-3-319-99608-0_47

2 Proposed System

The architecture for the proposed automatic screening system is presented in Fig. 1. The optic nerve originates in the brightest region of retinal image and it act as a main region to detect the retinal diseases using the ratio of cup and disc and the ratio between Optic rim & center of the Optic Disc. To verify the quality of the segmentation a base is needed that provides a correct marking of the region to be segmented and is an important prerequisite for the quantitative evaluation of the excavation caused by the loss of nerve fibers (ganglion cells) can be established through the relation between the diameter of the papilla and the center of the optical disk [2] extraction of color and entropy resources in the segmented area [3]. Many others methods of image segmentation for the purpose of diagnosis of the glaucomatous papilla have been recently proposed [4–8]. The controller agent receives smartphone photos of the optic disc and another eventually supplementary data from the anamnesis and visual field, and stores them in a database. One or more analyst agents are created and receives the photo ID and another complementary data from the anamnesis and visual field.

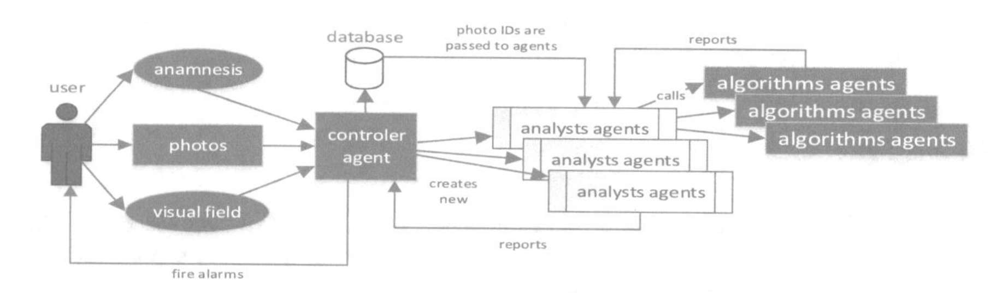

Fig. 1. Architecture for the proposed automatic screening system.

Each analyst agent can do two things: (1) call several algorithms in parallel for lesion detection and wait for the result (2) do a pre-analysis and call the algorithms that it considers most appropriate, for example, taking into account factors such as sharpness, number of pixels per inch, etc., call a filter to improve the image (e.g. before calling the algorithms), or call other anamnesis and visual field data. The algorithms can parallelize the problem by dividing the photo into several segments and integrating relevant data from the anamnesis (considered risk factors). The algorithms report to the analyst agent who will analyze various responses and send the final report to the controlling agent. Here the reports are classified in one of three different classes: (1) optic papilla without glaucomatous characteristics (2) optic papilla contains glaucomatous characteristics (3) result of doubtful diagnosis. In the case of (1) an alarm is immediately fired to inform the user. In the others situation, user can request notification.

3 Expected Results

Digital image processing is an area of study with great potential that is increasingly contributing to diagnostic accuracy in the health area, in an efficient and expensive way, and several diagnoses can be made. Glaucoma is a multifactorial etiology syndrome in which clinical reasoning proves to be essential, automated systems that use patterns of image processing of the optic disc associated with other clinical data are likely to benefit most by increasing sensitivity and specificity of the method (not the apparatus), especially in the screening of suspected cases.

Many algorithms for the diagnosis of the glaucomatous papilla have been recently proposed (e.g. [8]). On the one hand, the methods yielded good results, accounting for an accuracy of over 83% when evaluated with a requirement threshold of 70% which is the classic value found in the literature [9]. On the other hand some studies show that the lack of a reference data set can be a limiting factor to the diagnosis of glaucoma [10]. It is expected that in multi-agent systems and the association of other data related to glaucomatous disease risk factors correlated with papilla images associated with the visual field may complement the diagnosis, especially in suspicious excavations. This will be great assistance to professionals with less experience and in distant places and thus expedite the diagnosis and treatment.

References

1. Lassen-Schmidt, B.C., Kuhnigk, J.-M., Konrad, O., van Ginneken, B., van Rikxoort, E.M.: Automatic and interactive segmentation of pulmonary lobes and nodules in chest CT images. Inst. Phys. Eng. Med. Phys. Med. Biol. **62** (2017)
2. GeethaRamani, R., Dhanapackiam, C.: Automatic localization and segmentation of optic disc in retinal fundus images through image processing techniques. In: 2014 International Conference on Recent Trends in Information Technology, pp. 1–5 (2014)
3. Claro, M., Santos, L., Silva, W., Araújo, F., Santana, A.D.A.: Automatic detection of glaucoma using disc optic segmentation and feature extraction. In: Proceedings of 2015 41st Latin American Computing Conference, CLEI 2015 (2015)
4. Viquez, K.D.C., Arandjelovic, O., Blaikie, A., Hwang, I.A.: Synthesising wider field images from narrow-field retinal video acquired using a low-cost direct ophthalmoscope (Arclight) attached to a smartphone. In: 2017 IEEE International Conference on Computer Vision Workshops (ICCVW), pp. 90–98 (2017)
5. Aquino, A., Gegúndez-Arias, M.E., Marín, D.: Detecting the optic disc boundary in digital fundus images using morphological, edge detection, and feature extraction techniques. IEEE Trans. Med. Imaging **29**(11), 1860–1869 (2010)
6. Thakkar, K., Chauhan, K., Sudhalkar, A., Gulati, R., Ophthalmologist, M.S.: Detection of glaucoma from retinal fundus images by analysing ISNT measurement and features of optic cup and blood vessels. Int. J. Eng. Technol. Sci. Res. IJETSR 4(7), 2394–3386 (2017). www.ijetsr.com. ISSN
7. Wong, D.W.K., et al.: Intelligent fusion of cup-to-disc ratio determination methods for glaucoma detection in ARGALI. In: 2009 Annual International Conference of the IEEE Engineering in Medicine and Biology Society, pp. 5777–5780 (2009)

8. Joshi, G.D., Sivaswamy, J., Krishnadas, S.R.: Optic disk and cup segmentation from monocular color retinal images for glaucoma assessment. IEEE Trans. Med. Imaging **30**(6), 1192–1205 (2011)
9. Santos, L.M., Araújo, F.H.D., Claro, M.L., Silva, W.L., Silva, R.R.V., Drumond, P.M.L.L.: Implementação e Comparação de um Método de Detecção e Segmentação Automática do Disco Óptico em Diferentes Bases de Imagens da Retina. In: XII Simpósio Brasileiro de Automação Inteligente (XII SBAI) (2015)
10. Maninis, K.-K., Pont-Tuset, J., Arbeláez, P., Van Gool, L.: Deep retinal image understanding. In: Medical Image Computing and Computer-Assisted Intervention – MICCAI 2016, pp. 140–148 (2016)

The Resources Placement Problem in a 5G Hierarchical SDN Control Plane

Irian Leyva-Pupo(✉), Cristina Cervelló-Pastor, and Alejandro Llorens-Carrodeguas

Department of Network Engineering, Universitat Politècnica de Catalunya (UPC), Esteve Terradas, 7, 08860 Castelldefels, Spain
{irian.leyva,cristina,alejandro.llorens}@entel.upc.edu

Abstract. In this paper, we address the SDN Controllers and Virtual Network Functions (VNFs) placement problem in 5G networks. To this aim, we propose an architecture for the 5G Control Plane and a method to determine the optimal placement of controllers and VNFs. The placement is determined according not only to latency and capacity requirements but also to type of Network Function (NF).

Keywords: SDN · NFV · CPP · VNFP · 5G

1 Introduction

SDN and Network Function Virtualization (NFV) have been defined as key drivers in the implementation of the 5G architecture. Their combination enables dynamic, flexible deployment and on-demand scaling of network functions, which are necessary for the development of the future Mobile Packet Core towards a 5G system [1]. However, to satisfy the 5G requirements while network costs are minimized, the SDN Controllers and VNFs placement should be carefully planned.

The Controller Placement Problem (CPP) was first defined in 2012 by Heller et al. [2]. Since then, this topic has been addressed in several ways. Despite the great variety of **related works**, just a few approaches this issue in 5G networks [3,4]. The authors in [3] study the CPP considering the uncertainty in cellular users location without deepening in details about the control plane architecture. By contrast, Kentini et al. in [4] assume an SDN-based virtual mobile network architecture and define an algorithm for the SGW Controller (SGW-C) placement with the purpose of reducing SGW relocations and the load in the SGW-C.

The VNF Placement (VNFP) problem in mobile networks has also been addressed in some research works such as [5–7]. In [5] the authors propose an algorithm to place VNFs of PGWs and SGWs on a given topology of distributed DataCenters taking into account criteria of QoE for mobile users and SGW relocations. However, they do not consider either the variability of network traffic or VNF resource requirements. Similarly, Bagaa et al. [6] propose an algorithm

S. Rodríguez et al. (Eds.): DCAI 2018, AISC 801, pp. 370–373, 2019.
https://doi.org/10.1007/978-3-319-99608-0_48

to create virtual instances of PGW and determine their placement based on geographical location and applications/services type. Moreover, both [5,6] limit the scope of their work to specific NFs. In [7] the authors address the placement of all the core network functions. Although, they do not consider latency constraint on the VNF nodes and end-to-end network.

Despite the wide variety of research works in the field of the CPP and VNFP in mobile networks, until now, both **problems** have been solved separately. Thus, in this work, we attempt to propose a method to jointly resolve them.

The **hypothesis** of our research is that 5G services requirement of ultra-low latency can be achieved if the core networks functionalities and the SDN controllers are optimally placed.

2 Proposal of the 5G Control Plane Architecture

The architecture proposed for the mobile core network is based on the 5G 3GPP standardized architecture [8], SDN and NFV, as shown in Fig. 1. A two-level hierarchy of SDN controllers bridges between the control and user planes, specifically, between the Session Management Functions (SMFs) and the User Plane Functions (UPFs). Its bottom layer is composed of Area Controllers (ACs) which are mainly responsible for UPFs control and flows management. While, the upper layer is formed by Global Controllers (GCs), in charge of managing and controlling the ACs, doing load balancing and keeping a global network view. Moreover, both the control plane NFs and the SDN controllers are virtualized, deployed and executed on an NFV Infrastructure (NFVI).

2.1 Method for SDN Controllers and VNFs Placement

Assuming this architecture, our main objective is to find the optimal placement of the SDN Controllers and the VNFs, the Access and Mobility Management Functions (AMFs) and SMFs mainly, in order to minimize the network response time to users' requests. To this end, the controllers and core network functionalities are deployed by following one of the 5G key requirements: latency lower than 10 ms in the control plane.

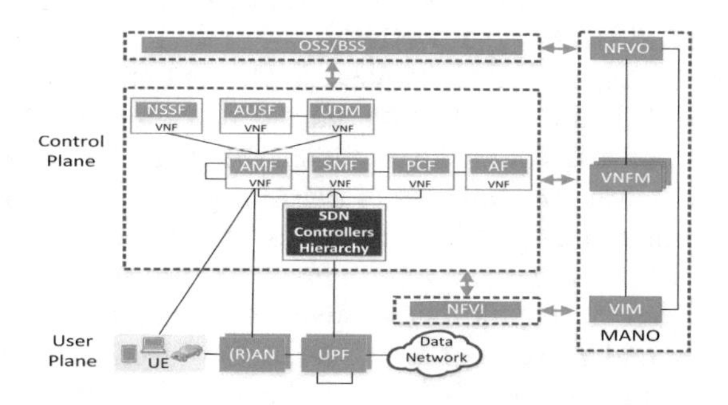

Fig. 1. Proposal of the 5G control plane architecture

Algorithm 1. Assignment of UPFs to ACs

Require: L_{req1}, $C_{AC_{max}}$, N_U, $D_{N_u X N_u}$:Delay matrix, T_{n_u}: UPFs traffic matrix
Ensure: K: Number of ACs, N_{AC}: Set of ACs, S_{u_k}: Sets of UPFs $\in AC_k$

1: $K \leftarrow 1$, $N_{AC} \leftarrow \emptyset$, $S_{u_k} \leftarrow \emptyset$
2: $U \leftarrow$ Furthest UPF in average to others UPFs
3: $S_u \leftarrow U$ # S_u: Set of candidate positions (n_{s_i}) for AC_k
4: $C_{AC_u} \leftarrow T_U$ # C_{AC_u}: Capacity used in the AC to manage S_u
Forming S_u with neighboring UPFs (n_{u_i}) to U, $n_{u_i} \in N_u$
5: **for all** $n_{u_i} \in N_u$ **do**
6: **if** $d(n_{u_i}, U) \leq L_{req1}$ **and** $C_{AC_u} + T_{n_{u_i}} \leq C_{AC_{max}}$ **then**
7: $S_u \leftarrow S_u + n_{u_i}$, $C_{AC_u} \leftarrow C_{AC_u} + T_{n_{u_i}}$
8: **end if**
9: **end for**
10: **for all** $n_{s_i} \in S_u$ **do**
11: $S_{ns_i} \leftarrow n_{s_i}$ # S_{ns_i}: Set of UPFs $\in$ AC placed in n_{s_i}
12: $C_{AC_{ns_i}} \leftarrow$ Capacity used in the AC to manage S_{ns_i}
Forming S_{ns_i} with neighboring UPFs (n_{nu_j}) to n_{s_i}, $n_{nu_j} \in N_u$
13: **for** $n_{nu_j} \in N_u$ **do**
14: **if** $d(n_{nu_j}, n_{s_i}) \leq L_{req1}$ **and** $C_{AC_{ns_i}} + T_{n_{nu_j}} \leq C_{AC_{max}}$ **then**
15: $S_{ns_i} \leftarrow S_{ns_i} + n_{nu_j}$, $C_{AC_{ns_i}} \leftarrow C_{AC_{ns_i}} + T_{n_{nu_j}}$
16: **end if**
17: **end for**
18: **if** $U \in S_{ns_i}$ **then**
19: **Evaluate** $F_{sel_1}(S_{ns_i})$ # F_{sel}: Funct. to select AC best placement
20: **Update** $S_{ns_{i_{best}}}$ # $S_{ns_{i_{best}}}$: S_{ns_i} with max. F_{sel_1}
21: **end if**
22: **end for**
23: $S_{u_k} \leftarrow S_{u_k} + S_{ns_{i_{best}}}$, $N_U \leftarrow N_U - S_{u_k}$
24: $N_{AC} \leftarrow N_{AC} + n_{s_i}$ # n_{s_i}: node with max. F_{sel_1}
25: **while** $N_U \neq \emptyset$ **do**
26: $U \leftarrow Z$, $K \leftarrow K + 1$ # Z: Nearest UPF to AC_k, $Z \in N_U$
27: **Go to** step 3
28: **end while**

The mobile network is modeled as a connected undirected graph G(N,E) where N is the set of nodes, E the links between them and $N_U \subset N$ and $N_R \subset N$ the sets of UPFs and (R)ANs, respectively.

Our method is composed of three phases. The aim of the first one is to find the optimal number of ACs and their placement, so that, the ACs response time to the UPFs is minimized. The UPFs are assigned to the ACs following criteria of latency (L_{req1}) and available capacity in the ACs ($C_{AC_{max}}$). Algorithm 1 shows the procedure to determine the best SDN Controller position. The SMFs are placed in the same nodes that the ACs in order to reduce propagation delays.

In the second phase, the AMFs placement is determined according to constraints of latency and AMFs capacity. The main objective of this phase is to minimize the AMF relocations. To this end, the user equipments (UEs) are classified according to their grade of mobility and three levels of AMFs are deployed to manage UEs with low, medium and high mobility patterns. Others 5G core NFs like the Unified Data Management (UDM) and the Police Control Function (PCF) are also placed in the upper level of AMFs.

Finally, in the third phase, the ACs are assigned to the Global Controllers with the main aim of minimizing the latency between GCs. Minimum latency between GCs is a key factor in order to keep network consistency and reduce controllers' response time. This phase is quite similar to the first one, but in this case, ACs are assigned instead of UPFs.

3 Reflections

Ultra-low latency is a key requirement of 5G networks to support services like vehicular connectivity and M2M communication. But, offering low response time while reducing network resources consumption and costs is a big challenge. By an optimal planning of SDN controllers and VNF placement these objectives can be achieved. Dynamic optimization of elements assignment and deployment of NFVI resources will be an important task for future works.

Acknowledgment. This work has been supported by the Ministerio de Economía y Competitividad of Spain under project TEC2016-76795-C6-1-R and AEI/FEDER, UE.

References

1. Nguyen, V.G., et al.: SDN/NFV-based mobile packet core network architectures: a survey. IEEE Commun. Surv. Tutor. **19**(3), 1567–1602 (2017)
2. Heller, B., Sherwood, R., McKeown, N.: The controller placement problem. In: Proceedings of the First Workshop on Hot Topics in Software Defined Networks, pp. 7–12. ACM (2012)
3. Abdel-Rahman, M.J., et al.: Robust controller placement and assignment in software-defined cellular networks. In: 2017 26th International Conference on ICCCN, pp. 1–9. IEEE (2017)
4. Ksentini, A., Bagaa, M., Taleb, T.: On using SDN in 5G: the controller placement problem. In: 2016 IEEE on GLOBECOM, pp. 1–6. IEEE (2016)
5. Taleb, T., Bagaa, M., Ksentini, A.: User mobility-aware virtual network function placement for virtual 5G network infrastructure. In: 2015 IEEE ICC, pp. 3879–3884. IEEE, June 2015
6. Bagaa, M., Taleb, T., Ksentini, A.: Service-aware network function placement for efficient traffic handling in carrier cloud. In: 2014 IEEE WCNC, pp. 2402–2407, April 2014
7. Baumgartner, A., Reddy, V.S., Bauschert, T.: Mobile core network virtualization: a model for combined virtual core network function placement and topology optimization. In: Proceedings of the 2015 1st IEEE Conference on NetSoft, pp. 1–9, April 2015
8. 3GPP: TS 23.501- system architecture for the 5G system; stage 2 (2017)

Computational Analysis of Multiple Instance Learning-Based Systems for Automatic Visual Inspection: A Doctoral Research Proposal

Eduardo-José Villegas-Jaramillo(✉) and Mauricio Orozco-Alzate

Facultad de Administración - Departamento de Informática y Computación, Universidad Nacional de Colombia - Sede Manizales, Campus La Nubia Km 7 vía al Magdalena, 170003 Manizales, Colombia
{ejvillegasj,morozcoa}@unal.edu.co

Abstract. The objective of this proposal is to select and analyze, functionally and computationally, a set of algorithms used for the detection of defects by automatic visual inspection, which make use of multiple instance learning and have the potential to be improved. From the analyses, modifications or updates, it is proposed to speed-up the response of the automatic visual inspection systems, allowing thereby, a decrease of the amount of undetected defective products in the production lines.

Keywords: Algorithm complexity · Classification algorithms · MIL Visual inspection

1 Problem Statement and Hypothesis

In the manufacturing industry, it is necessary to achieve high quality standards, seeking to minimize both: the amount of undetected defective objects and the rejection time in the production lines. For this reason, production processes require precise, effective and besides, automated controls, since quality control is traditionally done manually by human experts, who are prone to perform it in a slow or erroneous manner, due —among others— to factors such as fatigue, anxiety, stress and adverse work conditions, such as high temperatures [9]. For these reasons, the automatic visual inspection systems (AVIS) have gained great relevance and importance, when used as elements of quality control.

The AVIS are composed of four stages: (i) the *acquisition* of digital images of the objects; (ii) the image *pre-processing*, by applying filters to remove distortions that obstruct the image recognition, e.g. noise, light reflections and spots; (iii) the *representation* of the pre-processed image, typically by extracting visual and/or structural features, whose values are stored in a single vector (feature vector) also called *instance* and; (iv) the *classification* stage, in which an algorithm —based on examples provided by an expert— assigns a class label to the representation of the object; for example: defective or non-defective [9].

S. Rodríguez et al. (Eds.): DCAI 2018, AISC 801, pp. 374–377, 2019.
https://doi.org/10.1007/978-3-319-99608-0_49

Sometimes, the knowledge of the problem is incomplete and the samples do not include precise information about the defect location —if any— inside the images. In such situations, it is more convenient to represent an image as a group of feature vectors (also called *bag* of instances) instead of using a single vector for the representation. The instances typically correspond to regions or parts of the image. The class label is associated to all of the vectors in the bag. This alternative representation, along with particular classification algorithms that can process bags, is called *multi-instance learning* (MIL) [7].

The computational analyses of a large part of the different MIL algorithms found in the literature, are not sufficiently detailed and rigorous, therefore it is important to perform them, not just to the classification stage —the most studied one— but also, to the representation stage, in order to guarantee the efficiency and accuracy levels required to fulfill the classification objectives. According to [1,6], the computational complexity is defined as the performance measure, with respect to time and space, and in terms of the size of the data inputs, that are required to achieve the results of an algorithm. Hence, we have adopted this definition in this proposal.

1.1 Hypothesis

The use of computational algorithms based on the MIL paradigm, which have a better performance in terms of efficiency and complexity, will allow the AVIS to perform a faster detection of defects. This acceleration will improve the decisions in the production line, and also will to simplify the necessary simulations for the design and tuning of the algorithms. A more detailed knowledge of the representation and classification algorithms, will allow a better understanding of their behavior and also, will be useful to make improvements to the different stages of the inspection system.

2 Related Works

Among the different related works found in the literature, those including analyses of the algorithms used in each one of the stages of the inspection process are of our special interest. In them, it has been found that the classification stage exhibits the greatest complexity, besides taking a long computational time. For this reason, the classification stage of the system has been the focus of most research efforts. Some of the most relevant works in this area are the following:

Paper [2] presents the foundations of the MIL paradigm, applied to drug activity prediction. In this work, it is proposed to represent the regions of interest by means of axis-parallel rectangles; besides, it is emphasized that MIL can be understood as a generalization of the traditional single instance learning, in which each bag can contain different instances.

The work proposed in [5] emphasizes that the complexity of the MIL algorithm is totally dependent on the number of instances in the training set. Therefore, the authors present a new MIL algorithm, called MILIS (Multiple

Instance Learning with Instance Selection), based on an adaptive selection of the instances, which decreases the complexity, thanks to the reduction of the number of selected instances.

The works using support vector machines (SVM), such as the one presented in [12], in which two non-parallel hyperplanes are calculated —a positive hyperplane for positive bags and a negative hyperplane for negative ones— solving in that way two SVM problems. Another approach is proposed in [13], which uses the ambiguous labels that result from supervised learning, as an input for a new approach called similarity-based multiple instance learning (SMILE).

Additionally, the work presented in [8], which proposes a supervised aggregated feature learning (SAFE) that combines, in a single framework, the local instance level and the information at the bag level, in order to do the classification task, through a model formulated with a least squares SVM. The high computational cost of the formulated algorithms —with $O(N^2)$ complexity, where N is the number of instances— implies that it is necessary to look for computational techniques that allow the acceleration of the algorithms.

In [4] the use of an algorithm based on the calculation of the likelihood applied to MIL is proposed, which computes a kernel density estimator (KDE). The results of the proposed algorithm, called MILKDE, are compared with those obtained with forty three other algorithms using five data sets, showing that the results are better than other similar algorithms described in the literature. Another work, presented in [3], is characterized by a fast selection of positive instances and it also emphasizes that the computational complexity is $O(n \cdot m)$, where n is the number of bags and m the number of instances. In spite of that, in [3] it is described that the test results are better than those found in the literature.

Regarding the national research community, the work developed by Mera [11] is our main reference. In this work, a methodology for defect detection in AVIS is developed, presenting the problem when there is a very large imbalance, between the number of images with defects and the number of those without defects. The author refers to the high computational cost of the MIL algorithms and the need to accelerate them, either with better analyses or through their parallelization.

3 Proposal

With this doctoral research proposal, we aim to explore a set of methods, techniques and algorithms, that are used for the recognition of objects, under the MIL paradigm, where the MIL algorithms are analyzed to implement computational techniques that, in turn, improve the efficiency and speed-up of the AVIS.

4 Evaluation Plan

As a starting point, the work by Mera et al. [10,11] will be considered. In order to test the analysis and improvement of the algorithms, we will consider the use of public data sets, related to problems of the industrial sector, and that are characterized or can be easily characterized, as bags of instances. Currently, the following data sets have been identified: glass inspection[1], suitcases, welds and fish fillets[2] as well as others such as those compiled in the following website: http://www.miproblems.org/.

Acknowledgments. The authors acknowledge the support to attend DCAI'18 Doctoral Consortium provided by "Convocatoria para la Movilidad Internacional de la Universidad Nacional de Colombia (UNAL) 2016 - 2018".

References

1. Brassard, G., Bratley, P.: Fundamentos de algoritmia. Prentice Hall, Montreal (1997)
2. Dietterich, T.G., Lathrop, R.H., Lozano-Pérez, T.: Solving the multiple instance problem with axis-parallel rectangles. Artif. Intell. **89**, 31–71 (1997)
3. Du, R., Wu, Q., He, X., Yang, J.: MIL-SKDE: multiple-instance learning with supervised kernel density estimation. Signal Process. **93**(6), 1471–1484 (2013)
4. Faria, A., Coelho, F., Silva, A., Rocha, H., Almeida, G., Lemos, A., Braga, A.: MILKDE: a new approach for multiple instance learning based on positive instance selection and kernel density estimation. Eng. Appl. Artif. Intell. **59**, 196–204 (2017)
5. Fu, Z., Robles-Kelly, A., Zhou, J.: MILIS: multiple instance learning with instance selection. IEEE Trans. Pattern Anal. Mach. Intell. **33**(5), 958–977 (2011)
6. Gelder, V., Baase, S.: Algoritmos Computacionales - Introducción al análisis y diseño, 3rd edn. Pearson Educación, México (2002)
7. Herrera, F., Ventura, S., Bello, R., Cornelis, C., Zafra, A., Sánchez Tarragó, D., Vluymans, S.: Multiple Instance Learning - Foundations and Algorithms. Springer (2016)
8. Langone, R., Suykens, J.A.: Supervised aggregated feature learning for multiple instance classification. Inf. Sci. **375**, 234–245 (2017)
9. Malamas, E.N., Petrakis, E.G.M., Zervakis, M., Petit, L., Legat, J.D.: A survey on industrial vision systems, applications and tools. Image Vis. Comput. **21**(2), 171–188 (2003)
10. Mera, C., Orozco-Alzate, M., Branch, J., Mery, D.: Automatic visual inspection: an approach with multi-instance learning. Comput. Ind. **83**, 46–54 (2016)
11. Mera, C.A.: Detección de Defectos en Sistemas de Inspección Visual Automática a través del Aprendizaje de Múltiples Instancias. Ph.D. thesis, Universidad Nacional de Colombia (2017)
12. Tian, Y., Qi, Z., Ju, X., Shi, Y., Liu, X.: Nonparallel support vector machines for pattern classification. IEEE Trans. Cybern. **44**(7), 1067–1079 (2013)
13. Xiao, Y., Liu, B., Hao, Z., Cao, L.: A similarity-based classification framework for multiple-instance learning. IEEE Trans. Cybern. **44**(4), 500–515 (2014)

[1] http://www.deltamaxautomazione.it/risolvi/.
[2] http://dmery.ing.puc.cl/index.php/material/gdxray/.

An Agent-Based Model for Energy Management of Smart Home: Residences' Satisfaction Approach

Mahoor Ebrahimi[1(✉)] and Amin Hajizade[2]

[1] Amirkabir University of Technology, Tehran, Iran
mahoor.ebrahimi@yahoo.com

[2] Department of Energy Technology, Aalborg University, Esbjerg, Denmark

Abstract. Reducing the cost of energy leads to reduction in resident's comfort. Therefore, it is necessary to consider both reduction in cost and resident's dissatisfaction. Smart buildings include different systems such as communication system (CS), sensing system (SS), grid data collecting system (GDCS), building energy management system (BEMS), hybrid system (HS), temporary service systems (TSS) and permanent service systems (PSS). Considering the aforementioned systems, an agent-based model for energy management of smart buildings is presented in this work to reduce the energy cost as well as the resident's dissatisfaction.

Keywords: Smart home · Multi-agent system
Home energy management system

1 Introduction

Considering the importance of reducing the cost of energy for smart buildings a solution is needed to make a use of consumption management. This can be done by using the Building Energy Management System (BEMS) which is one of the main components of smart homes. Thus, the amount of energy consumed during the peak-hours should be shaved when electricity price is high, or the consumption can be shifted to other hours [1]. Moreover, Electric Vehicles (EVs) are used as energy storage systems to reduce the cost according to their potential of charging and discharging (buying and selling electricity) in the smart home [2]. However, it is noticeable that the above solutions may reduce resident's comfort. In this work, a solution is proposed to manage consumption in order to reduce the cost of energy with taking into account the residents' comfort.

2 Proposed Scheme

Energy production and consumption in smart home can be managed by the BEMS which is a decision-making agent of the smart home. The BEMS allows residents to apply several energy management mechanisms that can change the way of energy

S. Rodríguez et al. (Eds.): DCAI 2018, AISC 801, pp. 378–384, 2019.
https://doi.org/10.1007/978-3-319-99608-0_50

production and consumption to achieve less energy cost and more residents' comfort. In this work, we present the multi agent-based structure for BEMS. There are several works that implement agents in the BEMS but none of does not consider a manner to consider residents' comfort in the BEMS [3–8]. Readers can refer to [9–70] to know more about multi-agent systems. Our proposed multi-agent BEMS consists of different agents that are described in the following.

The smart home is equipped with sensors and operators to collect the data and send control commands according to the resident's activities and expectations. Sensors are including in the Sensing System (SS) that measures the parameter needed for making appropriate decisions such as inside temperature of different rooms and outside temperature. The purpose of the services at the smart home is increasing the comfort and security of the residents. As highlighted before, several load management mechanisms can be used to provide an optimal management decisions based on the criteria given to the BEMS through the existence of communication network in smart home. In this way, the Communication System (CS) conveys different signals from sensors and grid data system to the BEMS, and from the BEMS to operators. Grid data collecting system (GDCS) obtains data related to grid's parameters such as hourly electricity price. The consumption management allows residents to save money by reducing consumption in hours of high energy price- e.g. peak hours of grid- or changing the start time of operations in services where the type of consumption is such that it can be adjusted to another hour- e.g. like a washing machine. Hence, there is a necessity to find a solution that optimizes both the cost and the acceptable level of satisfaction of the residents. In this work, we want to find the solution by multi-gent system.

In addition to the above solutions, we want to assess the impact of Plug-in Hybrid Electric Vehicle (PHEV) as a hybrid system (HS). The battery consumes electricity to be charged and also it can be discharged to provide electricity for smart home. It can assist the BMES to increase energy efficiency considering satisfaction of residents.

Smart home residents use their services to improve their comfort. These are service systems. These systems can be divided into two categories: permanent service systems (PSS) and temporary service systems. Permanent services are services that are necessary to be used over a certain period of time in a day. For instance, the home heating system should operate all day along to provide an appropriate temperature. PSSs participate in the cost reduction problem by reducing the amount of their energy consumption. For example, the energy cost is reduced by decreasing the energy consumption of the heating system. On the other hand, the temperature of the house also decreases which impacts negatively on residents' satisfaction of the services.

The second category are temporary service systems (TSS) whose operation takes place in a limited and certain time- e.g. dishwasher and washing machine. TSSs participate in the cost reduction problem by moving their operation time. For example, the washing machine can be used in the late hours of the night where the price of electricity is lower. In this case, the use of services in a time that is different with the preferences of residents causes their dissatisfaction.

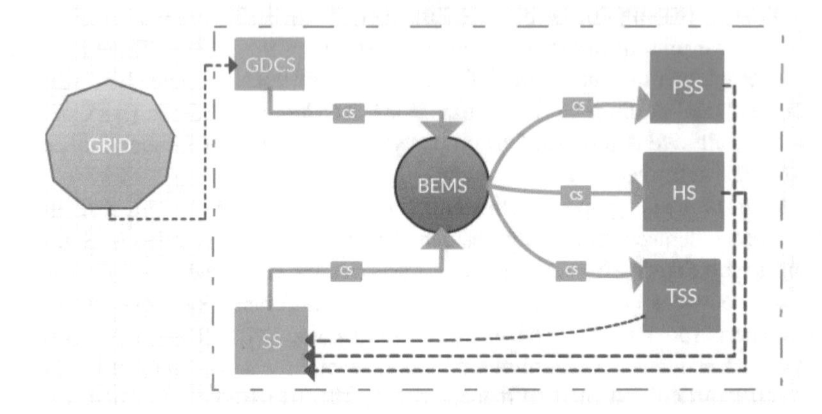

Fig. 1. General structure for multi agent-based smart home system.

Figure 1 shows the relation between different systems in smart home simply. SS and GDS obtains the information and inputs that are needed for BEMS and then this inputs will be sent to BEMS by CS. BEMS makes an appropriate decision and send it to the TSS, PSSS and HS agents.

References

1. Nassaj, A., Shahrtash, S.M.: An accelerated preventive agent based scheme for post-disturbance voltage control and loss reduction. IEEE Early Access Artic. (2018)
2. Nassaj, A., Shahrtash, S.M.: A predictive agent-based scheme for post-disturbance voltage control. Int. J. Electr. Power Energy Syst. **98**, 189–198 (2018)
3. Wooldridge, M.J.: An Introduction to Multi Agent Systems. Wiley, New York (2009)
4. Shokri Gazafroudi, A., Pinto, T., Prieto-Castrillo, F., Prieto, J., Corchado, J.M., Jozi, A., Vale, Z., Venayagamoorthy, G.K.L.: Organization-based multi-agent structure of the smart home electricity system. In: IEEE Congress on Evolutionary Computation (CEC), June 2017
5. Shokri Gazafroudi, A., Prieto-Castrillo, F., Pinto, T., Jozi, A., Vale, Z.: Economic evaluation of predictive dispatch model in MAS-based smart home. In: 15th International Conference on Practical Applications of Agents and Multi-Agent Systems (PAAMS), June 2017
6. Shokri Gazafroudi, A., De Paz, J.F., Prieto-Castrillo, F., Villarrubia, G., Talari, S., Shafie-khah, M., Catalão, J.P.S.: A review of multi-agent based energy management systems. In: 8th International Symposium on Ambient Intelligence (ISAmI), June 2017
7. Shokri Gazafroudi, A., Prieto-Castrillo, F., Pinto, T., Corchado, J.M.: Organization-based multi-agent system of local electricity market: bottom-up approach. In: 15th International Conference on Practical Applications of Agents and Multi-Agent Systems (PAAMS), June 2017
8. Shokri Gazafroudi, A., Abrishambaf, O., Jozi, A., Pinto, T., Preito-Castrillo, F., Corchado, J. M., Vale, Z.: Energy flexibility assessment of a multi agent-based smart home electricity system. In: 17th edn. of the IEEE International Conference on Ubiquitous Wireless Broadband (ICUWB), September 2017

9. Chamoso, P., Rivas, A., Martín-Limorti, J.J., Rodríguez, S.: A hash based image matching algorithm for social networks. In: Advances in Intelligent Systems and Computing, vol. 619, pp. 183–190 (2018)
10. Sittón, I., Rodríguez, S.: Pattern extraction for the design of predictive models in Industry 4.0. In: International Conference on Practical Applications of Agents and Multi-Agent Systems, pp. 258–261 (2017)
11. García, O., Chamoso, P., Prieto, J., Rodríguez, S., De La Prieta, F.: A serious game to reduce consumption in smart buildings. In: Communications in Computer and Information Science, vol. 722, pp. 481–493 (2017)
12. Palomino, C.G., Nunes, C.S., Silveira, R.A., González, S.R., Nakayama, M.K.: Adaptive agent-based environment model to enable the teacher to create an adaptive class. In: Advances in Intelligent Systems and Computing, vol. 617 (2017)
13. Canizes, B., Pinto, T., Soares, J., Vale, Z., Chamoso, P., Santos, D.: Smart city: a GECAD-BISITE energy management case study. In: 15th International Conference on Practical Applications of Agents and Multi-Agent Systems PAAMS 2017. Trends in Cyber-Physical Multi-Agent Systems, vol. 2, pp. 92–100 (2017)
14. Chamoso, P., de La Prieta, F., Eibenstein, A., Santos-Santos, D., Tizio, A., Vittorini, P.: A device supporting the self management of tinnitus. In: Lecture Notes in Computer Science (including subseries Lecture Notes in Artificial Intelligence and Lecture Notes in Bioinformatics) (LNCS), vol. 10209, pp. 399–410 (2017)
15. Román, J.A., Rodríguez, S., de da Prieta, F.: Improving the distribution of services in MAS. In: Communications in Computer and Information Science, vol. 616 (2016)
16. Buciarelli, E., Silvestri, M., González, S.R.: Decision economics. In: Commemoration of the Birth Centennial of Herbert A. Simon 1916–2016 (Nobel Prize in Economics 1978): Distributed Computing and Artificial Intelligence, 13th International Conference. Advances in Intelligent Systems and Computing, vol. 475. Springer (2016)
17. Li, T., Sun, S., Bolić, M., Corchado, J.M.: Algorithm design for parallel implementation of the SMC-PHD filter. Sig. Process. **119**, 115–127 (2016)
18. Lima, A.C.E.S., De Castro, L.N., Corchado, J.M.: A polarity analysis framework for Twitter messages. Appl. Math. Comput. **270**, 756–767 (2015)
19. Redondo-Gonzalez, E., De Castro, L.N., Moreno-Sierra, J., Maestro De Las Casas, M.L., Vera-Gonzalez, V., Ferrari, D.G., Corchado, J.M.: A cluster analysis. BioMed Res. Int. (2015)
20. Li, T., Sun, S., Corchado, J.M., Siyau, M.F.: Random finite set-based Bayesian filters using magnitude-adaptive target birth intensity. In: FUSION 2014 - 17th International Conference on Information Fusion (2014)
21. Prieto, J., Alonso, A.A., de la Rosa, R., Carrera, A.: Adaptive framework for uncertainty analysis in electromagnetic field measurements. Radiat. Prot. Dosim. (2014). https://doi.org/10.1093/rpd/ncu260
22. Chamoso, P., Raveane, W., Parra, V., González, A.: UAVs applied to the counting and monitoring of animals. In: Advances in Intelligent Systems and Computing, vol. 291, pp. 71–80 (2014)
23. Pérez, A., Chamoso, P., Parra, V., Sánchez, A.J.: Ground vehicle detection through aerial images taken by a UAV. In: 2014 17th International Conference on Information Fusion (FUSION) (2014)
24. Choon, Y.W., Mohamad, M.S., Deris, S., Illias, R.M., Chong, C.K., Chai, L.E., Corchado, J. M.: Differential bees flux balance analysis with OptKnock for in silico microbial strains optimization. PLoS ONE **9**(7), e102744 (2014)

25. Li, T., Sun, S., Corchado, J.M., Siyau, M.F.: A particle dyeing approach for track continuity for the SMC-PHD filter. In: FUSION 2014 - 17th International Conference on Information Fusion (2014)
26. García Coria, J.A., Castellanos-Garzón, J.A., Corchado, J.M.: Intelligent business processes composition based on multi-agent systems. Expert Syst. Appl. **41**(4), 1189–1205 (2014)
27. Prieto, J., Mazuelas, S., Bahillo, A., Fernández, P., Lorenzo, R.M., Abril, E.J.: Accurate and robust localization in harsh environments based on V2I communication. In: Vehicular Technologies - Deployment and Applications. INTECH Open Access Publisher (2013)
28. De La Prieta, F., Navarro, M., García, J.A., González, R., Rodríguez, S.: Multi-agent system for controlling a cloud computing environment. Lecture Notes in Computer Science (including subseries Lecture Notes in Artificial Intelligence and Lecture Notes in Bioinformatics) (LNAI), vol. 8154 (2013)
29. Tapia, D.I., Fraile, J.A., Rodríguez, S., Alonso, R.S., Corchado, J.M.: Integrating hardware agents into an enhanced multi-agent architecture for Ambient Intelligence systems. Inf. Sci. **222**, 47–65 (2013)
30. Prieto, J., Mazuelas, S., Bahillo, A., Fernandez, P., Lorenzo, R.M., Abril, E.J.: Adaptive data fusion for wireless localization in harsh environments. IEEE Trans. Signal Process. **60**(4), 1585–1596 (2012)
31. Muñoz, M., Rodríguez, M., Rodríguez, M.E., Rodríguez, S.: Genetic evaluation of the class III dentofacial in rural and urban Spanish population by AI techniques. In: Advances in Intelligent and Soft Computing (AISC), vol. 151 (2012)
32. Costa, Â., Novais, P., Corchado, J.M., Neves, J.: Increased performance and better patient attendance in an hospital with the use of smart agendas. Log. J. IGPL **20**(4), 689–698 (2012)
33. García, E., Rodríguez, S., Martín, B., Zato, C., Pérez, B.: MISIA: middleware infrastructure to simulate intelligent agents. In: Advances in Intelligent and Soft Computing, vol. 91 (2011)
34. Rodríguez, S., De La Prieta, F., Tapia, D.I., Corchado, J.M.: Agents and computer vision for processing stereoscopic images. In: Lecture Notes in Computer Science (including subseries Lecture Notes in Artificial Intelligence and Lecture Notes in Bioinformatics) (LNAI), vol. 6077 (2010)
35. Rodríguez, S., Gil, O., De La Prieta, F., Zato, C., Corchado, J.M., Vega, P., Francisco, M.: People detection and stereoscopic analysis using MAS. In: Proceedings of INES 2010 - 14th International Conference on Intelligent Engineering Systems (2010)
36. Prieto, J., Mazuelas, S., Bahillo, A., Fernández, P., Lorenzo, R.M., Abril, E.J.: On the minimization of different sources of error for an RTT-based indoor localization system without any calibration stage. In: 2010 International Conference on Indoor Positioning and Indoor Navigation (IPIN), pp. 1–6 (2010)
37. Rodríguez, S., Tapia, D.I., Sanz, E., Zato, C., De La Prieta, F., Gil, O.: Cloud computing integrated into service-oriented multi-agent architecture. In: IFIP Advances in Information and Communication Technology (AICT), vol. 322 (2010)
38. Corchado, J., Fyfe, C., Lees, B.: Unsupervised learning for financial forecasting. In: Proceedings of the IEEE/IAFE/INFORMS 1998 Conference on Computational Intelligence for Financial Engineering (CIFEr) (Cat. No. 98TH8367), pp. 259–263 (1998)
39. Durik, B.O.: Organisational metamodel for large-scale multi-agent systems: first steps towards modelling organisation dynamics. ADCAIJ Adv. Distrib. Comput. Artif. Intell. J. Salamanca **6**(3) (2017)
40. Bremer, J., Lehnhoff, S.: Decentralized coalition formation with agent-based combinatorial heuristics. ADCAIJ Adv. Distrib. Comput. Artif. Intell. J. Salamanca **6**(3), 29–44 (2017)
41. Cardoso, R.C., Bordini, R.H.: A multi-agent extension of a hierarchical task network planning formalism. ADCAIJ Adv. Distrib. Comput. Artif. Intell. J. Salamanca **6**(2), 5–17 (2017)

42. Gonçalves, E., Cortés, M., De Oliveira, M., Veras, N., Falcão, M., Castro, J.: An analysis of software agents, environments and applications school: retrospective, relevance, and trends. ADCAIJ Adv. Distrib. Comput. Artif. Intell. J. Salamanca **6**(2), 19–32 (2017)
43. Teixeira, E.P., Gonçalves, E.M., Adamatti, D.F.: Ulises: a agent-based system for timbre classification. ADCAIJ Adv. Distrib. Comput. Artif. Intell. J. Salamanca **6**(2), 29–40 (2017)
44. Souza de Castro, L.F., Vaz Alves, G., Pinz Borges, A.: Using trust degree for agents in order to assign spots in a smart parking. ADCAIJ Adv. Distrib. Comput. Artif. Intell. J. Salamanca **6**(2), 45–55 (2017)
45. Cunha, R., Billa, C., Adamatti, D.: Development of a graphical tool to integrate the prometheus AEOlus methodology and Jason platform. ADCAIJ Adv. Distrib. Comput. Artif. Intell. J. Salamanca **6**(2), 57–70 (2017)
46. Rincón, J., Poza, J.L., Posadas, J.L., Julián, V., Carrascosa, C.: Adding real data to detect emotions by means of smart resource artifacts in MAS. In: ADCAIJ Adv. Distrib. Comput. Artif. Intell. J. Salamanca **5**(4), 85–92 (2016)
47. Villavicencio, C.P., Schiaffino, S., Díaz-Pace, J.A., Monteserin, A.: A group recommendation system for movies based on MAS. ADCAIJ Adv. Distrib. Comput. Artif. Intell. J. Salamanca **5**(3), 1–12 (2016)
48. Briones, A.G., Chamoso, P., Barriuso, A.: Review of the main security problems with multi-agent systems used in E-commerce applications. ADCAIJ Adv. Distrib. Comput. Artif. Intell. J. Salamanca **5**(3), 55–61 (2016)
49. Carbó, J., Molina, J.M., Patricio, M.A.: Asset management system through the design of a Jadex agent system. ADCAIJ Adv. Distrib. Comput. Artif. Intell. J. Salamanca **5**(2), 1–14 (2016)
50. Santos, G., Pinto, T., Vale, Z., Praça, I., Morais, H.: Enabling communications in heterogeneous multi-agent systems: electricity markets ontology. ADCAIJ Adv. Distrib. Comput. Artif. Intell. J. Salamanca **5**(2), 15–42 (2016)
51. Murciego, Á.L., González, G.V., Barriuso, A.L., De La Iglesia, D.H., Herrero, J.R.: Multi agent gathering waste system. ADCAIJ Adv. Distrib. Comput. Artif. Intell. J. Salamanca **4** (4), 9–22 (2015)
52. Barriuso, A.L., De La Prieta, F., Murciego, Á.L., Hernández, D., Herrero, J.R.: JOUR-MAS: a multi-agent system approach to help journalism management. ADCAIJ Adv. Distrib. Comput. Artif. Intell. J. Salamanca **4**(4) (2015)
53. De La Iglesia, D.H., González, G.V., Barriuso, A.L., Murciego, Á.L., Herrero, J.R.: Monitoring and analysis of vital signs of a patient through a multi-agent application system. ADCAIJ Adv. Distrib. Comput. Artif. Intell. J. Salamanca **4**(3), 19–30 (2015)
54. Gallego, J.Á.R., González, S.R.: Improvement in the distribution of services in multi-agent systems with SCODA. ADCAIJ Adv. Distrib. Comput. Artif. Intell. J. Salamanca **4**(3), 31–46 (2015)
55. Chamoso, P., De La Prieta, F.: Simulation environment for algorithms and agents evaluation. ADCAIJ Adv. Distrib. Comput. Artif. Intell. J. Salamanca **4**(3) (2015)
56. González, A., Ramos, J., De Paz, J.F., Corchado, J.M.: Obtaining relevant genes by analysis of expression arrays with a multi-agent system. ADCAIJ Adv. Distrib. Comput. Artif. Intell. J. Salamanca **3**(3) (2014)
57. Faia, R., Pinto, T., Vale, Z.: Dynamic fuzzy clustering method for decision support in electricity markets negotiation. ADCAIJ Adv. Distrib. Comput. Artif. Intell. J. Salamanca **5** (1), 23–35 (2016)
58. Silva, A., Oliveira, T., Neves, J., Novais, P.: Treating colon cancer survivability prediction as a classification problem. ADCAIJ Adv. Distrib. Comput. Artif. Intell. J. Salamanca **5**(1) (2016)

59. Sánchez, D.L., Arrieta, A.G.: Preliminary results on nonparametric facial occlusion detection. ADCAIJ Adv. Distrib. Comput. Artif. Intell. J. Salamanca **5**(1), 51–61 (2016)
60. Chamoso, P., Pérez-Ramos, H., García-García, Á.: ALTAIR: supervised methodology to obtain retinal vessels caliber. ADCAIJ Adv. Distrib. Comput. Artif. Intell. J. Salamanca **3**(4), 48–57 (2014)
61. Cofini, V., De La Prieta, F., Di Mascio, T., Gennari, R., Vittorini, P.: Design Smart Games with requirements, generate them with a Click, and revise them with a GUIs. ADCAIJ Adv. Distrib. Comput. Artif. Intell. J. Salamanca **1**(3), 55–68 (2012)
62. Kushch, S., Castrillo, F.P.: A review of the applications of the Block-chain technology in smart devices and distributed renewable energy grids. ADCAIJ Adv. Distrib. Comput. Artif. Intell. J. Salamanca **6**(3) (2017)
63. Pinto, A., Costa, R.: Hash-chain-based authentication for IoT. ADCAIJ Adv. Distrib. Comput. Artif. Intell. J. Salamanca **5**(4) (2016)
64. García-Valls, M.: Prototyping low-cost and flexible vehicle diagnostic systems. ADCAIJ Adv. Distrib. Comput. Artif. Intell. J. Salamanca **5**(4), 93–103 (2016)
65. Fernández-Fernández, A., Cervelló-Pastor, C., Ochoa-Aday, L.: Energy-aware routing in multiple domains software-defined networks. ADCAIJ Adv. Distrib. Comput. Artif. Intell. J. Salamanca **5**(3), 13–19 (2016)
66. Koskimaki, H., Siirtola, P.: Accelerometer vs. electromyogram in activity recognition. ADCAIJ Adv. Distrib. Comput. Artif. Intell. J. Salamanca **5**(3), 31–42 (2016)
67. Chamoso, P., De La Prieta, F., Villarrubia, G.: Intelligent system to control electric power distribution networks. ADCAIJ Adv. Distrib. Comput. Artif. Intell. J. Salamanca **4**(4), 1–8 (2015)
68. Herrero, J.R., Villarrubia, G., Barriuso, A.L., Hernández, D., Lozano, Á., De La Serna González, M.A.: Wireless controller and smartphone based interaction system for electric bicycles. ADCAIJ Adv. Distrib. Comput. Artif. Intell. J. Salamanca **4**(4), 59–68 (2015)
69. Fernández-Isabel, A., Fuentes-Fernández, R.: Simulation of road traffic applying model-driven engineering. ADCAIJ Adv. Distrib. Comput. Artif. Intell. J. Salamanca **4**(2), 1–24
70. Chamoso, P., De La Prieta, F.: Swarm-based smart city platform: a traffic application. ADCAIJ Adv. Distrib. Comput. Artif. Intell. J. Salamanca **4**(2), 89–98 (2015)

A Novel Agent-Based Platform for Wide-Area Monitoring and Control in Power Systems

Amin Nassaj(✉)

NRI - Niroo Research Institute, Tehran, Iran
aminnasaj@gmail.com

Abstract. In this paper, a novel agent-based platform for wide-area monitoring and control of power systems is proposed, which maintains the security of network in a wide area manner. The proposed algorithm upon detecting any disturbance, employs multi-agent system (MAS) in order to coordinate the employment of controlling devices for returning the condition of system to the steady behavior. Also, via wide-area monitoring and control system (WAMCS) approach, the system will be able to prevent the propagation of large disturbances by processing wide-area information gathered from related local points.

Keywords: Multi-agent system (MAS)
Wide-area monitoring and control system (WAMCS)
Phasor measurement unit (PMU)

1 Introduction

Wide area monitoring and control system (WAMCS) is a platform for control targets, having a global management on power system contingencies with more effective contributions. WAMCS can prevent the propagation of large disturbances by processing wide-area information gathered from all/chosen local points. Indeed, WAMCS became enabled when synchronized measurement technology (SMT) detected its application in phasor measurement units (PMUs). These devices present voltage and current phasors and frequency information, synchronized with high rate and accuracy to a common time reference made by global positioning system (GPS). More key reasons for rationalization of WAMCS have been addressed in [1, 2].

Besides, in recent years, a distributed control concept named as multi agent system (MAS) has been presented by artificial intelligence researchers. By employing MAS, a complex problem is divided into some easier sub-problems and each sub-problem is appointed to an agent; while agents upon a common environment are in negotiation, cooperation, and coordination to solving problems [3]. With the development of the MAS strategy, many power engineers have endeavored to employ it to power system monitoring and control schemes [4–8]. To this end, this article introduces a novel agent-based platform for wide-area monitoring and control system in power networks.

S. Rodríguez et al. (Eds.): DCAI 2018, AISC 801, pp. 385–387, 2019.
https://doi.org/10.1007/978-3-319-99608-0_51

2 Proposed Scheme

Multi-agent systems are systems combined of multiple interacting smart elements, known as agents. The agents can be categorized into two types, including Intelligent Agent (IA) and Reactive Agent (RA) [3]. Intelligent agents are computer systems with two main abilities. First, they are capable of resolving for themselves what they need to do in order to assure their design targets. Second, they are able of negotiating with other IAs. The main characteristics of an IA are negotiation, cooperation, and coordination. In power system, the host computers in substations can treat as IAs and their collection in one of the common forms of organization founds a multi-agent system. In more detail, an MAS is a collection of IAs in companion with RAs; where the former ones negotiate and coordinate to resolve a particular problem by managing the actions assigned to RAs [1, 2].

Therefore, as a whole, the system includes two layers. The first layer presents physical equipment of power system. In the second layer, agents are set in companion with some parts of system such as, generators, transformers, busbars, loads, and switches.

Here, IAs are assumed to be those host computers fixed in a substation with the ability of receiving synchronized voltage and current phasors by PMU, processing received data, taking rational decisions, and dispatching commands to actuators devices. IAs can also negotiates with other IAs to convince the others to help it in attaining its assigned aims. Then, commands of IAs are applied to each of RAs in its territory, with the hope that system returns back into normal and safe condition. On the other hand, managing the positions and the connectivity of actuators devices (including tap changer transformers, capacitor banks, reactors and switches) are assumed to be conducted by Reactive Agents.

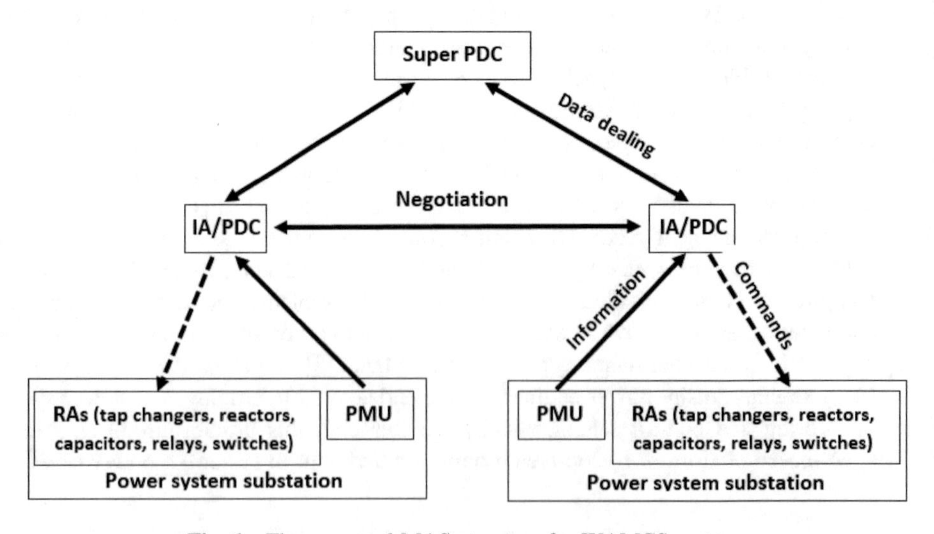

Fig. 1. The proposed MAS structure for WAMCS system.

The proposed multi agent system should be made in companion with WAMCS infrastructure, as shown in Fig. 1. Generally, there are phasor measurement units that measure and submit data at a pre-defined frame rate to phasor data concentrators (local PDCs), which have the same role and position as IAs. The data collected in this medium are processed by the associated IA, continuously, as well as those delivered to super PDC to creating a backup from data and doing central/common computing and the results being distributed back to all IAs.

References

1. Nassaj, A., Shahrtash, S.M.: An accelerated preventive agent based scheme for post-disturbance voltage control and loss reduction. Early Access IEEE Trans. Power Syst. (2018)
2. Nassaj, A., Shahrtash, S.M.: A predictive agent-based scheme for post-disturbance voltage control. Int. J. Electr. Power Energy Syst. **98**, 189–198 (2018)
3. Wooldridge, M.J.: An Introduction to Multi Agent Systems. Wiley, New York (2009)
4. Gazafroudi, A.S., Pinto, T., Prieto-Castrillo, F., Prieto, J., Corchado, J.M., Jozi, A., Vale, Z., Venayagamoorthy, G.K.: Organization-based multi-agent structure of the smart home electricity system. IEEE Congress on Evolutionary Computation (CEC), June 2017
5. Gazafroudi, A.S., Prieto-Castrillo, F., Pinto, T., Jozi, A., Vale, Z.: Economic evaluation of predictive dispatch model in MAS-based smart home. In: 15th International Conference on Practical Applications of Agents and Multi-Agent Systems (PAAMS), June 2017
6. Gazafroudi, A.S., De Paz, J.F., Prieto-Castrillo, F., Villarrubia, G., Talari, S., Shafie-khah, M., Catalão, J.P.S.: A review of multi-agent based energy management systems. In: 8th International Symposium on Ambient Intelligence (ISAmI), June 2017
7. Gazafroudi, A.S., Prieto-Castrillo, F., Pinto, T., Corchado, J.M.: Organization-based multi-agent system of local electricity market: bottom-up approach. In: 15th International Conference on Practical Applications of Agents and Multi-Agent Systems (PAAMS), June 2017
8. Gazafroudi, A.S., Abrishambaf, O., Jozi, A., Pinto, T., Preito-Castrillo, F., Corchado, J.M., Vale, Z.: Energy flexibility assessment of a multi agent-based smart home electricity system. In: 17th Edition of the IEEE International Conference on Ubiquitous Wireless Broadband (ICUWB), September 2017

An Agent-Based Model for Optimal Voltage Control and Power Quality by Electrical Vehicles in Smart Grids

Amirabbas Hadizade(✉)

Sharif University of Technology, Tehran, Iran
amirabbas.hadizade@yahoo.com

Abstract. The electric power industry is the main part of Science development, and today, with the advent of technology, the demand for electric power has been expanded. On the other hand, smart grids are developing heavily. One of the notable features of these networks is the presence of a plug-in hybrid electric vehicle (PHEV). The addition of these cars to the network has its own advantages and disadvantages. One of the most important issues in smart grids is network management and control of critical system parameters. In this paper the effect of these cars on the grid is investigated. These vehicles impose an increase in production capacity in the uncontrolled charge mode. They also have the ability to inject power into the network and can assist the grid at peak consumption time, leading to peak shaving in daily load curve. Our main goal is to provide a way to manage the charge and discharge of these vehicles using the agent based model, in order to control the voltage of the system buses.

Keywords: Optimal voltage control · PHEV · Smart grid · Agent based model
Power quality

1 Introduction

Nowadays, the power industry faces not only the supply of energy resources demanded by the industry, but also the reduction of the destructive effects, consumers have on the grid, as well as the important issues facing the industry. Smart grids are one of the solutions presented for this challenge, which has a lot of profits and efficiencies. In smart grids and in the distribution system, it can control the behavior of some controllable loads that leads to improve the voltage and current waveforms, as well as the power quality. One of the kind of controllable loads in the smart grids is plug-in electric vehicle (PEV). Charging electric vehicles has significant effects on the distribution network because these vehicles consume a lot of electrical energy, and this power demand can lead to large and unpleasant peaks in power consumption. From a power grid perspective, power losses during charging are an economic concern and should be minimized, and the transformer and feeder should avoid overloading. Not only the power loss issue, but also attention to power quality (e.g., voltage profile, voltage and current unbalance, harmonics, etc.) is also required by the distribution network as well as the consumers [1, 2]. Voltage deviation is a power quality concern. The excessive

S. Rodríguez et al. (Eds.): DCAI 2018, AISC 801, pp. 388–394, 2019.
https://doi.org/10.1007/978-3-319-99608-0_52

deviation causes reliability problems that should be avoided to ensure the desired performance of the appliances. Therefore, controlled charge is proposed to minimize power losses and maximize the power factor. It should be noted that uncontrolled charging of the electric vehicles reduces the efficiency of the distribution network. On the other hand, if the energy stored in electric vehicle that is not consumed during the day, it can inject and sell this stored energy according to the network voltage control pattern at peak times.

2 Proposed Scheme

Voltage Unbalancing (VU) is one of the major problems in power quality in LV distribution networks. In electrical networks, we try to divide the same load across all three phase distribution feeders. Although the voltages are balanced on the power side, voltages at the consumer level can be unbalanced due to different impedances and unequal distribution of single-phase loads.

One of the most important features of electric vehicles is their flexibility in connecting to power networks. With the advent of electric vehicles instead of gasoline, a new type of load is added to the power grid. Since PEVs are large load units or single phase generators, they can increase the grid VU. Although the VU, with respect to PEVs, has been discussed in the papers [3, 4], the effects of charging or discharging levels, the location of PEVs and functional states of V2G and G2V have not been completely analyzed for VU analysis.

Electric car batteries are like energy storage resources, so electric cars can connect to the power grid at optimal time for the network. This means that electric vehicles can be charged at low load times, usually at night. Thus, these electric cars can be recharged in low-cost hours, and sell the energy to grid in the peak hours of electricity consumption and high electrical energy costs. With the aid of electric vehicles, the daily load curve can be balanced and the power grid does not require the participation of small and expensive gas units. So, the electric vehicles can be called “controllable load”. The behavior of this series of loads can be modeled by a number of probabilistic functions, and according to statistical data, the level of battery capacity of different electric vehicles, the traveled distance throughout the day of each vehicle, the enter and exit time of each vehicle. On the other hand, since the purpose of this study is to control the optimal voltage and the power quality of the network, these can be managed in order to achieve this target. Smart grid in the presence of electric vehicles is composed of several components and can be investigated by multi-agent systems method. Consumers have an electric car connected to the power grid like other consumers. Also these consumers who have the ability to interact with the power grid are connected to the control center, and due to the network’s purpose, the center provides conditions for cooperation with subscribers, with the help of optimizing target functions. Readers can refer to [5–71] to know more about multi-agent systems. The schematic of this system is shown in Fig. 1.

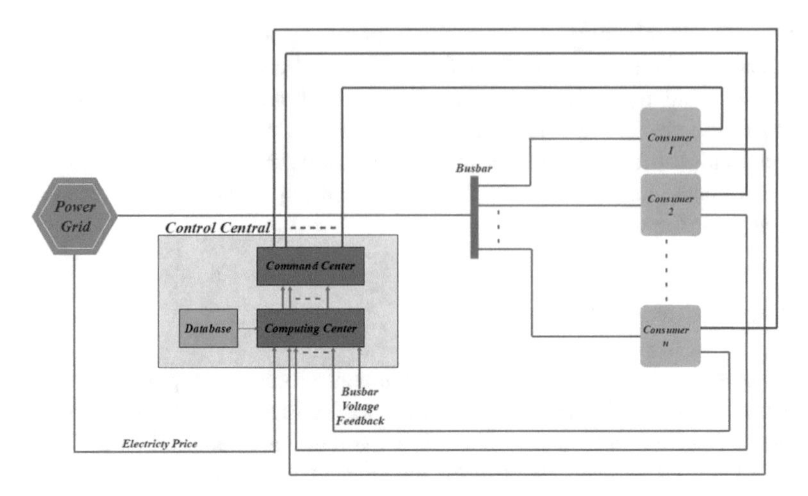

Fig. 1. General structure for smart grid in presence of electrical vehicles

When cars return home, they can notify the network control system of their next day to make the network optimize the defined parameters. The distribution network control center has a computing center. The information is sent by the users is collected in this unit and the buses voltage feedback will be transferred to the center and also by the aid of information in the database of the center-daily voltage and power curves in the past months - commands will be sent to connect the vehicles to the network for charging and discharging.

In order to optimize the smart grid parameters, it is necessary to optimize the power losses in V2G situation, voltage profile, electricity price, daily power curve and etc. Therefore, the first priority for selling power to the network should be the consumption of the nearest consumer to minimize the cost of losses. Now commands are sent from the command center to consumers.

In this optimization, not only the power losses and control the distribution network voltage can be reduced, but also the cost to consumers as well as the network will be minimized. Another advantage of this method is peak shaving of the daily power of the network, which prevents the commissioning of power supplies to provide consumers with the power they need.

References

1. Nassaj, A., Shahrtash, S.M.: An accelerated preventive agent based scheme for post-disturbance voltage control and loss reduction. IEEE Early Access Articles (2018)
2. Nassaj, A., Shahrtash, S.M.: A predictive agent-based scheme for post-disturbance voltage control. Int. J. Electr. Power Energy Syst. **98**, 189–198 (2018)
3. Fernández, L.P., Román, T.G.S., Cossent, R., Domingo, C.M., Frías, P.: Assessment of the impact of plug-in electric vehicles on distributionnetworks. IEEE Trans. Power Syst. **26**(1), 206–213 (2011)

4. Hashemi-Dezaki, H., et al.: Risk management of smart grids based onmanaged charging of PHEVs and vehicle-to-grid strategy using Monte Carlosimulation. Energy Convers. Manag. **100**, 262–276 (2015)
5. Shokri Gazafroudi, A., Pinto, T., Prieto-Castrillo, F., Prieto, J., Corchado, J.M., Jozi, A., Vale, Z., Venayagamoorthy, G.K.: Organization-based multi-agent structure of the smart home electricity system. In: IEEE Congress on Evolutionary Computation (CEC), June 2017
6. Shokri Gazafroudi, A., Prieto-Castrillo, F., Pinto, T., Jozi, A., Vale, Z.: Economic evaluation of predictive dispatch model in MAS-based smart home. In: 15th International Conference on Practical Applications of Agents and Multi-Agent Systems (PAAMS), June 2017
7. Shokri Gazafroudi, A., De Paz, J.F., Prieto-Castrillo, F., Villarrubia, G., Talari, S., Shafie-khah, M., Catalão, J.P.S.: A review of multi-agent based energy management systems. In: 8th International Symposium on Ambient Intelligence (ISAmI), June 2017
8. Shokri Gazafroudi, A., Prieto-Castrillo, F., Pinto, T., Corchado, J.M.: Organization-based multi-agent system of local electricity market: bottom-up approach. In: 15th International Conference on Practical Applications of Agents and Multi-Agent Systems (PAAMS), June 2017
9. Shokri Gazafroudi, A., Abrishambaf, O., Jozi, A., Pinto, T., Preito-Castrillo, F., Corchado, J. M., Vale, Z.: Energy flexibility assessment of a multi agent-based smart home electricity system. In: 17th edition of the IEEE International Conference on Ubiquitous Wireless Broadband (ICUWB), September 2017
10. Chamoso, P., Rivas, A., Martín-Limorti, J.J., Rodríguez, S.: A hash based image matching algorithm for social networks. In: Advances in Intelligent Systems and Computing, vol. 619, pp. 183–190 (2018)
11. Sittón, I., Rodríguez, S.: Pattern extraction for the design of predictive models in industry 4.0. In: International Conference on Practical Applications of Agents and MultiAgent Systems, pp. 258–261 (2017)
12. García, O., Chamoso, P., Prieto, J., Rodríguez, S., De La Prieta, F.: A serious game to reduce consumption in smart buildings. In: Communications in Computer and Information Science, vol. 722, pp. 481–493 (2017)
13. Palomino, C.G., Nunes, C.S., Silveira, R.A., González, S.R., Nakayama, M.K.: Adaptive agent-based environment model to enable the teacher to create an adaptive class. Adv. Intell. Syst. Comput. vol. 617 (2017)
14. Canizes, B., Pinto, T., Soares, J., Vale, Z., Chamoso, P., Santos, D.: Smart City: a GECAD-BISITE energy management case study. In: 15th International Conference on Practical Applications of Agents and Multi-Agent Systems PAAMS 2017, Trends in CyberPhysical Multi-Agent Systems, vol. 2, pp. 92–100 (2017)
15. Chamoso, P., de La Prieta, F., Eibenstein, A., Santos-Santos, D., Tizio, A., Vittorini, P.: A device supporting the self management of tinnitus. In: Lecture Notes in Computer Science (including subseries Lecture Notes in Artificial Intelligence and Lecture Notes in Bioinformatics). LNCS, vol. 10209, pp. 399–410 (2017)
16. Román, J.A., Rodríguez, S., de da Prieta, F.: Improving the distribution of services in MAS. Commun. Comput. Inf. Sci. **616** (2016)
17. Buciarelli, E., Silvestri, M., González, S.R.: Decision economics. In: 13th International Conference Commemoration of the Birth Centennial of Herbert A. Simon 1916–2016 (Nobel Prize in Economics 1978): Distributed Computing and Artificial Intelligence. Advances in Intelligent Systems and Computing, vol. 475. Springer (2016)
18. Li, T., Sun, S., Bolić, M., Corchado, J.M.: Algorithm design for parallel implementation of the SMC-PHD filter. Sig. Process. **119**, 115–127 (2016)
19. Lima, A.C.E.S., De Castro, L.N., Corchado, J.M.: A polarity analysis framework for Twitter messages. Appl. Math. Comput. **270**, 756–767 (2015)

20. Redondo-Gonzalez, E., De Castro, L.N., Moreno-Sierra, J., Maestro De Las Casas, M.L., Vera-Gonzalez, V., Ferrari, D.G., Corchado, J.M.: Bladder carcinoma data with clinical risk factors and molecular markers: a cluster analysis. BioMed Res. Int. (2015)
21. Li, T., Sun, S., Corchado, J.M., Siyau, M.F.: Random finite set-based Bayesian filters using magnitude-adaptive target birth intensity. In: FUSION 2014 - 17th International Conference on Information Fusion (2014)
22. Prieto, J., Alonso, A.A., de la Rosa, R., Carrera, A.: Adaptive framework for uncertainty analysis in electromagnetic field measurements. Radiat. Prot. Dosimetry (2014). ncu260
23. Chamoso, P., Raveane, W., Parra, V., González, A.: UAVs applied to the counting and monitoring of animals. In: Advances in Intelligent Systems and Computing, vol. 291, pp. 71–80 (2014)
24. Pérez, A., Chamoso, P., Parra, V., Sánchez, A.J.: Ground vehicle detection through aerial images taken by a UAV. In: 2014 17th International Conference on Information Fusion (FUSION) (2014)
25. Choon, Y.W., Mohamad, M.S., Deris, S., Illias, R.M., Chong, C.K., Chai, L.E., Corchado, J. M.: Differential bees flux balance analysis with OptKnock for in silico microbial strains optimization. PLoS ONE **9**(7) (2014)
26. Li, T., Sun, S., Corchado, J.M., Siyau, M.F.: A particle dyeing approach for track continuity for the SMC-PHD filter. In: FUSION 2014 - 17th International Conference on Information Fusion (2014)
27. García Coria, J.A., Castellanos-Garzón, J.A., Corchado, J.M.: Intelligent business processes composition based on multi-agent systems. Exp. Syst. Appl. **41**(4 PART 1), 1189–1205 (2014)
28. Prieto, J., Mazuelas, S., Bahillo, A., Fernández, P., Lorenzo, R.M., Abril, E.J.: Accurate and robust localization in harsh environments based on V2I communication. In: Vehicular Technologies - Deployment and Applications. INTECH Open Access Publisher (2013)
29. De La Prieta, F., Navarro, M., García, J.A., González, R., Rodríguez, S.: Multiagent system for controlling a cloud computing environment. In: Lecture Notes in Computer Science (including subseries Lecture Notes in Artificial Intelligence and Lecture Notes in Bioinformatics). LNAI, vol. 8154 (2013)
30. Tapia, D.I., Fraile, J.A., Rodríguez, S., Alonso, R.S., Corchado, J.M.: Integrating hardware agents into an enhanced multi-agent architecture for Ambient Intelligence systems. Inf. Sci. **222**(47–65), 5 (2013)
31. Prieto, J., Mazuelas, S., Bahillo, A., Fernandez, P., Lorenzo, R.M., Abril, E.J.: Adaptive data fusion for wireless localization in harsh environments. IEEE Trans. Signal Process. **60**(4), 1585–1596 (2012)
32. Muñoz, M., Rodríguez, M., Rodríguez, M.E., Rodríguez, S.: Genetic evaluation of the class III dentofacial in rural and urban Spanish population by AI techniques. In: Advances in Intelligent and Soft Computing. AISC, vol. 151 (2012)
33. Costa, Â., Novais, P., Corchado, J.M., Neves, J.: Increased performance and better patient attendance in an hospital with the use of smart agendas. Logic J. IGPL (2012)
34. García, E., Rodríguez, S., Martín, B., Zato, C., Pérez, B.: MISIA: middleware infrastructure to simulate intelligent agents. In: Advances in Intelligent and Soft Computing, vol. 91 (2011)
35. Rodríguez, S., De La Prieta, F., Tapia, D.I., Corchado, J.M.: Agents and computer vision for processing stereoscopic images. In: Lecture Notes in Computer Science (including subseries Lecture Notes in Artificial Intelligence and Lecture Notes in Bioinformatics). LNAI, vol. 6077 (2010)
36. Rodríguez, S., Gil, O., De La Prieta, F., Zato, C., Corchado, J.M., Vega, P., Francisco, M.: People detection and stereoscopic analysis using MAS. In: Proceedings of INES 2010 - 14th International Conference on Intelligent Engineering Systems (2010)

37. Prieto, J., Mazuelas, S., Bahillo, A., Fernández, P., Lorenzo, R.M., Abril, E.J.: On the minimization of different sources of error for an RTT-based indoor localization system without any calibration stage. In: 2010 International Conference on Indoor Positioning and Indoor Navigation (IPIN), pp. 1–6 (2010)
38. Rodríguez, S., Tapia, D.I., Sanz, E., Zato, C., De La Prieta, F., Gil, O.: Cloud computing integrated into service-oriented multi-agent architecture. In: IFIP Advances in Information and Communication Technology. AICT, vol. 322 (2010)
39. Corchado, J., Fyfe, C., Lees, B.: Unsupervised learning for financial forecasting. In: Proceedings of the IEEE/IAFE/INFORMS 1998 Conference on Computational Intelligence for Financial Engineering (CIFEr) (Cat. No. 98TH8367), pp. 259–263 (1998)
40. Durik, B.O.: Organisational metamodel for large-scale multi-agent systems: first steps towards modelling organisation dynamics. ADCAIJ Adv. Distrib. Comput. Artif. Intell. J. **6** (3) (2017)
41. Bremer, J., Lehnhoff, S.: Decentralized coalition formation with agent based combinatorial heuristics. ADCAIJ Adv. Distrib. Comput. Artif. Intell. J. **6**(3) (2017)
42. Cauê Cardoso, R., Bordini, R.H.: A multi-agent extension of a hierarchical task network planning formalism. ADCAIJ Adv. Distrib. Comput. Artif. Intell. J. **6**(2) (2017)
43. Gonçalves, E., Cortés, M., De Oliveira, M., Veras, N., Falcão, M., Castro, J.: An analysis of software agents, environments and applications school: retrospective, relevance, and trends. ADCAIJ Adv. Distrib. Comput. Artif. Intell. J. **6**(2) (2017)
44. Teixeira, E.P., Goncalves, E.M.N., Adamatti, D.F.: Ulises: a agent based system for timbre classification. ADCAIJ Adv. Distrib. Comput. Artif. Intell. J. **6**(2) (2017)
45. De Castro, L.F.S., Vaz Alves, G., Borges, A.P.: Using trust degree for agents in order to assign spots in a Smart Parking. ADCAIJ Adv. Distrib. Comput. Artif. Intell. J. **6**(2.6) (2017)
46. Cunha, R., Billa, C., Adamatti, D.: Development of a graphical tool to integrate the prometheus AEOlus methodology and Jason platform. ADCAIJ Adv. Distrib. Comput. Artif. Intell. J. **6**(2) (2017)
47. Rincón, J., Poza, J.L., Posadas, J.L., Julián, V., Carrascosa, C.: Adding real data to detect emotions by means of smart resource artifacts in MAS. ADCAIJ Adv. Distrib. Comput. Artif. Intell. J. **5**(4) (2016)
48. Villavicencio, C.P., Schiaffino, S., Andrés Díaz-Pace, J., Monteserin, A.: A group recommendation system for movies based on MAS. ADCAIJ Adv. Distrib. Comput. Artif. Intell. J. **5**(3) (2016)
49. Briones, A.G., Chamoso, P., Barriuso, A.: Review of the main security problems with multi-agent systems used in e-commerce applications. ADCAIJ Adv. Distrib. Comput. Artif. Intell. J. **5**(3) (2016)
50. Carbó, J., Molina, J.M., Patricio, M.A.: Asset management system through the design of a Jadex agent system. ADCAIJ Adv. Distrib. Comput. Artif. Intell. J. **5**(2) (2016)
51. Santos, G., Pinto, T., Vale, Z., Praça, I., Morais, H.: Enabling communications in heterogeneous multi-agent systems: electricity markets ontology. ADCAIJ Adv. Distrib. Comput. Artif. Intell. J. **5**(2) (2016)
52. Murciego, Á.L., González, G.V., Barriuso, A.L., De La Iglesia, D.H., Herrero, J.R.: Multi agent gathering waste system. ADCAIJ Adv. Distrib. Comput. Artif. Intell. J. **4**(4) (2015)
53. Barriuso, A.L., De La Prieta, F., Murciego, Á.L., Hernández, D., Herrero, J.R.: JOUR-MAS: a multi-agent system approach to help journalism management. ADCAIJ Adv. Distrib. Comput. Artif. Intell. J. **4**(4) (2015)
54. De La Iglesia, D.H., González, G.V., Barriuso, A.L., Murciego, Á.L., Herrero, J.R.: Monitoring and analysis of vital signs of a patient through a multi-agent application system. ADCAIJ Adv. Distrib. Comput. Artif. Intell. J. **4**(3) (2015)

55. Gallego, J.Á.R., González, S.R.: Improvement in the distribution of services in multi-agent systems with SCODA. ADCAIJ Adv. Distrib. Comput. Artif. Intell. J. **4**(3) (2015)
56. Chamoso, P., De La Prieta, F.: Simulation environment for algorithms and agents evaluation. ADCAIJ Adv. Distrib. Comput. Artif. Intell. J. **4**(3) (2015)
57. González, A., Ramos, J., De Paz, J.F., Corchado, J.M.: Obtaining relevant genes by analysis of expression arrays with a multi-agent system. ADCAIJ Adv. Distrib. Comput. Artif. Intell. J. **3**(3) (2014)
58. Faia, R., Pinto, T., Vale, Z.: Dynamic fuzzy clustering method for decision support in electricity markets negotiation. ADCAIJ Adv. Distrib. Comput. Artif. Intell. J. **5**(1) (2016)
59. Silva, A., Oliveira, T., Neves, J., Novais, P.: Treating colon cancer survivability prediction as a classification problem. ADCAIJ Adv. Distrib. Comput. Artif. Intell. J. **5**(1.7) (2016)
60. Sánchez, D.L., Arrieta, A.G.: Preliminary results on nonparametric facial occlusion detection. ADCAIJ Adv. Distrib. Comput. Artif. Intell. J. **5**(1) (2016)
61. Chamoso, P., Pérez-Ramos, H., García-García, Á.: ALTAIR: supervised methodology to obtain retinal vessels caliber. ADCAIJ Adv. Distrib. Comput. Artif. Intell. J. **3**(4) (2014)
62. Cofini, V., De La Prieta, F., Di Mascio, T., Gennari, R., Vittorini, P.: Design smart games with requirements, generate them with a click, and revise them with a GUIs. ADCAIJ Adv. Distrib. Comput. Artif. Intell. J. **1**(3) (2012)
63. Kushch, S., Castrillo, F.P.: A review of the applications of the Blockchain technology in smart devices and distributed renewable energy grids. ADCAIJ Adv. Distrib. Comput. Artif. Intell. J. **6**(3) (2017)
64. Pinto, A., Costa, R.: Hash-chain-based authentication for IoT. ADCAIJ Adv. Distrib. Comput. Artif. Intell. J. **5**(4) (2016)
65. García-Valls, M.: Prototyping low-cost and flexible vehicle diagnostic systems. ADCAIJ Adv. Distrib. Comput. Artif. Intell. J. **5**(4) (2016)
66. Fernández-Fernández, A., Cervelló-Pastor, C., Ochoa-Aday, L.: Energy-aware routing in multiple domains software-defined networks. ADCAIJ Adv. Distrib. Comput. Artif. Intell. J. **5**(3) (2016)
67. Koskimaki, H., Siirtola, P.: Accelerometer vs. electromyogram in activity recognition. ADCAIJ Adv. Distrib. Comput. Artif. Intell. J. **5**(3) (2016)
68. Chamoso, P., De La Prieta, F., Villarrubia, G.: Intelligent system to control electric power distribution networks. ADCAIJ Adv. Distrib. Comput. Artif. Intell. J. **4**(4) (2015)
69. Herrero, J.R., Villarrubia, G., Barriuso, A.L., Hernández, D., Lozano, Á., De La Serna González, M.A.: Wireless controller and smartphone based interaction system for electric bicycles. ADCAIJ Adv. Distrib. Comput. Artif. Intell. J. **4**(4) (2015)
70. Fernández-Isabel, A., Fuentes-Fernández, R.: Simulation of road traffic applying model-driven engineering. ADCAIJ Adv. Distrib. Comput. Artif. Intell. J. **4**(2) (2015)
71. Chamoso, P., De La Prieta, F.: Swarm-based smart city platform: a traffic application. ADCAIJ Adv. Distrib. Comput. Artif. Intell. J. **4**(21) (2015)

Stock Recommendation Platform Based on the Environment. INSIDER

Elena Hernández Nieves(✉)

BISITE Digital Innovation Hub, University of Salamanca, Edificio Multiusos I+D+I, 37007 Salamanca, Spain
elenahn@ieee.org

Abstract. The research presented in this paper, focuses on an investment recommendation system for businesses in order to provide investment related suggestions. For this purpose, it is identified different factors that could be extracted from the internet and from the information provided by the users. Currently, the research is in its initial stage, it has been reviewed the literature on data based techniques for investment recommendations, which will provide a complete overview of the methodologies, techniques and recent developments in this field. Once the state of the art has been reviewed, the platform model developed through a virtual organization of agents, called INSIDER.

Keywords: Recommender system · Virtual organization of agents
Hybrid A.I. algorithm · Investment decisions

1 Proposal

Social networks collect a large amount of events that affect the stock market, internal developments of companies such as mergers, acquisitions, frauds; global events such as wars, terrorism and natural disasters or macroeconomic policies that generate movements over inflation, interest rates, etc. In this article is addressed a new way of make investment portfolio recommendations taking into account the environment to improve the decision making: an investment recommendation system for businesses in order to provide investment related suggestions.

In the field of management, the most commonly used approaches are optimization and machine learning. Li and Hoi [31] applied machine learning as online decision support system, Wang (2015) [54] applied a different model to stock operations, Hadavandi *et al.* en [26], applied neural networks and integrated genetic fuzzy systems to predict performance on stock markets. In this research it is proposed a Virtual Organization of agents with new human-agent interaction modules which allow to improve user experience. The proposal consisted in adding Agent-based Virtual Organizations for investment recommendation. VOs were implemented with the aim of creating a light, well-structured, scalable and user-adapted system that integrate different capabilities, of which the most notable are autonomy, reactivity, proactivity, learning, ubiquitous distributed communication and most importantly the intelligence of all their elements. This distributed design is going to facilitate subsequent development and allow for future modifications and extensions.

S. Rodríguez et al. (Eds.): DCAI 2018, AISC 801, pp. 395–399, 2019.
https://doi.org/10.1007/978-3-319-99608-0_53

It is being considered to perform a sample analysis of the user's characteristics to identify the level of risk assumed. Once a significant sample has been obtained, the data collected could be from the IBEX35 history stock market. The hybrid algorithm HBP-PSO [25, 29] will be applied and the results obtained will be shown.

References

1. Baruque, B., Corchado, E., Mata, A., Corchado, J.M.: A forecasting solution to the oil spill problem based on a hybrid intelligent system. Inf. Sci. **180**(10), 2029–2043 (2010). https://doi.org/10.1016/j.ins.2009.12.032
2. Durik, B.O.: Organisational metamodel for large-scale multi-agent systems: first steps towards modelling organisation dynamics. ADCAIJ Adv. Distrib. Comput. Artif. Intell. J. **6** (3), Salamanca (2017)
3. Buciarelli, E., Silvestri, M., González, S.R.: Decision economics. In: Commemoration of the Birth Centennial of Herbert A. Simon 1916–2016 (Nobel Prize in Economics 1978): Distributed Computing and Artificial Intelligence, 13th International Conference. Advances in Intelligent Systems and Computing, vol. 475. Springer (2016)
4. Chamoso, P., Rivas, A., Martín-Limorti, J.J., Rodríguez, S.: A hash based image matching algorithm for social networks. In: Advances in Intelligent Systems and Computing, vol. 619, pp. 183–190 (2018). https://doi.org/10.1007/978-3-319-61578-3_18
5. Choon, Y.W., Mohamad, M.S., Deris, S., Illias, R.M., Chong, C.K., Chai, L.E., Corchado, J. M.: Differential bees flux balance analysis with OptKnock for in silico microbial strains optimization. PLoS ONE **9**(7) (2014). https://doi.org/10.1371/journal.pone.0102744
6. Corchado, J.A., Aiken, J., Corchado, E.S., Lefevre, N., Smyth, T.: Quantifying the Ocean's CO2 budget with a CoHeL-IBR system. In: Advances in Case-Based Reasoning, Proceedings, vol. 3155, pp. 533–546
7. Corchado, J.M., Aiken, J.: Hybrid artificial intelligence methods in oceanographic forecast models. IEEE Trans. Syst. Man Cybern. Part C-Appl. Rev. **32**(4), 307–313 (2002). https://doi.org/10.1109/tsmcc.2002.806072
8. Corchado, J.M., Fyfe, C.: Unsupervised neural method for temperature forecasting. Artif. Intell. Eng. **13**(4), 351–357 (1999). https://doi.org/10.1016/S0954-1810(99)00007-2
9. Corchado, J.M., Borrajo, M.L., Pellicer, M.A., Yáñez, J.C.: Neuro-symbolic system for business internal control. In: Industrial Conference on Data Mining, pp. 1–10 (2004). https://doi.org/10.1007/978-3-540-30185-1_1
10. Corchado, J.M., Corchado, E.S., Aiken, J., Fyfe, C., Fernandez, F., Gonzalez, M.: Maximum likelihood Hebbian learning based retrieval method for CBR systems. In: Lecture Notes in Computer Science (Including Subseries Lecture Notes in Artificial Intelligence and Lecture Notes in Bioinformatics), vol. 2689, pp. 107–121 (2003). https://doi.org/10.1007/3-540-45006-8_11
11. Corchado, J.M., Pavón, J., Corchado, E. S., Castillo, L.F.: Development of CBR-BDI agents: a tourist guide application. In: Lecture Notes in Computer Science (Including Subseries Lecture Notes in Artificial Intelligence and Lecture Notes in Bioinformatics), vol. 3155, pp. 547–559 (2004). https://doi.org/10.1007/978-3-540-28631-8
12. Corchado, J., Fyfe, C., Lees, B.: Unsupervised learning for financial forecasting. In: Proceedings of the IEEE/IAFE/INFORMS 1998 Conference on Computational Intelligence for Financial Engineering (CIFEr) (Cat. No. 98TH8367), pp. 259–263 (1998). https://doi.org/10.1109/CIFER.1998.690316

13. Costa, Â., Novais, P., Corchado, J.M., Neves, J.: Increased performance and better patient attendance in a hospital with the use of smart agendas. Logic J. IGPL **20**(4), 689–698 (2012). https://doi.org/10.1093/jigpal/jzr021
14. De La Prieta, F., Navarro, M., García, J.A., González, R., Rodríguez, S.: Multi-agent system for controlling a cloud computing environment. In: Lecture Notes in Computer Science (Including Subseries Lecture Notes in Artificial Intelligence and Lecture Notes in Bioinformatics), LNAI, vol. 8154 (2013). https://doi.org/10.1007/978-3-642-40669-0_2
15. Gonçalves, E., Cortés, M., De Oliveira, M., Veras, N., Falcão, M., Castro, J.: An analysis of software agents, environments and applications school: retrospective, relevance, and trends. ADCAIJ: Adv. Distrib. Comput. Artif. Intell. J. **6**(2), Salamanca (2017)
16. Fdez-Riverola, F., Corchado, J.M.: CBR based system for forecasting red tides. Knowl. Based Syst. **16**(5–6 SPEC.), 321–328 (2003). https://doi.org/10.1016/S0950-7051(03)00034-0
17. Fdez-Rtverola, F., Corchado, J.M.: FSfRT: forecasting system for red tides. Appl. Intell. **21**(3), 251–264 (2004). https://doi.org/10.1023/B:APIN.0000043558.52701.b1
18. Fernández-Riverola, F., Díaz, F., Corchado, J.M.: Reducing the memory size of a fuzzy case-based reasoning system applying rough set techniques. IEEE Trans. Syst. Man Cybern. Part C Appl. Rev. **37**(1), 138–146 (2007). https://doi.org/10.1109/TSMCC.2006.876058
19. Fyfe, C., Corchado, J.: A comparison of kernel methods for instantiating case based reasoning systems. Adv. Eng. Inf. **16**(3), 165–178 (2002). https://doi.org/10.1016/S1474-0346(02)00008-3
20. Fyfe, C., Corchado, J.M.: Automating the construction of CBR systems using kernel methods. Int. J. Intell. Syst. **16**(4), 571–586 (2001). https://doi.org/10.1002/int.1024
21. Coria, J.A.G., Castellanos-Garzón, J.A., Corchado, J.M.: Intelligent business processes composition based on multi-agent systems. Expert Syst. Appl. **41**(4 Part 1), 1189–1205 (2014). https://doi.org/10.1016/j.eswa.2013.08.003
22. Glez-Bedia, M., Corchado, J.M., Corchado, E.S., Fyfe, C.: Analytical model for constructing deliberative agents. Int. J. Eng. Intell. Syst. Electr. Eng. Commun. **10**(3) (2002)
23. Glez-Peña, D., Díaz, F., Hernández, J.M., Corchado, J.M., Fdez-Riverola, F.: geneCBR: a translational tool for multiple-microarray analysis and integrative information retrieval for aiding diagnosis in cancer research. BMC Bioinf. **10** (2009). https://doi.org/10.1186/1471-2105-10-187
24. Isaza, G., Mejía, M.H., Castillo, L.F., Morales, A., Duque, N.: Network management using multi-agents system. ADCAIJ: Adv. Distrib. Comput. Artif. Intell. J. **1**(3), Salamanca (2012)
25. Han, F., Gu, T.Y., Ju, S.G.: An improved hybrid algorithm based on PSO and BP for feedforward neural networks. JDCTA: Int. J. Digital Content Technol. Appl. **5**(2), 106–115 (2011)
26. Havandi, E., Shavandi, H., Ghanbari, A.: Integration of genetic fuzzy systems and artificial neural networks for stock price forecasting. Knowl. Syst. **23**(8), 800–808 (2010)
27. Hüllermeier, E., Minor, M. (eds.): Case-Based Reasoning Research and Development: 22nd International Conference, ICCBR 2014, Cork, Ireland, 29 September–1 October 2014. Proceedings, vol. 8765. Springer (2015)
28. Bremer, J., Lehnhoff, S.: Decentralized coalition formation with agent-based combinatorial heuristics. ADCAIJ Adv. Distrib. Comput. Artif. Intell. J. **6**(3), Salamanca (2017)
29. Kaastra, I., Boyd, M.: Designing a neural network for forecasting financial and economic time series. Neurocomputing **10**(3), 215–236 (1996)
30. Laza, R., Pavn, R., Corchado, J.M.: A reasoning model for CBR_BDI agents using an adaptable fuzzy inference system. In: Lecture Notes in Computer Science (Including Subseries Lecture Notes in Artificial Intelligence and Lecture Notes in Bioinformatics), vol. 3040, pp. 96–106. Springer, Heidelberg (2004)

31. Li, B., Hoi, S.C.: Online portfolio selection: a survey. ACM Comput. Surv. (CSUR) **46**(3), 35 (2014)
32. Li, T., Sun, S., Bolić, M., Corchado, J.M.: Algorithm design for parallel implementation of the SMC-PHD filter. Signal Process. **119**, 115–127 (2016). https://doi.org/10.1016/j.sigpro.2015.07.013
33. Li, T., Sun, S., Corchado, J.M., Siyau, M.F.: A particle dyeing approach for track continuity for the SMC-PHD filter. In: FUSION 2014 - 17th International Conference on Information Fusion (2014). https://www.scopus.com/inward/record.uri?eid=2-s2.0-84910637583&partnerID=40&md5=709eb4815eaf544ce01a2c21aa749d8f
34. Li, T., Sun, S., Corchado, J.M., Siyau, M.F.: Random finite set-based Bayesian filters using magnitude-adaptive target birth intensity. In: FUSION 2014 - 17th International Conference on Information Fusion (2014). https://www.scopus.com/inward/record.uri?eid=2-s2.0-84910637788&partnerID=40&md5=bd8602d6146b014266cf07dc35a681e0
35. Li, T.-C., Su, J.-Y., Liu, W., Corchado, J.M.: Approximate Gaussian conjugacy: parametric recursive filtering under nonlinearity, multimodality, uncertainty, and constraint, and beyond. Front. Inf. Technol. Electr. Eng. **18**(12), 1913–1939 (2017)
36. Lima, A.C.E.S., De Castro, L.N., Corchado, J.M.: A polarity analysis framework for Twitter messages. Appl. Math. Comput. **270**, 756–767 (2015). https://doi.org/10.1016/j.amc.2015.08.059
37. Mata, A., Corchado, J.M.: Forecasting the probability of finding oil slicks using a CBR system. Expert Syst. Appl. **36**(4), 8239–8246 (2009). https://doi.org/10.1016/j.eswa.2008.10.003
38. Méndez, J.R., Fdez-Riverola, F., Díaz, F., Iglesias, E.L., Corchado, J.M.: A comparative performance study of feature selection methods for the anti-spam filtering domain. In: Lecture Notes in Computer Science (Including Subseries Lecture Notes in Artificial Intelligence and Lecture Notes in Bioinformatics), LNAI, vol. 4065, pp. 106–120 (2006). https://www.scopus.com/inward/record.uri?eid=2-s2.0-33746435792&partnerID=40&md5=25345ac884f61c182680241828d448c5
39. Méndez, J.R., Fdez-Riverola, F., Iglesias, E.L., Díaz, F., Corchado, J.M.: Tracking concept drift at feature selection stage in SpamHunting: an anti-spam instance-based reasoning system. In: Lecture Notes in Computer Science (Including Subseries Lecture Notes in Artificial Intelligence and Lecture Notes in Bioinformatics), LNAI, vol. 4106, pp. 504–518 (2006). https://www.scopus.com/inward/record.uri?eid=2-s2.0-33750974465&partnerID=40&md5=f468552f565ecc3af2d3ca6336e09cc2
40. Morente-Molinera, J.A., Kou, G., González-Crespo, R., Corchado, J.M., Herrera-Viedma, E.: Solving multi-criteria group decision making problems under environments with a high number of alternatives using fuzzy ontologies and multi-granular linguistic modelling methods. Knowl. Based Syst. **137**, 54–64 (2017)
41. Chamoso, P., De La Prieta, F.: Simulation environment for algorithms and agents evaluation. ADCAIJ Adv. Distrib. Comput. Artif. Intell. J. **4**(3), Salamanca (2015)
42. Palomino, C.G., Nunes, C.S., Silveira, R.A., González, S.R., Nakayama, M.K.: Adaptive agent-based environment model to enable the teacher to create an adaptive class. In: Advances in Intelligent Systems and Computing, vol. 617 (2017). https://doi.org/10.1007/978-3-319-60819-8_3
43. Pinto, T., Gazafroudi, A.S., Prieto-Castrillo, F., Santos, G., Silva, F., Corchado, J.M., Vale, Z.: Reserve costs allocation model for energy and reserve market simulation. In: 2017 19th International Conference on Intelligent System Application to Power Systems, ISAP 2017, Art. no. 8071410 (2017)

44. Cardoso, R.C., Bordini, R.H.: A multi-agent extension of a hierarchical task network planning formalism. ADCAIJ Adv. Distrib. Comput. Artif. Intell. J. **6**(2), Salamanca (2017)
45. Redondo-Gonzalez, E., De Castro, L.N., Moreno-Sierra, J., De Las Casas, M.L.M., Vera-Gonzalez, V., Ferrari, D.G., Corchado, J.M.: Bladder carcinoma data with clinical risk factors and molecular markers: a cluster analysis. BioMed Res. Int. (2015). https://doi.org/10.1155/2015/168682
46. Silveira, R., Da Silva Bitencourt, G.K., Gelaim, T.Â., Marchi, J., De La Prieta, F.: Towards a model of open and reliable cognitive multiagent systems: dealing with trust and emotions. ADCAIJ Adv. Distrib. Comput. Artif. Intell. J. **4**(3), Salamanca (2015)
47. Rodríguez, S., De La Prieta, F., Tapia, D.I., Corchado, J.M.: Agents and computer vision for processing stereoscopic images. In: Lecture Notes in Computer Science (Including Subseries Lecture Notes in Artificial Intelligence and Lecture Notes in Bioinformatics), LNAI, vol. 6077 (2010). https://doi.org/10.1007/978-3-642-13803-4_12
48. Rodríguez, S., Gil, O., De La Prieta, F., Zato, C., Corchado, J.M., Vega, P., Francisco, M.: People detection and stereoscopic analysis using MAS. In: INES 2010 - 14th International Conference on Intelligent Engineering Systems, Proceedings (2010). https://doi.org/10.1109/INES.2010.5483855
49. Rodríguez, S., Tapia, D.I., Sanz, E., Zato, C., De La Prieta, F., Gil, O.: Cloud computing integrated into service-oriented multi-agent architecture. IFIP Adv. Inf. Commun. Technol. (AICT) **322** (2010). https://doi.org/10.1007/978-3-642-14341-0_29
50. Román, J.A., Rodríguez, S., de da Prieta, F.: Improving the distribution of services in MAS. Commun. Comput. Inf. Sci. **616** (2016). https://doi.org/10.1007/978-3-319-39387-2_4
51. Sittón, I., Rodríguez, S.: Pattern extraction for the design of predictive models in industry 4.0. In: International Conference on Practical Applications of Agents and Multi-agent Systems, pp. 258–261 (2017)
52. Tapia, D.I., Corchado, J.M.: An ambient intelligence based multi-agent system for alzheimer health care. Int. J. Ambient Comput. Intell. **1**(1), 15–26 (2009). https://doi.org/10.4018/jaci.2009010102
53. Tapia, D.I., Fraile, J.A., Rodríguez, S., Alonso, R.S., Corchado, J.M.: Integrating hardware agents into an enhanced multi-agent architecture for ambient intelligence systems. Inf. Sci. **222**, 47–65 (2013). https://doi.org/10.1016/j.ins.2011.05.002
54. Wang, L.X.: Dynamical models of stock prices based on technical trading rules part I: the models. IEEE Trans. Fuzzy Syst. **23**(4), 787–801 (2015)
55. Wang, X., Li, T., Sun, S., Corchado, J.M.: A survey of recent advances in particle filters and remaining challenges for multitarget tracking. Sensors (Switzerland), **17**(12), Art. no. 2707 (2017)

Interaction Analysis in Asperger Through Virtual Media of Text Communication

Manuel A. Arroyave[1(✉)], Luis F. Castillo[2,3], Gustavo A. Isaza[2], and Manuel G. Bedia[4]

[1] Fac. de Ingeniería, Dpto. de Sistemas e Informática, Univ. de Caldas, Manizales 170001, Colombia
manuel.arroyavehenao@gmail.com
[2] Fac. de Ingeniería, Dpto. de Sistemas e Informática, GITIR Grupo Investigación Tecnologías Información y Redes, Universidad de Caldas, Manizales 170001, Colombia
[3] Facultad de Ingeniería y Arquitectura, GTA en Innovación y Desarrollo Tecnológico, Universidad Nacional de Colombia Sede Manizales, Departamento de Ingenierìa Industrial, Campus la Nubia, Manizales 170001, Colombia
[4] Dpto. de Informática e Ingeniería de Sistemas, Zaragoza, Spain

Abstract. Currently the use of technologies by individuals with special mental conditions has shown significant impact. Social interaction is the foundation for the natural development and a component that requires different cognitive abilities. The study of the linguistic component through virtual media is a task that is approached from different methodologies allowing the extraction of additional information to the interaction process and fundamental for the work with therapists, patients and relatives. This article presents the development of a computational platform for the analysis of social interactions in terms of quantity and content of the information with the goal of approach the intervention and monitoring of mental pathologies. Indices to measure the cognitive development of an individual and Information Recuperation techniques has been adapted to the analysis context with the purpose of determine characteristics and find patterns in the linguistic and developmental components of the person. We evaluate interactions with 10 patients, 6 diagnosed with Asperger and 4 under Attention Deficit and Hyperactivity Disorder finding evidence of characteristics for these pathologies like abrupt change and redundancy towards conversation topics. The results suggest that these systems of computer mediated communication allow the discovering of therapeutic keys and boosting the person cognitive development.

Keywords: Computational platform · Social interaction · Asperger

1 Introduction

In areas like neuroscience or cognitive psychology, and in general in areas related to the study of the learning process and cognitive development, are various the studies related to the integration of ICT as a way to promote the learning mechanisms, reasoning, communication and language; specifically oriented to contribute in platforms of communication for people with special conditions in the vital development [1].

S. Rodríguez et al. (Eds.): DCAI 2018, AISC 801, pp. 400–408, 2019.
https://doi.org/10.1007/978-3-319-99608-0_54

Diverse technological tools are constantly implemented and oriented to satisfy learning and communication needs in kids and adults with special development conditions [2, 3]. They are presented as systems to allow a person under some disorder to establish communication with other people and, in more complex scenarios, to transmit their wishes and fundamental needs.

Complex linguistic abilities in human beings have to be tackled from diverse perspectives due to the close relationship with other vital components of general development for the person [4, 5]. Due to the considerable number of patients and the information that is generated for every one of them, is important to have means that allow the storage, data processing and consequent deployment of results. The use of specialized algorithms for data processing and visualization, allows a fast and efficient way to evaluate the interventions in favor of supporting decisions making over diagnostics and interactions between patients and specialists [6].

The study of the cognitive capacities in a specific individual is a complex task that requires the intervention of diverse disciplines, so a classification based on behaviors observation and theories may not be reliable because each condition is unique for every person [1] and variables that can provide a precise approach to every particular situation from diverse points of view are sometimes omitted [7].

Diverse studies have shown the relation between different mental pathologies and language development, affecting both fields mutually when one of the deficit is present [8]. Likewise, different tools have been developed with the aim to assist patients under some mental condition with the requirement of their communicative needs such as the case of Augmented and Collaborative Communications (ACC) [1, 9], that generally correspond to pictographic systems by means of which the individual can make known his needs. However, although they support in a significant way the communications with patients [10], they not reflect a recovery that increase in time within the person basic communication channels (speak, read, write). By the same way, the observable results of these investigations are obtained through the patient abstraction with respect to the use of technologies to evaluate and qualify by offline media the tools usage; this is, through surveys, questionnaires and post-fact data analysis to the interventions [9, 11].

For the mentioned above, we have built a platform for the constant analysis of text that establishes through the communication and the representation of the results by visualization algorithms, what makes it possible to establish a real time monitoring of the conversations held in a virtual media, giving the possibility to determine diverse characteristics of these, such as rhythm, fluency, variability, information quantity, recurrence to topics, among others, with the aim of determine the responds to diverse stimulus as they are provided allowing the identification of unique characteristics for each person. This is the basis of the Computer Mediated Communication (CMC), in which the temporality and information quantity that is being shared can be controlled to study the different moments of the interaction.

2 Methods and Materials

A software system accessible through the web and in test over mobile devices has been developed allowing the capture and load of data, additional to the analysis process and results deployment. The data to analyze are directly represented by the texts exchanged through send messages over the platform between users. Diverse measures have been adapted to identify communication characteristics like quantity and size of the messages, determine the function of the used terms and correlate the interactions based on those terms and their relevance for each interaction. The results have been shown in a graphic interface to speed up the data analysis and interpretation.

2.1 Architecture of the Platform

The use of a messaging server was required for user management and the message exchange within the platform. For this purpose, Openfire, an open source server under apache license and that operates based on the open protocol for messaging XMPP was elected as the best option due to the completeness about management and administration, easy integration with other systems developed in different languages and the available resources to expand and adapt its capacities. The interactive component that is in permanent contact with the person has been developed in PhoneGap, a framework for mobile applications development that allows compatibility with desktop web browsers and generation of the system for diverse platforms like Android and iOS. This platform works under the html5 standard with support for scripting languages JavaScript and CSS labels. The graphic component is related to a JavaScript library called d3.js that presents good integration with the deployment model of PhoneGap. The communication between Openfire and PhoneGap is made by an external component called Strophe.js, that contains the necessary tools for login, status indication, contacts deployment and message exchange. The analysis component is implemented entirely in Java and is represented by a Web Service operating under a Glassfish server. The java service contains the indices implementation and models for the data analysis. The system storage is working on a MySQL Database engine that accept connections from messaging server and his extra components, and the processing service for information extraction and consequent results generation. The Physical deployment of the components are shown in Fig. 1.

2.2 Preprocessing

It is common in this type of software applications to apply a text normalization that allows the reduction of processing complexity, besides of eliminating the noise that can interfere in the obtained results. The elimination of stop-words or words without content like articles, pronouns, and others; help to significantly reduce the space of analysis allowing to focus the efforts on those symbols that can contribute with major significance according to the analysis objectives. In the same way, a reduction of the symbols to their lemmas allows the reduction of complexity and size of the used vector models, ultimately contributing in the processing times and in a better clustering and deployment of results.

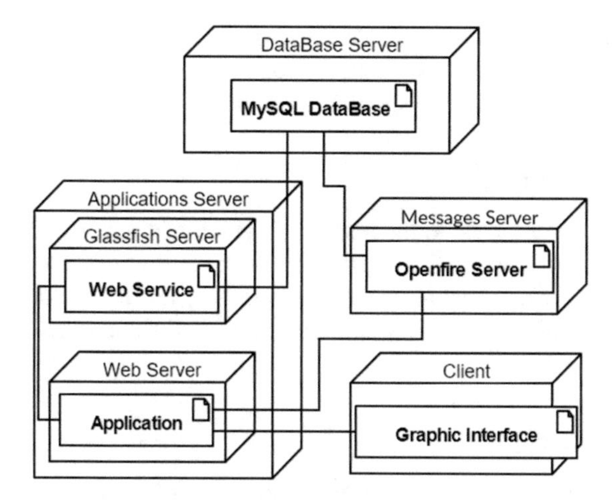

Fig. 1. Physical deployment of components

2.3 Ordered Distribution and Average Length of the Statement (ALS) in Words

To determine the quantity of information present in the interaction, an ordered distribution has been designed in which number of lines or messages per communication session are compared, additional to the adaptation of the ALS index to determine the average length of the statement in words. To obtain the comparison between communications and the number of messages, a count of the messages must be made, grouping them by conversation. The results of the above are captured in a line graph as shown in the Fig. 3, in which the "x" axis represents the communications within the interaction, and the "y" axis the number of messages for every one of those. The ALS index is calculated based on the amount of words in the communication, value that is divided by the amount of communications present in the interactions.

2.4 Term Frequency – Inverse Document Frequency TFIDF

For the communicative process of study, the terms in a communication session are analyzed in reference with the other communications in the interaction. The TFIDF index is calculated in various steps:

Term Frequency TF. Is calculated for every term in every communication. For this, the number of times that the term appears in the text portion analyzed is extracted and then divided for the number of words in that text portion [12].

Inverse Document Frequency IDF. The IDF is calculated based on the number of communications that contain the analyzed term. For this, the total amount of communications in the analyzed interaction is obtained and divided by the number of communications that contain the term in analysis. Finally, the logarithm of the quotient is taken.

TFIDF. The product over the results of the above indices reflects the obtained TFIDF value for the analyzed term in respect to the communication and interaction in which it is presented [13]. The results are show in a bar graph like in Fig. 2a.

2.5 Space Vector Model

A dictionary with every word in the interaction is initially created. This is represented as a bag of words, a structure that represents the content of the interaction. For every communication, a vector of equal size of the dictionary is created and in every position of it, where a word of the dictionary is found, its TFIDF value calculated previously is assigned.

2.6 Cosine Similarity

In a general way, the cosine similarity is calculated with the dot product between the vectors that represents the document to compare, all of this divided by the product of the Euclidean lengths of both vectors [12].

In this way and having built the SVM previously for the communications in an interaction, is possible to operate over the resulting vectors for every present text portion, and in this way apply the formula for the similarity of the vectors mentioned above. A way of representing the above in a graphic interface is through an unguided graph or a cooccurrence matrix like in Fig. 2b. where the nodes represent the diverse communications in the interaction and the edges indicate if a relation exists and the grade of this.

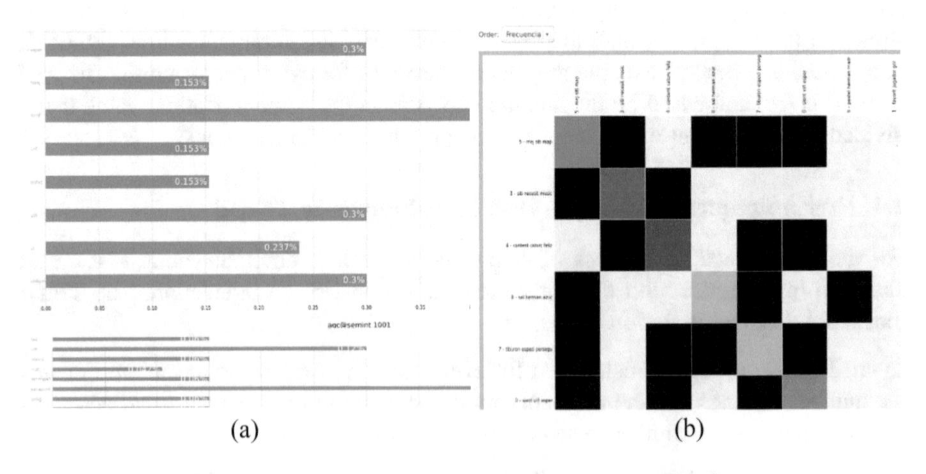

(a) (b)

Fig. 2. a. TFIDF representation b. Cooccurrence Matrix

3 Results

Figure 3 represents 4 interactions loaded of Asperger's Syndrome in terms of number of lines by scene, allowing us to evidence the following communication moments: With the patients denominated age and bfg; and in contrast with the content of the dialogue, concrete and coherent answers to the questions are evidenced, with no intention to explain or complement with additional contents; which in the figure can be evidenced because it represents a kind of linearity in the interaction with a media number of lines per scene in respect to the other interactions so they server as a reference point for the rest of the analysis since by the content of them can be classified as normal conversations. Analyzing the interaction with a patient indexed by dfc, it is shown how he get dispersed at the beginning of the conversation complementing answers with real life events, in addition that describes what is being shown to him in pictures without the specialist having asked the question; so, it implies a major number of lines for the initial interactions in the conversation. Although initiates in a pressured way wanting to get ahead the events, once the control of the interaction is taken, the kid answer correctly and with coherent assertions about the shown facts, generating conclusions based on details on the pictures.

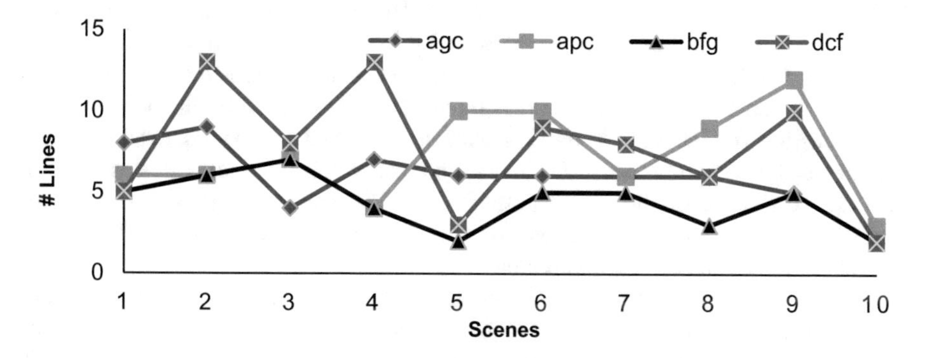

Fig. 3. Average length of sentence and ordered distribution that compares communications and amount of messages

Parallelly it can be seen in the graphic analysis with the picks at the beginning of the interactions coupling to the reference interactions (agc y bfg) as it develops, however being always the calculated results over what we have considered as normal, indicating that the patient may be a little bit more intense or extroverted on his behaviors or expressions. For the patient marked as apc exists evidence about the emission of incoherent sounds and get distracted along conversation bringing topics that do not fit in the discussion; the interviewer must call for attention permanently over the patient accenting this act over the intermediate and end of the conversation, turning into a more receptive patient about the questions and answering them in a more coherent way.

In a parallel way the graphic allows the identification of how the interaction grow in content as it develops, and how the initial stages could have been affected in terms of linguistic production due to the distractions or personal thoughts of the patient. Once the interviewer catches the patient attention, the interaction flows in a natural way, however he keeps making sounds and the calls for attention make to increase the resulting number of lines per interaction scene.

The figure reflects how two interactions in special represents more content being these identified for patients that presents a higher level of dispersion in their communicative process, although it is observed that the answer are extended with additional sentences and explications which can be seen as some grade of imagination, extraversion or access to memories that in general can be taken as positive aspects for the individual development. It is noted how an effortless process of call for attention can help to align the individual in the communicative process and in the development of the interaction.

In the other side, the lower graphic presentations show interactions with concrete answers to stimulus (questions based on illustrations), facilitating the interaction whit the subjects without need of corrective actions for the normal development of the session.

An important observation is made based on abrupt return to interest topics by the patients diagnosed with autistic spectrum disorder, in this case with Asperger's syndrome. In the conclusions of the study made by Verano [14] and having into account that two of the datasets analyzed were from hers work, it is observed situations in both interactions that are perceived as abrupt return to interest topics by the patients when the interaction is proposed. In relation with the observed for experimental subject 1 communication is more evident the change of topics returning to own interests, that is for example when at the beginning of the conversation they dialogue about dinosaurs, and the interviewer tries to change the conversation topic, the patient suggests that what he likes or remember are the dinosaurs.

This situation is analyzed by our system through the cosine similarity index that allow us to identify when and in what grade different portions of the communication can be related between them. For the experimental subject 1 case, the graphic results in Fig. 2b shows a series of relations between different scenes or interactions in the communication, indicating that in some moment along the interaction topics where introduced or related terms were used that in the same way were present in other times during the whole conversation, while in the experimental subject 2 conversation analysis no evidence was seen about relations between interventions, causing a natural development of the interaction talking about the topics introduced in the moment [14].

4 Conclusions

From the analysis and the different results it is possible to reach the following conclusions: (1) the adaptation of Information Retrieval techniques like TFIDF in social interactions contexts over mental pathologies is applicable for the identification of characteristic patterns in these, for this specific case in autistic conditions; (2) no evidence was reported about symptomatology of the autistic spectrum in the results obtained for ADHD in reference to recurrence to interest topics, which allows the upgrade of knowledge about the historically reported relation between these two pathologies and (3) Visualizations of interactions according to different purposes of analysis allow the deduction of characteristics of the communication and the individuals that interact, that are no evident during the act of interaction. Visualization of information allows the discovering of hidden relations among the data, upgrading the processing and interventions times.

Acknowledgment. Universidad de Caldas and Universidad Nacional de Colombia Sede Manizales project code 36715 Computational prototype for the fusion and analysis of large volumes of data in IoT (Internet of Things) environments based on Machine Learning techniques and secure architectures between sensors, to characterize the behavior and interaction of users in a Connected Home ecosystem.

References

1. Fernández-López, Á., Rodríguez-Fórtiz, M.J., Rodríguez-Almendros, M.L., Martínez-Segura, M.J.: Mobile learning technology based on iOS devices to support students with special education needs. Comput. Educ. **61**, 77–90 (2013). https://doi.org/10.1016/j.compedu.2012.09.014
2. Martin, S., Sutcliffe, P., Griffiths, F., et al.: Effectiveness and impact of networked communication interventions in young people with mental health conditions: a systematic review. Patient Educ. Couns. **85** (2011). https://doi.org/10.1016/j.pec.2010.11.014
3. van der Aa, C., Pollmann, M., Plaat, A., van der Gaag, R.: Computer-mediated communication in adults with high-functioning autism spectrum disorders and controls. Res. Autism Spectr. Disord. **23**, 15–27 (2016). https://doi.org/10.1016/j.rasd.2015.11.007
4. Rogers, C.R., Nulty, K.L., Betancourt, M.A., DeThorne, L.S.: Causal effects on child language development: a review of studies in communication sciences and disorders. J. Commun. Disord. **57**, 3–15 (2015). https://doi.org/10.1016/j.jcomdis.2015.06.004
5. Gabriela, M., Elizabeth, C., Estela, C.: Dimensiones del lenguaje: las dificultades del lenguaje en el nivel inicial. En: X Congreso Nacional y II Congreso Internacional "REPENSAR LA NIÑEZ EN EL SIGLO XXI". Mendoza - Argentina, p. 9 (2008)
6. Mahmud, S., Iqbal, R., Doctor, F.: Cloud enabled data analytics and visualization framework for health-shocks prediction. Futur. Gener. Comput. Syst. **65**, 169–181 (2016). https://doi.org/10.1016/j.future.2015.10.014
7. Shipstead, Z., Hicks, K.L., Engle, R.W.: J. Appl. Res. Mem. Cogn. **1**, 217–219 (2012). https://doi.org/10.1016/j.jarmac.2012.07.005
8. Stephanie, D., Julie, F.: Exploring links between language and cognition in autism spectrum disorders: Complement sentences, false belief, and executive functioning. J. Commun. Disord. **54**, 15–31 (2015). https://doi.org/10.1016/j.jcomdis.2014.12.001
9. Campigotto, R., McEwen, R., Demmans Epp, C.: Especially social: exploring the use of an iOS application in special needs classrooms. Comput. Educ. **60**, 74–86 (2013). https://doi.org/10.1016/j.compedu.2012.08.002
10. Hutchins, T.L., Prelock, P.A.: Using communication to reduce challenging behaviors in individuals with autism spectrum disorders and intellectual disability. Child Adolesc. Psychiatr. Clin. N Am. **23**, 41–55 (2014)
11. Davis-McShan, M.L.: Impact of Computer-Mediated Communication Duration on Adolescent Social Self-E cacy, Social Anxiety, and Depression. Walden University (2015)

12. Manning, C.D., Raghavan, P., Schutze, H.: An Introduction to Information Retrieval. Cambridge (2009)
13. Saint-Georges, C., Mahdhaoui, A., Chetouani, M., et al.: Do parents recognize autistic deviant behavior long before diagnosis? taking into account interaction using computational methods. PLoS ONE **6**, 13 (2011). https://doi.org/10.1371/journal.pone.0022393
14. Verano, P.L.Y.: El tópico perseverante en el discurso de niños diagnosticados con síndrome de asperger. Pontificia Universidad Católica del Perú PUCP (2013)

A Self-organized Multiagent System for Industry 4.0

Inés Sittón Candanedo(✉)

Grupo GITCE, Universidad Tecnológica de Panamá, Panama City, Panama
isittonc@gmail.com

Abstract. Industry 4.0 has revolutionized the recent years because the requirements in all domains of manufacturing, production or sales are dynamics and uncertainty and with them the challenges such as emerging technologies, great volumes of data and to make decisions in real time. This paper describes the advantage of a self-organized multiagent system to addresses the problem of data and how process them in Industry 4.0 environment.

Keywords: Artificial intelligence · Multiagent system · Self-organized Industry 4.0

1 Introduction

Artificial intelligence together with cloud computing and the emerging technologies such as Internet of Things (IoT), big data, ciber physical systems or wireless sensors networks help to implement the smart factory of the Industry 4.0 paradigm, nevertheless, all of them collect a great volume of data and nowadays, how process and store them is the companies challenge [47]. This research addresses this challenge from the perspective of multiagent systems.

For several authors [1–20] the agents and multiagent systems (MAS) have characteristics such as: cooperation, organization, communication and distribution, which makes them capable of offering the analysis, modeling or design of solutions for the Industry 4.0 environment. In recent decades the multiagent systems were been used in a several fields with success, such as: logistics, robotics, manufacturing, electronic commerce, bioinformatics, telecommunications, smart cities and energy are some of them [21–46].

2 Proposal

In industry environment is possible to mention the application of MAS solutions in: Chrysler factory in Stuttgart or US Navy to control the heating, ventilation and air condition (HVAC) systems in their ships, they are just some examples. The objective of this self-organized multiagent system is helping to make autonomous decisions and

S. Rodríguez et al. (Eds.): DCAI 2018, AISC 801, pp. 409–413, 2019.
https://doi.org/10.1007/978-3-319-99608-0_55

cooperation where, machines and products communicate as intelligent agents in the Industry 4.0 environment [48–50]. The aim of future work will be to presenting a case study that tests this novel model and represent it in a simulated framework, containing a set of different tasks.

Acknowledgments. I. Sittón Candanedo has been supported by IFARHU – SENACYT scholarship program (Government of Panama).

References

1. Adam, E., Grislin-Le Strugeon, E., Mandiau, R.: MAS architecture and knowledge model for vehicles data communication. ADCAIJ Adv. Distrib. Comput. Artif. Intell. J. **1**(1) (2012)
2. Baruque, B., Corchado, E., Mata, A., Corchado, J.M.: A forecasting solution to the oil spill problem based on a hybrid intelligent system. Inf. Sci. **180**(10), 2029–2043 (2010). https://doi.org/10.1016/j.ins.2009.12.032
3. Buciarelli, E., Silvestri, M., González, S.R.: Decision economics, in commemoration of the birth centennial of Herbert A. Simon 1916–2016 (Nobel Prize in Economics 1978). In: Distributed Computing and Artificial Intelligence, 13th International Conference. Advances in Intelligent Systems and Computing, vol. 475. Springer (2016)
4. Chamoso, P., Rivas, A., Martín-Limorti, J.J., Rodríguez, S.: A hash based image matching algorithm for social networks. In: Advances in Intelligent Systems and Computing, vol. 619, pp. 183–190 (2018)
5. Choon, Y.W., Mohamad, M.S., Deris, S., Illias, R.M., Chong, C.K., Chai, L.E., Corchado, J. M.: Differential bees flux balance analysis with OptKnock for in silico microbial strains optimization. PLoS ONE **9**(7) (2014)
6. Corchado, J.A., Aiken, J., Corchado, E.S., Lefevre, N., Smyth, T.: Quantifying the Ocean's CO2 budget with a CoHeL-IBR system. In: Proceedings of Advances in Case-Based Reasoning, vol. 3155, pp. 533–546 (2004)
7. Corchado, J.M., Aiken, J.: Hybrid artificial intelligence methods in oceanographic forecast models. IEEE Trans. Syst. Man Cybern. Part C-Appl. Rev. **32**(4), 307–313 (2002). https://doi.org/10.1109/tsmcc.2002.806072
8. Corchado, J.M., Fyfe, C.: Unsupervised neural method for temperature forecasting. Artif. Intell. Eng. **13**(4), 351–357 (1999). https://doi.org/10.1016/S0954-1810(99)00007-2
9. Corchado, J.M., Borrajo, M.L., Pellicer, M.A., Yáñez, J.C.: Neuro-symbolic system for business internal control. In: Industrial Conference on Data Mining, pp. 1–10 (2004)
10. Corchado, J.M., Corchado, E.S., Aiken, J., Fyfe, C., Fernandez, F., Gonzalez, M.: Maximum likelihood Hebbian learning based retrieval method for CBR systems. In: Lecture Notes in Computer Science (including subseries Lecture Notes in Artificial Intelligence and Lecture Notes in Bioinformatics), vol. 2689, pp. 107–121 (2003). https://doi.org/10.1007/3-540-45006-8_11
11. Corchado, J.M., Pavón, J., Corchado, E.S., Castillo, L.F.: Development of CBR-BDI agents: a tourist guide application. In: Lecture Notes in Computer Science (including subseries Lecture Notes in Artificial Intelligence and Lecture Notes in Bioinformatics), vol. 3155, pp. 547–559 (2004)
12. Corchado, J., Fyfe, C., Lees, B.: Unsupervised learning for financial forecasting. In: Proceedings of the IEEE/IAFE/INFORMS 1998 Conference on Computational Intelligence for Financial Engineering (CIFEr) (Cat. No. 98TH8367), pp. 259–263 (1998). https://doi.org/10.1109/CIFER.1998.690316

13. Costa, Â., Novais, P., Corchado, J.M., Neves, J.: Increased performance and better patient attendance in an hospital with the use of smart agendas. Logic J. IGPL **20**(4), 689–698 (2012)
14. De La Prieta, F., Navarro, M., García, J.A., González, R., Rodríguez, S.: Multi-agent system for controlling a cloud computing environment. In: Lecture Notes in Computer Science (including subseries Lecture Notes in Artificial Intelligence and Lecture Notes in Bioinformatics). LNAI, vol. 8154 (2013)
15. Fdez-Riverola, F., Corchado, J.M.: CBR based system for forecasting red tides. Knowl.-Based Syst. **16**(5–6 SPEC.), 321–328 (2003). https://doi.org/10.1016/S0950-7051(03)00034-0
16. Fdez-Rtverola, F., Corchado, J.M.: FSfRT: forecasting system for red tides. Appl. Intell. **21**(3), 251–264 (2004). https://doi.org/10.1023/B:APIN.0000043558.52701.b1
17. Fernández-Riverola, F., Díaz, F., Corchado, J.M.: Reducing the memory size of a Fuzzy case-based reasoning system applying rough set techniques. IEEE Trans. Syst. Man Cybern. Part C Appl. Rev. **37**(1), 138–146 (2007)
18. Fyfe, C., Corchado, J.: A comparison of Kernel methods for instantiating case based reasoning systems. Adv. Eng. Inform. **16**(3), 165–178 (2002). https://doi.org/10.1016/S1474-0346(02)00008-3
19. Fyfe, C., Corchado, J.M.: Automating the construction of CBR systems using kernel methods. Int. J. Intell. Syst. **16**(4), 571–586 (2001). https://doi.org/10.1002/int.1024
20. García Coria, J.A., Castellanos-Garzón, J.A., Corchado, J.M.: Intelligent business processes composition based on multi-agent systems. Expert Syst. Appl. **41**(4 PART 1), 1189–1205 (2014). https://doi.org/10.1016/j.eswa.2013.08.003
21. García, E., Rodríguez, S., Martín, B., Zato, C., Pérez, B.: MISIA: Middleware infrastructure to simulate intelligent agents. In: Advances in Intelligent and Soft Computing, vol. 91 (2011). https://doi.org/10.1007/978-3-642-19934-9_14
22. Glez-Bedia, M., Corchado, J.M., Corchado, E.S., Fyfe, C.: Analytical model for constructing deliberative agents. Int. J. Eng. Intell. Syst. Electr. Eng. Commun. **10**(3) (2002)
23. Glez-Peña, D., Díaz, F., Hernández, J.M., Corchado, J.M., Fdez-Riverola, F.: geneCBR: a translational tool for multiple-microarray analysis and integrative information retrieval for aiding diagnosis in cancer research. BMC Bioinform. **10** (2009). https://doi.org/10.1186/1471-2105-10-187
24. González Briones, A., Chamoso, P., Barriuso A.: Review of the main security problems with multi-agent systems used in e-commerce applications. ADCAIJ Adv. Distrib. Comput. Artif. Intell. J. **5** (2016)
25. Isaza, G., Mejía, M., Castillo, L.F., Morales, A., Duque, N.: Network management using multi-agents system. ADCAIJ Adv. Distrib. Comput. Artif. Intell. J. **1**(3) (2012)
26. Kou, G., González-Crespo, R., Corchado, J.M., Herrera-Viedma, E.: Solving multi-criteria group decision making problems under environments with a high number of alternatives using fuzzy ontologies and multi-granular linguistic modelling methods. Knowl.-Based Syst. **137**, 54–64 (2017)
27. Laza, R., Pavn, R., Corchado, J.M.: A reasoning model for CBR_BDI agents using an adaptable fuzzy inference system. In: Lecture Notes in Computer Science (including subseries Lecture Notes in Artificial Intelligence and Lecture Notes in Bioinformatics), vol. 3040, pp. 96–106. Springer, Heidelberg (2004)
28. Li, T., Sun, S., Bolić, M., Corchado, J.M.: Algorithm design for parallel implementation of the SMC-PHD filter. Sig. Process. **119**, 115–127 (2016). https://doi.org/10.1016/j.sigpro.2015.07.013

29. Li, T., Sun, S., Corchado, J.M., Siyau, M.F.: A particle dyeing approach for track continuity for the SMC-PHD filter. In: FUSION 2014 - 17th International Conference on Information Fusion (2014)
30. Li, T., Sun, S., Corchado, J.M., Siyau, M.F.: Random finite set-based Bayesian filters using magnitude-adaptive target birth intensity. In: FUSION 2014 - 17th International Conference on Information Fusion (2014)
31. Li, T.-C., Su, J.-Y., Liu, W., Corchado, J.M.: Approximate Gaussian conjugacy: parametric recursive filtering under nonlinearity, multimodality, uncertainty, and constraint, and beyond. Front. Inf. Technol. Electr. Eng. **18**(12), 1913–1939 (2017)
32. Lima, A.C.E.S., De Castro, L.N., Corchado, J.M.: A polarity analysis framework for Twitter messages. Appl. Math. Comput. **270**, 756–767 (2015). https://doi.org/10.1016/j.amc.2015.08.059
33. Mata, A., Corchado, J.M.: Forecasting the probability of finding oil slicks using a CBR system. Expert Syst. Appl. **36**(4), 8239–8246 (2009). https://doi.org/10.1016/j.eswa.2008.10.003
34. Méndez, J.R., Fdez-Riverola, F., Díaz, F., Iglesias, E.L., Corchado, J.M.: A comparative performance study of feature selection methods for the anti-spam filtering domain. In: Lecture Notes in Computer Science (Including Subseries Lecture Notes in Artificial Intelligence and Lecture Notes in Bioinformatics). LNAI, vol. 4065, pp. 106–120 (2006)
35. Méndez, J.R., Fdez-Riverola, F., Iglesias, E.L., Díaz, F., Corchado, J.M.: Tracking concept drift at feature selection stage in SpamHunting: an anti-spam instance-based reasoning system. In: Lecture Notes in Computer Science (Including Subseries Lecture Notes in Artificial Intelligence and Lecture Notes in Bioinformatics). LNAI, vol. 4106, pp. 504–518. Morente-Molinera, J.A. (2006)
36. Omatu, S., Wada, T., Chamoso, P.: Odor classification using agent technology. DCAIJ Adv. Distrib. Comput. Artif. Intell. J. **2**(4) (2013)
37. Palomino, C.G., Nunes, C.S., Silveira, R.A., González, S.R., Nakayama, M.K.: Adaptive agent-based environment model to enable the teacher to create an adaptive class. In: Advances in Intelligent Systems and Computing, vol. 617 (2017). https://doi.org/10.1007/978-3-319-60819-8_3
38. Peñaranda, C., Agüero, J., Carrascosa, C., Rebollo, M., Julián, V.: An agent-based approach for a smart transport system. ADCAIJ Adv. Distrib. Comput. Artif. Intell. J. **5**(2) (2016)
39. Pinto, T., Gazafroudi, A.S., Prieto-Castrillo, F., Santos, G., Silva, F., Corchado, J.M., Vale, Z.: Reserve costs allocation model for energy and reserve market simulation. In: 2017 19th International Conference on Intelligent System Application to Power Systems, ISAP 2017, art. no. 8071410 (2017)
40. Redondo-Gonzalez, E., De Castro, L.N., Moreno-Sierra, J., Maestro De Las Casas, M.L., Vera-Gonzalez, V., Ferrari, D.G., Corchado, J.M.: Bladder carcinoma data with clinical risk factors and molecular markers: a cluster analysis. BioMed Res. Int. (2015). https://doi.org/10.1155/2015/168682
41. Rodríguez, S., De La Prieta, F., Tapia, D.I., Corchado, J.M.: Agents and computer vision for processing stereoscopic images. In: Lecture Notes in Computer Science (including subseries Lecture Notes in Artificial Intelligence and Lecture Notes in Bioinformatics). LNAI, vol. 6077 (2010). https://doi.org/10.1007/978-3-642-13803-4_12
42. Rodríguez, S., Gil, O., De La Prieta, F., Zato, C., Corchado, J.M., Vega, P., Francisco, M.: People detection and stereoscopic analysis using MAS. In: Proceedings of INES 2010 - 14th International Conference on Intelligent Engineering Systems (2010). https://doi.org/10.1109/INES.2010.5483855

43. Rodríguez, S., Tapia, D.I., Sanz, E., Zato, C., De La Prieta, F., Gil, O.: Cloud computing integrated into service-oriented multi-agent architecture. In: IFIP Advances in Information and Communication Technology. AICT, vol. 322 (2010). https://doi.org/10.1007/978-3-642-14341-0_29
44. Román Gallego, J.A.,, Rodríguez González, S.: Improvement in the distribution of services in multi-agent systems with SCODA. ADCAIJ Adv. Distrib. Comput. Artif. Intell. J. **4**(3) (2015)
45. Román, J.A., Rodríguez, S., de da Prieta, F.: Improving the distribution of services in MAS. Commun. Comput. Inf. Sci. **616** (2016). https://doi.org/10.1007/978-3-319-39387-2_4
46. Santos, G., Pinto, T., Vale, Z., Praça, I., Morais, H.: Enabling communications in heterogeneous multi-agent systems: electricity markets ontology. ADCAIJ Adv. Distrib. Comput. Artif. Intell. J. **5**(2) (2016)
47. Sittón, I., Rodríguez, S.: Pattern extraction for the design of predictive models in industry 4.0. In: International Conference on Practical Applications of Agents and Multi-Agent Systems, pp. 258–261 (2017)
48. Tapia, D.I., Corchado, J.M.: An ambient intelligence based multi-agent system for alzheimer health care. Int. J. Ambient Comput. Intell. **1**(1), 15–26 (2009). https://doi.org/10.4018/jaci.2009010102
49. Tapia, D.I., Fraile, J.A., Rodríguez, S., Alonso, R.S., Corchado, J.M.: Integrating hardware agents into an enhanced multi-agent architecture for Ambient Intelligence systems. Inf. Sci. **222**, 47–65 (2013). https://doi.org/10.1016/j.ins.2011.05.002
50. Wang, X., Li, T., Sun, S., Corchado, J.M.: A survey of recent advances in particle filters and remaining challenges for multitarget tracking. Sensors **17**(12), art. no. 2707 (2017)

Blockchain-Based Distributed Cooperative Control Algorithm for WSN Monitoring

Roberto Casado-Vara(✉)

BISITE Digital Innovation Hub,
University of Salamanca, Edificio Multiusos I+D+i, 37007 Salamanca, Spain
rober@usal.es
https://bisite.usal.es/es

Abstract. The management of heterogeneous distributed sensor networks requires new solutions to address the problem of data quality and false data detection in Wireless Sensor Networks (WSN). In this paper, we present a nonlinear cooperative control algorithm based on game theory and blockchain. Here, a new model is proposed for the automatic processing and management of data in heterogeneous distributed wireless sensor networks stored in a blockchain. We apply our algorithm for improving temperature data quality in indoor surfaces.

Keywords: Wireless Sensors Network · Blockchain · Game theory
Nonlinear models and systems · Data quality · False data detection
Nonlinear cooperative control

1 Introduction

Wireless Sensor Networks have become important in the last years and nowadays are present in practically all the sectors of our society. Their great capacity to gather data may facilitate the construction of smart environments, allowing for a flexible analysis of processes that occur in the environment and the services offered to users [1–9]. Distributed sensor networks and depending on the network topology and sensor neighbourhood are also presented. In our work, coalitions of neighbours are created by using clustering techniques. This distributed and self-organized (overall temperature equilibrium arises from local game interactions between sensors of an initially disordered temperatures system) game is designed to provide reliability and robustness to the data collected by a WSN [10–13]. It identifies sensors gathering defective or inaccurate measurements and detects areas with similar temperatures. This article tackles the problem of WSN data reliability from the point of view of game theory and probability, which is a novelty approach in this field [14–17].

In this paper, we present an algorithm that will ensure the robustness and reliability of the data collected by WSN and stored in a blockchain [18–22]. In our

S. Rodríguez et al. (Eds.): DCAI 2018, AISC 801, pp. 414–417, 2019.
https://doi.org/10.1007/978-3-319-99608-0_56

approach we apply game theory (GT) to data stored in the blockchain. Our game is distributed and self-organized so that it can work in a WSN regardless of the number of sensors, the architecture of the WSN or the type of sensors to which the game is applied [23–27]. The design of the game fulfills the following needs: it is capable of recognizing the neighborhood in which it is implemented [28–30]. The game also identifies the possible coalitions that can be formed between the neighbors. Finally, the temperature is determined by the winning coalition for the sensor to which the game has been applied. The algorithm also has a high reliability as it uses blockchain to store data and create a log [31,32].

2 Conclusion

This work proposes a distributed and self-organized cooperative algorithm using game theory. The algorithm has been applied to the data collected by a WSN in an indoor surface. The main goal of the game is to improve the robust control of the WSN by consensus temperature monitoring. Furthermore, the submitted work achievements ensure data quality, false data detection (i.e., inaccurate sensors) and temperature data optimization to improve energy efficiency in cooperative WSNs. The most significant results obtained in this work are listed below.

This paper provides a novel, blockchain-based, distributed and self-organized algorithm, which allows to self-correct temperature data collected by the sensors according to their surrounding temperatures. We also address some interesting results and some quite promising industry applications. In our future work, we will extend the game to larger topological manifolds and we will study these manifolds dynamically.

Acknowledgments. This paper has been funded by the European Regional Development Fund (FEDER) within the framework of the Interreg program V-A Spain-Portugal 2014-2020 (PocTep) grant agreement No 0123_IOTEC_3_E (project IOTEC).

References

1. Li, T., Sun, S., Bolič, M., Corchado, J.M.: Algorithm design for parallel implementation of the SMC-PHD filter. Signal Process. **119**, 115–127 (2016). https://doi.org/10.1016/j.sigpro.2015.07.013
2. Lima, A.C.E.S., De Castro, L.N., Corchado, J.M.: A polarity analysis framework for Twitter messages. Appl. Math. Comput. **270**, 756–767 (2015). https://doi.org/10.1016/j.amc.2015.08.059
3. Redondo-Gonzalez, E., De Castro, L.N., Moreno-Sierra, J., De Las, M., Casas, M.L., Vera-Gonzalez, V., Ferrari, D.G., Corchado, J.M.: Bladder carcinoma data with clinical risk factors and molecular markers: a cluster analysis. BioMed Res. Int. **2015**, 168682 (2015). https://doi.org/10.1155/2015/168682
4. Li, T., Sun, S., Corchado, J.M., Siyau, M.F.: Random finite set-based Bayesian filters using magnitude-adaptive target birth intensity. In: FUSION 2014 - 17th International Conference on Information Fusion (2014)

5. Choon, Y.W., Mohamad, M.S., Deris, S., Illias, R.M., Chong, C.K., Chai, L.E., Corchado, J.M.: Differential bees flux balance analysis with OptKnock for in silico microbial strains optimization. PLoS ONE **9**(7) (2014). https://doi.org/10.1371/journal.pone.0102744
6. Li, T., Sun, S., Corchado, J.M., Siyau, M.F.: A particle dyeing approach for track continuity for the SMC-PHD filter. In: FUSION 2014 - 17th International Conference on Information Fusion (2014)
7. García Coria, J.A., Castellanos-Garzón, J.A., Corchado, J.M.: Intelligent business processes composition based on multi-agent systems. Expert Syst. Appl. **41**(4 PART 1), 1189–1205 (2014). https://doi.org/10.1016/j.eswa.2013.08.003
8. Tapia, D.I., Fraile, J.A., Rodríguez, S., Alonso, R.S., Corchado, J.M.: Integrating hardware agents into an enhanced multi-agent architecture for ambient intelligence systems. Inf. Sci. **222**, 47–65 (2013). https://doi.org/10.1016/j.ins.2011.05.002
9. Costa, A., Novais, P., Corchado, J.M., Neves, J.: Increased performance and better patient attendance in an hospital with the use of smart agendas. Logic J. IGPL **20**(4), 689–698 (2012). https://doi.org/10.1093/jigpal/jzr021
10. García, E., Rodriguez, S., Martin, B., Zato, C., Perez, B.: MISIA: middleware infrastructure to simulate intelligent agents. Advances in Intelligent and Soft Computing, vol. 91 (2011)
11. Rodriguez, S., De La Prieta, F., Tapia, D.I., Corchado, J.M.: Agents and computer vision for processing stereoscopic images. Lecture Notes in Computer Science (including subseries Lecture Notes in Artificial Intelligence and Lecture Notes in Bioinformatics), vol. 6077 LNAI (2010)
12. Rodriguez, S., Gil, O., De La Prieta, F., Zato, C., Corchado, J.M., Vega, P., Francisco, M.: People detection and stereoscopic analysis using MAS. In: INES 2010 - 14th International Conference on Intelligent Engineering Systems, Proceedings (2010). https://doi.org/10.1109/INES.2010.5483855
13. Baruque, B., Corchado, E., Mata, A., Corchado, J.M.: A forecasting solution to the oil spill problem based on a hybrid intelligent system. Inf. Sci. **180**(10), 2029–2043 (2010). https://doi.org/10.1016/j.ins.2009.12.032
14. Tapia, D.I., Corchado, J.M.: An ambient intelligence based multi-agent system for Alzheimer health care. Int. J. Ambient Comput. Intell. **1**(1), 15–26 (2009). https://doi.org/10.4018/jaci.2009010102
15. Mata, A., Corchado, J.M.: Forecasting the probability of finding oil slicks using a CBR system. Expert Syst. Appl. **36**(4), 8239–8246 (2009). https://doi.org/10.1016/j.eswa.2008.10.003
16. Glez-Peña, D., Diaz, F., Hernandez, J.M., Corchado, J.M., Fdez-Riverola, F.: geneCBR: a translational tool for multiple-microarray analysis and integrative information retrieval for aiding diagnosis in cancer research. BMC Bioinform. **10** (2009). https://doi.org/10.1186/1471-2105-10-187
17. Fernandez-Riverola, F., Diaz, F., Corchado, J.M.: Reducing the memory size of a fuzzy case-based reasoning system applying rough set techniques. IEEE Trans. Syst. Man Cybern. Part C Appl. Rev. **37**(1), 138–146 (2007). https://doi.org/10.1109/TSMCC.2006.876058
18. Mendez, J.R., Fdez-Riverola, F., Diaz, F., Iglesias, E.L., Corchado, J.M.: A comparative performance study of feature selection methods for the anti-spam filtering domain. Lecture Notes in Computer Science (Including Subseries Lecture Notes in Artificial Intelligence and Lecture Notes in Bioinformatics), LNAI, vol. 4065, pp. 106–120 (2006)

19. Mendez, J.R., Fdez-Riverola, F., Iglesias, E.L., Diaz, F., Corchado, J.M.: Tracking concept drift at feature selection stage in SpamHunting: an anti-spam instance-based reasoning system. Lecture Notes in Computer Science (Including Subseries Lecture Notes in Artificial Intelligence and Lecture Notes in Bioinformatics), LNAI, vol. 4106, pp. 504–518 (2006)
20. Fdez-Rtverola, F., Corchado, J.M.: FSfRT: forecasting system for red tides. Appl. Intell. **21**(3), 251–264 (2004). https://doi.org/10.1023/B:APIN.0000043558.52701.b1
21. Corchado, J.M., Pavon, J., Corchado, E.S., Castillo, L.F.: Development of CBR-BDI agents: a tourist guide application. Lecture Notes in Computer Science (including subseries Lecture Notes in Artificial Intelligence and Lecture Notes in Bioinformatics), vol. 3155, pp. 547–559 (2004). https://doi.org/10.1007/978-3-540-28631-8
22. Laza, R., Pavn, R., Corchado, J.M.: A reasoning model for CBR-BDI agents using an adaptable fuzzy inference system. Lecture Notes in Computer Science (including subseries Lecture Notes in Artificial Intelligence and Lecture Notes in Bioinformatics), vol. 3040, pp. 96–106. Springer, Heidelberg (2004)
23. Corchado, J.A., Aiken, J., Corchado, E.S., Lefevre, N., Smyth, T.: Quantifying the Ocean's CO2 budget with a CoHeL-IBR system. In: Advances in Case-Based Reasoning, Proceedings, vol. 3155, pp. 533–546 (2004)
24. Corchado, J.M., Borrajo, M.L., Pellicer, M.A., Yáñez, J.C.: Neuro-symbolic system for business internal control. In: Industrial Conference on Data Mining, pp. 1–10 (2004)
25. Corchado, J.M., Corchado, E.S., Aiken, J., Fyfe, C., Fernandez, F., Gonzalez, M.: Maximum likelihood hebbian learning based retrieval method for CBR systems. Lecture Notes in Computer Science (including subseries Lecture Notes in Artificial Intelligence and Lecture Notes in Bioinformatics), vol. 2689, pp. 107–121 (2003)
26. Fdez-Riverola, F., Corchado, J.M.: CBR based system for forecasting red tides. Knowl. Based Syst. **16**(5-6), 321–328 (2003). https://doi.org/10.1016/S0950-7051(03)00034-0
27. Glez-Bedia, M., Corchado, J.M., Corchado, E.S., Fyfe, C.: Analytical model for constructing deliberative agents. Int. J. Eng. Intell. Syst. Electr. Eng. Commun. **10**(3), 173 (2002)
28. Corchado, J.M., Aiken, J.: Hybrid artificial intelligence methods in oceanographic forecast models. IEEE Trans. Syst. Man Cybern. Part C Appl. Rev. **32**(4), 307–313 (2002). https://doi.org/10.1109/tsmcc.2002.806072
29. Fyfe, C., Corchado, J.: A comparison of Kernel methods for instantiating case based reasoning systems. Adv. Eng. Inform. **16**(3), 165–178 (2002). https://doi.org/10.1016/S1474-0346(02)00008-3
30. Fyfe, C., Corchado, J.M.: Automating the construction of CBR systems using kernel methods. Int. J. Intell. Syst. **16**(4), 571–586 (2001). https://doi.org/10.1002/int.1024
31. Li, T.-C., Su, J.-Y., Liu, W., Corchado, J.M.: Approximate Gaussian conjugacy: parametric recursive filtering under nonlinearity, multimodality, uncertainty, and constraint, and beyond. Front. Inf. Technol. Electron. Eng. **18**(12), 1913–1939 (2017)
32. Casado-Vara, R., Prieto-Castrillo, F., Corchado, J.M.: A game theory approach for cooperative control to improve data quality and false data detection in WSN. Int. J. Robust Nonlinear Control

New Approach to Power System Grid Security with a Blockchain-Based Model

Roberto Casado-Vara(✉)

BISITE Digital Innovation Hub,
University of Salamanca, Edificio Multiusos I+D+i, 37007 Salamanca, Spain
rober@usal.es
https://bisite.usal.es/es

Abstract. There are many power system grid carrying energy from power plants to consumer. In order to manage these huge grid are all monitored by WSN and controlled by a large cluster of computers. However, the problem is how to ensure that all the data transmitted by the grid is authentic and has not been modified in any way. This paper presents a model for increasing communication security in the power system grid. For this purpose, blockchain is used to store all communication data, blockchain is distributed, secure and reliable. The main goal is that information is protected in the blockchain against modification attempts. In addition, power source can be authenticated and tracked from the consumer to the source.

Keywords: Power system · Grid · Blockchain · Track · Authenticate

1 Introduction

Grid security in the power system grid is currently vulnerable to the coordinated attack of hackers [1–3]. They can join in the grid and modify some data, although over time the electric companies find and fix it [4,5]. Another possible point of improvement in the power system grid is to be able to authenticate and track the energy generated from renewable sources and that generated from fossil fuels [6–11]. The paper main goal allow to updating the power system grid in smart grid are to save all the information in the blockchain [12–14]. This improvement is intended to achieve several things [15,16]. First thing is to secure the information in the grid communications [17–20]. Next, one of the most important points today is the production and distribution of renewable energies. Here is presented one of the most important problems [21,22], the electrons that produce renewable energies are the same as those produced by fossil fuels [23]. However, if we use the blockchain, authentication certificates and origin to the energy generated can be included [24,25]. This energy and these certificates will be available in the blockchain and anyone who uses the grid and has access will be able to verify the source of their energy [26–28].

S. Rodríguez et al. (Eds.): DCAI 2018, AISC 801, pp. 418–421, 2019.
https://doi.org/10.1007/978-3-319-99608-0_57

In this paper we address a new model for power systems grid. Blockchain give us a new way to update old power systems grid into smarter grids [29]. In the new model we propose use smart devices in some of the electric poles. These smart devices use a Wireless Sensor Network (WSN) to communicate with each other and share information with the blockchain [30]. The smart poles will generate events with the information they produce and send it to the miners which link it to the blockchain [31]. In addition, it is expected to obtain an authentication and energy tracking log with the blockchain data. The previous results obtained in the monitoring and control of WSN allow to improve data quality collected by smart devices.

2 Conclusion

Our paper proposes a new model for a power system grid. This new model uses blockchain to provide some advantages over current market models. It provides more security, since the data is cryptographically protected inside the blockchain. It also provides a reliable way to track electricity from the source to the end consumer. Finally, they allow you to authenticate each grid element as everything is encrypted within the blockchain. In the other hand, this model can be used to track the energy generated by renewable sources, and the energy generated by fossil fuels. The novelty of the model proposed in this paper is that smart devices are added to the current power system grid to update it to smarter grids. Therefore, the information collected by the WSN formed by smart devices is stored in a blockchain. In addition, several proprietary development algorithms are added that will take care of tasks such as eliminating duplicate information, controlling the intake of information, generating random numerical models to add more security to communications.

Acknowledgments. This paper has been funded by the European Regional Development Fund (FEDER) within the framework of the Interreg program V-A Spain-Portugal 2014-2020 (PocTep) grant agreement No 0123_IOTEC_3_E (project IOTEC).

References

1. Coria, J.A.G., Castellanos-Garzon, J.A., Corchado, J.M.: Intelligent business processes composition based on multi-agent systems. Expert Syst. Appl. **41**(4 PART 1), 1189–1205 (2016). https://doi.org/10.1016/j.eswa.2013.08.003
2. Tapia, D.I., Fraile, J.A., Rodríguez, S., Alonso, R.S., Corchado, J.M.: Integrating hardware agents into an enhanced multi-agent architecture for ambient intelligence systems. Inf. Sci. **222**, 47–65 (2013). https://doi.org/10.1016/j.ins.2011.05.002
3. Costa, Â., Novais, P., Corchado, J.M., Neves, J.: Increased performance and better patient attendance in an hospital with the use of smart agendas. Logic J. IGPL **20**(4), 689–698 (2012). https://doi.org/10.1093/jigpal/jzr021
4. Rodríguez, S., De La Prieta, F., Tapia, D.I., Corchado, J.M.: Agents and computer vision for processing stereoscopic images. Lecture Notes in Computer Science (including subseries Lecture Notes in Artificial Intelligence and Lecture Notes in Bioinformatics), LNAI, vol. 6077 (2010)

5. Durik, B.O.: Organisational metamodel for large-scale multi-agent systems: first steps towards modelling organisation dynamics. Adv. Distrib. Comput. Artif. Intell. J. (ADCAIJ), **6**(3), 17 (2017)
6. Becerra-Bonache, L., Lopez, M.D.J.: Linguistic models at the crossroads of agents, learning and formal languages. Adv. Distrib. Comput. Artif. Intell. J. (ADCAIJ) **3**(4), 67 (2014)
7. Rodríguez, S., Gil, O., De La Prieta, F., Zato, C., Corchado, J.M., Vega, P., Francisco, M.: People detection and stereoscopic analysis using MAS. In: INES 2010 - 14th International Conference on Intelligent Engineering Systems, Proceedings (2010). https://doi.org/10.1109/INES.2010.5483855
8. Baruque, B., Corchado, E., Mata, A., Corchado, J.M.: A forecasting solution to the oil spill problem based on a hybrid intelligent system. Inf. Sci. **180**(10), 2029–2043 (2010). https://doi.org/10.1016/j.ins.2009.12.032
9. Tapia, D.I., Corchado, J.M.: An ambient intelligence based multi-agent system for Alzheimer health care. Int. J. Ambient. Comput. Intell. **1**(1), 15–26 (2009). https://doi.org/10.4018/jaci.2009010102
10. Mata, A., Corchado, J.M.: Forecasting the probability of finding oil slicks using a CBR system. Expert Syst. Appl. **36**(4), 8239–8246 (2009). https://doi.org/10.1016/j.eswa.2008.10.003
11. Glez-Peña, D., Díaz, F., Hernández, J.M., Corchado, J.M., Fdez-Riverola, F.: geneCBR: a translational tool for multiple-microarray analysis and integrative information retrieval for aiding diagnosis in cancer research. BMC Bioinform. **10** (2009). https://doi.org/10.1186/1471-2105-10-187
12. Fernández-Riverola, F., Díaz, F., Corchado, J.M.: Reducing the memory size of a Fuzzy case-based reasoning system applying rough set techniques. IEEE Trans. Syst. Man Cybern. Part C Appl. Rev. **37**(1), 138–146 (2007). https://doi.org/10.1109/TSMCC.2006.876058
13. Méndez, J.R., Fdez-Riverola, F., Díaz, F., Iglesias, E.L., Corchado, J.M.: A comparative performance study of feature selection methods for the anti-spam filtering domain. Lecture Notes in Computer Science (including subseries Lecture Notes in Artificial Intelligence and Lecture Notes in Bioinformatics), LNAI, vol. 4065, pp. 106–120 (2006)
14. Méndez, J.R., Fdez-Riverola, F., Iglesias, E.L., Díaz, F., Corchado, J.M.: Tracking concept drift at feature selection stage in SpamHunting: an anti-spam instance-based reasoning system. Lecture Notes in Computer Science (including subseries Lecture Notes in Artificial Intelligence and Lecture Notes in Bioinformatics), LNAI, vol. 4106, pp. 504–518 (2006)
15. Fdez-Rtverola, F., Corchado, J.M.: FSfRT: forecasting system for red tides. Appl. Intell. **21**(3), 251–264 (2004). https://doi.org/10.1023/B:APIN.0000043558.52701.b1
16. Corchado, J.M., Pavón, J., Corchado, E.S., Castillo, L.F.: Development of CBR-BDI agents: a tourist guide application. Lecture Notes in Computer Science (including subseries Lecture Notes in Artificial Intelligence and Lecture Notes in Bioinformatics), vol. 3155, pp. 547–559 (2004). https://doi.org/10.1007/978-3-540-28631-8
17. Laza, R., Pavn, R., Corchado, J.M.: A reasoning model for CBR-BDI agents using an adaptable fuzzy inference system. Lecture Notes in Computer Science (including subseries Lecture Notes in Artificial Intelligence and Lecture Notes in Bioinformatics), vol. 3040, pp. 96–106. Springer, Heidelberg (2004)
18. Corchado, J.A., Aiken, J., Corchado, E.S., Lefevre, N., Smyth, T.: Quantifying the Ocean's CO2 budget with a CoHeL-IBR system. In: Advances in Case-Based Reasoning, Proceedings, vol. 3155, pp. 533–546 (2004)

19. Corchado, J.M., Borrajo, M.L., Pellicer, M.A., Yáñez, J.C.: Neuro-symbolic system for business internal control. In: Industrial Conference on Data Mining, pp. 1–10 (2004)
20. Li, T., Sun, S., Bolić, M., Corchado, J.M.: Algorithm design for parallel implementation of the SMC-PHD filter. Signal Process. **119**, 115–127 (2016). https://doi.org/10.1016/j.sigpro.2015.07.013
21. Lima, A.C.E.S., De Castro, L.N., Corchado, J.M.: A polarity analysis framework for Twitter messages. Appl. Math. Comput. **270**, 756–767 (2015). https://doi.org/10.1016/j.amc.2015.08.059
22. Redondo-Gonzalez, E., De Castro, L.N., Moreno-Sierra, J., Maestro De Las Casas, M.L., Vera-Gonzalez, V., Ferrari, D.G., Corchado, J.M.: Bladder carcinoma data with clinical risk factors and molecular markers: a cluster analysis. BioMed Res. Int. (2015). https://doi.org/10.1155/2015/168682
23. Li, T., Sun, S., Corchado, J.M., Siyau, M.F.: Random finite set-based Bayesian filters using magnitude-adaptive target birth intensity. In: FUSION 2014 - 17th International Conference on Information Fusion (2014)
24. Choon, Y.W., Mohamad, M.S., Deris, S., Illias, R.M., Chong, C.K., Chai, L.E., Corchado, J.M.: Differential bees flux balance analysis with OptKnock for in silico microbial strains optimization. PLoS ONE **9**(7) (2014). https://doi.org/10.1371/journal.pone.0102744
25. Li, T., Sun, S., Corchado, J.M., Siyau, M.F.: A particle dyeing approach for track continuity for the SMC-PHD filter. In: FUSION 2014 - 17th International Conference on Information Fusion (2014)
26. Corchado, J.M., Corchado, E.S., Aiken, J., Fyfe, C., Fernandez, F., Gonzalez, M.: Maximum likelihood hebbian learning based retrieval method for CBR systems. Lecture Notes in Computer Science (including subseries Lecture Notes in Artificial Intelligence and Lecture Notes in Bioinformatics), vol. 2689, pp. 107–121 (2014)
27. Fdez-Riverola, F., Corchado, J.M.: CBR based system for forecasting red tides. Knowl. Based Syst. **16**(5–6), 321–328 (2003). https://doi.org/10.1016/S0950-7051(03)00034-0
28. Glez-Bedia, M., Corchado, J.M., Corchado, E.S., Fyfe, C.: Analytical model for constructing deliberative agents. Int. J. Eng. Intell. Syst. Electr. Eng. Commun. **10**(3), 173 (2002)
29. Corchado, J.M., Aiken, J.: Hybrid artificial intelligence methods in oceanographic forecast models. IEEE Trans. Syst. Man Cybern. Part C Appl. Rev. **32**(4), 307–313 (2002). https://doi.org/10.1109/tsmcc.2002.806072
30. Fyfe, C., Corchado, J.: A comparison of Kernel methods for instantiating case based reasoning systems. Adv. Eng. Inform. **16**(3), 165–178 (2002). https://doi.org/10.1016/S1474-0346(02)00008-3
31. Casado-Vara, R., Prieto-Castrillo, F., Corchado, J.M.: A game theory approach for cooperative control to improve data quality and false data detection in WSN. Int. J. Robust Nonlinear Control

Stochastic Approach for Prediction of WSN Accuracy Degradation with Blockchain Technology

Roberto Casado-Vara(✉)

BISITE Digital Innovation Hub,
University of Salamanca, Edificio Multiusos I+D+i, 37007 Salamanca, Spain
rober@usal.es
https://bisite.usal.es/es

Abstract. Nowadays, Wireless Sensors Network (WSN) sensors lose accuracy in their measurements. This address two problems that have a direct influence on monitoring and control that WSN performs. The first one is data collected by the WSN from inaccurate sensors. And secondly, high maintenance cost of WSN if it is not known exactly which sensors are inaccurate. In this paper we propose a stochastic model using Blockchain to predict the degradation of sensor accuracy, knowing its current state. The expected results are the prediction with a high degree of accuracy that sensors will be inaccurate in the near future in order to perform proper maintenance and maintain data quality.

Keywords: Blockchain · Markov chains · WSN · Non-linear control
Inaccurate sensors

1 Introduction

A maintenance policy has become an essential point within a framework of designing business strategies aimed at improving maintenance operations [1–5], reducing the quantity and frequency of interventions and optimizing costs in maintenance service and business production. All decision processes involved with maintenance are naturally stochastic due to the large number of uncertain environmental factors that affect the service life and variability of each component [6–10]. It is therefore desirable to simulate and predict these processes within stochastic models. However, it is the accuracy of the templates that determines the validity of the templates. When a system model with uncertainty is developed, one can choose to either ignore it, recognize it implicitly or model it explicitly [11–14]. Markov's chains are a mathematical model that explicitly allow incorporating inaccuracies into decision-making models [15–18], especially if the system involves probabilities and human subjectivity. Proposed model is addressed to show how Markov's chains can be used to represent and predict the degree of degradation of any equipment or system [19–24].

S. Rodríguez et al. (Eds.): DCAI 2018, AISC 801, pp. 422–425, 2019.
https://doi.org/10.1007/978-3-319-99608-0_58

This model allow to justify feasibility of this method to give a vision in the selection of maintenance policies and therefore, planning budgets and overall performance of systems [25–27]. Expected results are all possible options of the different precision states in which the sensors will find themselves at a chosen future time. Once all the possible states are obtained [28], a study will be carried out based on probabilities conditioned by the initial states of the sensors and by the probabilities of degradation associated with each of the states of precision [29]. The final results expected are the most likely states in which all sensors can be found considering the uncertainty of the entire WSN [11,30].

2 Conclusion

Degradation process of the precision of sensors of a WSN uncertainty is introduced. This allows to know the status of each sensor in the WSN. We address all possible states of sensors using stochastic processes and blockchain. From these calculations, and using the theory of random walks, it is possible to predict all the possible states that can occur among all the possible combinations of future sensor states [29,30]. Finally, using conditioned probabilities dependent on the initial conditions of each sensor, different random walks with the smallest probabilities are ruled out. In the other hand, the advantage of this model is that it can predict which sensors will fail, making maintenance easier. In addition, this will increase the quality of WSN data. Finally, along with other work that I am doing in parallel will be able to increase the energy efficiency of the infrastructure monitored by WSN. Novelty of this paper is that uncertainty is added to the calculations for predicting WSN degradation. Using stochastic processes and blockchain, a new model is created that knowing the present state of the sensors allows predicting future states in which the WSN sensors will be found with a certain probability. This can generate a new industrial maintenance model.

Acknowledgments. This paper has been funded by the European Regional Development Fund (FEDER) within the framework of the Interreg program V-A Spain-Portugal 2014-2020 (PocTep) grant agreement No 0123_IOTEC_3_E (project IOTEC).

References

1. Glez-Peña, D., Díaz, F., Hernández, J.M., Corchado, J.M., Fdez-Riverola, F.: geneCBR: a translational tool for multiple-microarray analysis and integrative information retrieval for aiding diagnosis in cancer research. BMC Bioinform. **10** (2009). https://doi.org/10.1186/1471-2105-10-187
2. Fernández-Riverola, F., Díaz, F., Corchado, J.M.: Reducing the memory size of a fuzzy case-based reasoning system applying rough set techniques. IEEE Trans. Syst. Man Cybern. Part C Appl. Rev. **37**(1), 138–146 (2007). https://doi.org/10.1109/TSMCC.2006.876058

3. Méndez, J.R., Fdez-Riverola, F., Díaz, F., Iglesias, E.L., Corchado, J.M.: A comparative performance study of feature selection methods for the anti-spam filtering domain. Lecture Notes in Computer Science (including subseries Lecture Notes in Artificial Intelligence and Lecture Notes in Bioinformatics), LNAI, vol. 4065, pp. 106–120 (2006)
4. Li, T., Sun, S., Bolić, M., Corchado, J.M.: Algorithm design for parallel implementation of the SMC-PHD filter. Signal Process. **119**, 115–127 (2016). https://doi.org/10.1016/j.sigpro.2015.07.013
5. Lima, A.C.E.S., De Castro, L.N., Corchado, J.M.: A polarity analysis framework for Twitter messages. Appl. Math. Comput. **270**, 756–767 (2015). https://doi.org/10.1016/j.amc.2015.08.059
6. Redondo-Gonzalez, E., De Castro, L.N., Moreno-Sierra, J., Maestro De Las Casas, M.L., Vera-Gonzalez, V., Ferrari, D.G., Corchado, J.M.: Bladder carcinoma data with clinical risk factors and molecular markers: a cluster analysis. BioMed Res. Int. (2015). https://doi.org/10.1155/2015/168682
7. Li, T., Sun, S., Corchado, J.M., Siyau, M.F.: Random finite set-based Bayesian filters using magnitude-adaptive target birth intensity. In: FUSION 2014 - 17th International Conference on Information Fusion (2014)
8. Choon, Y.W., Mohamad, M.S., Deris, S., Illias, R.M., Chong, C.K., Chai, L.E., Corchado, J.M.: Differential bees flux balance analysis with OptKnock for in silico microbial strains optimization. PLoS ONE **9**(7) (2014). https://doi.org/10.1371/journal.pone.0102744
9. Li, T., Sun, S., Corchado, J.M., Siyau, M.F.: A particle dyeing approach for track continuity for the SMC-PHD filter. In: FUSION 2014 - 17th International Conference on Information Fusion (2014)
10. García Coria, J.A., Castellanos-Garzón, J.A., Corchado, J.M.: Intelligent business processes composition based on multi-agent systems. Expert Syst. Appl. **41**(4 PART 1), 1189–1205 (2014). https://doi.org/10.1016/j.eswa.2013.08.003
11. Tapia, D.I., Fraile, J.A., Rodríguez, S., Alonso, R.S., Corchado, J.M.: Integrating hardware agents into an enhanced multi-agent architecture for ambient intelligence systems. Inf. Sci. **222**, 47–65 (2013). https://doi.org/10.1016/j.ins.2011.05.002
12. Costa, Â., Novais, P., Corchado, J.M., Neves, J.: Increased performance and better patient attendance in an hospital with the use of smart agendas. Logic J. IGPL **20**(4), 689–698 (2012). https://doi.org/10.1093/jigpal/jzr021
13. García, E., Rodríguez, S., Martín, B., Zato, C., Pérez, B.: MISIA: middleware infrastructure to simulate intelligent agents. Advances in Intelligent and Soft Computing, vol. 91 (2012)
14. Rodríguez, S., De La Prieta, F., Tapia, D.I., Corchado, J.M.: Agents and computer vision for processing stereoscopic images. Lecture Notes in Computer Science (including subseries Lecture Notes in Artificial Intelligence and Lecture Notes in Bioinformatics), LNAI, vol. 6077 (2010)
15. Rodríguez, S., Gil, O., De La Prieta, F., Zato, C., Corchado, J.M., Vega, P., Francisco, M.: People detection and stereoscopic analysis using MAS. In: INES 2010 - 14th International Conference on Intelligent Engineering Systems, Proceedings (2010). https://doi.org/10.1109/INES.2010.5483855
16. Baruque, B., Corchado, E., Mata, A., Corchado, J.M.: A forecasting solution to the oil spill problem based on a hybrid intelligent system. Inf. Sci. **180**(10), 2029–2043 (2010). https://doi.org/10.1016/j.ins.2009.12.032
17. Tapia, D.I., Corchado, J.M.: An ambient intelligence based multi-agent system for Alzheimer health care. Int. J. Ambient Comput. Intell. **1**(1), 15–26 (2009). https://doi.org/10.4018/jaci.2009010102

18. Mata, A., Corchado, J.M.: Forecasting the probability of finding oil slicks using a CBR system. Expert Syst. Appl. **36**(4), 8239–8246 (2009). https://doi.org/10.1016/j.eswa.2008.10.003
19. Méndez, J.R., Fdez-Riverola, F., Iglesias, E.L., Díaz, F., Corchado, J.M.: Tracking concept drift at feature selection stage in SpamHunting: an anti-spam instance-based reasoning system. Lecture Notes in Computer Science (including subseries Lecture Notes in Artificial Intelligence and Lecture Notes in Bioinformatics), LNAI, vol. 4106, pp. 504–518 (2006)
20. Fdez-Rtverola, F., Corchado, J.M.: FSfRT: forecasting system for red tides. Appl. Intell. **21**(3), 251–264 (2004). https://doi.org/10.1023/B:APIN.0000043558.52701.b1
21. Corchado, J.M., Pavón, J., Corchado, E.S., Castillo, L.F.: Development of CBR-BDI agents: a tourist guide application. Lecture Notes in Computer Science (including subseries Lecture Notes in Artificial Intelligence and Lecture Notes in Bioinformatics), vol. 3155, pp. 547–559 (2004). https://doi.org/10.1007/978-3-540-28631-8
22. Laza, R., Pavn, R., Corchado, J.M.: A reasoning model for CBR-BDI agents using an adaptable fuzzy inference system. Lecture Notes in Computer Science (including subseries Lecture Notes in Artificial Intelligence and Lecture Notes in Bioinformatics), vol. 3040, pp. 96–106. Springer, Heidelberg (2004)
23. Corchado, J.A., Aiken, J., Corchado, E.S., Lefevre, N., Smyth, T.: Quantifying the Ocean's CO2 budget with a CoHeL-IBR system. In: Advances in Case-Based Reasoning, Proceedings, vol. 3155, pp. 533–546 (2004)
24. Corchado, J.M., Borrajo, M.L., Pellicer, M.A., Yáñez, J.C.: Neuro-symbolic system for business internal control. In: Industrial Conference on Data Mining, pp. 1–10 (2004)
25. Corchado, J.M., Corchado, E.S., Aiken, J., Fyfe, C., Fernandez, F., Gonzalez, M.: Maximum likelihood hebbian learning based retrieval method for CBR systems. Lecture Notes in Computer Science (including subseries Lecture Notes in Artificial Intelligence and Lecture Notes in Bioinformatics), vol. 2689, pp. 107–121 (2003)
26. Fdez-Riverola, F., Corchado, J.M.: CBR based system for forecasting red tides. Knowl. Based Syst. **16**(5–6), 328 (2003). https://doi.org/10.1016/S0950-7051(03)00034-0
27. Glez-Bedia, M., Corchado, J.M., Corchado, E.S., Fyfo, C.: Analytical model for constructing deliberative agents. Int. J. Eng. Intell. Syst. Electr. Eng. Commun. **10**(3), 173 (2002)
28. Corchado, J.M., Aiken, J.: Hybrid artificial intelligence methods in oceanographic forecast models. IEEE Trans. Syst. Man Cybern. Part C Appl. Rev. **32**(4), 307–313 (2002). https://doi.org/10.1109/tsmcc.2002.806072
29. Becerra-Bonache, L., Lopez, M.D.J.: Linguistic models at the crossroads of agents, learning and formal languages. Adv. Distrib. Comput. Artif. Intell. J. (ADCAIJ) **3**(4), 67 (2014)
30. Casado-Vara, R., Prieto-Castrillo, F., Corchado, J.M.: A game theory approach for cooperative control to improve data quality and false data detection in WSN. Int. J. Robust Nonlinear Control

Implementation of Automated Pipelines to Generate Knowledge on Challenging Biological Queries

Noé Vázquez[1,2(✉)]

[1] ESEI - Escuela Superior de Ingeniería Informática, Universidad de Vigo, Edificio Politécnico, Campus Universitario As Lagoas s/n, 32004 Ourense, Spain
nvazquezg@gmail.com
[2] CINBIO - Centro de Investigaciones Biomédicas, Universidad de Vigo, Vigo, Spain

Abstract. The main objective of this work is the design and implementation of a reduced set of automated pipelines able to integrate a wide range of existing bioinformatics applications and libraries with the goal of delivering an easy-to-use resource, which can be further used to provide different answers to complex biological questions mainly related with nucleotide and amino acid sequences.

Keywords: Automated pipeline · Software integration
Nucleotide and amino acid sequences · Biological queries

1 Introduction and Motivation

In the last decade, advances in both bioinformatics and computational biology have provided the availability of a huge number of theoretically unrelated tools, applications, and libraries focussed on very distant areas [1]. However, relatively early on, different but complementary approaches were successfully used for integrating existing resources and techniques with the goal of generating even more significant results from a biological perspective [2]. In this line, data and text mining frameworks [3, 4], ensemble alternatives [5], case-based reasoning systems [6] and even agent-based approaches [7] were presented as appropriate envelopes to tackle different biological problems.

However, even the most basic studies require the use of different but complementary tools being executed sequentially or in parallel, which makes it necessary to automatize the analysis pipeline with ad-hoc disposable scripts [8] or to execute it manually. This situation dramatically reduces the reproducibility of the underlying study, as other researcher must repeat the analysis pipeline manually or create their own ad-hoc scripts. Moreover, both alternatives are error prone, as any misunderstanding of the original workflow will result in a different pipeline. In such a situation, the availability of easy-to-install, cross-platform software applications with good Graphical User Interfaces (GUI) that implement the analysis pipelines, not only significantly increases the analysis reproducibility, but also encourages that other research groups

S. Rodríguez et al. (Eds.): DCAI 2018, AISC 801, pp. 426–430, 2019.
https://doi.org/10.1007/978-3-319-99608-0_59

may consider doing similar large-scale analyses, especially if a background in informatics is not a requirement to perform the work.

In this context, the main objective of the proposed research is the identification and generation of a set of automated pipelines able to integrate a wide range of previously developed bioinformatics tools, applications and libraries. As an example, the first step of this doctoral work was the development of BDBM (Blast Database Management, http://sing-group.org/BDBM/), in a collaboration between the Molecular Evolution (ME, http://evolution.ibmc.up.pt/) and the Next Generation Computer System (SING, http://sing-group.org/) groups. This tool provides a GUI that allows the creation of high quality sequence datasets using the several state-of-the-art tools for sequence processing. By using BDBM the user can perform several common tasks such as doing BLAST alignments, getting the open reading frames of the sequences in a FASTA file or changing the format of a FASTA file through a graphical interface. All these tasks make use of several known tools such as EMBOSS [9], BEDTools [10], and NCBI's BLAST [11], Splign and Compart [12], and ProSplign and ProCompart [13]. In addition, BDBM helps the final user to keep his sequence files ordered in a file repository.

In this line, although existing software applications have already eased the work significantly, there is still room for optimization and improvement in them, which is the main objective of this proposal. The experience of the SING group developing bioinformatics tools [14–17] and the experience of the ME group in the definition of biological pathways [18, 19] will be essential for the success of the proposed work.

2 Background and Related Work

Specifically related with the processing and management of nucleotide and amino acid sequences, in a first collaboration between the previously mentioned ME and SING research groups, the ADOPS tool [20] was developed with the main goal of simplifying the identification of positively selected sites in genomic sequences. Since then, several complementary tools are being developed as part of this doctoral work, namely, the aforementioned BDBM (https://www.sing-group.org/BDBM), B+ [21], EvoPPI (http://evoppi.i3s.up.pt/), and SEDA (https://www.sing-group.org/seda/).

The ADOPS software is a Java desktop application, but it requires the installation of third-party software in order to work. Since some of the required software is only available for Linux operating systems, ADOPS can only be used in such environments and, therefore, it may be difficult for researchers working in the medical and life sciences fields to use it. In order to overcome this limitation, as part of this doctoral work, a VirtualBox image with the fully installed software that can be used in an operating system was created. However, as this is not an optimal solution for large-scale projects, an even more user-friendly ADOPS version that uses Docker containers was created, where the required third-party software are already installed, making it truly cross-platform and easy-to-install.

B+ is an online database that complements the ADOPS software, as it is specifically designed to store the results of studies done with the ADOPS software. On one side, it provides researchers using ADOPS with an easy way to share the results of their studies, even when they are under development. On the other side, users using B+ will find a web interface designed to explore the results of large-scale analysis, that may be comprised of hundreds or thousands of results obtained with ADOPS. In this respect, when exploring a study, users can quickly identify the genes where positively selected sites where identified and view the results for the genes in a web interface that mimics the ADOPS GUI.

EvoPPI is an open source web application tool that allows users to compare the protein interactions reported in two different interactomes. When interactomes belong to different species, a versatile BLAST search approach is used to identify orthologous/paralogous genes. This tool integrates several protein-protein interaction (PPI) databases, such as BioGRID [22], FlyBase [23], HIPPIE [24], HomoMINT [25], Interactome3D [26], mentha [27] or PINA [28]. In addition, it implements two custom algorithms that integrate the interactions coming from different interactomes. The algorithm applied depends on whether the interactomes belong to the same or different species.

Finally, SEDA is a multiplatform desktop Java application that has a plugin-based architecture, thus new functions can be easily added to it through the creation of new plugins. Currently, SEDA already allows users to go from hundreds or even thousands of genome annotations to a useful FASTA file containing the sequences of interest in a couple of days, and this is the likely reason why this recently developed software application, that is not yet published, has been already downloaded more than one thousand times. Nevertheless, the incorporation of new plugins such as one for the identification of orthologous genes in different genomes using a two-way BLAST approach will significantly speed up the preparation of the data for bacterial projects.

3 Conclusions

The proposed research project (improving ADOPS and implementing BDBM, B+, EvoPPI, and SEDA) is clearly transdisciplinary, requiring the participation of research teams with a background in biological/health sciences and research teams with a background in informatics. When research teams with expertise in these two research fields collaborate, as in this proposal, it is possible to efficiently perform analyses using whole genome data, using hundreds of individuals, and the biological implications of those analyses understood.

Nowadays, no single research group is able to perform all relevant analyses on their own. Therefore, the availability of easy-to-install, multiplatform software applications with good and user-friendly GUIs, as those to be here improved, is important, especially because this software were developed having in mind that users may not have a background in informatics. It should be noted that some of the options implemented in BDBM, SEDA and even in ADOPS are quite general, and thus such software applications are likely to be used by a wider public than those interested in the identification of amino acid sites under diversifying selection. By depositing the ADOPS projects at B+, researchers can see all the details of the projects that have been performed, reuse the data and reproduce results. Finally, the EvoPPI tool provides a way to compare and explore the PPI in different species.

References

1. Edwards, D., Stajich, J., Hansen, D. (eds.): Bioinformatics: tools and applications. Springer (2009)
2. Matos, S., Araújo, H., Oliveira, J.L.: Biomedical literature exploration through latent semantics. Adv. Distrib. Comput. Artif. Intell. J., ADCAIJ **2**(2), 65 (2013)
3. Ramos, J., Castellanos-Garzón, J.A., de Paz, J.F., Corchado, J.M.: A data mining framework based on boundary-points for gene selection from DNA-microarrays: pancreatic Ductal Adenocarcinoma as a case study. Eng. Appl. Artif. Intell. **70**, 92–108 (2018)
4. Santos, A., Nogueira, R., Lourenço, A.: Applying a text mining framework to the extraction of numerical parameters from scientific literature in the biotechnology domain. Adv. Distrib. Comput. Artif. Intell. J., ADCAIJ **1**(1), 1–8 (2012)
5. Castellanos-Garzón, J.A., Ramos, J., López-Sánchez, D., F. de Paz, J., Corchado, J.M.: An ensemble framework coping with instability in the gene selection process. Interdiscip. Sci.: Comput. Life Sci. **10**, 12–23 (2018)
6. Ramos, J., López-Sánchez, D., Castellanos-Garzón, J.A., de Paz, J.F., Corchado, J.M.: A CBR framework with gradient boosting based feature selection for lung cancer subtype classification. Comput. Biol. Med. **86**, 98–106. (2017)
7. Ramos, J., Castellanos-Garzón, J.A., González-Briones, A., de Paz, J.F., Corchado, J.M.: An agent-based clustering approach for gene selection in gene expression microarray. Interdiscip. Sci. Comput. Life Sci .**9**, 1–13 (2017)
8. Leipzig, J.: A review of bioinformatic pipeline frameworks. Brief. Bioinform. **18**(3:1), 530–536 (2017)
9. Rice, P., Longden, I., Bleasby, A.: EMBOSS: the European molecular biology open software suite. Trends Genet. **16**(6), 276–277 (2000)
10. Quinlan, A.R., Hall, I.M.: BEDTools: a flexible suite of utilities for comparing genomic features. Bioinformatics **26**(6), 841–842 (2010)
11. Altschul, S.F., Gish, W., Miller, W., Myers, E.W., Lipman, D.J.: Basic local alignment search tool. J. Mol. Biol. **215**, 403–410 (1990)
12. Kapustin, Y., Souvorov, A., Tatusova, T., Lipman, D.: Splign: algorithms for computing spliced alignments with identification of paralogs. Biol. Direct **3**, 20 (2008)
13. Kiryutin, B., Souvorov, A., Tatusova, T.: ProSplign - protein to genomic alignment tool. manuscript in preparation. https://www.ncbi.nlm.nih.gov/sutils/static/prosplign/prosplign.html
14. Lourenço, A., Carreira, R.C., Glez-Peña, D., Méndez, J.R., Carneiro, S.A., Rocha, L.M., Díaz, F., Ferreira, E.C., Rocha, I.P., Fdez-Riverola, F., Rocha, M.: BioDR: semantic indexing networks for biomedical document retrieval. Expert Syst. Appl. **37**(4), 3444–3453 (2010)
15. Glez-Peña, D., Díaz, F., Hernández, J.M., Corchado, J.M., Fdez-Riverola, F.: geneCBR: a translational tool for multiple-microarray analysis and integrative information retrieval for aiding diagnosis in cancer research. BMC Bioinform. **10**, 187 (2009)
16. López-Fernández, H., Reboiro-Jato, M., Pérez Rodríguez, J.A., Fdez-Riverola, F., Glez-Peña, D.: The artificial intelligence workbench: a retrospective review. Adv. Distrib. Comput. Artif. Intell. J. (ADCAIJ) **5**(1), 73–85 (2016)
17. López-Fdez, H., Santos, H.M., Capelo, J.L., Fdez-Riverola, F., Glez-Peña, D., Reboiro-Jato, M.: Mass-up: an all-in-one open software application for MALDI-TOF mass spectrometry knowledge discovery. BMC Bioinform. **16**, 318 (2015)

18. Araújo, A.R., Reis, M., Rocha, H., Aguiar, B., Morales-Hojas, R., Macedo-Ribeiro, S., Fonseca, N.A., Reboiro-Jato, D., Reboiro-Jato, M., Fdez-Riverola, F., Vieira, C.P., Vieira, J.: The drosophila melanogaster methuselah gene: a novel gene with ancient functions. PLoS ONE **8**(5), e63747 (2013)
19. Aguiar, B., Vieira, B., Cunha, A.E., Fonseca, N.A., Reboiro-Jato, D., Reboiro-Jato, M., Fdez-Riverola, F., Raspé, O., Vieira, C.P.: Patterns of evolution at the gametophytic self-incompatibility Sorbus aucuparia (Pyrinae) S pollen genes support the non-self recognition by multiple factors model. J. Exp. Bot. **64**(1), 2423–2434 (2013)
20. Reboiro-Jato, D., Reboiro-Jato, M., Fdez-Riverola, F., Vieira, C.P., Fonseca, N.A., Vieira, J:. ADOPS - automatic detection of positively selected sites. J. Integr. Bioinform. **9**(3), 18–32 (2016)
21. Vázquez, N., Vieira, C.P., Amorim, B.S.R., Torres, A., López-Fernández, H., Fdez-Riverola, F., Sousa, J.L.R., Reboiro-Jato, M., Vieira, J.: Large scale analyses and visualization of adaptive amino acid changes projects. Interdiscip. Sci. Comput. Life Sci. **10**, 24 (2018)
22. Chatr-Aryamontri, A., Oughtred, R., Boucher, L., Rust, J., Chang, C., Kolas, N.K., O'Donnell, L., Oster, S., Theesfeld, C., Sellam, A., Stark, C., Breitkreutz, B.J., Dolinski, K., Tyers, M.: The BioGRID interaction database: 2017 update. Nucleic Acids Res. **45**(Database issue), D369–D379 (2017)
23. Gramates, L.S., Marygold, S.J., dos Santos, G., Urbano, J-M., Antonazzo, G., Matthews, B. B., Rey, A.J., Tabone, C.J., Crosby, M.A., Emmert, D.B., Falls, K., Goodman, J.L., Hu, Y., Ponting, L., Schroeder, A.J., Strelets, V.B., Thurmond, J., Zhou, P., The FlyBase Consortium: FlyBase at 25: looking to the future. Nucleic Acids Res. **45**(D1), D663–D671 (2017)
24. Persico, M., Ceol, A., Gavrila, C., Hoffmann, R., Florio, A., Cesareni, G.: HomoMINT: an inferred human network based on orthology mapping of protein interactions discovered in model organisms. BMC Bioinform. **6**(Suppl 4), S21 (2005)
25. Mosca, R., Céol, A., Aloy, P.: Interactome3D: adding structural details to protein networks. Nat. Methods **10**(1), 47–53 (2013)
26. Calderone, A., Castagnoli, L., Cesareni, G.: mentha: a resource for browsing integrated protein-interaction networks. Nat. Methods **10**, 690 (2013)
27. Alanis-Lobato, G., Andrade-Navarro, M.A., Schaefer, M.H.: HIPPIE v2.0: enhancing meaningfulness and reliability of protein–protein interaction networks. Nucleic Acids Res. **45**, D408–D414 (2017)
28. Wu, J., Vallenius, T., Ovaska, K., Westermarck, J., Makela, T.P., Hautaniemi, S.: Integrated network analysis platform for protein-protein interactions. Nat. Methods **6**, 75–77 (2009)

A Data Mining Approach Applied to Wireless Sensor Neworks in Greenhouses

José A. Castellanos-Garzón[1(✉)], Yeray Mezquita Martín[1], José Luis Jaimes S.[2], and Santiago M. López G.[2]

[1] BISITE Digital Innovation Hub, University of Salamanca, Edificio Multiusos I+D+i, 37007 Salamanca, Spain
{jantonio,yeraymm}@usal.es

[2] Instituto Universitario de Estudios de la Ciencia y la Tecnología, University of Salamanca, Salamanca, Spain
{al08985,slopez}@usal.es

Abstract. This research presents an innovative multi-agent system based on virtual organizations. It has been designed to manage the information collected by wireless sensor networks for knowledge discovery and decision making in greenhouses. The developed multi-agent system allowed us to take decisions on the basis of the analysis of the historical data obtained from sensors. The proposed approach improves the efficiency of greenhouses by optimizing the use of resources.

Keywords: Agents · Adaptivity · IoT · Greenhouse · Crop optimization Data mining

1 Introduction

Today, the agricultural industry seeks to maximize economic profit by using the least amount of resources possible, this is achieved with precision agriculture technology. Wireless sensor networks (WSNs) can be used to collect information on plant growth and the environmental conditions that affect it [1–4]. Optimal plant growth depends on several parameters such as irrigation, soil moisture, humidity, temperature, light radiation, pH and CO_2 level.

The aim of this research is to optimize the use of resources [43–52] in greenhouses by means of a multi-agent system (MAS) connected to an available technology (ZigBee, Wi-Fi, Bluetooth, etc.). The aim of this architecture is to yield high quality crops in the greenhouse with the use of minimal resources [35–42]. Through the measurement of parameters, it is possible for example, to irrigate only the part of the crops that need it. Moreover, we can regulate the amount of light radiation and fertilizers that each plant receives.

In order to use resources only when and where they are needed, we propose a MAS for knowledge discovery from the historical data collected by WSNs, see Fig. 1. The platform is subdivided into virtual organizations, which perform different roles such as, WSN data collection, the transformation of data into differen formats, data mining techniques for the identification of patterns and relationships [5–20] and an interface

S. Rodríguez et al. (Eds.): DCAI 2018, AISC 801, pp. 431–436, 2019.
https://doi.org/10.1007/978-3-319-99608-0_60

for user interaction. This architecture allows the patform to manage and fuse heterogeneous information that comes from different types of sensors. Once data have been stored by the MAS, knowledge discovery and/or prediction algorithms can be applied to find out when and where different types of resources are needed [21–34], in this way we obtain a precision agriculture system. Upon the implementation of the developed framework, the users of the system can monitor the greenhouse and estimate the required resources and the quality of crops. Optimization techniques such as those applied here have already had success in other areas of application [53, 54].

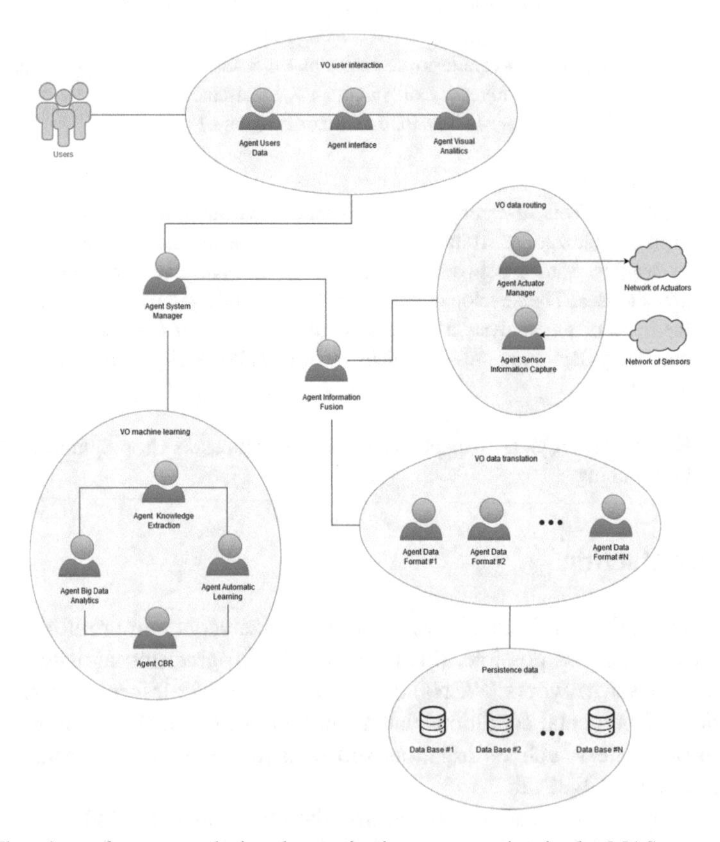

Fig. 1. Flowchart of agents and virtual organizations cooperating in the MAS to reach global goals.

2 Conclusions

The aim of this research was to improve the quality of the crops grown in greenhouses while optimizing the consumption of resources. The above goal has been achieved by obtaining data from a WSN deployed in the greenhouse and storing it in a database. The stored historical data are later analyzed to discover knowledge about the environment and support the user in decision making. Data analysis allowed us to predict the outcome of certain patterns and act accordingly by using resources only when and where they are needed.

The innovation of our approach not only lies in the collection and monitoring of the data collected by sensors, the proposed MAS is also capable of transforming and storing data adequatley. In addition, intelligent algorithms analyze all the information stored in the database and extract the necessary knowledge to support decision making in the process of greenhouse optimization.

Acknowledgements. This work has been supported by the project "IOTEC: Development of Technological Capacities around the Industrial Application of Internet of Things (IoT)". 0123_IOTEC_3_E. Project financed with FEDER funds, Interreg Spain-Portugal (PocTep).

References

1. Shrouf, F., Miragliotta, G.: Energy management based on Internet of Things: practices and framework for adoption in production management. J. Clean. Prod. **100**, 235–246 (2015)
2. Fang, S., Da Xu, L., Zhu, Y., Ahati, J., Pei, H., Yan, J., Liu, Z.: An integrated system for regional environmental monitoring and management based on Internet of Things. IEEE Trans. Industr. Inf. **10**(2), 1596–1605 (2014)
3. Sicari, S., Rizzardi, A., Grieco, L.A., Coen-Porisini, A.: Security, privacy and trust in Internet of Things: the road ahead. Comput. Netw. **76**, 146–164 (2015)
4. Mumtaz, S., Alsohaily, A., Pang, Z., Rayes, A., Tsang, K.F., Rodriguez, J.: Massive Internet of Things for industrial applications: addressing wireless IIoT connectivity challenges and ecosystem fragmentation. IEEE Ind. Electron. Mag. **11**(1), 28–33 (2017)
5. Li, T., Sun, S., Bolić, M., Corchado, J.M.: Algorithm design for parallel implementation of the SMC-PHD filter. Sig. Process. **119**, 115–127 (2016). https://doi.org/10.1016/j.sigpro.2015.07.013
6. Lima, A.C.E.S., De Castro, L.N., Corchado, J.M.: A polarity analysis framework for Twitter messages. Appl. Math. Comput. **270**, 756–767 (2015). https://doi.org/10.1016/j.amc.2015.08.059
7. Redondo-Gonzalez, E., De Castro, L.N., Moreno-Sierra, J., Maestro De Las Casas, M.L., Vera-Gonzalez, V., Ferrari, D.G., Corchado, J.M.: Bladder carcinoma data with clinical risk factors and molecular markers: a cluster analysis. BioMed Res. Int. (2015). https://doi.org/10.1155/2015/168682
8. Li, T., Sun, S., Corchado, J.M., Siyau, M.F.: Random finite set-based Bayesian filters using magnitude-adaptive target birth intensity. In: FUSION 2014 - 17th International Conference on Information Fusion (2014). https://www.scopus.com/inward/record.uri?eid=2-s2.0-84910637788&partnerID=40&md5=bd8602d6146b014266cf07dc35a681e0
9. Choon, Y.W., Mohamad, M.S., Deris, S., Illias, R.M., Chong, C.K., Chai, L.E., Omatu, S., Corchado, J.M.: Differential bees flux balance analysis with OptKnock for in silico microbial strains optimization. PLoS ONE **9**(7) (2014). https://doi.org/10.1371/journal.pone.0102744
10. Li, T., Sun, S., Corchado, J.M., Siyau, M.F.: A particle dyeing approach for track continuity for the SMC-PHD filter. In: FUSION 2014 - 17th International Conference on Information Fusion (2014). https://www.scopus.com/inward/record.uri?eid=2-s2.0-84910637583&partnerID=40&md5=709eb4815eaf544ce01a2c21aa749d8f
11. Coria, J.A.G., Castellanos-Garzón, J.A., Corchado, J.M.: Intelligent business processes composition based on multi-agent systems. Expert Syst. Appl. **41**(4 PART 1), 1189–1205 (2014). https://doi.org/10.1016/j.eswa.2013.08.003

12. Tapia, D.I., Fraile, J.A., Rodríguez, S., Alonso, R.S., Corchado, J.M.: Integrating hardware agents into an enhanced multi-agent architecture for ambient intelligence systems. Inf. Sci. **222**, 47–65 (2013). https://doi.org/10.1016/j.ins.2011.05.002
13. Costa, Â., Novais, P., Corchado, J.M., Neves, J.: Increased performance and better patient attendance in a hospital with the use of smart agendas. Logic J. IGPL **20**(4), 689–698 (2012). https://doi.org/10.1093/jigpal/jzr021
14. García, E., Rodríguez, S., Martín, B., Zato, C., Pérez, B.: MISIA: middleware infrastructure to simulate intelligent agents. Advances in Intelligent and Soft Computing, vol. 91 (2011). https://doi.org/10.1007/978-3-642-19934-9_14
15. Rodríguez, S., De La Prieta, F., Tapia, D.I., Corchado, J.M.: Agents and computer vision for processing stereoscopic images. Lecture Notes in Computer Science (including subseries Lecture Notes in Artificial Intelligence and Lecture Notes in Bioinformatics), LNAI, vol. 6077 (2010). https://doi.org/10.1007/978-3-642-13803-4_12
16. Rodríguez, S., Gil, O., De La Prieta, F., Zato, C., Corchado, J.M., Vega, P., Francisco, M.: People detection and stereoscopic analysis using MAS. In: INES 2010 - 14th International Conference on Intelligent Engineering Systems, Proceedings (2010). https://doi.org/10.1109/INES.2010.5483855
17. Baruque, B., Corchado, E., Mata, A., Corchado, J.M.: A forecasting solution to the oil spill problem based on a hybrid intelligent system. Inf. Sci. **180**(10), 2029–2043 (2010). https://doi.org/10.1016/j.ins.2009.12.032
18. Tapia, D.I., Corchado, J.M.: An ambient intelligence based multi-agent system for alzheimer health care. International J. Ambient Comput. Intell. **1**(1), 15–26 (2009). https://doi.org/10.4018/jaci.2009010102
19. Mata, A., Corchado, J.M.: Forecasting the probability of finding oil slicks using a CBR system. Expert Syst. Appl. **36**(4), 8239–8246 (2009). https://doi.org/10.1016/j.eswa.2008.10.003
20. Glez-Peña, D., Díaz, F., Hernández, J.M., Corchado, J.M., Fdez-Riverola, F.: geneCBR: a translational tool for multiple-microarray analysis and integrative information retrieval for aiding diagnosis in cancer research. BMC Bioinform. **10** (2009). https://doi.org/10.1186/1471-2105-10-187
21. Fernández-Riverola, F., Díaz, F., Corchado, J.M.: Reducing the memory size of a Fuzzy case-based reasoning system applying rough set techniques. IEEE Trans. Syst. Man Cybern. Part C Appl. Rev. **37**(1), 138–146 (2007). https://doi.org/10.1109/TSMCC.2006.876058
22. Méndez, J.R., Fdez-Riverola, F., Díaz, F., Iglesias, E.L., Corchado, J.M.: A comparative performance study of feature selection methods for the anti-spam filtering domain. Lecture Notes in Computer Science (including subseries Lecture Notes in Artificial Intelligence and Lecture Notes in Bioinformatics), LNAI, vol. 4065, 106–120 (2006). https://www.scopus.com/inward/record.uri?eid=2-s2.0-33746435792&partnerID=40&md5=25345ac884f61c182680241828d448c5
23. Méndez, J.R., Fdez-Riverola, F., Iglesias, E.L., Díaz, F., Corchado, J.M.: Tracking concept drift at feature selection stage in SpamHunting: an anti-spam instance-based reasoning system. Lecture Notes in Computer Science (including subseries Lecture Notes in Artificial Intelligence and Lecture Notes in Bioinformatics), LNAI, vol. 4106, pp. 504–518 (2006). https://www.scopus.com/inward/record.uri?eid=2-s2.0-33750974465&partnerID=40&md5=f468552f565ecc3af2d3ca6336e09cc2
24. Fdez-Rtverola, F., Corchado, J.M.: FSfRT: forecasting system for red tides. Appl. Intell. **21**(3), 251–264 (2004). https://doi.org/10.1023/B:APIN.0000043558.52701.b1

25. Corchado, J.M., Pavón, J., Corchado, E.S., Castillo, L.F.: Development of CBR-BDI agents: a tourist guide application. Lecture Notes in Computer Science (including subseries Lecture Notes in Artificial Intelligence and Lecture Notes in Bioinformatics), vol. 3155, pp. 547–559 (2004). https://doi.org/10.1007/978-3-540-28631-8
26. Laza, R., Pavn, R., Corchado, J.M.: A reasoning model for CBR_BDI agents using an adaptable fuzzy inference system. Lecture Notes in Computer Science (including subseries Lecture Notes in Artificial Intelligence and Lecture Notes in Bioinformatics), vol. 3040, pp. 96–106. Springer, Heidelberg (2004)
27. Corchado, J.A., Aiken, J., Corchado, E.S., Lefevre, N., Smyth, T.: Quantifying the Ocean's CO2 budget with a CoHeL-IBR system. In: Advances in Case-Based Reasoning, Proceedings, vol. 3155, pp. 533–546 (2004)
28. Corchado, J.M., Borrajo, M.L., Pellicer, M.A., Yáñez, J.C.: Neuro-symbolic system for business internal control. In: Industrial Conference on Data Mining, pp. 1–10 (2004). https://doi.org/10.1007/978-3-540-30185-1_1
29. Corchado, J.M., Corchado, E.S., Aiken, J., Fyfe, C., Fernandez, F., Gonzalez, M.: Maximum likelihood hebbian learning based retrieval method for CBR systems. Lecture Notes in Computer Science (including subseries Lecture Notes in Artificial Intelligence and Lecture Notes in Bioinformatics), vol. 2689, pp. 107–121 (2003). https://doi.org/10.1007/3-540-45006-8_11
30. Fdez-Riverola, F., Corchado, J.M.: CBR based system for forecasting red tides. Knowl. Based Syst. **16**(5–6), 321–328 (2003). https://doi.org/10.1016/S0950-7051(03)00034-0
31. Glez-Bedia, M., Corchado, J.M., Corchado, E.S., Fyfe, C.: Analytical model for constructing deliberative agents. Int. J. Eng. Intell. Syst. Electr. Eng. Commun. **10**(3), 173 (2002)
32. Corchado, J.M., Aiken, J.: Hybrid artificial intelligence methods in oceanographic forecast models. IEEE Trans. Syst. Man Cybern. Part C-Appl. Rev. **32**(4), 307–313 (2002). https://doi.org/10.1109/tsmcc.2002.806072
33. Fyfe, C., Corchado, J.: A comparison of Kernel methods for instantiating case based reasoning systems. Adv. Eng. Inform. **16**(3), 165–178 (2002). https://doi.org/10.1016/S1474-0346(02)00008-3
34. Fyfe, C., Corchado, J.M.: Automating the construction of CBR systems using kernel methods. Int. J. Intell. Syst. **16**(4), 571–586 (2001). https://doi.org/10.1002/int.1024
35. Li, T.-C., Su, J.-Y., Liu, W., Corchado, J.M.: Approximate Gaussian conjugacy: parametric recursive filtering under nonlinearity, multimodality, uncertainty, and constraint, and beyond. Front. Inf. Technol. Electron. Eng. **18**(12), 1913–1939 (2017)
36. Wang, X., Li, T., Sun, S., Corchado, J.M.: A survey of recent advances in particle filters and remaining challenges for multitarget tracking. Sensors (Switzerland), **17**(12), Article no. 2707 (2017)
37. Morente-Molinera, J.A., Kou, G., González-Crespo, R., Corchado, J.M., Herrera-Viedma, E.: Solving multi-criteria group decision making problems under environments with a high number of alternatives using fuzzy ontologies and multi-granular linguistic modelling methods. Knowl. Based Syst. **137**, 54–64 (2017)
38. Pinto, T., Gazafroudi, A.S., Prieto-Castrillo, F., Santos, G., Silva, F., Corchado, J.M., Vale, Z.: Reserve costs allocation model for energy and reserve market simulation. In: 2017 19th International Conference on Intelligent System Application to Power Systems, ISAP 2017, art. no. 8071410 (2017)
39. Chamoso, P., Rivas, A., Martín-Limorti, J.J., Rodríguez, S.: A hash based image matching algorithm for social networks. Advances in Intelligent Systems and Computing, vol. 619, pp. 183–190 (2018). https://doi.org/10.1007/978-3-319-61578-3_18

40. Sittón, I., Rodríguez, S.: Pattern extraction for the design of predictive models in Industry 4.0. In: International Conference on Practical Applications of Agents and Multi-Agent Systems, pp. 258–261 (2017)
41. García, O., Chamoso, P., Prieto, J., Rodríguez, S., De La Prieta, F.: A serious game to reduce consumption in smart buildings. Communications in Computer and Information Science, vol. 722, pp. 481–493 (2017). https://doi.org/10.1007/978-3-319-60285-1_41
42. Palomino, C.G., Nunes, C.S., Silveira, R.A., González, S.R., Nakayama, M.K.: Adaptive agent-based environment model to enable the teacher to create an adaptive class. Advances in Intelligent Systems and Computing, vol. 617 (2017). https://doi.org/10.1007/978-3-319-60819-8_3
43. Canizes, B., Pinto, T., Soares, J., Vale, Z., Chamoso, P., Santos, D.: Smart city: A GECAD-BISITE energy management case study. In: 15th International Conference on Practical Applications of Agents and Multi-Agent Systems PAAMS 2017, Trends in Cyber-Physical Multi-Agent Systems, vol. 2, pp. 92–100 (2017). https://doi.org/10.1007/978-3-319-61578-3_9
44. Chamoso, P., de La Prieta, F., Eibenstein, A., Santos-Santos, D., Tizio, A., Vittorini, P.: A device supporting the self management of tinnitus. Lecture Notes in Computer Science (including subseries Lecture Notes in Artificial Intelligence and Lecture Notes in Bioinformatics), LNCS, vol. 10209, pp. 399–410 (2017). https://doi.org/10.1007/978-3-319-56154-7_36
45. Román, J.A., Rodríguez, S., de da Prieta, F.: Improving the distribution of services in MAS. Communications in Computer and Information Science, vol. 616 (2016). https://doi.org/10.1007/978-3-319-39387-2_4
46. Durik, B.O.: Organisational metamodel for large-scale multi-agent systems: first steps towards modelling organisation dynamics. Adv. Distrib. Comput. Artif. Intell. J. (ADCAIJ) **6**(3), 17 (2017)
47. Bremer, J., Lehnhoff, S.: Decentralized coalition formation with agent-based combinatorial heuristics. Adv. Distrib. Comput. Artif. Intell. J. (ADCAIJ) **6**(3), 29 (2017)
48. Cardoso, R.C., Bordini, R.H: A multi-agent extension of a hierarchical task network planning formalism. Adv. Distrib. Comput. Artif. Intell. J. (ADCAIJ) **6**(2), 5 (2017)
49. Gonçalves, E., Cortés, M., Oliveira, M.D., Veras, N., Falcão, M., Castro, J.: An analysis of software agents, environments and applications school: retrospective, relevance, and trends. Adv. Distrib. Comput. Artif. Intell. J. (ADCAIJ) **6**(2), 19 (2017)
50. Teixeira, E.P., Goncalves, E.M.N., Adamatti, D.F.: Ulises: a agent-based system for timbre classification. Adv. Distrib. Comput. Artif. Intell. J. (ADCAIJ) **7**(1), 29 (2017)
51. de Castro, L.F.S., Alves, G.V., Borges, A.P.: Using trust degree for agents in order to assign spots in a Smart Parking. Adv. Distrib. Comput. Artif. Intell. J. (ADCAIJ) **6**(2) (2017)
52. Cunha, R., Billa, C., Adamatti, D.: Development of a graphical tool to integrate the Prometheus AEOlus methodology and Jason platform. Adv. Distrib. Comput. Artif. Intell. J. (ADCAIJ) **6**(2), 57 (2017)
53. Ramos, J., Castellanos-Garzón, J.A., González-Briones, A., de Paz, J.F., Corchado, J.M.: An agent-based clustering approach for gene selection in gene expression microarray. Interdiscip. Sci. Comput. Life Sci. **9**(1), 1–13 (2017)
54. Castellanos-Garzón, J.A., Ramos, J.: A gene selection approach based on clustering for classification tasks in colon cancer. Adv. Distrib. Comput. Artif. Intell. Journal. **4**(3), 1–10 (2015)

Administration 4.0: The Challenge of Institutional Competitiveness as a Requisite for Development

Pedro T. Nevado-Batalla Moreno(✉)

University of Salamanca, Campus Miguel de Unamuno, 37071 Salamanca, Spain
pnevado@usal.es

Abstract. Public Administration must live up to the standards of the new environment 4.0. This economic-industrial paradigm is concerned with the whole society. The need for modernization in Public Administration brings to light the directly proportional relationship between institutional and economic competitiveness.

Keywords: Public administration · Competitiveness citizens · Development eGovernment policy · Technology

Process 4.0 [48] is currently associated with the evolution of the industrial sector and has a broad effect on private activities. However, the mere talking or writing of these processes [21] has become a superfluous effort. The concept synthesis of 'Internet of Things' has permeated our society [1, 2, 26, 27] and Public Administration cannot be indifferent to this reality.

We should consider the fact that the modernization of Public Administration is not a new idea at all, it has been with us since decades and takes a temporal logic [14] approach; its aim is to adapt to the reality [29, 53], to what is already available.

This idea of modernity is not as simple as it seems. It may be straightforward in the case of a private organization but not in the case of a public one, such as Public Administration. This is because Public Administration is subject to political variables and contingencies which can change palpable decisions and objectives that require a fully reasonable consensus. Moreover, conflicting situations may lead to significant delays. For this reason, it has been verified on many occasions that Public Administration is always one step behind of the society and is subject to a continuous process of modernization [7, 43, 50].

The norm in all types of private organizations is to follow technical rules which regulate the development [41, 46, 47, 52] of a given activity or sector. However, in Public Administration such conduct receives many objections from those that direct it, due to their inherently political viewpoint. In other words, Public Administration finds itself in a context of governments and political interests which modulate its objectives and decisions linked by more than those and they can be unquestionable or aligned with the needs and expectations of citizens or economic operators.

Beyond any doubt, we should propose that Public Administration be part of horizon 4.0. This objective should not be considered an option, neither an assumption of a largely supported idea. This is a fundamental objective that will prevent Public Administration from getting outplaced in the process of change and development [12]

S. Rodríguez et al. (Eds.): DCAI 2018, AISC 801, pp. 437–443, 2019.
https://doi.org/10.1007/978-3-319-99608-0_61

which is already taking place. Industry 4.0 is a good example of this development. In a highly critical and strongly questioning environment [26, 40], Public Administration is at risk of delegitimization if it does not set the objective 4.0 at the same level as our society and the sectors in which it plays a role. Administration is not a differentiated organization and it does not have to fight for space nor a market. We are dealing with a public organization that sustains a social [5, 45] and democratic model of law, covering all social activities, particularly economic ones, as well as determinants of economic development and the quality of life of the nation [20, 24, 32, 33, 42].

It is not possible to separate Administration and society. Some may think that we would be better off without Administration, that there would be no harm in not having administrative structures nor public employees, but they are wrong. Good Administration brings development, richness, employment and social strength. For this reason, it is so important for Administration to follow the agile flow of our society's economic and industrial activity. And today's path to success for Administration is called 4.0.

The aim to reconcile Public and Political Administration must be a challenge 4.0 [54], as the role of the government cannot be fulfilled correctly without the help of Public Administration if it does not move at the same pace as the rest of the society and, in particular, if it is not at the same level as economic operators [23, 28, 35].

This idea has been well understood in the European Union, although concept 4.0 had not been explicitly stated. The EU updated its Administration with technological advances and tools that are already providing a good standard of life and a significant level of progress in the most developed societies.

In fact, "Communication from the Commission to the European Parliament, The Council, The European Economic and Social Committee and The Committee Of The Regions [44, 49] EU eGovernment Action Plan 2016–2020 Accelerating the digital transformation of government" [17, 18, 22] this is expressed in the first lines of its preamble:

> *eGovernment supports administrative processes, improves the quality of the services and increases internal public-sector efficiency. Digital public services reduce administrative burden on businesses and citizens by making their interactions with public administrations faster and efficient, more convenient and transparent, and less costly. In addition, using digital technologies as an integrated part of governments' modernisation strategies can unlock further economic and social benefits for society as a whole. The digital transformation of government is a key element to the success of the Single Market.*

However, the issue is not as simple as in private sectors. Although, Administration is concerned with public service and material supply activities, in which the fitting of elements that characterize objective 4.0 is fairly simple, the other part of administrative activities requires careful thought. In these activities, the cognitive skills of a human being: analysis, motivation, evaluation, judgement, choice, decision, still cannot be substituted by a machine [25, 36]. We can state, with no prejudice towards the advances in artificial intelligence and the attribution of administrative competencies to a machine, that at present this possibility seems to be quite far away from us [3].

Beyond these concerns however, we are certain that the incorporation of objective 4.0 in Administration is unquestionable. All this entails a twofold challenge.

Firstly, we are concerned with the tools of work in Administration, both for its material supply activity as well as help in the adoption of decisions.

The instruments and tools associated with an environment 4.0 can decisively help in the basic activities of Public Administration: organization, direction, prevention, intervention, coordination etc. This will facilitate the understanding of administrative activity in its entire cycle, from the decision to the execution, follow-up and evaluation. Administration will be like authentic science that is governed by stringent rules based on rationality, efficiency and efficacy, these are far away from political purposes or the zest to always obtain gains, even if it is at the expense of the common good "political bonus".

Secondly, Public Administration must create a favorable environment for businesses and economic operators and it must provide the required support and motivation. The generation of richness and employment must be a priority objective of Public Administration and the Government alike.

In conclusion, both viewpoints can be found in this highly institutional organization, favoring a highly competitive society and economic sector. Thus, a directly proportional link is established between institutional and economic competitiveness, as shown in very well-known studies, like "Doing business" conducted by World Bank [16, 40, 51].

The majority of public and private operators can certainly be subscribed to these ideas as they are general, and their correctness and benefit is difficult to reproach.

As Saint Thomas said, the good is not only to be known, you must want to put it in practice and you must want to achieve it, and this implies effort which in the case of Public Administration is highly required.

Three requirements should be met if we want to achieve domestic benefit and create an environment that favours economic activity, institutional competitiveness and its advantageous economic effects. However, these three requirements must be achieved jointly, it is not possible to compensate the absence or weakness of one with the strength of another [31, 39].

Furthermore, all efforts towards modernization in Public Administration require strong will and determination, not only on the side of superior managing organs. In the "bottom up" process it is necessary that these initiatives be assumed by ordinary citizens [6, 8, 11].

This willingness must be upheld at all times. The beginning of a continuity of public policies must become a reality in Public Administration. Changes must be made according to consistent logic based on long-term planning [9, 13, 19]. Consistency and long-term planning is at times incompatible with the deadlines established by political changes which usually are not consistent with what has been done before. This practice condemns Public Administration to the terrible Penelope public syndrome, in a way that in each political phase it weaves, unweaves and weaves the fabric again. As a result, public resources are wasted, valuable time is lost and the there is no logic to our reality. All this, of course, is at the expense of our citizens.

Finally, it is necessary to make a long-term, timely and reasonable economic investment to cope with a threefold need: staff, material resources and maintenance. This is a complicated issue as investing in institutional competitiveness means investing in actions that are invisible to the majority of citizens, although the effects of such investments are clearly tangible and highly beneficial. This is a conflicting situation which may be solved with the elaboration of programs or theoretical proposals that unfortunately do not tend to receive the necessary financial support and if they receive it, it is not provided for a sufficient period of time and does not result beneficial for the organization [14, 30, 38].

There must be a strong conviction of the need to achieve these standards, if we truly want to establish a proportional relationship between institutional and economic competitiveness under a common denominator of objective 4.0 [34, 37, 39].

In conclusion, challenge 4.0 represents a "manifest destiny" to transforming the society and its institutions. It will provide citizens with greater advantages and quality [15] of life. Public Administration must be the guarantor of a prosperous and growing environment. The organization and resources of Administration will be rationalized through a rigorous and reliable environment 4.0, in which the achievement of public objectives will be faster and more adequate and efficient [4].

References

1. Gazafroudi, A.S., Pinto, T., Prieto-Castrillo, F., Prieto, J., Corchado, J.M., Jozi, A., Vale, Z., Venayagamoorthy, G.K.: Organization-based multi-agent structure of the smart home electricity system. In: 2017 IEEE Congress on Evolutionary Computation (CEC), pp. 1327–1334. IEEE (2017)
2. Gazafroudi, A.S., Prieto-Castrillo, F., Pinto, T., Corchado, J.M.: Organization-based multi-agent system of local electricity market: bottom-up approach. In: International Conference on Practical Applications of Agents and Multi-Agent Systems, pp. 281–283. Springer (2017)
3. Baruque, B., Corchado, E., Mata, A., Corchado, J.M.: A forecasting solution to the oil spill problem based on a hybrid intelligent system. Inf. Sci. **180**(10), 2029–2043 (2010). https://doi.org/10.1016/j.ins.2009.12.032
4. Nihan, C.E.: Healthier? More efficient? Fairer? An overview of the main ethical issues raised by the use of ubicomp in the workplace. Adv. Distrib. Comput. Artif. Intell. (ADCAIJ) **2**(1), 29 (2013). ISSN 2255-2863
5. Chamoso, P., Rivas, A., Martín-Limorti, J.J., Rodríguez, S.: A hash based image matching algorithm for social networks. Advances in Intelligent Systems and Computing, vol. 619, pp. 183–190 (2018). https://doi.org/10.1007/978-3-319-61578-3_18
6. Choon, Y.W., Mohamad, M.S., Deris, S., Illias, R.M., Chong, C.K., Chai, L.E., Omatu, S., Corchado, J.M.: Differential bees flux balance analysis with OptKnock for in silico microbial strains optimization. PLoS ONE **9**(7) (2014). https://doi.org/10.1371/journal.pone.0102744
7. Corchado, J.A., Aiken, J., Corchado, E.S., Lefevre, N., Smyth, T.: Quantifying the Ocean's CO2 budget with a CoHeL-IBR system. In: Advances in Case-Based Reasoning, Proceedings, vol. 3155, pp. 533–546 (2004)
8. Corchado, J.M., Aiken, J.: Hybrid artificial intelligence methods in oceanographic forecast models. IEEE Trans. Syst. Man Cybern. Part C Appl. Rev. **32**(4), 307–313 (2002). https://doi.org/10.1109/tsmcc.2002.806072
9. Corchado, J.M., Fyfe, C.: Unsupervised neural method for temperature forecasting. Artif. Intell. Eng. **13**(4), 351–357 (1999). https://doi.org/10.1016/S0954-1810(99)00007-2
10. Corchado, J.M., Borrajo, M.L., Pellicer, M.A., Yáñez, J.C.: Neuro-symbolic system for business internal control. In: Industrial Conference on Data Mining, pp. 1–10. https://doi.org/10.1007/978-3-540-30185-1_1
11. Corchado, J.M., Corchado, E.S., Aiken, J., Fyfe, C., Fernandez, F., Gonzalez, M.: Maximum likelihood hebbian learning based retrieval method for CBR systems. Lecture Notes in

Computer Science (including subseries Lecture Notes in Artificial Intelligence and Lecture Notes in Bioinformatics), vol. 2689, pp. 107–121 (2003). https://doi.org/10.1007/3-540-45006-8_11

12. Corchado, J.M., Pavón, J., Corchado, E.S., Castillo, L.F.: Development of CBR-BDI agents: a tourist guide application. Lecture Notes in Computer Science (including subseries Lecture Notes in Artificial Intelligence and Lecture Notes in Bioinformatics), vol. 3155, pp. 547–559 (2004). https://doi.org/10.1007/978-3-540-28631-8
13. Corchado, J., Fyfe, C., Lees, B.: Unsupervised learning for financial forecasting. In: Proceedings of the IEEE/IAFE/INFORMS 1998 Conference on Computational Intelligence for Financial Engineering (CIFEr) (Cat. No. 98TH8367), pp. 259–263 (1998). https://doi.org/10.1109/CIFER.1998.690316
14. Costa, Â., Novais, P., Corchado, J.M., Neves, J.: Increased performance and better patient attendance in an hospital with the use of smart agendas. Logic J. IGPL **20**(4), 689–698 (2012). https://doi.org/10.1093/jigpal/jzr021
15. Martínez-Martín, E., Escrig, M.T., Pobil, A.P.D.: A qualitative acceleration model based on intervals. Adv. Distrib. Comput. Artif. Intell. (ADCAIJ) **2**(2), 17 (2013). ISSN 2255-2863
16. Fdez-Riverola, F., Corchado, J.M.: CBR based system for forecasting red tides. Knowl. Based Syst. **16**(5–6), 321–328 (2003). https://doi.org/10.1016/S0950-7051(03)00034-0
17. Fdez-Rtverola, F., Corchado, J.M.: FSfRT: forecasting system for red tides. Appl. Intell. **21** (3), 251–264 (2004). https://doi.org/10.1023/B:APIN.0000043558.52701.b1
18. Fernández-Riverola, F., Díaz, F., Corchado, J.M.: Reducing the memory size of a fuzzy case-based reasoning system applying rough set techniques. IEEE Trans. Syst. Man Cybern. Part C Appl. Rev. **37**(1), 138–146 (2007). https://doi.org/10.1109/TSMCC.2006.876058
19. Fyfe, C., Corchado, J.: A comparison of Kernel methods for instantiating case based reasoning systems. Adv. Eng. Inform. **16**(3), 165–178 (2002). https://doi.org/10.1016/S1474-0346(02)00008-3
20. Fyfe, C., Corchado, J.M.: Automating the construction of CBR systems using kernel methods. Int. J. Intell. Syst. **16**(4), 571–586 (2001). https://doi.org/10.1002/int.1024
21. Coria, J.A.G., Castellanos-Garzón, J.A., Corchado, J.M.: Intelligent business processes composition based on multi-agent systems. Expert Syst. Appl. **41**(4 PART 1), 1189–1205 (2014). https://doi.org/10.1016/j.eswa.2013.08.003
22. García, E., Rodríguez, S., Martín, B., Zato, C., Pérez, B.: MISIA: middleware infrastructure to simulate intelligent agents. Advances in Intelligent and Soft Computing, vol. 91 (2011). https://doi.org/10.1007/978-3-642-19934-9_14
23. García, O., Chamoso, P., Prieto, J., Rodríguez, S., De La Prieta, F.: A serious game to reduce consumption in smart buildings. Communications in Computer and Information Science, vol. 722, pp. 481–493 (2017). https://doi.org/10.1007/978-3-319-60285-1_41
24. Glez-Bedia, M., Corchado, J.M., Corchado, E.S., Fyfe, C.: Analytical model for constructing deliberative agents. Int. J. Eng. Intell. Syst. Electr. Eng. Commun. **10**(3) (2002)
25. Glez-Peña, D., Díaz, F., Hernández, J.M., Corchado, J.M., Fdez-Riverola, F.: geneCBR: a translational tool for multiple-microarray analysis and integrative information retrieval for aiding diagnosis in cancer research. BMC Bioinform. **10** (2009). https://doi.org/10.1186/1471-2105-10-187
26. Palanca, J., Del Val, E., García-Fornes, A., Billhardt, H., Corchado, J.M., Julian, V.: Designing a goal-oriented smart-home environment. Inf. Syst. Front. **20**(1), 125–142 (2017)
27. Rodríguez-Fernandez, J., Pinto, T., Silva, F., Praca, I., Vale, Z., Corchado, J.M.: Bilateral contract prices estimation using a Q-learning based approach. In: 2017 IEEE Symposium Series on Computational Intelligence (SSCI), pp. 1–6 (2017)

28. Macek, K., Rojicek, J., Kontes, G., Rovas, D.V.: Black-box optimization for buildings and its enhancement by advanced communication infrastructure. Adv. Distrib. Comput. Artif. Intell. (ADCAIJ) **2**(2), 53 (2013). ISSN 2255-2863
29. Laza, R., Pavn, R., Corchado, J.M.: A reasoning model for CBR_BDI agents using an adaptable fuzzy inference system. Lecture Notes in Computer Science (including subseries Lecture Notes in Artificial Intelligence and Lecture Notes in Bioinformatics), vol. 3040, pp. 96–106. Springer, Heidelberg (2004)
30. Li, T., De la Prieta Pintado, F., Corchado, J.M., Bajo, J.: Multi-source homogeneous data clustering for multi-target detection from cluttered background with misdetection. Appl. Soft Comput. J. **60**, 436–446 (2017)
31. Li, T., Sun, S., Bolić, M., Corchado, J.M.: Algorithm design for parallel implementation of the SMC-PHD filter. Sig. Process. **119**, 115–127 (2016). https://doi.org/10.1016/j.sigpro.2015.07.013
32. Li, T., Sun, S., Corchado, J.M., Siyau, M.F.: A particle dyeing approach for track continuity for the SMC-PHD filter. In: FUSION 2014 - 17th International Conference on Information Fusion (2014). https://www.scopus.com/inward/record.uri?eid=2-s2.0-84910637583&partnerID=40&md5=709eb4815eaf544ce01a2c21aa749d8f
33. Li, T., Sun, S., Corchado, J.M., Siyau, M.F.: Random finite set-based Bayesian filters using magnitude-adaptive target birth intensity. In: FUSION 2014 - 17th International Conference on Information Fusion (2014). https://www.scopus.com/inward/record.uri?eid=2-s2.0-84910637788&partnerID=40&md5=bd8602d6146b014266cf07dc35a681e0
34. Li, T.-C., Su, J.-Y., Liu, W., Corchado, J.M.: Approximate Gaussian conjugacy: parametric recursive filtering under nonlinearity, multimodality, uncertainty, and constraint, and beyond. Front. Inf. Technol. Electron. Eng. **18**(12), 1913–1939 (2017)
35. Lima, A.C.E.S., De Castro, L.N., Corchado, J.M.: A polarity analysis framework for Twitter messages. Appl. Math. Comput. **270**, 756–767 (2015). https://doi.org/10.1016/j.amc.2015.08.059
36. Mata, A., Corchado, J.M.: Forecasting the probability of finding oil slicks using a CBR system. Expert Syst. Appl. **36**(4), 8239–8246 (2009). https://doi.org/10.1016/j.eswa.2008.10.003
37. Méndez, J.R., Fdez-Riverola, F., Díaz, F., Iglesias, E.L., Corchado, J.M.: A comparative performance study of feature selection methods for the anti-spam filtering domain. Lecture Notes in Computer Science (including subseries Lecture Notes in Artificial Intelligence and Lecture Notes in Bioinformatics), LNAI, vol. 4065, pp. 106–120 (2006). https://www.scopus.com/inward/record.uri?eid=2-s2.0-33746435792&partnerID=40&md5=25345ac884f61c182680241828d448c5
38. Méndez, J.R., Fdez-Riverola, F., Iglesias, E.L., Díaz, F., Corchado, J.M.: Tracking concept drift at feature selection stage in SpamHunting: An anti-spam instance-based reasoning system. Lecture Notes in Computer Science (including subseries Lecture Notes in Artificial Intelligence and Lecture Notes in Bioinformatics), LNAI, vol. 4106, pp. 504–518 (2006). https://www.scopus.com/inward/record.uri?eid=2-s2.0-33750974465&partnerID=40&md5=f468552f565ecc3af2d3ca6336e09cc2
39. Teixido, M., Palleja, T., Tresanchez, M., Font, D., Moreno, J., Fernández, A., Palacín, J., Rebate, C.: Optimization of the virtual mouse HeadMouse to foster its classroom use by children with physical disabilities. Adv. Distrib. Comput. Artif. Intell. (ADCAIJ) **2**(4), 1–8 (2013)
40. Morente-Molinera, J.A., Kou, G., González-Crespo, R., Corchado, J.M., Herrera-Viedma, E.: Solving multi-criteria group decision making problems under environments with a high number of alternatives using fuzzy ontologies and multi-granular linguistic modelling methods. Knowl. Based Syst. **137**, 54–64 (2017)

41. García, Ó., Prieto, J., Alonso, R.S., Corchado, J.M.: A framework to improve energy efficient behaviour at home through activity and context monitoring. Sensors **17**(8), 1749 (2017)
42. Redondo-Gonzalez, E., De Castro, L.N., Moreno-Sierra, J., Maestro De Las Casas, M.L., Vera-Gonzalez, V., Ferrari, D.G., Corchado, J.M.: Bladder carcinoma data with clinical risk factors and molecular markers: a cluster analysis. BioMed Res. Int. (2015). https://doi.org/10.1155/2015/168682
43. Rodríguez, S., De La Prieta, F., Tapia, D.I., Corchado, J.M.: Agents and computer vision for processing stereoscopic images. Lecture Notes in Computer Science (including subseries Lecture Notes in Artificial Intelligence and Lecture Notes in Bioinformatics), LNAI, vol. 6077 (2010). https://doi.org/10.1007/978-3-642-13803-4_12
44. Rodríguez, S., Gil, O., De La Prieta, F., Zato, C., Corchado, J.M., Vega, P., Francisco, M.: People detection and stereoscopic analysis using MAS. In: INES 2010 - 14th International Conference on Intelligent Engineering Systems, Proceedings (2010). https://doi.org/10.1109/INES.2010.5483855
45. Romero, S., Fardoun, H.M., Penichet, V.M.R., Gallud, J.A.: Tweacher: new proposal for online social networks impact in secondary education. Adv. Distrib. Comput. Artif. Intell. (ADCAIJ) **2**(1), 9 (2013). ISSN 2255-2863
46. Gazafroudi, A.S., Pinto, T., Castrillo, F.P., Rodríguez, J.M.C., Abrishambaf, O., Jozi, A., Vale, Z.: Energy flexibility assessment of a multi agent-based smart home energy system. In: 2017 IEEE 17th International Conference on Ubiquitous Wireless Broadband (ICUWB), Salamanca (2017)
47. Shokri Gazafroudi, A., Prieto Castrillo, F., Pinto, T., Prieto Tejedor, J., Corchado Rodríguez, J.M., Bajo Pérez, J.: Energy flexibility management based on predictive dispatch model of domestic energy management system. Energies **10**(9), 1397 (2017)
48. Sittón, I., Rodríguez, S.: Pattern extraction for the design of predictive models in Industry 4.0. In: International Conference on Practical Applications of Agents and Multi-Agent Systems, pp. 258–261 (2017)
49. Tapia, D.I., Corchado, J.M.: An ambient intelligence based multi-agent system for alzheimer health care. International J. Ambient Comput. Intell. **1**(1), 15–26 (2009). https://doi.org/10.4018/jaci.2009010102
50. Tapia, D.I., Fraile, J.A., Rodríguez, S., Alonso, R.S., Corchado, J.M.: Integrating hardware agents into an enhanced multi-agent architecture for Ambient Intelligence systems. Inf. Sci. **222**, 47–65 (2013). https://doi.org/10.1016/j.ins.2011.05.002
51. Oliveira, T., Neves, J., Novais, P.: Guideline formalization and knowledge representation for clinical decision support. Adv. Distrib. Comput. Artif. Intell. (ADCAIJ) **1**(2), 1–11 (2012). ISSN 2255-2863
52. Li, T., Corchado, J.M., Prieto, J.: Convergence of distributed flooding and its application for distributed Bayesian filtering. IEEE Trans. Signal Inf. Process. Over Netw. **3**(3), 580–591 (2017)
53. Li, T., Sun, S.: Online adapting the magnitude of target birth intensity in the PHD Filter. Adv. Distrib. Comput. Artif. Intell. J. **2**(4), 31 (2013). ISSN 2255-2863
54. Wang, X., Li, T., Sun, S., Corchado, J.M.: A survey of recent advances in particle filters and remaining challenges for multitarget tracking. Sensors (Switzerland), **17**(12), Article no. 2707 (2017)

Programmed Physical Activity for the Elderly as a Motor of Active Ageing

Galo Sánchez(✉) and Valeria Lobina

E.U. Magisterio de Zamora, Universidad de Salamanca,
Campus Viriato, Zamora, Spain
galo@usal.es

Abstract. Our population is ageing, and this makes it necessary to implement initiatives that encourage sports activities among people over the age of 65. According to the concept of active ageing, keeping this age group active will allow to improve its health. This research aims to create an ageing model. To this end, we developed a platform for the capture of data and used it to conduct our study. To validate the proposed hypotheses, we examined the correlation between the data captured by the platform with medical data.

Keywords: Active ageing · e-health

1 Introduction

The European Union had declared 2012 as the International Year of Active Ageing and Solidarity between Generations. Since then, there has been considerable interest in improving the quality of life of elderly people. As life expectancy increases and birth rates decline, the elderly live longer and become more active. The western society has a growing obligation to be more attentive to the needs of people over the age of 65 and to provide them with more resources.

Population forecasts show that there will be major social, cultural and economic changes in the short and medium terms. The elderly will be able to spend their free time engaged in a wide range of activities that will make them active players in the society they live in. Governments will have to use their creativity and economic resources to increase the elderly's options for social and cultural participation. Organized physical activities (OPA) are central to the delivery of these opportunities, as they are vital for maintaining body mobility, affective and social relationships, and improving the quality of life.

2 State of the Art

The pace at which Europe's population is ageing has led to the institutional design of strategic lines of action in every EU country. In Spain, several initiatives seek to offer a comprehensive and well-defined framework for action [8].

In comparison to previous generations, todays' elders are more dynamic and play a leading role in our society. It is a silent but steady revolution that is changing the social structure and its socio-economic potential. Typically, the people in this social group are

S. Rodríguez et al. (Eds.): DCAI 2018, AISC 801, pp. 444–450, 2019.
https://doi.org/10.1007/978-3-319-99608-0_62

healthy, they take care of their nutrition and physical activity, and they want to maintain their independence. The policies for equality and solidarity, and protection of well-being, are obligations that are made explicit in our Constitution and public authorities must fulfil them by promoting collective and efficient activities among the population.

The World Health Organization (WHO) defines active ageing as "the process of optimizing opportunities for health, participation and security in order to enhance quality of life as people age" [6]. Nowadays, the daily and moderate practice of physical activity is considered one of the most important preventive health measures that is key to active ageing. However, the lack of physical activity among the population, sedentary lifestyles and associated diseases constitute a serious public health problem that must be combated. For this reason, it is important to promote activity among older people and progress towards social awareness and global participation. The programs of public institutions focus intensively on this issue and encourage healthy lifestyles which involve daily physical activity [1].

3 Study

In the Autonomous Community of Castile and León, the problems of ageing and depopulation are very acute [2]. Political and social initiatives seem to be ineffective in slowing down this population drain. Our research project focuses on the possibilities for active ageing offered by the rural municipalities of the province of Salamanca. In particular, we look at the physical activities organised for the elderly. In the unfavourable context of depopulation and ageing, we would like to analyse the level of commitment to and involvement in physical culture.

We will focus our study on active programs and their stability over time, the teachers who teach them and their professional qualification levels as well as the users' degree of satisfaction with the programs and their direct influence on the health and well-being of the participants. This study will be conducted using the data obtained from the *Depends on You (Depende de ti)* program, the Council of Salamanca and the municipalities' town councils and social action managers. From the analysis of this information we want to create a Guide to good practice, providing useful guidelines on the planning and organization of these activities, to institutions and private management companies alike.

The study includes 3000 program participants all of them over the age of 60; 40 instructors and more than 300 municipalities. We want to delve deeper into the effects of physical activity (physical maintenance mode or Yoga mode) on this group of people. This is a research topic is of increasing interest to the scientific community [3], in both population studies that provide direct evidence and partial studies that focus on the positive influence of physical activity on the well-being and daily development of the elderly.

The indicators tell us that in the last ten years, the levels of physical activity have risen the most in the older age groups. A multidisciplinary team of experts from the University of Salamanca made it possible to examine in-depth the motor, physical and expressive development of the body and the mobility of the elderly. To ensure the success of the project, the team members are experts in fields such as education and culture, physical activity, depopulation, ageing and new rural dwellers. Our interests are specific, we want to learn about the body itself and its movement capabilities in physical activity and sport. We want to learn about improving physical fitness and its influence on the elderly's well-being, personality development and the positive impact on affectivity and socialization.

4 Proposal

To conduct the study, a technological platform [9–23] has been designed to capture data from multiple heterogeneous sources [24–52], such as questionnaires and interviews, planned workshops and monitoring of class sessions, discussion groups and open forums to gather as much information as possible. During the months of April to November 2018, the study will be conducted in more than 100 locations.

The specific objective of this research is to create our collaboration plan with educational technology professionals, aimed at designing a platform for the fusion of medical data and the individualized monitoring of physical activity sessions performed in the Yoga modality. This will allow us to correlate continued practice with the improvement of general physical condition (specifically flexibility, agility and body balance), to demonstrate its preventive value in falls and hip fractures.

Accurate data on regular physical practice in a group of people between the ages of 65 and 85, could help us demonstrate the importance of a specific set of yoga exercises that help prevent falls and hip fractures.

An explanatory, clear and very visual chart of ten basic exercises will be made. The purpose of these exercises is to strengthen static joints and give elasticity to the supporting muscles in order to improve physical condition and balance. There will be a set of twelve sessions which will explain how to do each exercise correctly, there will also be a final learning assessment to ensure that the exercise routine is followed by each participant in a timely manner and that they are able to do the exercise independently.

Technology will be used to do a follow-up and compare participants with the control group.

5 Conclusions

The objective of the institutions must be to improve the quality of life of elderly people through active ageing. All programs should be highly humanized, preventive and respectful of each individual. For this reason, our research draws from the health value of physical activity in elderly people, serving as a medical support resource in the prevention of daily risk situations.

The collaborative work of primary health care services and the reports of social action managers on physical activities organized by physical education professionals, will have a major impact on the quality of life and social participation of older people.

The extraction of data from physical activity practices will allow to improve people's health. Well-being must be an important part of the policies for promotion and quality of life in the 21st century.

References

1. Barrio Aliste, J.M.: De los problemas a los retos de la población rural de Castilla y León. Encrucijadas, Revista crítica de Ciencias Sociales **6**, 117–128 (2013)
2. Gómez-Limón Rodríguez, J.A., Atance Muñiz, I., Rico González, M.: Percepción pública del problema de la despoblación del medio rural en Castilla y León. Ager, Revista de estudios sobre despoblación y desarrollo rural **6**, 9–60 (2007)
3. Guillén García, F., Castro Sánchez, J.J., Guillén García, M.A.: Calidad de vida, salud y ejercicio físico: una aproximación al tema desde una perspectiva psicosocial. Revista Psicología del deporte **12**, 91–110 (1997)
4. López Fernández, F.J.: Acción social en España. Centros, servicios y establecimientos de servicios sociales. ACCI, Madrid (2014)
5. Martínez de Haro, V. (coord.): Actividad física, salud y calidad de vida. UAM, Fundación Estudiantes, Madrid (2010)
6. OMS (ed.): Active Ageing: A Policy Framework. Organización Mundial de la Salud (2002). http://goo.gl/oG5w8M. Recuperado 18 Feb 2018
7. VVAA: Plan Integral para la actividad física y el deporte en personas mayores. Ministerio de Cultura (2009)
8. VVAA: Libro blanco del envejecimiento activo. CSD, Ministerio de Educación y Cultura (2010)
9. Li, T., Sun, S., Bolić, M., Corchado, J.M.: Algorithm design for parallel implementation of the SMC-PHD filter. Sig. Process. **119**, 115–127 (2016). https://doi.org/10.1016/j.sigpro.2015.07.013
10. Lima, A.C.E.S., De Castro, L.N., Corchado, J.M.: A polarity analysis framework for Twitter messages. Appl. Math. Comput. **270**, 756–767 (2015). https://doi.org/10.1016/j.amc.2015.08.059
11. Redondo-Gonzalez, E., De Castro, L.N., Moreno-Sierra, J., Maestro De Las Casas, M.L., Vera-Gonzalez, V., Ferrari, D.G., Corchado, J.M.: A cluster analysis. BioMed Res. Int. (2015). https://doi.org/10.1155/2015/168682
12. Li, T., Sun, S., Corchado, J.M., Siyau, M.F.: Random finite set-based Bayesian filters using magnitude-adaptive target birth intensity. In: FUSION 2014 - 17th International Conference on Information Fusion (2014). https://www.scopus.com/inward/record.uri?eid=2-s2.0-84910637788&partnerID=40&md5=bd8602d6146b014266cf07dc35a681e0
13. Choon, Y.W., Mohamad, M.S., Deris, S., Illias, R.M., Chong, C.K., Chai, L.E., Omatu, S., Corchado, J.M.: Differential bees flux balance analysis with OptKnock for in silico microbial strains optimization. PLoS ONE **9**(7) (2014). https://doi.org/10.1371/journal.pone.0102744
14. Li, T., Sun, S., Corchado, J.M., Siyau, M.F.: A particle dyeing approach for track continuity for the SMC-PHD filter. In: FUSION 2014 - 17th International Conference on Information Fusion (2014). https://www.scopus.com/inward/record.uri?eid=2-s2.0-84910637583&partnerID=40&md5=709eb4815eaf544ce01a2c21aa749d8f
15. García Coria, J.A., Castellanos-Garzón, J.A., Corchado, J.M.: Intelligent business processes composition based on multi-agent systems. Expert Syst. Appl. **41**(4 PART 1), 1189–1205 (2014). https://doi.org/10.1016/j.eswa.2013.08.003

16. Tapia, D.I., Fraile, J.A., Rodríguez, S., Alonso, R.S., Corchado, J.M.: Integrating hardware agents into an enhanced multi-agent architecture for Ambient Intelligence systems. Inf. Sci. **222**, 47–65 (2013). https://doi.org/10.1016/j.ins.2011.05.002
17. Costa, Â., Novais, P., Corchado, J.M., Neves, J.: Increased performance and better patient attendance in an hospital with the use of smart agendas. Log. J. IGPL **20**(4), 689–698 (2012). https://doi.org/10.1093/jigpal/jzr021
18. García, E., Rodríguez, S., Martín, B., Zato, C., Pérez, B.: MISIA: middleware infrastructure to simulate intelligent agents. In: Advances in Intelligent and Soft Computing, vol. 91 (2011). https://doi.org/10.1007/978-3-642-19934-9_14
19. Rodríguez, S., De La Prieta, F., Tapia, D.I., Corchado, J.M.: Agents and computer vision for processing stereoscopic images. In: Lecture Notes in Computer Science (including subseries Lecture Notes in Artificial Intelligence and Lecture Notes in Bioinformatics) (LNAI), vol. 6077 (2010). https://doi.org/10.1007/978-3-642-13803-4_12
20. Rodríguez, S., Gil, O., De La Prieta, F., Zato, C., Corchado, J.M., Vega, P., Francisco, M.: People detection and stereoscopic analysis using MAS. In: Proceedings of INES 2010 - 14th International Conference on Intelligent Engineering Systems (2010). https://doi.org/10.1109/INES.2010.5483855
21. Baruque, B., Corchado, E., Mata, A., Corchado, J.M.: A forecasting solution to the oil spill problem based on a hybrid intelligent system. Inf. Sci. **180**(10), 2029–2043 (2010). https://doi.org/10.1016/j.ins.2009.12.032
22. Tapia, D.I., Corchado, J.M.: An ambient intelligence based multi-agent system for alzheimer health care. Int. J. Ambient. Comput. Intell **1**(1), 15–26 (2009). https://doi.org/10.4018/jaci.2009010102
23. Mata, A., Corchado, J.M.: Forecasting the probability of finding oil slicks using a CBR system. Expert Syst. Appl. **36**(4), 8239–8246 (2009). https://doi.org/10.1016/j.eswa.2008.10.003
24. Glez-Peña, D., Díaz, F., Hernández, J.M., Corchado, J.M., Fdez-Riverola, F.: geneCBR: a translational tool for multiple-microarray analysis and integrative information retrieval for aiding diagnosis in cancer research. BMC Bioinform. **10** (2009). https://doi.org/10.1186/1471-2105-10-187
25. Fernández-Riverola, F., Díaz, F., Corchado, J.M.: Reducing the memory size of a Fuzzy case-based reasoning system applying rough set techniques. IEEE Trans. Syst. Man Cybern. Part C Appl. Rev. **37**(1), 138–146 (2007). https://doi.org/10.1109/TSMCC.2006.876058
26. Méndez, J.R., Fdez-Riverola, F., Díaz, F., Iglesias, E.L., Corchado, J.M.: A comparative performance study of feature selection methods for the anti-spam filtering domain. In: Lecture Notes in Computer Science (Including Subseries Lecture Notes in Artificial Intelligence and Lecture Notes in Bioinformatics) (LNAI), vol. 4065, pp. 106–120 (2006). https://www.scopus.com/inward/record.uri?eid=2-s2.0-33746435792&partnerID=40&md5=25345ac884f61c182680241828d448c5
27. Fdez-Rtverola, F., Corchado, J.M.: FSfRT: forecasting system for red tides. Appl. Intell. **21** (3), 251–264 (2004). https://doi.org/10.1023/B:APIN.0000043558.52701.b1
28. Corchado, J.M., Pavón, J., Corchado, E.S., Castillo, L.F.: Development of CBR-BDI agents: a tourist guide application. In: Lecture Notes in Computer Science (including subseries Lecture Notes in Artificial Intelligence and Lecture Notes in Bioinformatics), vol. 3155, pp. 547–559 (2004). https://doi.org/10.1007/978-3-540-28631-8
29. Corchado, J.A., Aiken, J., Corchado, E.S., Lefevre, N., Smyth, T.: Quantifying the Ocean's CO2 budget with a CoHeL-IBR system. In: Proceedings of Advances in Case-Based Reasoning, vol. 3155, pp. 533–546 (2004)

30. Corchado, J.M., Borrajo, M.L., Pellicer, M.A., Yáñez, J.C.: Neuro-symbolic system for business internal control. In: Industrial Conference on Data Mining, pp. 1–10 (2004). https://doi.org/10.1007/978-3-540-30185-1_1
31. Corchado, J.M., Corchado, E.S., Aiken, J., Fyfe, C., Fernandez, F., Gonzalez, M.: Maximum likelihood hebbian learning based retrieval method for CBR systems. In: Lecture Notes in Computer Science (including subseries Lecture Notes in Artificial Intelligence and Lecture Notes in Bioinformatics), vol. 2689, pp. 107–121 (2003). https://doi.org/10.1007/3-540-45006-8_11
32. Fdez-Riverola, F., Corchado, J.M.: CBR based system for forecasting red tides. Knowl.-Based Syst. **16**(5–6 SPEC.), 321–328 (2003). https://doi.org/10.1016/S0950-7051(03)00034-0
33. Glez-Bedia, M., Corchado, J.M., Corchado, E.S., Fyfe, C.: Analytical model for constructing deliberative agents. Int. J. Eng. Intell. Syst. Electr. Eng. Commun. **10**(3), 173–185 (2002)
34. Corchado, J.M., Aiken, J.: Hybrid artificial intelligence methods in oceanographic forecast models. IEEE Trans. Syst. Man Cybern. Part C Appl. Rev. **32**(4), 307–313 (2002). https://doi.org/10.1109/tsmcc.2002.806072
35. Fyfe, C., Corchado, J.: A comparison of Kernel methods for instantiating case based reasoning systems. Adv. Eng. Inform. **16**(3), 165–178 (2002). https://doi.org/10.1016/S1474-0346(02)00008-3
36. Fyfe, C., Corchado, J.M.: Automating the construction of CBR systems using kernel methods. Int. J. Intell. Syst. **16**(4), 571–586 (2001). https://doi.org/10.1002/int.1024
37. Corchado, J.M., Fyfe, C.: Unsupervised neural method for temperature forecasting. Artif. Intell. Eng. **13**(4), 351–357 (1999). https://doi.org/10.1016/S0954-1810(99)00007-2
38. Corchado, J., Fyfe, C., Lees, B.: Unsupervised learning for financial forecasting. In: Proceedings of the IEEE/IAFE/INFORMS 1998 Conference on Computational Intelligence for Financial Engineering (CIFEr) (Cat. No. 98TH8367), pp. 259–263 (1998). https://doi.org/10.1109/CIFER.1998.690316
39. Li, T.-C., Su, J.-Y., Liu, W., Corchado, J.M.: Approximate Gaussian conjugacy: parametric recursive filtering under nonlinearity, multimodality, uncertainty, and constraint, and beyond. Front. Inf. Technol. Electron. Eng. **18**(12), 1913–1939 (2017)
40. Wang, X., Li, T., Sun, S., Corchado, J.M.: A survey of recent advances in particle filters and remaining challenges for multitarget tracking. Sens. (Switz.) **17**(12) (2017). Article No. 2707
41. Morente-Molinera, J.A., Kou, G., González Crespo, R., Corchado, J.M., Herrera-Viedma, E.: Solving multi-criteria group decision making problems under environments with a high number of alternatives using fuzzy ontologies and multi-granular linguistic modelling methods. Knowl. Based Syst. **137**, 54–64 (2017)
42. Pinto, T., Gazafroudi, A.S., Prieto-Castrillo, F., Santos, G., Silva, F., Corchado, J.M., Vale, Z.: Reserve costs allocation model for energy and reserve market simulation. In: 2017 19th International Conference on Intelligent System Application to Power Systems, ISAP 2017 (2017). Article No. 8071410
43. Lim, S.Y., Mohamad, M.S., Chai, L.E., Deris, S., Chan, W.H., Omatu, S., Corchado, J.M., Sjaugi, M.F., Zainuddin, M.M., Rajamohan, G., Ibrahim, Z., Yusof, Z.M.: Investigation of the effects of imputation methods for gene regulatory networks modelling using dynamic bayesian networks. In: DCAI 2016, pp. 413–421 (2016)
44. Fernandes, F., Gomes, L., Morais, H., Silva, M.R., Vale, Z.A., Corchado, J.M.: Dynamic energy management method with demand response interaction applied in an office building. In: PAAMS (Special Sessions), pp. 69–82 (2016)
45. Dang, N.C., de la Prieta, F., Corchado, J.M., Moreno, M.N.: Framework for retrieving relevant contents related to fashion from online social network data. In: PAAMS (Special Sessions), pp. 335–347 (2016)

46. Chamoso, P., de la Prieta, F., de Paz, J.F., Corchado, J.M.: Swarm agent-based architecture suitable for internet of things and smartcities. In: DCA 2015, pp. 21–29 (2015)
47. Omatu, S., Wada, T., Rodríguez, S., Chamoso, P., Corchado, J.M.: Multi-agent technology to perform odor classification. In: ISAmI 2014, pp. 241–252 (2014)
48. Román, J.Á., Rodríguez, S., Corchado, J.M.: Improving intelligent systems: specialization. In: PAAMS (Workshops), pp. 378–385 (2014)
49. Tapia, D.I., García, Ó., Alonso, R.S., Guevara, F., Catalina, J., Bravo, R.A., Corchado, J.M.: Evaluating the n-core polaris real-time locating system in an indoor environment. In: PAAMS (Workshops), pp. 29–37 (2012)
50. Tapia, D.I., Alonso, R.S., García, Ó., Corchado, J.M.: HERA: hardware-embedded reactive agents platform. In: PAAMS (Special Sessions), pp. 249–256 (2011)
51. Batista, Vivian Fad López Aguilar, R., Alonso, L., García, María Nand Moreno Corchado, J. M.: Data mining for grammatical inference with bioinformatics criteria. In: HAIS (2), pp. 53–60 (2010)
52. Mata, A., Lancho, B.P., Corchado, J.M.: Forest fires prediction by an organization based system. In: PAAMS, pp. 135–144 (2010)

Blockchain Technology for Luggage Tracking

Alberto Rodríguez Ludeiro(✉)

BISITE Digital Innovation Hub, University of Salamanca.
Edificio Multiusos I+D+I, 37007 Salamanca, Spain
albertoludeiro@usal.es

Abstract. Lost luggage is one of the main fears when boarding a plane, especially on long flights. Around 10,000 suitcases are lost every day at airports around the world. The proposed solution is to locate much faster and more efficiently the lost object. There are currently a multitude of players involved in this process, who could synchronize by sharing information, which would save airlines costs. This way, the customer and the airline can know where the luggage is at any time.

Keywords: Blockchain · Tracking · Sharing information

1 Introduction

Lost luggage is one of the main fears when we board a plane, especially on long flights where we have to make stopovers and change aircraft. During 2016, more than 26 million bags are lost, damaged, delayed or stolen across 200 countries. There were 8.83 baggage mismanaged for every thousand passengers. Of the total number of parts that caused complaints from passengers, 82.2% were delays, 12.9% damaged luggage and 4.2% were stolen or lost.

The aim of this project is to improve the tracking of luggage, saving time and trying to avoid its loss. All this through the development of a multi-platform application and the use of barcodes that are currently used in luggage, which means a very low cost for the airline. It is important to bear in mind that this application must be easily understood and used by any type of user.

This project was based on the idea that we can have a platform that integrates and coordinates two different faces, the blockchain and a mobile device. The main development of the system was done according to a client-server architecture, where the client is a mobile application and the server a set of Blockchain and a RESTful API (Fig. 1).

The results of the implementation of this platform are that its use saves a lot of money in addition to the time that would otherwise be lost. Airports have a comprehensive luggage tracking system. Users can search easily and in real time for their suitcase.

S. Rodríguez et al. (Eds.): DCAI 2018, AISC 801, pp. 451–456, 2019.
https://doi.org/10.1007/978-3-319-99608-0_63

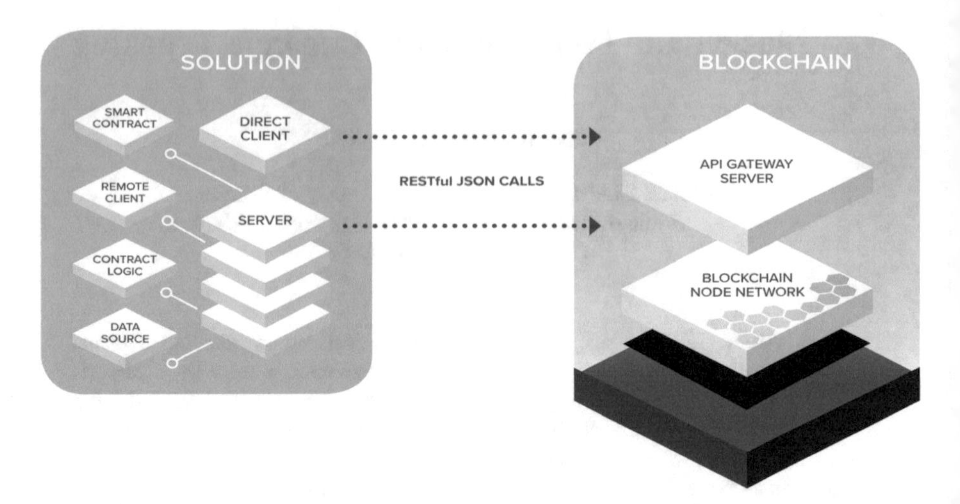

Fig. 1. Blockchain architecture

2 Conclusions

This research aims to improve the conventional baggage tracking method, allowing users to have control of where their luggage is and save time. It is also a great improvement for companies because they can increase their reputation by reducing the number of lost luggage, as well as reducing the resulting subsidies.

The blockchain is the most revolutionary technology we will know in our time and will mark the beginning of a new "Internet of value". With this project, it is possible to use this new technology to achieve a new functionality unexplored to date.

Acknowledgements. This work has been supported by project "IOTEC: Development of Technological Capacities around the Industrial Application of Internet of Things (IoT)". 0123_IOTEC_3_E. Project financed with FEDER funds, Interreg Spain-Portugal (PocTep).

References

1. Li, T., Sun, S., Bolić, M., Corchado, J.M.: Algorithm design for parallel implementation of the SMC-PHD filter. Sig. Process. **119**, 115–127 (2016). https://doi.org/10.1016/j.sigpro.2015.07.013
2. Lima, A.C.E.S., De Castro, L.N., Corchado, J.M.: A polarity analysis framework for Twitter messages. Appl. Math. Comput. **270**, 756–767 (2015). https://doi.org/10.1016/j.amc.2015.08.059. Redondo-Gonzalez, E., De Castro, L.N., Moreno-Sierra, J., Maestro De Las Casas, M.L., Vera-Gonzalez, V., Ferrari, D.G., Corchado, J.M.: Bladder carcinoma data with clinical risk factors and molecular markers: a cluster analysis. BioMed Res. Int. **2015** (2015). https://doi.org/10.1155/2015/168682

3. Li, T., Sun, S., Corchado, J.M., Siyau, M.F.: Random finite set-based Bayesian filters using magnitude-adaptive target birth intensity. In: FUSION 2014 - 17th International Conference on Information Fusion (2014). https://www.scopus.com/inward/record.uri?eid=2-s2.0-84910637788&partnerID=40&md5=bd8602d6146b014266cf07dc35a681e0. Choon, Y.W., Mohamad, M.S., Deris, S., Illias, R.M., Chong, C.K., Chai, L.E., Corchado, J.M.: Differential bees flux balance analysis with OptKnock for in silico microbial strains optimization. PLoS ONE **9**(7) (2014). https://doi.org/10.1371/journal.pone.0102744
4. Li, T., Sun, S., Corchado, J.M., Siyau, M.F.: A particle dyeing approach for track continuity for the SMC-PHD filter. In: FUSION 2014 - 17th International Conference on Information Fusion (2014). https://www.scopus.com/inward/record.uri?eid=2-s2.0-84910637583&partnerID=40&md5=709eb4815eaf544ce01a2c21aa749d8f
5. García Coria, J.A., Castellanos-Garzón, J.A., Corchado, J.M.: Intelligent business processes composition based on multi-agent systems. Expert Syst. Appl. **41**(4 PART 1), 1189–1205 (2014). https://doi.org/10.1016/j.eswa.2013.08.003. Tapia, D.I., Fraile, J.A., Rodríguez, S., Alonso, R.S., Corchado, J.M.: Integrating hardware agents into an enhanced multi-agent architecture for Ambient Intelligence systems. Inf. Sci. **222**, 47–65 (2013). https://doi.org/10.1016/j.ins.2011.05.002. Costa, Â., Novais, P., Corchado, J.M., Neves, J.: Increased performance and better patient attendance in an hospital with the use of smart agendas. Logic J. IGPL **20**(4), 689–698 (2012). https://doi.org/10.1093/jigpal/jzr021
6. García, E., Rodríguez, S., Martín, B., Zato, C., Pérez, B.: MISIA: middleware infrastructure to simulate intelligent agents. In: Advances in Intelligent and Soft Computing, vol. 91 (2011). https://doi.org/10.1007/978-3-642-19934-9_14
7. Rodríguez, S., De La Prieta, F., Tapia, D.I., Corchado, J.M.: Agents and computer vision for processing stereoscopic images. In: Lecture Notes in Computer Science (Including Subseries Lecture Notes in Artificial Intelligence and Lecture Notes in Bioinformatics) (LNAI), vol. 6077 (2010). https://doi.org/10.1007/978-3-642-13803-4_12
8. Rodríguez, S., Gil, O., De La Prieta, F., Zato, C., Corchado, J.M., Vega, P., Francisco, M.: People detection and stereoscopic analysis using MAS. In: INES 2010 - 14th International Conference on Intelligent Engineering Systems, Proceedings (2010). https://doi.org/10.1109/INES.2010.5483855. Baruque, B., Corchado, E., Mata, A., Corchado, J.M.: A forecasting solution to the oil spill problem based on a hybrid intelligent system. Inf. Sci. **180**(10), 2029–2043 (2010). https://doi.org/10.1016/j.ins.2009.12.032
9. Tapia, D.I., Corchado, J.M.: An ambient intelligence based multi-agent system for alzheimer health care. Int. J. Ambient Comput. Intell. **1**(1), 15–26 (2009). https://doi.org/10.4018/jaci.2009010102. Mata, A., Corchado, J.M.: Forecasting the probability of finding oil slicks using a CBR system. Expert Syst. Appl. **36**(4), 8239–8246 (2009). https://doi.org/10.1016/j.eswa.2008.10.003
10. Glez-Peña, D., Díaz, F., Hernández, J.M., Corchado, J.M., Fdez-Riverola, F.: geneCBR: a translational tool for multiple-microarray analysis and integrative information retrieval for aiding diagnosis in cancer research. BMC Bioinf. **10** (2009). https://doi.org/10.1186/1471-2105-10-187
11. Fernández-Riverola, F., Díaz, F., Corchado, J.M.: Reducing the memory size of a fuzzy case-based reasoning system applying rough set techniques. IEEE Trans. Syst. Man Cybern. Part C Appl. Rev. **37**(1), 138–146 (2007). https://doi.org/10.1109/TSMCC.2006.876058
12. Méndez, J.R., Fdez-Riverola, F., Díaz, F., Iglesias, E.L., Corchado, J.M.: A comparative performance study of feature selection methods for the anti-spam filtering domain. In: Lecture Notes in Computer Science (Including Subseries Lecture Notes in Artificial Intelligence and Lecture Notes in Bioinformatics), LNAI, vol. 4065, pp. 106–120 (2006). https://www.scopus.com/inward/record.uri?eid=2-s2.0-33746435792&partnerID=40&md5=25345ac884f61c182680241828d448c5

13. Méndez, J.R., Fdez-Riverola, F., Iglesias, E.L., Díaz, F., Corchado, J.M.: Tracking concept drift at feature selection stage in SpamHunting: an anti-spam instance-based reasoning system. In: Lecture Notes in Computer Science (Including Subseries Lecture Notes in Artificial Intelligence and Lecture Notes in Bioinformatics), LNAI, vol. 4106, 504–518 (2006). https://www.scopus.com/inward/record.uri?eid=2-s2.0-33750974465&partnerID=40&md5=f468552f565ecc3af2d3ca6336e09cc2
14. Fdez-Rtverola, F., Corchado, J.M.: FSfRT: forecasting system for red tides. Appl. Intell. **21** (3), 251–264 (2004). https://doi.org/10.1023/B:APIN.0000043558.52701.b1
15. Corchado, J.M., Pavón, J., Corchado, E.S., Castillo, L.F.: Development of CBR-BDI agents: a tourist guide application. In: Lecture Notes in Computer Science (including subseries Lecture Notes in Artificial Intelligence and Lecture Notes in Bioinformatics), vol. 3155, pp. 547–559 (2004). https://doi.org/10.1007/978-3-540-28631-8
16. Laza, R., Pavn, R., Corchado, J.M.: A reasoning model for CBR_BDI agents using an adaptable fuzzy inference system. In: Lecture Notes in Computer Science (Including Subseries Lecture Notes in Artificial Intelligence and Lecture Notes in Bioinformatics), vol. 3040, pp. 96–106. Springer, Heidelberg (2004)
17. Corchado, J.A., Aiken, J., Corchado, E.S., Lefevre, N., Smyth, T.: Quantifying the Ocean's CO2 budget with a CoHeL-IBR system. In: Advances in Case-Based Reasoning, Proceedings, vol. 3155, pp. 533–546 (2004)
18. Corchado, J.M., Borrajo, M.L., Pellicer, M.A., Yáñez, J.C.: Neuro-symbolic system for business internal control. In: Industrial Conference on Data Mining, pp. 1–10 (2004). https://doi.org/10.1007/978-3-540-30185-1_1
19. Corchado, J.M., Corchado, E.S., Aiken, J., Fyfe, C., Fernandez, F., Gonzalez, M.: Maximum likelihood Hebbian learning based retrieval method for CBR systems. In: Lecture Notes in Computer Science (Including Subseries Lecture Notes in Artificial Intelligence and Lecture Notes in Bioinformatics), vol. 2689, pp. 107–121 (2003). https://doi.org/10.1007/3-540-45006-8_11
20. Fdez-Riverola, F., Corchado, J.M.: CBR based system for forecasting red tides. Knowl. Based Syst. **16**(5–6 SPEC), 321–328 (2003). https://doi.org/10.1016/S0950-7051(03)00034-0
21. Glez-Bedia, M., Corchado, J.M., Corchado, E.S., Fyfe, C.: Analytical model for constructing deliberative agents. Int. J. Eng. Intell. Syst. Electr. Eng. Commun. **10**(3) (2002)
22. Corchado, J.M., Aiken, J.: Hybrid artificial intelligence methods in oceanographic forecast models. IEEE Trans. Syst. Man Cybern. Part C Appl. Rev. **32**(4), 307–313 (2002). https://doi.org/10.1109/tsmcc.2002.806072
23. Li, T.-C., Su, J.-Y., Liu, W., Corchado, J.M.: Approximate Gaussian conjugacy: parametric recursive filtering under nonlinearity, multimodality, uncertainty, and constraint, and beyond. Front. Inf. Technol. Electron. Eng. **18**(12), 1913–1939 (2017)
24. Wang, X., Li, T., Sun, S., Corchado, J.M.: A survey of recent advances in particle filters and remaining challenges for multitarget tracking. Sensors (Switzerland) **17**(12) (2017). Art. no. 2707
25. Morente-Molinera, J.A., Kou, G., González-Crespo, R., Corchado, J.M., Herrera-Viedma, E.: Solving multi-criteria group decision making problems under environments with a high number of alternatives using fuzzy ontologies and multi-granular linguistic modelling methods. Knowl. Based Syst. **137**, 54–64 (2017)

26. Pinto, T., Gazafroudi, A.S., Prieto-Castrillo, F., Santos, G., Silva, F., Corchado, J.M., Vale, Z.: Reserve costs allocation model for energy and reserve market simulation. In: 2017 19th International Conference on Intelligent System Application to Power Systems, ISAP 2017 (2017). Art. no. 8071410
27. Fyfe, C., Corchado, J.: A comparison of Kernel methods for instantiating case based reasoning systems. Adv. Eng. Inf. **16**(3), 165–178 (2002). https://doi.org/10.1016/S1474-0346(02)00008-3
28. Chamoso, P., Rivas, A., Martín-Limorti, J.J., Rodríguez, S.: A hash based image matching algorithm for social networks. In: Advances in Intelligent Systems and Computing, vol. 619, pp. 183–190 (2018). https://doi.org/10.1007/978-3-319-61578-3_18
29. Sittón, I., Rodríguez, S.: Pattern extraction for the design of predictive models in industry 4.0. In: International Conference on Practical Applications of Agents and Multi-agent Systems, pp. 258–261 (2017)
30. García, O., Chamoso, P., Prieto, J., Rodríguez, S., De La Prieta, F.: A serious game to reduce consumption in smart buildings. In: Communications in Computer and Information Science, vol. 722, pp. 481–493 (2017). https://doi.org/10.1007/978-3-319-60285-1_41
31. Palomino, C.G., Nunes, C.S., Silveira, R.A., González, S.R., Nakayama, M.K.: Adaptive agent-based environment model to enable the teacher to create an adaptive class. In: Advances in Intelligent Systems and Computing, vol. 617 (2017). https://doi.org/10.1007/978-3-319-60819-8_3
32. Canizes, B., Pinto, T., Soares, J., Vale, Z., Chamoso, P., Santos, D.: Smart city: a GECAD-BISITE energy management case study. In: 15th International Conference on Practical Applications of Agents and Multi-agent Systems PAAMS 2017, Trends in Cyber-Physical Multi-Agent Systems, vol. 2, pp. 92–100 (2017). https://doi.org/10.1007/978-3-319-61578-3_9
33. Chamoso, P., de La Prieta, F., Eibenstein, A., Santos-Santos, D., Tizio, A., Vittorini, P.: A device supporting the self management of tinnitus. In: Lecture Notes in Computer Science (Including Subseries Lecture Notes in Artificial Intelligence and Lecture Notes in Bioinformatics) (2017)
34. Prieto, J., Mazuelas, S., Bahillo, A., Fernández, P., Lorenzo, R.M., Abril, E.J.: On the minimization of different sources of error for an RTT-based indoor localization system without any calibration stage. In: 2010 International Conference on Indoor Positioning and Indoor Navigation (IPIN), pp. 1–6 (2010)
35. del Rey, ÁM., Batista, F.K., Dios, A.Q.: Malware propagation in Wireless Sensor Networks: global models vs Individual-based models. ADCAIJ: Adv. Distrib. Comput. Artif. Intell. J. Salamanca **6**(3) (2017)
36. Kushch, S., Castrillo, F.P.: A review of the applications of the Block-chain technology in smart devices and distributed renewable energy grids. ADCAIJ: Adv. Distrib. Comput. Artif. Intell. J. Salamanca **6**(3) (2017)
37. Pinto, A., Costa, R.: Hash-chain-based authentication for IoT. ADCAIJ: Adv. Distrib. Comput. Artif. Intell. J. Salamanca **5**(4) (2016)
38. García-Valls, M.: Prototyping low-cost and flexible vehicle diagnostic systems. ADCAIJ: Adv. Distrib. Comput. Artif. Intell. J. Salamanca **5**(4) (2016)
39. Fernández-Fernández, A., Cervelló-Pastor, C., Ochoa-Aday, L.: Energy-aware routing in multiple domains software-defined networks. ADCAIJ: Adv. Distrib. Comput. Artif. Intell. J. Salamanca **5**(3) (2016)

40. Koskimaki, H., Siirtola, P.: Accelerometer vs. electromyogram in activity recognition. ADCAIJ: Adv. Distrib. Comput. Artif. Intell. J. Salamanca **5**(3) (2016)
41. Herrero, J.R., Villarrubia, G., Barriuso, A.L., Hernández, D., Lozano, Á., De La Serna González, M.A.: Wireless controller and smartphone based interaction system for electric bicycles. ADCAIJ: Adv. Distrib. Comput. Artif. Intell. J. Salamanca **4**(4) (2015)
42. OMS (ed.): Active Ageing: A Policy Framework. Organización Mundial de la Salud (2002). http://goo.gl/oG5w8M. Accessed 18 Feb 2018
43. VVAA: Plan Integral para la actividad física y el deporte en personas mayores. Ministerio de Cultura (2009)
44. VVAA: Libro blanco del envejecimiento activo, CSD. Ministerio de Educación y Cultura (2010)

Multiagent System for Semantic Categorization of Places Mean the Use of Distributed Surveillance Cameras

José Carlos Rangel and Cristian Pinzón(✉)

Grupo de Investigación ROBOTSiS,
Universidad Tecnológica de Panamá, Santiago de Veraguas, Panama
{jose.rangel,Cristian.pinzon}@utp.ac.pa

Abstract. Surveillance systems are quite common in almost every building. The current dimension of these systems is huge and involves a great deal of hardware and human resources for achieving these objectives. This paper proposes the use of an agent-based architecture for helping in the categorization of the places where these are deployed. Proposal uses a deep learning model for evaluating the images captured by the cameras and then label the zone where the camera is located.

Keywords: Software-Agents · Semantic-Categorization · Deep-Learning

1 Introduction

Nowadays the necessity of security on several kind of areas demands the development of robust surveillance systems that have to be able to manage the differences between these areas in order to execute their respective security or behavior protocols. Usually this systems are compose of a camera network that is constantly acquiring video on a real-time fashion of the scenes.

Determination of the kind of places the camera is pointing to, is a task usually carried out manually by a human operator. Therefore, for system deployed in huge areas with a great deal of divisions or different sections, this task would take a lot of time and effort. Consequently, the existence of a systems that automatically categorize the environment where each camera is positioned, would help to a reduce the time used for the initial deployment of the system.

This paper proposes a method for automatically assign a semantic category to the rooms of a building, the system merge multi-agent techniques with deep learning methods for accomplish the semantic categorization of the places. The proposal takes as input images captured with the cameras available within a surveillance network.

S. Rodríguez et al. (Eds.): DCAI 2018, AISC 801, pp. 457–464, 2019.
https://doi.org/10.1007/978-3-319-99608-0_64

2 Related Works

The semantic classification or categorization consist on assigning a semantic concept to places, this concept highly describes the function of the place and helps the human for understanding the area. Nowadays, the semantic classification of places is one of the popular topics in the literature, several works addressed this issue using traditional computer vision methods or taking advantage of the recent advanced on deep learning techniques.

Recently, the use of deep learning techniques, specifically the Convolutional Neural Networks (CNN), have become the standard for scene understanding by mean of the images. In the works presented in [1, 2], authors defined a semantic descriptor generated with a classification system based on CNN pre-trained models. The descriptor is tested for evaluating their description for the generation of topological maps inside university buildings respectively.

Multi-agent architecture has been tested in several scenarios [3, 4] with the results that this approach is able to optimize the accomplishment of the task which the system is responsible for.

3 Proposal

Our proposal consist on the development of a multi-agent architecture that will be on charge of the automatic categorization of several different room inside a building. Then, the proposal consist on a categorization procedure with a multi-agent architecture.

3.1 Semantic Categorization

We make the most of Semantic Descriptors generated with pre-trained CNN Models in order to automatically assign a category to an area under video-vigilance.

This involves labeling each image captured by the cameras, with a set of semantic terms. With this aim, we will use previously-trained deep learning models, CNNs in our case, in order to assign a probability value to each semantic label which that could be recognized by the model.

This generates a semantic descriptor (Eq. 1), as proposed in [2] for every image analyzed by the CNN model. Let $L = \{l_1, \ldots, l_{|L|}\}$ be the set of $|L|$ predefined lexical labels of a CNN model, and I an input image. Then, the generated descriptor can be defined as:

$$d_{CNN}(I) = \left[p_I(l_1), \ldots, p_I\left(l_{|L|}\right)\right] \quad (1)$$

where $p_I(l_i)$ denotes the probability of describing the image I using the i-th label in L.

Once the images belonging to a room we will apply a merging and averaging procedure over the labels of each image. We use a similar notation to represent the descriptor of a room/area R of the environment, which is composed of a set of $|N|$ images $\left(N = \{I_1, \ldots, I_{|N|}\}\right)$. The descriptor of R (Eq. 2) is defined as the vector of the average label probability of its $|N|$ images, and the corresponding vector of standard deviations. More formally:

$$d_{CNN}(R) = \left(\left[\bar{p}_R(l_1), \ldots, \bar{p}_R\left(l_{|L|}\right)\right], \left[\sigma_R(l_1), \ldots, \sigma_R\left(l_{|L|}\right)\right]\right) \quad (2)$$

where:

$$\bar{p}_R(l_i) = \frac{1}{|N|}\sum_{j=1}^{|N|} p_j(l_i) \quad (3)$$

and

$$\sigma_R^2(l_i) = \frac{1}{|N|}\sum_{j=1}^{|N|} \left(p_j(l_i) - \bar{p}_j(l_i)\right)^2 \quad (4)$$

Then, the category for the room will be determinate by the label or semantic terms with the higher mean value as showed in Eq. 5

$$Semantic\ \ Category = argmax[\bar{p}_R(l_i) \cdot \sigma_R(l_i)], \quad i = 1, \ldots, |L| \quad (5)$$

3.2 Multi-agent Architecture

Agents are characterized by their autonomy; which gives them the ability to work independently in real-time environments [3, 5–50]. Furthermore, when they are integrated within a multi-agent system they can also offer collaborative and distributed assistance in carrying out tasks [3].

The proposed system will be implemented using a multi-agent architecture in order to successfully accomplish the different task of the problem. The defined agent would be responsible for the recollection, management, labeling image and the categorization of the zones. The function of each agent is defined as follows and it is showed on Fig. 1.

- **Tagger Agent:** The task of this agent is to generate the semantic descriptor for the images which have been sent to him. This agent works on a server fashion, working only on demand (when an image arrives). The agent receive images from the *Camera Agent* and implements the CNN pre-trained model for classifying the images.
- **Camera Agent:** This agent is in charge of capturing the images of the rooms. There will be a *Camera Agent* for each camera in the surveillance network. Therefore, each agent only captures images belonging to their physical location. Images are send to *Tagger Agent* and when the whole location have been capture the agent lets to know this to the *Management Agent*.

- **Categorizer Agent:** Once all the images of a room have been captured and labeled, this agent analysis the whole set of semantic descriptor of the room and then applies the merging and averaging procedure for selecting the category of the room. This agent waits an order of the *Management Agent* for beginning to work.
- **Management Agent:** Takes care of the supervision of the correct performance of the system. This agent receives a signal from the *Camera Agent* indicating the finish of capturing procedure for its camera, then sends an order to the *Categorizer Agent* for beginning the descriptors analysis in order to produce the category of the room.

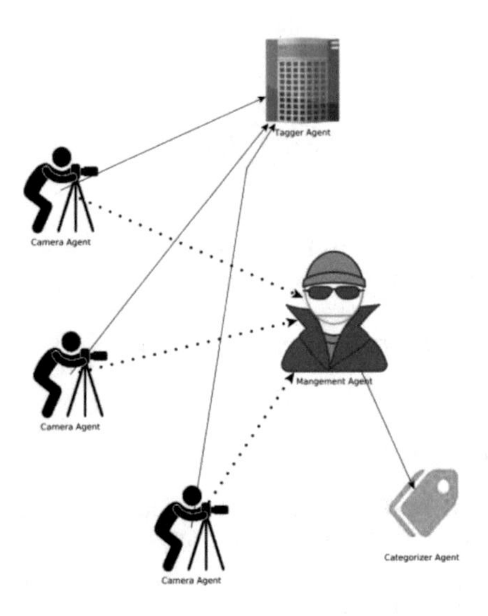

Fig. 1. Agent's architecture of the proposal

4 Tools

In order to develop this project a set of distributed agents is required. In this case it is proposed to use the JADE agent platform [51]. JADE (Java Agent Development Framework) is a software framework that significantly facilitates the development of agent-based applications in compliance with the FIPA specifications. The Deep Learning component of the project will a pre-trained classification model, built with the Convolutional Architecture for Fast Feature Embedding (Caffe) [52]. This framework provides the tools for training and testing classification models using images as the input. For this proposal we selected the Places365-GoogLeNet [53] model, which is able to produce 365 different labels related to indoors and outdoors locations. In this approximation we plane to use IP cameras for testing the system in a real environment.

5 Conclusions

The presented architecture deals with the categorization problems for surveillance systems. The proposal make the most of a multi-agent architecture for communicating several cameras and classifying the captured images. The architecture includes the use of deep learning models for assigning the semantic category to a room.

The systems allows to easily adapt the solution to others kind of environments by mean of the classifier model. Even, the use of agents allows for the addition of new camera modules without changing the entire system.

References

1. Scene classification based on semantic labeling. Adv. Robot
2. LexToMap: lexical-based topological mapping. Adv. Robot. (2016)
3. Corchado, J.M., Bajo, J., Abraham, A.: GerAmi: improving healthcare delivery in geriatric residences. IEEE Intell. Syst. **23**(2), 19–25 (2008)
4. Carrascosa, C., Bajo, J., Julian, V., Corchado, J.M., Botti, V.: Hybrid multi-agent architecture as a real-time problem-solving model. Expert Syst. Appl. **34**(1), 2–17 (2008)
5. Li, T., De la Prieta Pintado, F., Corchado, J.M., Bajo, J.: Multi-source homogeneous data clustering for multi-target detection from cluttered background with misdetection. Appl. Soft Comput. **60**, 436–446 (2017)
6. Morente-Molinera, J.A., Kou, G., González-Crespo, R., Corchado, J.M., Herrera-Viedma, E.: Solving multi-criteria group decision making problems under environments with a high number of alternatives using fuzzy ontologies and multi-granular linguistic modelling methods. Knowl.-Based Syst. (2017)
7. Wang, X., Li, T., Sun, S., Corchado, J.M.: A survey of recent advances in particle filters and remaining challenges for multitarget tracking. Sensors **17**(12) (2017)
8. Li, T.-C., Su, J.-Y., Liu, W., Corchado, J.M.: Approximate Gaussian conjugacy: parametric recursive filtering under nonlinearity, multimodality, uncertainty, and constraint, and beyond. Front. Inf. Technol. Electron. Eng. **18**, 1913–1939 (2017)
9. Oliveira, T., Neves, J., Novais, P.: Guideline formalization and knowledge representation for clinical decision support. ADCAIJ Adv. Distrib. Comput. Artif. Intell. J. **1**(2) (2012)
10. Nihan, C.E.: Healthier? More Efficient? Fairer? an overview of the main ethical issues raised by the use of Ubicomp in the workplace. ADCAIJ Adv. Distrib. Comput. Artif. Intell. J. **2**(1) (2013)
11. Macek, K., Rojicek, J., Kontes, G., Rovas, D.V.: Black-box optimization for buildings and its enhancement by advanced communication infrastructure. ADCAIJ Adv. Distrib. Comput. Artif. Intell. J. **2**(2) (2013)
12. Romero, S., Fardoun, H.M., Penichet, V.M.R., Gallud, J.A.: Tweacher: new proposal for online social networks impact in secondary education. ADCAIJ Adv. Distrib. Comput. Artif. Intell. J. **2**(1) (2013)
13. Martinez-Martin, E., Escrig, M.T., Del Pobil, A.P.: A qualitative acceleration model based on intervals. ADCAIJ Adv. Distrib. Comput. Artif. Intell. J. **2**(2) (2013)

14. Li, T., Sun, S.: Online adapting the magnitude of target birth intensity in the PHD filter. ADCAIJ Adv. Distrib. Comput. Artif. Intell. J. **2**(4) (2013)
15. Teixido, M., et al.: Optimization of the virtual mouse HeadMouse to foster its classroom use by children with physical disabilities. ADCAIJ Adv. Distrib. Comput. Artif. Intell. J. **2**(4) (2013)
16. García, O., Chamoso, P., Prieto, J., Rodríguez, S., De La Prieta, F.: A serious game to reduce consumption in smart buildings. Commun. Comput. Inf. Sci. **722**, 481–493 (2017)
17. Sittón, I., Rodríguez, S.: Pattern extraction for the design of predictive models in industry 4. 0 (2017)
18. Chamoso, P., Rivas, A., Martín-Limorti, J.J., Rodríguez, S.: A hash based image matching algorithm for social networks. In: Advances in Intelligent Systems and Computing, vol. 619, pp. 183–190 (2018)
19. Corchado, J., Fyfe, C., Lees, B.: Unsupervised learning for financial forecasting. In: Proceedings of the IEEE/IAFE/INFORMS 1998 Conference on Computational Intelligence for Financial Engineering (CIFEr) (Cat. No, 1998), pp. 259–263 (1998)
20. Corchado, J.M., Fyfe, C.: Unsupervised neural method for temperature forecasting. Artif. Intell. Eng. **13**(4), 2–7 (1810)
21. Fyfe, C., Corchado, J.M.: Automating the construction of CBR systems using kernel methods. Int. J. Intell. Syst. **16**(4) (2001)
22. Fyfe, C., Corchado, J.: A comparison of Kernel methods for instantiating case based reasoning systems. Adv. Eng. Inform. **16**(3) (2002)
23. Corchado, J.M., Aiken, J.: Hybrid artificial intelligence methods in oceanographic forecast models. IEEE Trans. Syst. Man Cybern. Part C-Appl. Rev. **32**(4) (2002)
24. Glez-Bedia, M., Corchado, J.M., Corchado, E.S., Fyfe, C.: Analytical model for constructing deliberative agents. Int. J. Eng. Intell. Syst. Electr. Eng. Commun. **10**(3) (2002)
25. Fdez-Riverola, F., Corchado, J.M.: CBR based system for forecasting red tides. Knowl.-Based Syst. 321–328 (2003)
26. Corchado, J.M., Corchado, E.S., Aiken, J., Fyfe, C., Fernandez, F., Gonzalez, M.: Maximum likelihood hebbian learning based retrieval method for CBR systems. In: Lecture Notes in Computer Science (including subseries Lecture Notes in Artificial Intelligence and Lecture Notes in Bioinformatics), vol. 2689, pp. 107–121 (2003)
27. Corchado, J.M., Borrajo, M.L., Pellicer, M.A., Yáñez, J.C.: Neuro-symbolic system for business internal control. In: Industrial Conference on Data Mining, pp. 1–10 (2004)
28. Corchado, J.A., Aiken, J., Corchado, E.S., Lefevre, N., Smyth, T.: Quantifying the ocean's CO2 budget with a CoHeL-IBR system. In: Proceedings of Advances in Case-Based Reasoning, vol. 3155, pp. 533–546 (2004)
29. Laza, R., Pavn, R., Corchado, J.M.: A reasoning model for CBR_BDI agents using an adaptable fuzzy inference system. In: Lecture Notes in Computer Science (including subseries Lecture Notes in Artificial Intelligence and Lecture Notes in Bioinformatics), vol. 3040, pp. 96–106 (2004)
30. Corchado, J.M., Pavón, J., Corchado, E.S., Castillo, L.F.: Development of CBR-BDI agents: a tourist guide application. In: Lecture Notes in Computer Science (including subseries Lecture Notes in Artificial Intelligence and Lecture Notes in Bioinformatics), vol. 3155, pp. 547–559 (2004)
31. Méndez, J.R., Fdez-Riverola, F., Iglesias, E.L., Díaz, F., Corchado, J.M.: Tracking concept drift at feature selection stage in SpamHunting: an anti-spam instance-based reasoning system. In: Lecture Notes in Computer Science (Including subseries Lecture Notes in Artificial Intelligence and Lecture Notes in Bioinformatics), vol. 4106 (2006)

32. Fdez-Rtverola, F., Corchado, J.M.: FSfRT: forecasting system for red tides. Appl. Intell. **21**(3) (2004)
33. Méndez, J.R., Fdez-Riverola, F., Díaz, F., Iglesias, E.L., Corchado, J.M.: A comparative performance study of feature selection methods for the anti-spam filtering domain. In: Lecture Notes in Computer Science (Including Subseries Lecture Notes in Artificial Intelligence and Lecture Notes in Bioinformatics), vol. 4065 (2006)
34. Fernández-Riverola, F., Díaz, F., Corchado, J.M.: Reducing the memory size of a fuzzy case-based reasoning system applying rough set techniques. IEEE Trans. Syst. Man Cybern. Part C Appl. Rev. **37**(1) (2006)
35. Glez-Peñaa, D., Díaz, F., Hernández, J.M., Corchado, J.M., Fdez-Riverola, F.: geneCBR: a translational tool for multiple-microarray analysis and integrative information retrieval for aiding diagnosis in cancer research. BMC Bioinform. **10** (2009)
36. Mata, A., Corchado, J.M.: Forecasting the probability of finding oil slicks using a CBR system. Expert Syst. Appl. **36**(4) (2008)
37. Tapia, D.I., Corchado, J.M.: An ambient intelligence based multi-agent system for alzheimer health care. Int. J. Ambient Comput. Intell. **1**(1), 15–26 (2009)
38. Baruque, B., Corchado, E., Mata, A., Corchado, J.M.: A forecasting solution to the oil spill problem based on a hybrid intelligent system. Inf. Sci. **180**(10) (2009)
39. Rodríguez, S., et al.: People detection and stereoscopic analysis using MAS. In: Proceedings of INES 2010-14th International Conference on Intelligent Engineering Systems (2010)
40. Rodríguez, S., de la Prieta, F., Tapia, D.I., Corchado, J.M.: Agents and computer vision for processing stereoscopic images. In: Lecture Notes in Computer Scienc. (including Subseries Lecture Notes in Artificial Intelligence and Lecture Notes in Bioinformatics), vol. 6077 (2010)
41. García, E., Rodríguez, S., Martín, B., Zato, C., Pérez, B.: MISIA: middleware infrastructure to simulate intelligent agents. Adv. Intell. Soft Comput. (2011)
42. Costa, Â., Novais, P., Corchado, J.M., Neves, J.: Increased performance and better patient attendance in an hospital with the use of smart agendas. Log. J. IGPL **20**(4) (2012)
43. Tapia, D.I., Fraile, J.A., Rodríguez, S., Alonso, R.S., Corchado, J.M.: Integrating hardware agents into an enhanced multi-agent architecture for ambient intelligence systems. Inf. Sci. **222** (2011)
44. Coria, J.A.G., Castellanos-Garzón, J.A., Corchado, J.M.: Intelligent business processes composition based on multi-agent systems. Expert Syst. Appl. (2013)
45. Li, T., Sun, S., Corchado, J.M., Siyau, M.F.: A particle dyeing approach for track continuity for the SMC-PHD filter. In: FUSION 2014-17th International Conference on Information Fusion (2014)
46. Choon, Y.W., et al.: Differential bees flux balance analysis with OptKnock for in silico microbial strains optimization. PLoS One **9**(7) (2014)
47. Redondo-Gonzalez, E., et al.: Bladder carcinoma data with clinical risk factors and molecular markers: a cluster analysis. BioMed Res. Int. (2015)
48. Li, T., Sun, S., Corchado, J.M., Siyau, M.F.: Random finite set-based Bayesian filters using magnitude-adaptive target birth intensity. In: FUSION 2014-17th International Conference on Information Fusion (2014)
49. Li, T., Sun, S.: Bolić, M., Corchado, J.M.: Algorithm design parallel implementation SMC-PHD filter. Sig. Process. **119**, 115–127 (2016)
50. Lima, A.C.E.S., de Castro, L.N., Corchado, J.M.: A polarity analysis framework for Twitter messages. Appl. Math. Comput. **270** (2015)

51. Abras, S., Kieny, C., Ploix, S., Wurtz, F.: MAS architecture for energy management: developing smart networks with JADE platform. In: 2013 IEEE International Conference on Smart Instrumentation, Measurement and Applications (ICSIMA), pp. 1–6 (2013)
52. Jia, Y., et al.: Caffe: convolutional architecture for fast feature embedding, June 2014
53. Zhou, B., Lapedriza, A., Khosla, A., Oliva, A., Torralba, A.: Places: a 10 million image database for scene recognition. IEEE Trans. Pattern Anal. Mach. Intell. (2017)

Customer Experience Management (CEM)

Samuel Gallego Chimeno(✉)

BISITE Digital Innovation Hub,
University of Salamanca. Edificio Multiusos I+D+I, 37007 Salamanca, Spain
samuelgch@usal.es

Abstract. The CEM Project arises from the need to create a web app that will enable users to work with the tools necessary for collecting and manipulating statistical data coming from surveys, campaigns and waves, etc. Primarily, our aim is to offer the customer a product that allows to collect data, in a simple and descriptive manner, for their subsequent interpretation with other types of tools that CEM also offers, such as its graphics module.

Keywords: Segment · List · Server · User · Project · Campaign Wave · Survey · Template · Client · Control panel · Management module

1 Introduction

CEM is a project that arose from the need for a tool that would unify the methods of collecting statistical data and would exceed the standards of other similar tools currently on the market. Its functional design as well as a clear and powerful interface will allow users to comprehend, manage and organize different data collection methods, such as (questionnaires, surveys, campaigns, waves, etc.) for the subsequent management and visualization of these data, also in the simplest and most graphic way possible.

2 Objectives

The core of the application is the management module, which is key to achieving this goal. Three types of profiles are proposed a priori, the administrator, the manager and the standard user. The role of each of them is defined in the system by their ability to create profiles in the system for other users. The administrator can create profiles for standard users and for managers. Managers on the other hand, can only create profiles for standard users, while standard users cannot create profiles. A fourth profile may also be defined uniquely for subscriber users, who have no power in the system, other than the right to receive e-mails containing surveys.

Returning to the concept of the previously described management module, and after having introduced the manager profile, we can state that the relationship between the two is absolute. The manager user in each company, will be in charge of using this module for three main purposes: adding subscribers, projects and campaigns.

S. Rodríguez et al. (Eds.): DCAI 2018, AISC 801, pp. 465–470, 2019.
https://doi.org/10.1007/978-3-319-99608-0_65

The subscribing users, as mentioned above, are the people who will receive surveys.

The project will be defined as a virtual space for the management of a specific CEM (indicators, users, campaigns and surveys). It encompasses campaigns. All information about the project's activities can be found on the web platform's home panel.

Campaigns will be associated with surveys, which will in turn be associated with a project and its indicators. To create a campaign, the first step is to assign it a name and a description. In addition, it can be anonymous (no previous subscriber list attached) or non-anonymous (it does have a list). At the end of the creation and configuration process, the campaign in question will then be disseminated.

Furthermore, the term wave is implicit in the concept of a campaign. Once a campaign has been configured and is available in the general panel, a wave can be assigned to it, which consists of a name, description, the associated survey that will be sent and a list of subscribers or a specific target group. Once all these aspects have been configured, in addition to other types of adjustments that are not so significant for this paper.

Some of the adjustments, such as parametrization and templates, are not described here because they are not important for the academic purposes of this article. Once all the aspects have been configured, the campaign will be visible in the general campaigns and waves panel, and therefore it will be possible to send a cycle.

The concept of lists and segments has also been addressed and it is of vital importance. As mentioned above it refers to groups of 'x' subscriber users, organized in lists, to whom the surveys will be sent by e-mail so that they can read and complete them properly. These lists can be assigned a name and a description in the same way as campaigns and waves. Once correctly configured, it will be possible to specify the name and e-mail address of the subscriber users to whom the surveys will be sent.

Surveys, on the other hand, are one of the fundamental pillars of the project. They are made up of a series of elements such as the title of the question, the issue in point, the answer block and the thank you text. Surveys are therefore used to compile the statistical data obtained from the opinion of the subscriber users (which will be the population to which the survey is applied).

On the other hand, it is also interesting to talk about the control panel, which, in a way, is also one of the main features of the CEM project. Once the statistical data have been collected, they can be interpreted visually together with campaigns, by means of interactive graphics, tables and crosses of variables, these characteristics are offered by this panel. It can also be used with the role of the three profiles described at the beginning: administrator, with access to the management of subscriber users and projects; manager, with access to campaigns, projects including surveys and subscriber users, or standard users (who will be called viewers), will simply have access to the results of a campaign.

To conclude, we would like to mention the interoperability feature of the CEM project, as well as describe, broadly speaking, its architecture and integration [1–33] with the CRM. The latter is achieved thanks to the design based on RESTful (http +JSON) API type web services; separating the API from the CRM, this interoperability between different systems is guaranteed. In addition, the web application on which the CEM project is based, was constructed with different programming languages for web development such as html, css and twig templates, as well as the increasingly used Symfony, which is a PHP framework that offers the possibility of optimizing and speeding up the development of this type of application [34–49] based on the architecture of the MVC (Model-View-Controller) and that separates, in a very efficient way, the business logic, the server logic and the presentation of the web application.

Acknowledgements. "This work has been supported by project "IOTEC: Development of Technological Capacities around the Industrial Application of Internet of Things (IoT)". 0123_IOTEC_3_E. Project financed with FEDER funds, Interreg Spain-Portugal (PocTep)."

References

1. Mata, A., Lancho, B.P., Corchado, J.M.: Forest fires prediction by an organization based system. In: PAAMS 2010, pp. 135–144 (2010)
2. Baruque, B., Corchado, E., Mata, A., Corchado, J.M.: A forecasting solution to the oil spill problem based on a hybrid intelligent system. Inf. Sci. **180**(10), 2029–2043 (2010). https://doi.org/10.1016/j.ins.2009.12.032
3. Choon, Y.W., Mohamad, M.S., Deris, S., Illias, R.M., Chong, C.K., Chai, L.E., Corchado, J.M.: Differential bees flux balance analysis with OptKnock for in silico microbial strains optimization. PLoS ONE **9**(7) (2014). https://doi.org/10.1371/journal.pone.0102744
4. Corchado, J.A., Aiken, J., Corchado, E.S., Lefevre, N., Smyth, T.: Quantifying the Ocean's CO2 budget with a CoHeL-IBR system. In: Proceedings of Advances in Case-Based Reasoning, vol. 3155, pp. 533–546 (2004)
5. Corchado, J.M., Aiken, J.: Hybrid artificial intelligence methods in oceanographic forecast models. IEEE Trans. Syst. Man Cybern. Part C-Appl. Rev. **32**(4), 307–313 (2002). https://doi.org/10.1109/tsmcc.2002.806072
6. Corchado, J.M., Fyfe, C.: Unsupervised neural method for temperature forecasting. Artif. Intell. Eng. **13**(4), 351–357 (1999). https://doi.org/10.1016/S0954-1810(99)00007-2
7. Corchado, J.M., Borrajo, M.L., Pellicer, M.A., Yáñez, J.C.: Neuro-symbolic system for business internal control. In: Industrial Conference on Data Mining, pp. 1–10 (2004). https://doi.org/10.1007/978-3-540-30185-1_1
8. Corchado, J.M., Corchado, E.S., Aiken, J., Fyfe, C., Fernandez, F., Gonzalez, M.: Maximum likelihood hebbian learning based retrieval method for CBR systems. In: Lecture Notes in Computer Science (including subseries Lecture Notes in Artificial Intelligence and Lecture Notes in Bioinformatics), vol. 2689, pp. 107–121 (2003). https://doi.org/10.1007/3-540-45006-8_11
9. Corchado, J.M., Pavón, J., Corchado, E.S., Castillo, L.F.: Development of CBR-BDI agents: a tourist guide application. In: Lecture Notes in Computer Science (including subseries Lecture Notes in Artificial Intelligence and Lecture Notes in Bioinformatics), vol. 3155, pp. 547–559 (2004). https://doi.org/10.1007/978-3-540-28631-8
10. Corchado, J., Fyfe, C., Lees, B.: Unsupervised learning for financial forecasting. In: Proceedings of the IEEE/IAFE/INFORMS 1998 Conference on Computational Intelligence for Financial Engineering (CIFEr) (Cat. No. 98TH8367), pp. 259–263 (1998). https://doi.org/10.1109/CIFER.1998.690316
11. Costa, Â., Novais, P., Corchado, J.M., Neves, J.: Increased performance and better patient attendance in an hospital with the use of smart agendas. Logic J. IGPL **20**(4), 689–698 (2012). https://doi.org/10.1093/jigpal/jzr021

12. Tapia, D.I., García, Ó., Alonso, R.S., Guevara, F., Catalina, J., Bravo, R.A., Corchado, J.M.: Evaluating the n-Core polaris real-time locating system in an indoor environment. In: PAAMS (Workshops), pp. 29–37 (2012)
13. Tapia, D.I., Alonso, R.S., García, Ó., Corchado, J.M.: HERA: hardware-embedded reactive agents platform. In: PAAMS (Special Sessions), pp. 249–256 (2011)
14. Fdez-Riverola, F., Corchado, J.M.: CBR based system for forecasting red tides. Knowl.-Based Syst. **16**(5–6 SPEC.), 321–328 (2003). https://doi.org/10.1016/S0950-7051(03)00034-0
15. Fdez-Rtverola, F., Corchado, J.M.: FSfRT: forecasting system for red tides. Appl. Intell. **21** (3), 251–264 (2004). https://doi.org/10.1023/B:APIN.0000043558.52701.b1
16. Fernández-Riverola, F., Díaz, F., Corchado, J.M.: Reducing the memory size of a Fuzzy case-based reasoning system applying rough set techniques. IEEE Trans. Syst. Man Cybern. Part C Appl. Rev. **37**(1), 138–146 (2007). https://doi.org/10.1109/TSMCC.2006.876058
17. Fernandes, F., Gomes, L., Morais, H., Silva, M.R., Vale, Z.A., Corchado, J.M.: Dynamic energy management method with demand response interaction applied in an office building. In: PAAMS (Special Sessions), pp. 69–82 (2016)
18. Fyfe, C., Corchado, J.: A comparison of Kernel methods for instantiating case based reasoning systems. Adv. Eng. Inform. **16**(3), 165–178 (2002). https://doi.org/10.1016/S1474-0346(02)00008-3
19. Fyfe, C., Corchado, J.M.: Automating the construction of CBR systems using kernel methods. Int. J. Intell. Syst. **16**(4), 571–586 (2001). https://doi.org/10.1002/int.1024
20. García Coria, J.A., Castellanos-Garzón, J.A., Corchado, J.M.: Intelligent business processes composition based on multi-agent systems. Exp. Syst. Appl. **41**(4 PART 1), 1189–1205 (2014). https://doi.org/10.1016/j.eswa.2013.08.003
21. García, E., Rodríguez, S., Martín, B., Zato, C., Pérez, B.: MISIA: middleware infrastructure to simulate intelligent agents. In: Advances in Intelligent and Soft Computing, vol. 91 (2011). https://doi.org/10.1007/978-3-642-19934-9_14
22. Glez-Bedia, M., Corchado, J.M., Corchado, E.S., Fyfe, C.: Analytical model for constructing deliberative agents. Int. J. Eng. Intell. Syst. Electr. Eng. Commun. **10**(3) (2002)
23. Oliver, M., Molina, J.P., Fernández-Caballero, A., González, P.: Collaborative computer-assisted cognitive rehabilitation system. ADCAIJ Adv. Distrib. Comput. Artif. Intell. J. **6**(3) (2017). ISSN: 2255-2863
24. Ueno, M., Suenaga, T., Isahara, H.: Classification of two comic books based on convolutional neural networks. ADCAIJ Adv. Distrib. Comput. Artif. Intell. J. **6**(1) (2017). ISSN: 2255-2863l
25. Blanco Valencia, X.P., Becerra, M.A., Castro Ospina, A.E., Ortega Adarme, M., Viveros Melo, D., Peluffo Ordóñez, D.H.: Kernel-based framework for spectral dimensionality reduction and clustering formulation: a theoretical study. Adv. Distrib. Comput. Artif. Intell. J. (2017). ISSN: 2255-2863
26. Bullón, J., Arrieta, A.G., Encinas, A.H., Dios, A.Q.: Manufacturing processes in the textile industry. Expert Systems for fabrics production. Adv. Distrib. Comput. Artif. Intell. J. **6**(1) (2017). ISSN: 2255-2863
27. Griol, D., Molina, J.M.: Simulating heterogeneous user behaviors to interact with conversational interfaces. ADCAIJ Adv. Distrib. Comput. Artif. Intell. J. **5**(4) (2016). ISSN: 2255-2863
28. Glez-Peña, D., Díaz, F., Hernández, J.M., Corchado, J.M., Fdez-Riverola, F.: geneCBR: a translational tool for multiple-microarray analysis and integrative information retrieval for aiding diagnosis in cancer research. BMC Bioinform. **10** (2009). https://doi.org/10.1186/1471-2105-10-187

29. Román, J.Á., Rodríguez, S., Corchado, J.M.: Improving intelligent systems: specialization. In: PAAMS (Workshops), pp. 378–385 (2014)
30. Li, T., Sun, S., Bolić, M., Corchado, J.M.: Algorithm design for parallel implementation of the SMC-PHD filter. Sig. Process. **119**, 115–127 (2016). https://doi.org/10.1016/j.sigpro.2015.07.013
31. Li, T., Sun, S., Corchado, J.M., Siyau, M.F.: A particle dyeing approach for track continuity for the SMC-PHD filter. In: FUSION 2014 - 17th International Conference on Information Fusion (2014). https://www.scopus.com/inward/record.uri?eid=2-s2.0-84910637583&partnerID=40&md5=709eb4815eaf544ce01a2c21aa749d8f
32. Li, T., Sun, S., Corchado, J.M., Siyau, M.F.: Random finite set-based Bayesian filters using magnitude-adaptive target birth intensity. In: FUSION 2014 - 17th International Conference on Information Fusion (2014). https://www.scopus.com/inward/record.uri?eid=2-s2.0-84910637788&partnerID=40&md5=bd8602d6146b014266cf07dc35a681e0
33. Li, T.-C., Su, J.-Y., Liu, W., Corchado, J.M.: Approximate Gaussian conjugacy: parametric recursive filtering under nonlinearity, multimodality, uncertainty, and constraint, and beyond. Front. Inf. Technol. Electr. Eng. **18**(12), 1913–1939 (2017)
34. Lima, A.C.E.S., De Castro, L.N., Corchado, J.M.: A polarity analysis framework for Twitter messages. Appl. Math. Comput. **270**, 756–767 (2015). https://doi.org/10.1016/j.amc.2015.08.059
35. Mata, A., Corchado, J.M.: Forecasting the probability of finding oil slicks using a CBR system. Expert Syst. Appl. **36**(4), 8239–8246 (2009). https://doi.org/10.1016/j.eswa.2008.10.003
36. Méndez, J.R., Fdez-Riverola, F., Díaz, F., Iglesias, E.L., Corchado, J.M.: A comparative performance study of feature selection methods for the anti-spam filtering domain. In: Lecture Notes in Computer Science (Including Subseries Lecture Notes in Artificial Intelligence and Lecture Notes in Bioinformatics). LNAI, vol. 4065, pp. 106–120 (2006). https://www.scopus.com/inward/record.uri?eid=2-s2.0-33746435792&partnerID=40&md5=25345ac884f61c182680241828d448c5
37. Morente-Molinera, J.A., Kou, G., González-Crespo, R., Corchado, J.M., Herrera-Viedma, E.: Solving multi-criteria group decision making problems under environments with a high number of alternatives using fuzzy ontologies and multi-granular linguistic modelling methods. Knowl.-Based Syst. **137**, 54–64 (2017)
38. Dang, N.C., de la Prieta, F., Corchado, J.M., Moreno, M.N.: Framework for retrieving relevant contents related to fashion from online social network data. In: PAAMS (Special Sessions), pp. 335–347 (2016)
39. Chamoso, P., de la Prieta, F., de Paz, F., Corchado, J.M.: Swarm agent-based architecture suitable for internet of things and smartcities. In: DCAI 2015, pp. 21–29 (2015)
40. Pinto, T., Gazafroudi, A.S., Prieto-Castrillo, F., Santos, G., Silva, F., Corchado, J.M., Vale, Z.: Reserve costs allocation model for energy and reserve market simulation. In: 2017 19th International Conference on Intelligent System Application to Power Systems, ISAP 2017, art. no. 8071410 (2017)
41. Redondo-Gonzalez, E., De Castro, L.N., Moreno-Sierra, J., Maestro De Las Casas, M.L., Vera-Gonzalez, V., Ferrari, D.G., Corchado, J.M.: Bladder carcinoma data with clinical risk factors and molecular markers: a cluster analysis. BioMed Res. Int. (2015). https://doi.org/10.1155/2015/168682
42. Rodríguez, S., De La Prieta, F., Tapia, D.I., Corchado, J.M.: Agents and computer vision for processing stereoscopic images. In: Lecture Notes in Computer Science (including subseries Lecture Notes in Artificial Intelligence and Lecture Notes in Bioinformatics). LNAI, vol. 6077 (2010). https://doi.org/10.1007/978-3-642-13803-4_12

43. Rodríguez, S., Gil, O., De La Prieta, F., Zato, C., Corchado, J.M., Vega, P., Francisco, M.: People detection and stereoscopic analysis using MAS. In: Proceedings of INES 2010 - 14th International Conference on Intelligent Engineering Systems (2010). https://doi.org/10.1109/INES.2010.5483855
44. Omatu, S., Wada, T., Rodríguez, S., Chamoso, P., Corchado, J.M.: Multi-agent technology to perform odor classification. In: ISAmI 2014, pp. 241–252 (2014)
45. Lim, S.Y., Mohamad, M.S., En Chai, L., Deris, S., Chan, W.H., Omatu, S., Corchado, J.M., Sjaugi, M.F., Mahfuz Zainuddin, M., Rajamohan, G., Ibrahim, Z., Md Yusof, Z.: Investigation of the effects of imputation methods for gene regulatory networks modelling using dynamic Bayesian networks. In: DCAI 2016, pp. 413–421 (2016)
46. Tapia, D.I., Corchado, J.M.: An ambient intelligence based multi-agent system for alzheimer health care. Int. J. Ambient Comput. Intell. **1**(1), 15–26 (2009). https://doi.org/10.4018/jaci.2009010102
47. Tapia, D.I., Fraile, J.A., Rodríguez, S., Alonso, R.S., Corchado, J.M.: Integrating hardware agents into an enhanced multi-agent architecture for Ambient Intelligence systems. Inf. Sci. **222**, 47–65 (2013). https://doi.org/10.1016/j.ins.2011.05.002
48. López Batista, V.F., Aguilar, R., Alonso, L., Moreno García, M.N., Corchado, J.M.: Data mining for grammatical inference with bioinformatics criteria. In: HAIS (2), pp. 53–60 (2010)
49. Wang, X., Li, T., Sun, S., Corchado, J.M.: A survey of recent advances in particle filters and remaining challenges for multitarget tracking. Sensors **17**(12), art. no. 2707 (2017)

Author Index

S. Rodríguez et al. (Eds.): DCAI 2018, AISC 801, pp. 471–473, 2019.
https://doi.org/10.1007/978-3-319-99608-0

Printed in the United States
By Bookmasters